Vespidae of the Middle East (Hymenoptera)

The cosmopolitan family Vespidae (Hymenoptera: Vespoidea) comprises approximately 5,000 described species worldwide, traditionally classified into six subfamilies: Stenogastrinae, Eumeninae *s.l.*, Euparagiinae, Masarinae, Polistinae, and Vespinae. Among vespid wasps, social behavior ranges from solitary nesting to highly eusocial societies that function as superorganisms. Vespid wasps play vital roles in terrestrial ecosystems and are recognized as efficient biological control agents, preying on caterpillars and other insects, including many agricultural pests. Additionally, by occupying top positions in terrestrial arthropod food webs and visiting flowers for nectar, they contribute to the pollination of fruit and vegetable crops. Some vespid species have become invasive pests, and/or are feared for their painful stings, which can be fatal to individuals allergic to their venom.

This book presents the first comprehensive treatment of the Vespidae of the Middle East, offering detailed information on 716 valid species and subspecies across 69 genera in four subfamilies: Eumeninae *s.l.*, Masarinae, Polistinae, and Vespinae. It provides complete taxonomic treatment of all subfamilies known from the region, along with insights into their general biology, ecology, and geographic distribution. The book also includes discussions on morphology, identification keys to genera and sometimes species, economic importance, diagnostic characters of each group, and their affinities with related groups. An extensive bibliography of over 800 references supports further study and, whenever possible, the information has been updated to reflect the most current knowledge.

As the definitive guide to the Vespidae of the Middle East, this book serves as a valuable resource for researchers studying wasps worldwide.

Vespidae of the Middle East (Hymenoptera)

Hornets, Paper Wasps, Pollen Wasps, Potter Wasps, and Yellowjackets

Edited by

Hassan Ghahari

Neveen S. Gadallah

Patrick K. Piekarski

CRC Press
Taylor & Francis Group
Boca Raton London New York

CRC Press is an imprint of the
Taylor & Francis Group, an **informa** business

Designed Cover Image: Author's own

First edition published 2027
by CRC Press
2385 NW Executive Center Drive, Suite 320, Boca Raton FL 33431

and by CRC Press
4 Park Square, Milton Park, Abingdon, Oxon, OX14 4RN

CRC Press is an imprint of Taylor & Francis Group, LLC

Library of Congress Cataloging-in-Publication Data
Names: Ghahari, Hassan editor | Gadallah, Neveen S. editor | Piekarski,
Patrick K. editor
Title: Vespidae of the Middle East (hymenoptera) : hornets, paper wasps,
pollen wasps, potter wasps, and yellowjackets / edited by Hassan Ghahari,
Neveen S. Gadallah, Patrick K. Piekarski.
Description: First edition. | Boca Raton, FL : CRC Press, 2026. | Includes
bibliographical references and index.
Identifiers: LCCN 2026001479 (print) | LCCN 2026001480 (ebook) | ISBN
9781041241461 hardback | ISBN 9781041241485 paperback |
ISBN 9781003740438 ebook
Subjects: LCSH: Vespidae–Middle East
Classification: LCC QL568.V5 V47 2026 (print) | LCC QL568.V5 (ebook)
LC record available at https://lccn.loc.gov/2026001479
LC ebook record available at https://lccn.loc.gov/2026001480

ISBN: 978-1-041-24146-1 (hbk)
ISBN: 978-1-041-24148-5 (pbk)
ISBN: 978-1-003-74043-8 (ebk)

DOI: 10.1201/9781003740438

Typeset in Times
by SPi Technologies India Pvt Ltd (Straive)

Dedication

This book is dedicated to all the researchers who have had roles in discovering the Vespidae of the Middle East.

Katamenes tauricus tauricus (de Saussure, 1855) (Eumeninae).

(Photo courtesy of Alexander Fateryga)

"It is interesting to contemplate an entangled bank clothed with many plants of many kinds with birds singing on the bushes, with various insects flitting about and with worms crawling through the damp earth and to reflect that these elaborately constructed forms so different from each other and dependent on each other and so complex a manner have all been produced by laws acting around us."

Charles Darwin [12 February 1809–19 April 1882; English naturalist, geologist and biologist]

Contents

About the Editors

Hassan Ghahari is currently working as a professor of entomology (insect systematics, and biological control) at Plant Protection Department, Yadegar-e-Imam Khomeini (RAH) Shahre Rey Branch, Islamic Azad University, Tehran, Iran. He received his B.S. degree in Plant Protection from Urmia University, M.Sc. in Agricultural Entomology from Isfahan University of Technology, and Ph.D. in Agricultural Entomology from Tehran Islamic Azad University. He is an accomplished scholar and expert in the field of insect systematics and biological control. He teaches several courses in entomology and has guided many M.Sc. and Ph.D. students. He has received several research grants from Islamic Azad University as well as many Iranian prestigious awards and coveted honors.

Dr. Ghahari has authored or co-authored more than 250 scientific papers in peer-reviewed international journals together with several authorized entomologists from different countries. He has served as editor for several books and co-authored over 150 book chapters for various reputable publishers (e.g., CABI, Elsevier, Springer, Magnolia Press, CRC Press, and IAU Press) on various topics related to insect systematics and biological control. He also is a member of the editorial board of many agricultural scientific journals in different countries.

Neveen S. Gadallah is professor of insect taxonomy in the Entomology Department, Faculty of Science, Cairo University, Egypt. She earned her B.S. in Entomology, as well as her M.Sc. and Ph.D. degrees in Insect Taxonomy, from Cairo University. Her research focuses on insect biodiversity and taxonomy, particularly within the order Hymenoptera in the Middle East. She has authored and co-authored over 150 scientific papers, monographs, books and book chapters published in peer-reviewed international journals and publishers, collaborating with numerous hymenopteran taxonomists worldwide. She has also served as an associate editor in many scientific journals.

Patrick K. Piekarski is a postdoctoral researcher in the Laboratory of Social Evolution and Behavior at The Rockefeller University, USA. He holds a B.Sc. (Honors) in Zoology from the University of Calgary, M.Sc. in Biology from the University of Central Florida, and Ph.D. in Biology from The Rockefeller University. His research spans Vespidae taxonomy and phylogenetics, animal social behavior, particularly in wasps and ants, as well as molecular biology and behavioral genetics. Dr. Piekarski has received several awards and fellowships, including the prestigious NSERC Canada Graduate Scholarship. He has authored or co-authored seven peer-reviewed journal articles, including a landmark study reconstructing the phylogeny of the Vespidae family.

He has presented his work at major scientific conferences and has been a member of several professional societies, including the American Genetics Association, the International Union for the Study of Social Insects, the Entomological Society of America, and the International Society of Hymenopterists. Driven by a fascination with insect societies, Dr. Piekarski investigates how genetics and the social environment shape behavior and morphology.

Contributors

Kees van Achterberg
Department of Entomology
National Natuurhistorisch Museum
Leiden, Netherlands

Libor Dvořák
Sumava NP and PLA
Kasperske Hory, Czech Republic

Neveen S. Gadallah
Entomology Department
Faculty of Science
Cairo University
Giza, Egypt

Hassan Ghahari
Department of Plant Protection
Yadegar-e-Imam Khomeini (RAH) Shahre Rey Branch
Islamic Azad University
Tehran, Iran

Patrick K. Piekarski
Laboratory of Social Evolution and Behavior
Rockefeller University
New York, New York

Alexander Somavilla
Instituto Nacional de Pesquisas da Amazõnia
Coordenação de Biodiversidade
Manaus, Brazil

Christopher K. Starr
Department of Life Sciences
University of the West Indies
St Augustine, Trinidad & Tobago

Erol Yildirim
Department of Plant Protection
Faculty of Agriculture
Ataturk University
Erzurum, Turkey

Preface

The cosmopolitan family Vespidae (Hymenoptera: Vespoidea) is an ecologically and economically significant group of insects. It includes wasps commonly known as potter wasps, paper wasps, pollen wasps, hover wasps, hornets, and yellowjackets. Members of this family function as pollinators, predators, and biocontrol agents. Despite their benefits, they are often considered nuisance pests in urban areas, and their stings can cause severe allergic reactions. Vespid wasps are distributed worldwide, with greater species diversity in tropical regions. While most species are solitary, many exhibit social behavior. In social species, adult females progressively provision their larvae with masticated insects. In contrast, solitary species typically provision their larvae with caterpillars or other insect prey placed in a cell by the adult female. Notably, pollen wasps differ by feeding their larvae with a mixture of pollen and nectar.

Vespid wasps play a vital role in ecosystems and can serve as valuable indicators of environmental conditions and habitat perturbation. Due to their predatory behavior, they have the potential to serve as effective biological control agents against agricultural pests, and some species have already been successfully employed in agroecosystems.

Solitary wasps nest in soil, pre-existing cavities (such as abandoned mud nests of other hymenopterans or beetle tunnels in wood), or in free-standing mud nests they construct themselves. Social wasps, on the other hand, often build elaborate paper nests and vigorously defend their colonies through stinging.

This family comprises more than 5000 described species and 250 genera across six subfamilies: Eumeninae, Masarinae, Polistinae, Vespinae, Euparagiinae, and Stenogastrinae, with the first four subfamilies occurring in the Middle East. The intricate behaviors, diverse morphologies, and ecological significance of Vespidae make them a compelling subject for scientific research.

Bees and wasps are both vital components of natural capital for human societies. However, while bees are generally well-regarded, wasps are often viewed unfavorably. Despite the ecological significance of both groups, bees are widely appreciated, whereas wasps are commonly met with aversion.

In this book, we document the diversity of vespid wasps in the Middle East and aim to highlight their broader ecological value, with the hope that they will come to be appreciated as much as bees.

Hassan Ghahari
[Tehran, Iran]
Neveen S. Gadallah
[Cairo, Egypt]
Patrick K. Piekarski
[New York, USA]

Acknowledgments

The first editor wishes to express sincere gratitude to several hymenopterists who kindly assisted in identifying vespid specimens from the Iranian fauna. Their invaluable contributions have significantly advanced recent efforts to document the Vespidae of Iran, and their work represents an important foundation for this project: Friedrich Wolfgang Gess (South Africa; 27 September 1936 – 6 August 2013), Josef Gusenleitner (Austria; 17 September 1929 – 6 July 2023), Libor Dvořák (Czech Republic), Michael E. Archer (UK), Erol Yildırim and Nil Bagriacik (Turkey). Additionally, the systematic work of Ebrahim Ebrahimi on Iranian Vespidae laid the groundwork for research on this family in Iran.

The authors would also like to thank the colleagues who edited various chapters or provided insightful comments: James M. Carpenter (Eumeninae), Sarah K. Gess (Masarinae), Fuki Saito-Morooka and Rainer Neumeyer (Polistinae), Erol Yildırim (Polistinae, Vespinae), and Christopher K. Starr (Chapter 6).

Many thanks are also due to Alexander V. Fateryga (Russia), Michael E. Archer (UK), James M. Carpenter (USA), Marco Selis (Italy), George R. Else (UK) and Erol Yildırim (Turkey) for providing essential literature, and to Shaaban Abd-Rabou (Egypt) and Hamid Sakenin (Iran) for loaning specimens for study.

This work was supported by Islamic Azad University, Yadegar-e-Imam Khomeini (RAH) Shahre Rey Branch (Iran), Cairo University (Egypt), and Rockefeller University (USA).

Finally, we would like to thank everyone else who provided us with advice, support, and assistance throughout the course of this study, as well as the staff at CRC Press for their cooperation and support.

Polistes metricus Say, 1831 (Polistinae) on fragrant sumac (Anacardiaceae). (Photo courtesy of Angella Moorehouse)

Dolichovespula sylvestris (Scopoli, 1763) (Vespinae). (Photo courtesy of Hana Křížinecká)

Pseudepipona beckeri (Morawitz, 1867) (Eumeninae). (Photo courtesy of Alexander Fateryga)

Celonites kozlovi (Kostylev, 1935) (Masarinae). (Photo courtesy of Alexander Fateryga)

Celonites ivanovi (Mauss & Fateryga, 2022) (Masarinae). (Photo courtesy of Alexander Fateryga)

Polistes aurifer (Saussure, 1853) (Polistinae) on pearly everlasting (Asteraceae). (Photo courtesy of Angella Moorehouse)

Paravespa rex (von Schulthess, 1924) (Eumeninae). (Photo courtesy of Alexander Fateryga)

> The wasps terrorize housewives, ruin picnics, and build large aerial nests that challenge fleet-footed stone-throwing boys the world over.
>
> (Evans and West-Eberhard, 1970)

Polistes nimpha (Christ, 1791) (Polistinae). (Photo courtesy of Alexander Fateryga)

No insect hangs its nest on threads as frail as those which will sustain the weight of human vanity.
(Edith Newbold Wharton; 24 January 1862 – 11 August 1937; American writer and designer)

Vespa analis (Fabricius, 1775) (Vespinae) on flowers of Banana (*Musa paradisiaca* L., Musaceae). (Photo courtesy of Christophe Barthelemy)

> The wing structure of the hornet, in relation to its weight, is not suitable for flight, but he does not know this and flies anyway.
>
> (Albert Einstein; 14 March 1879 – 18 April 1955; A German physicist)

1 Introduction to the Family Vespidae (Hymenoptera: Vespoidea)

Hassan Ghahari, Neveen S. Gadallah, Christopher K. Starr, Patrick K. Piekarski, and Kees van Achterberg

Concerning the generation of animals akin to them, as hornets and wasps, the facts in all cases are similar to a certain extent, but are devoid of the extraordinary features which characterize bees; this we should expect, for they have nothing divine about them as the bees have.

(Aristotle; 384–322 BC; Ancient Greek philosopher and polymath)

HYMENOPTERA: ECOLOGICAL AND ECONOMIC ROLES

Hymenoptera is one of the largest and most megadiverse insect orders, notable not only for its vast number of species, but also for its wide range of lifestyles (Austin and Dowton, 2000). It includes wasps, sawflies, bees, and ants (Aguiar *et al.*, 2013), highlighting the ecological, agricultural, and cultural importance of the group (Gauld and Bolton, 1988).

The order encompasses a wide array of forms, such as large-bodied Vespidae that provision their young with larvae of crop pests, wingless Chrysididae that act as egg parasitoids, and tiny Mymaridae, commonly called "fairy-flies" due to their feather-like wings.

Arguably, Hymenoptera is best known for a few social families, which include species that build elaborate colonies and are often noticed for their ability to sting, bite, and infest human dwellings. Despite these negative associations, they play essential roles as biological control agents, primary pollinators of food crops, and producers of honey (Gauld and Bolton, 1988; Hanson, 1995; Austin and Dowton, 2000; Michener, 2000).

Hymenoptera is one of four insect orders with more than 100,000 described species, the others being Coleoptera, Lepidoptera, and Diptera. Although Hymenoptera ranks a distant second to Coleoptera in terms of the number of described species worldwide, largely due to under-sampling in tropical regions, it is often the most species-rich in surveyed temperate regions (Gauld and Bolton, 1988; Goulet and Huber, 1993; LaSalle and Gauld, 1993; Forbes *et al.*, 2018).

For example, in the Netherlands, there are 6089 species of Hymenoptera, compared with the 5887 species of Diptera, and 4163 spp. of Coleoptera (Peeters *et al.*, 2024). In many tropical arthropod samples, Hymenoptera appears to rival beetles in diversity, particularly among the lesser-studied parasitoid groups (Noyes, 1989; Gaston, 1991; Whitfield, 1998).

Currently, over 153,000 species of Hymenoptera have been described (Peters *et al.*, 2017), and estimates of the total number of existing species range from 300,000 to 3,000,000 (Gaston, 1991; Aguiar *et al.*, 2013), with a commonly cited estimate around 1,000,000 (Sharkey, 2007). This suggests that only about one-tenth of species have been formally described. However, estimating the

DOI: 10.1201/9781003740438-1

true number of species remains challenging, as most megadiverse regions of the world have yet to be thoroughly studied and inventoried (LaSalle and Gauld, 1993; Rasnitsyn and Quicke, 2002; Grimaldi and Engel, 2005).

The order Hymenoptera is known from fossils dating back at least 220 million years ago (Mya) (Rasnitsyn, 1969; Kukalova-Peck, 1991; Labandeira and Sepkoski, 1993). There is also indirect evidence, such as damage patterns on fossilized plants, that suggests the order may have originated as early as 300 Mya (Whitfield, 1998). Molecular dating analyses suggest that the most recent common ancestor of all extant Hymenoptera dates to ~281 Mya (239–329 Mya; Peters *et al.*, 2017). The diversification of Hymenoptera may have been driven, in part, by the concurrent radiation of flowering plants and their associated insect fauna. However, identifying the critical events in this co-evolutionary relationship remains challenging with the currently available data. Several parasitoid lineages had already emerged by 160 Mya, and all superfamilies of Hymenoptera were present by the late Cretaceous (65–70 Mya) (Gauld and Bolton, 1988; Gauld and Hanson, 1995; Rasnitsyn, 1980, 1988; Whitfield, 1998).

Hymenoptera is divided into two suborders: Symphyta and Apocrita. Symphyta are mostly phytophagous and are considered the most primitive members of the order. The Apocrita include two groups: the paraphyletic **Parasitica** and the monophyletic **Aculeata**. Parasitica are primarily parasitoids, although some have secondarily become phytophagous. In contrast, Aculeata exhibits a broader range of lifestyles, including predation, pollination, and parasitism, and encompasses all eusocial Hymenoptera (Austin and Dowton, 2000; Rasplus *et al.*, 2010).

The monophyly of the Aculeata is most notably supported by a key morphological feature: the modification of the female reproductive system, in which the egg is laid through an opening separated from the ovipositor (Brothers and Carpenter, 1993; Gadallah and Brothers, 2020). In Aculeata, the ovipositor has lost its original function as an egg-laying organ and has been transformed into a stinger used to envenomate hosts, prey, or enemies. It is secondarily reduced or lost in the stingless bees (Brothers and Carpenter, 1993; Whitfield, 1998) and some other groups. The ability to paralyze or incapacitate hosts or prey was likely already present in the common ancestor of aculeates and ichneumonoids. Some aculeates have further specialized their venom to induce pain and allergic reactions in vertebrates (Piek, 1986).

Many members of the Vespoidea and Apoidea excavate burrows or construct nests from various materials in which they deposit hosts or prey to provision their offspring and protect them (Evans and West-Eberhard, 1970). The evolution of progressive larval provisioning with prey, combined with nest defense using the stinger, preadapted certain groups toward social living. This also facilitated a shift in some lineages to using seeds, pollen, and nectar as food for larvae, instead of insect hosts or prey (O'Toole and Raw, 1991; Hanson and Gauld, 1995; Whitfield, 1998).

Ecologically and economically, few insect groups are as important to humanity as the Hymenoptera (Rasplus *et al.*, 2010). The order exhibits a wide spectrum of natural life histories and is found in virtually all terrestrial habitats, where its members play numerous ecological roles (Goulet and Huber, 1993; Whitfield, 1998; Huber, 2009).

Members of Aculeata, including ants, bees, and wasps, provide essential ecosystem services, notably pollination in both natural and managed ecosystems (Gallai *et al.*, 2009; Gess and Roosenschoon, 2016; Ghahari, 2017). Meanwhile, parasitoid Hymenoptera regulate populations of phytophagous insects (Tscharntke *et al.*, 2007) and serve as effective biological control agents against agricultural and forest pests (Bale *et al.*, 2008; Brodeur and Boivin, 2004; Vilhelmsen and Turrisi, 2011).

Hymenoptera often outperform other insect orders in pollination and parasitism, likely due to the high proportion of species that specialize in these roles and their frequent host specificity. Additionally, some phytophagous Hymenoptera have intimate associations with their host plants (Nyman *et al.*, 2006) and certain groups, such as Diprionidae, are considered major forest pests (De Somviele *et al.*, 2004).

Invasive ants can cause significant economic and ecological damage (Holway *et al.*, 2002; Lach and Thomas, 2008), and Hymenoptera stings are capable of inducing severe allergic reactions, including anaphylaxis (Flabbee *et al.*, 2008; Klotz *et al.*, 2009).

The importance of Hymenoptera to basic research often stems from their distinctive, and sometimes unique, biological features, the study of which can yield valuable insights with broad applications beyond pollination alone (LaSalle and Gauld, 1993; Hanson, 1995; Grissell, 2010).

One well-known feature is their haplodiploid method of sex determination, in which unfertilized eggs develop into males and fertilized eggs into females. This system has been hypothesized to promote the evolution of eusociality and complex social behaviors due to its effects on relatedness among individuals (Hanson, 1995; Linksvayer and Wade, 2005; Heimpel and de Boer, 2008; Rasplus *et al.*, 2010; Beukeboom and Perrin, 2014).

The evolutionary history of Hymenoptera includes multiple independent origins of eusociality, more than any other animal order, and a remarkable radiation of parasitoid lineages. As such, the order remains a rich source of comparative research in evolutionary biology and ecology (Rasnitsyn, 1980; Davis *et al.*, 2010; Sharkey *et al.*, 2012; Peters *et al.*, 2017).

FAMILY VESPIDAE

Vespid wasps are typically medium-sized, though they can range from relatively small (body length 2.5 mm) to very large (up to 50 mm). They are recognized by several key morphological features: the inner eye orbits often display a pronounced notch opposite to the antennal socket; the pronotum extends laterally back to the tegulae, with a deep oval or triangular dorsal emargination; the forewings fold longitudinally when at rest (with the exception of Stenogastrinae, Euparagiinae, and part of Masarinae); and the first discoidal cell of the forewing is longer than half the wing length. Their body coloration typically includes yellow or white spots and bands on a black background, although some species may be black with reddish markings, or entirely black-brown or yellow (Kurzenko, 1995, 2012; Antropov and Fateryga, 2017).

This is a cosmopolitan family, though its greatest species richness is found in tropical regions (Brothers and Finnamore, 1993; Dvořák and Straka, 2007). The ongoing study of vespid wasps is essential for improving our understanding of their life history, behavior, habitat preferences, and ecological roles (Brock *et al.*, 2021). Vespid wasps commonly prey on caterpillars and other insects, including many agricultural pests, to feed their larvae (Das and Gupta, 1989; Goulet and Huber, 1993). This predatory behavior makes them effective biological control agents in pest management programs (Gould and Jeanne, 1984; Gess, 1996; Pham, 2014). In addition to their role as predators, some vespid wasps also contribute to pollination, visiting a wide variety of flowers to collect nectar for energy (Kojima, 1993; Carpenter and Wenzel, 1999; Khuat *et al.*, 2004). These ecological functions suggest that vespid wasps may serve as valuable bioindicators of environmental conditions and habitat disturbance (Carpenter, 1991; Pham, 2014).

Females of subfamily Eumeninae (potter wasps) construct nests and provision them with paralyzed prey, typically larvae of Lepidoptera, Curculionidae, and Chrysomelidae (Junior *et al.*, 2012), as food for their developing offspring (Krombein *et al.*, 1979; O'Neill, 2001; Johnson *et al.*, 2023). The female stings and paralyzes the prey before placing it in the nest cell. The larvae then develop inside the cell, feeding on the immobilized prey, and emerge as adults by breaking out of the sealed cell (Hermann, 1982; Carpenter, 1991; Cowan, 1991; Matthews and Matthews, 2018). Several studies have reported that potter wasps are effective generalist predators of economically significant Lepidoptera pests (Buschini and Buss, 2010; Udayakumar *et al.*, 2022; Johnson *et al.*, 2023).

Vespidae belong to the superfamily Vespoidea, which has traditionally included nine other families: Bradynobaenidae, Formicidae (i.e., ants), Mutillidae (i.e., velvet ants), Pompilidae (i.e., spider wasps), Rhopalosomatidae, Sapygidae, Scoliidae, Sierolomorphidae, and Tiphiidae (Brothers and Finnamore, 1993; Aguiar *et al.*, 2013).

A revised classification of families and subfamilies within Vespoidea was proposed by Pilgrim *et al.* (2008) based on molecular analyses, whereby several families were elevated to the superfamily level. Furthermore, recent molecular phylogenies of Hymenoptera suggest Vespoidea is paraphyletic (Peters *et al.*, 2017; Blaimer *et al.*, 2023). Nonetheless, the traditional classification presented by Aguiar *et al.* (2013), which retains all these families under the single superfamily Vespoidea, continues to be used by some hymenopterists (e.g., Sharkey *et al.*, 2012; Schmid-Egger and Schmidt, 2021). This group is one of the three aculeate superfamilies, Apoidea, Chrysidoidea, and Vespoidea, and include many of the most conspicuous wasp groups, among them two of the most commonly encountered aculeates: the highly social ants and paper-nesting wasps. The social behaviors exhibited by ants and vespids have played a particularly significant role in shaping theories on the transitional stages of social evolution (Wilson, 1971; Hölldobler and Wilson, 1990; Pilgrim *et al.*, 2008; Rehan and Toth, 2015).

The family Rhopalosomatidae, comprising fewer than 100 species, all presumed to be ectoparasitoids of crickets (Lohrmann and Engel, 2017), is considered the sister group of Vespidae (Pilgrim *et al.*, 2008; Peters *et al.*, 2017; Blaimer *et al.*, 2023).

The Vespidae comprises more than 5465 species worldwide, 58 of which are known only from fossils (Perrard *et al.*, 2017), across 255 extant and 13 extinct genera. The family is traditionally divided into six subfamilies: Eumeninae, Euparagiinae, Masarinae, Polistinae, Stenogastrinae, and Vespinae (Carpenter, 1982; Pickett and Carpenter, 2010; Piekarski *et al.*, 2014), along with two extinct subfamilies represented exclusively by fossils (Priorvespinae and Protovespinae) (Aguiar *et al.*, 2013; Antropov and Fateryga, 2017).

The life history and behaviors of four subfamilies are discussed in the following chapters. Since Euparagiinae and Stenogastrinae are not found in the Middle East, we provide only a brief description of these groups here.

The subfamily Euparagiinae is a small, relict group of vespids that includes a single extant genus, *Euparagia* Cresson, 1879, comprising 10 known species found exclusively in the southwest of North America. Females excavate vertical burrows in the ground and construct mud turrets or "chimneys" above the entrance. Each nest contains several cells, which are provisioned with larvae of Curculionidae (Carpenter, 1982; Bohart, 1989; Carpenter and Kimsey, 2009). These wasps exhibit nesting behavior similar to those of the eumenine genus *Odynerus* (Mauss, 2007).

The subfamily Stenogastrinae (commonly known as hover wasps), includes seven genera and 63 species of medium-sized (11–21 mm long), slender wasps characterized by a very long and narrow first gastral segment. These wasps are restricted in their distribution to the Oriental region and the island of New Guinea. Hover wasps are eusocial, though their colonies are typically small, rarely comprising more than 10 adults (Field *et al.*, 2000). Nests are often founded by a single female, although in some species, colonies may be established by multiple females (Wenzel, 1998; Turillazzi, 2012). These social wasps display allomaternal care, with daughters frequently remaining in their natal nest to help rear their siblings. Notably, there is no morphological differentiation between reproductive and non-reproductive individuals (i.e., helpers).

Unlike most other social vespids, their nests lack a primary petiole, and the ventrally open brood cells are attached directly to the substratum – such as thin tree branches, vine tendrils, loose strands of thatch under roof tops, or more rarely, under stones. The cells are made of chewed plant material, sometimes mixed with soil, and are gradually lengthened as the larvae grow. Stenogastrines usually build their nest in sheltered environments, such as under cliff overhangs, in caves, beneath small bridge vaults, or inside water pipelines, generally dark, humid places (Wenzel, 1998; Hunt and Toth, 2017; Varghese and Girish Kumar, 2023).

A unique feature of hover wasps is their secretion of a white gelatinous substance from the Dufour's gland. This secretion serves as a substrate for egg deposition and larval development. In some species, it is also used to construct protective barriers, known as ant guards, to deter ant predation (Turillazzi, 1991, 2012; Field, 2021).

The generic classification of Eumeninae has long been problematic and is marked by a complex taxonomic history (Hermes *et al.*, 2014). Richards (1962) proposed a classification for the then-recognized "Eumenidae," dividing it into three subfamilies: Discoeliinae (= Zethinae), Eumeninae, and Raphiglossinae. In 1982, Carpenter restructured the group by recognizing a single family, Vespidae, with a broad Eumeninae subfamily that encompassed the three subfamilies proposed by Richards (1962). He did not provide any tribal classification and rejected the use of the subfamily names Raphiglossinae and Zethinae. However, according to modern taxonomic treatments (Banks *et al.*, 2017; Piekarski *et al.*, 2018; Mauss *et al.*, 2019; Luo *et al.*, 2022; Benda *et al.*, 2024), eumenine wasps should be divided into at least three distinct subfamilies: Eumeninae *s.str.*, Raphiglossinae, and Zethinae (Fateryga and Proshchalykin, 2024). In addition, Gayellinae (formerly the tribe Gayellini within Masarinae) was elevated to subfamily rank (Piekarski *et al.*, 2018). Altogether, based on recent molecular phylogenetic studies, three former taxa have now been elevated to subfamily status: Gayellinae, Raphiglossinae, and Zethinae (Banks *et al.*, 2017; Piekarski *et al.*, 2018), bringing the total number of recognized vespid subfamilies to nine. However, in this book, we bin the Raphiglossinae, Zethinae and Eumeninae *s.str.* under the label of Eumeninae *sensu lato*.

IMPORTANCE OF VESPIDAE

Most vespid species are solitary and many play an important role as a biological control agent by protecting crops; they feed their larvae with various harmful insects such as lepidopteran caterpillars, orthopterans, dipterans, and other pests (Figure 1.1) (Soares and Lara, 1994; Donovan, 2003; Grimaldi and Engel, 2005; Fateryga, 2009; Prezoto *et al.*, 2019). Additionally, many vespids serve as potential pollinators of various fruit and vegetable crops (Figure 1.2) (Ghahari, 2017; Carpenter and Gadallah, 2020). Yellowjackets scavenge dead insects, earthworms and other carrion. Some species have become invasive pests, having been recorded in new regions (Matsuura and Sakagami, 1973; Akre and Davis, 1978; Otis *et al.*, 2023). The predation of wasps on honey and honeybees

FIGURE 1.1 *Tropidodynerus interruptus* (Brullé, 1832) (Eumeninae) – carrying a Coleoptera larvae to the nest.
(Photo courtesy of Alexander Fateryga)

FIGURE 1.2 *Vespula maculifrons* (Buysson, 1905) (Vespinae) on calico aster (Asteraceae).

(Photo courtesy of Angella Moorehouse)

was known since at least the sixteenth century (Edwards, 1980). Certain species of *Vespa* that hunt honey bees can cause serious damage to apiculture (Monceau *et al.*, 2014; Choi and Kwon, 2015). The economic importance of social wasps to beekeepers has been well documented in some cases. For example, as few as eight workers of *Vespa mandarinia* Smith can destroy an entire hive within a few hours (De Jong, 1979; Matsuura, 1988). Moreover, species of *Vespa* and *Vespula* may occasionally cause severe damage to fruit crops, particularly grapes and pears (Al-Mahdawi and Al-Kihani, 2011; Cranshaw *et al.*, 2011). Females are feared due to their painful stings, and certain species of Polistinae and Vespinae can cause certain medical risks due to their aggressive attacks, which reportedly result in hundreds of fatalities annually in various countries (Spradbery, 1973; Akre, 1982; Antropov and Fateryga, 2017). Additionally, these wasps can cause problems in pastry and confectionery shops, hinder fruit harvesting, and attack castration wounds or other injuries on livestock (Rasnitsyn and Quicke, 2002; Lester, 2018).

Vespid wasps have also been regarded as beneficial insects, both as a source of food (the larvae and pupae) and in traditional medicine. *Brachygastra mellifica* (Say, 1837) is a traditional source of honey in Central America, while honey produced by some species of *Belonogaster* de Saussure, 1854, is used in traditional medicine in Africa. Additionally, the larvae and pupae of *Vespa* are part of the traditional diet of local populations in South Eastern Asia (Evans and West-Eberhard, 1970; Edwards, 1980; Antropov and Fateryga, 2017).

Vespid venom contains three major proteins that act as allergens, the most important of which are antigen 5, phospholipase, and hyaluronidase (Dongol *et al.*, 2014). It also includes a wide variety of vasoactive amines and peptides (Dongol *et al.*, 2014), along with several other proteins that remain incompletely characterized (Fitzgerald and Flood, 2006; Kolarich *et al.*, 2007). Some components of vespid venom act effectively against mammals, for example, kinins, which contribute to lowered blood pressure. This suggests a possible evolutionary interaction between the feeding habits of mammals and the remarkable aggressiveness of certain vespid species (Matsuura and Yamane, 1990).

Vespid venoms are a complex mixture composed of low-molecular weight amines, low-molecular weight peptides, and high-molecular weight enzymes (Chen *et al.*, 2010; Lee *et al.*, 2016; Luo *et al.*, 2022). The substances believed to cause the sharp pain during a sting include amines such as histamine and serotonin, as well as low-molecular weight kinins (wasp kinins) (Edery *et al.*, 1978; Schmidt, 1982). Ongoing research are exploring the potential use of toxic components from some species' venom in the treatment of neuralgia (Matsuura and Yamane, 1990; Karsai and Hunt, 2002; Donovan, 2003). Vespid venoms are also of interest due to their ability to trigger severe allergic reactions (Kolarich *et al.*, 2007; Gong *et al.*, 2019; Sharma and Upadhyay, 2021).

Public aversion to wasps is widespread, largely stemming from deep-rooted cultural bias and a general lack of recognition of their ecological and economic importance (Sumner *et al.*, 2018). This negative perception is primarily shaped by a small subset of wasps—namely, the vespines—which represent less than 1% of aculeate (stinging) wasp species (Carpenter and Kojima, 1997). To improve public understanding and appreciation of wasps, it is essential that scientists place greater emphasis on these insects and undertake the necessary research to quantify their contributions to ecosystem services. Unlike bees, which are widely recognized for their multi-billion-dollar role in pollination and ecosystem functioning, wasps currently lack comprehensive estimates of their ecological and economic value, particularly in their role as natural pest controllers and regulators of arthropod populations (Sumner *et al.*, 2018).

BIOLOGY OF VESPIDAE

The family Vespidae (yellowjackets, hornets, paper wasps, potter wasps, hover wasps, and pollen wasps) exhibits remarkable diversity in behavior and life history across species. Among social species, the behavioral plasticity of individual wasps can be particularly exceptional (Cowan, 1991; West-Eberhard, 2005; Kelstrup *et al.*, 2023). Nearly 80% of all species in the family are solitary wasps, although some lineages of eumenines and masarines have evolved subsocial behavior, including progressive provisioning, and in some cases, quasi-communal nest-sharing (Carpenter, 1988; West-Eberhard, 2005; Fateryga *et al.*, 2021; Kelstrup *et al.*, 2023). In species that are cooperative breeders and/or superorganismal, colonies consist of queens, workers, and males. In both subsocial and social species, larvae are typically fed progressively by adult females with masticated insects (Goulet and Huber, 1993). Potter wasps (Eumeninae) capture caterpillars or the larvae of beetles or sawflies, whereas pollen wasps (Masarinae) provision their nests with a mixture of pollen and nectar, similar to bees. Paper wasps (Polistinae), along with the highly social yellowjackets and hornets (Vespinae), construct their nests from chewed wood ("paper"). A few species are cleptoparasitic, living in the nests of other species within the same genus (Evans, 1977; Goulet and Huber, 1993; Gess, 1996; Gogala, 2022).

BEHAVIOR OF VESPIDAE

A derived trait of vespid wasps is the habit of laying an egg into an empty nest cell. In the ancestral, solitary condition, the cell is subsequently filled with provisions and typically sealed before the egg hatches, a behavior known as mass provisioning, which represents the ancestral state in the family (Cowan, 1991). In subsocial and social species, however, the cell is not sealed before the egg hatches, allowing for interactions between the larva and its mother and/or other adult caregivers.

Progressive provisioning is characteristic of all the social wasps (Stenogastrinae, Polistinae and Vespinae), approximately 17 species from various genera within Eumeninae *s.l.* (West-Eberhard, 2005), and at least one species of Masarinae (Cowan, 1991). Colony organization varies both among and within the social subfamilies (Carpenter, 1989) but typically includes at least one queen and multiple female workers who forego direct reproduction to invest their time and energy in nest-building, brood care, foraging, and nest defense (Spradbery, 1973; Matsuura and Yamane, 1990; Richter, 2000). A common behavioral trait across the social subfamilies is the simultaneous rearing and provisioning of multiple offspring. A *Liostenogaster flavolineata* (Stenogastrinae) lone foundress typically provisions an average of five cells simultaneously (Field *et al.*, 2000), whereas a polistine or vespine foundress often provisions 10–30 offspring at a time (Reeve, 1991; Mead *et al.*, 1994).

Nesting Behavior of Vespidae

Many solitary bees and wasps use holes in dead wood for offspring production. The majority cannot bore into the wood themselves and are therefore dependent on wood-boring insects to create suitable nesting cavities (Figure 1.3). Studies using artificial trap nests have shown that each species of bee or wasp has specific requirements regarding the hole size (Budriene *et al.*, 2004; Westerfelt *et al.*, 2015), and an increase in the availability of holes can lead to an increase in nest numbers (Tscharntke *et al.*, 1998). However, to the best of our knowledge, no study has yet mapped the distribution of natural potential nesting holes across the landscape (Westerfelt *et al.*, 2015).

Stenogastrine nests are constructed from mud and vegetable fibers, while vespines and polistines generally use chewed plant fibers mixed with saliva to form a paper-like substance known as carton (Carpenter and Gadallah, 2020). The nest material typically consists of wood or foliage that is chewed and processed by the insect (Borror *et al.*, 1989). Solitary vespid wasps may nest in burrows

FIGURE 1.3 *Vespa crabro* (Linnaeus, 1758) (Vespinae) – standing guard at the entrance hole of the nest cavity.

(Photo courtesy of Alexander Fateryga)

FIGURE 1.4 *Polistes mongolicus* (du Buysson, 1911) (Polistinae) – nest consisting of a single open comb. **(Photo courtesy of Alexander Fateryga)**

in the ground or dead wood, or build free-standing nests of mud attached to rocks or plants. In species of independent-founding Polistinae (*Mischocyttarus*, *Polistes*, and most Ropalidiini) (Figure 1.4), paper nests are open and consist of a single comb supported by a single pedicel, a pattern known as stelocyttarous. In contrast, nests of the subfamily Vespinae and most swarm-founding polistines (Epiponini and some Ropalidiini) contain multiple combs surrounded by an envelope. Some epiponines lack a pedicel supporting the comb (i.e., astelocyttarous) and instead, have sessile combs (Budriene *et al.*, 2004; Westerfelt *et al.*, 2015; Noll *et al.*, 2021).

The open-comb nest structure of paper wasps leaves the brood exposed to predators and nest parasites, prompting strong defensive reactions by workers in response to nest disturbances. In contrast, the highly eusocial yellowjackets and hornets build paper nests consisting of stacked combs enclosed within a spherical paper sheath. These species exhibit extremely aggressive and intimidating defensive behaviors when disturbances occur near their nests. Wasp colonies are vulnerable to a wide range of natural enemies. Consequently, social wasps have evolved a variety of primary (nest construction) and secondary (coordinated behavioral responses) defenses to protect the colony and reduce the risk of predation (Richards and Richards, 1951; Edwards, 1980; Wenzel, 1991; Detoni *et al.*, 2021).

Among social vespids, several species are obligate nest parasites (e.g., *Dolichovespula adulterina*, *D. omissa*, *Polistes atrimandibularis*, *P. sulcifer*, *Vespula austriaca*) that do not build their own nests but instead usurp the nests of closely related species. Obligate parasites lack a worker caste, and their first generation consists solely of males and females capable of mating. Other species, such as

Vespa dybowskii and *Polistes nimpha* may be facultative nest parasites: they can usurp the nests of related species but are also capable of founding their own colonies (Archer, 1999, 2006, 2008; Cervo, 2006).

In social species, larvae are progressively fed with masticated insects, and occasionally even carrion (Brothers and Finnamore, 1993; Gogala, 2022). In temperate regions, colonies typically last for only a single season. Only the queens overwinter, and in spring, each queen, or sometimes a group of related queens, establishes a new colony. The queen rears the first brood alone, which develops into workers. Once mature, these workers take over the duties of the colony, while the queen's role is then largely limited to egg-laying (Archer, 2008).

Eusocial wasps are often divided by some authors into two groups: the primitively eusocial paper wasps, which lack morphological differences between castes, and the highly eusocial yellowjackets and hornets, which exhibit distinct morphological differences between queens and workers (Hunt, 2007; Boomsma and Gawne, 2018; Breed and Moore, 2022). However, several instances of morphologically distinct castes in "primitively eusocial" paper wasps have been reported, suggesting that preimaginal caste biasing can occur to some extent even when workers remain reproductively plastic (Piekarski *et al.*, 2018 and references therein).

The mechanisms of queen dominance in social vespids are not fully understood, although they likely involve pheromonal and/or physical control over workers. In the highly social *Vespula vulgaris*, queen pheromones directly suppress ovarian development in adult workers and mark queen-laid eggs, allowing workers to recognize and eliminate haploid eggs laid by other workers (Oi *et al.*, 2015; Oi *et al.*, 2020). In contrast, in primitively eusocial wasps, reproductive division of labor is generally believed to be regulated through aggressive dominance behavior (Oi *et al.*, 2019), although in some species, pheromonal control appears to play a more prominent role (Mitra *et al.*, 2011).

Unfortunately, the nesting habits and biology of many social wasps remain poorly understood, as taxonomic studies often rely solely on museum specimens. Detailed natural history research is essential to uncover the ecological and biological intricacies of these social wasps (Richards, 1971; Hunt and Toth, 2017; Varghese and Girish Kumar, 2023).

DIAGNOSIS OF VESPIDAE

Species of the family Vespidae are predominantly black or brown, often with yellow or white bands and, in some cases, reddish markings. Some species are entirely dark brown or yellow, while a few, such as certain Malagasy Ropalidiini, may even be green (Richards, 1982; Goulet and Huber, 1993).

Vespidae can readily distinguished by the following combination of characteristics: eyes kidney-shaped, with the inner margins usually emarginate; pronotum produced laterally above anterior margin of the tegula, with a concavely oval, distinctly U-shaped posterior margin; meso- and metacoxae contiguous; wings fully developed in both sexes and folded longitudinally at rest (except in Stenogastrinae, Euparagiinae, and some Masarinae); forewing with first discoidal cell longer than half the wing length (except Gayellini); hindwing with an indistinct claval lobe and with or without a jugal lobe; posterior (inner) spur of metatibia weakly modified into a calcar; metasoma either sessile or petiolate; first metasomal segment without a true node, though sometimes approaching it; metasoma constricted between the first and second segments; male hypopygium simple and exposed (Richards, 1985; Brothers and Finnamore, 1993; Douwes *et al.*, 2012; Antropov and Fateryga, 2017). Sexual dimorphism is slight, and both sexes are macropterous (Brothers and Finnamore, 1993). An illustrative figure is provided to facilitate a clearer understanding of the morphological characteristics of the Vespidae family (see Figure 1.5).

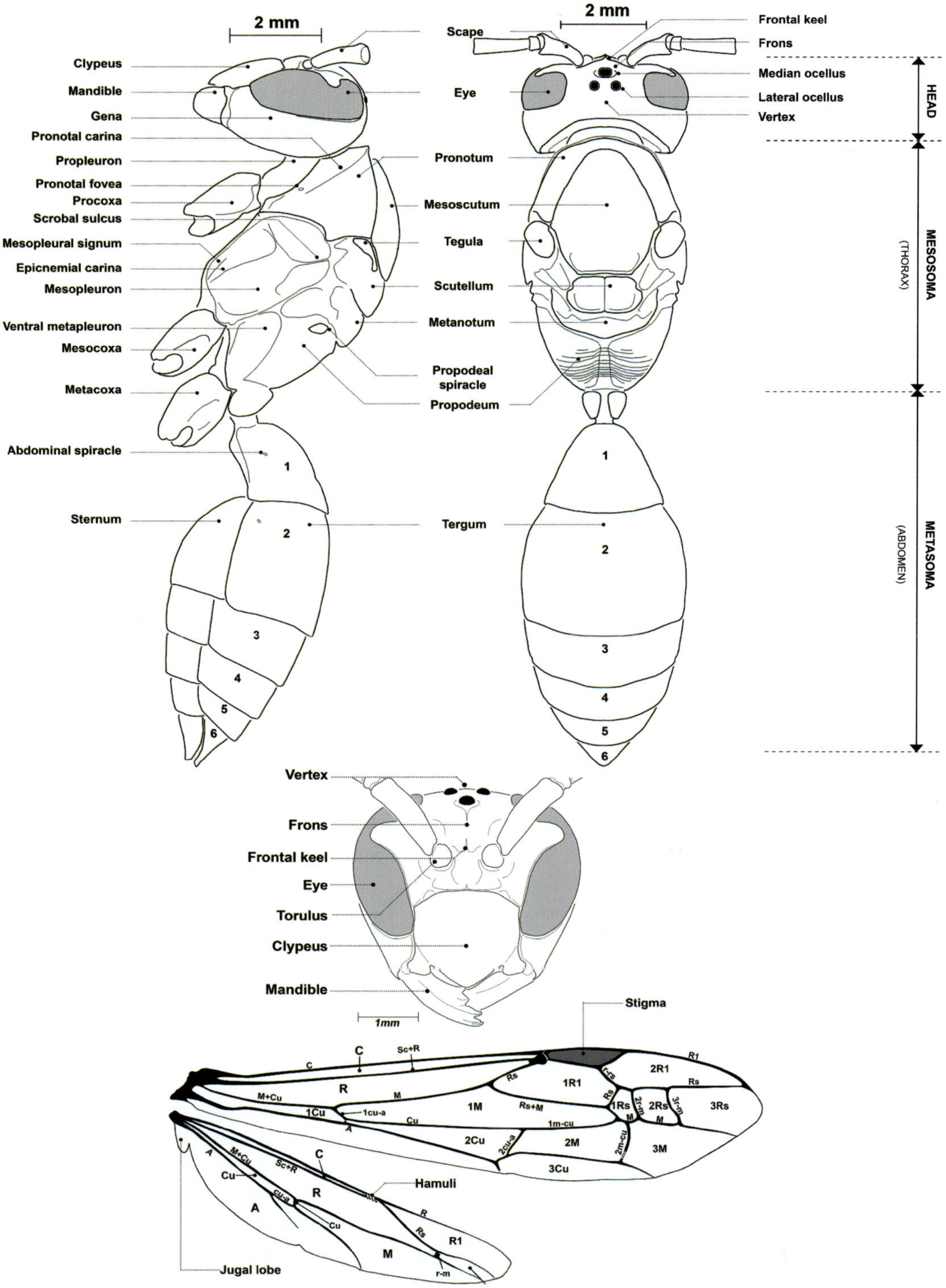

FIGURE 1.5 Morphological structures of Vespidae; Habitus – lateral and dorsal view; Head – frontal view; Wing venation; Wings cell: C Costal, Sc Subcostal, R Radial, Cu Cubital, 1Cu First Cubital, 2Cu Second cubital, 3Cu Third cubital, A Anal, 1R1 First radial 1, 2R1 Second radial 2, 1Rs First radial sector, 2Rs Third radial sector, 3Rs Third radial sector, M Medial, 1M First medial, 2M Second medial, 3M Third medial.

(Adapted from Barthélémy, 2021)

Identification key to subfamilies of Vespidae of the Middle East (Adapted from Carpenter, 1982; Brothers and Finnamore, 1993)

1. Forewing with two submarginal cells; apex of marginal cell of forewing widely separated from anterior margin of wing; apical free section of radial vein at least as long as width of pterostigma; mesoscutum without lateral longitudinal carina opposite tegula, and without projecting lobe at posterolateral apex; middle femur without basal ring (trochanter)..**Masarinae**

— Forewing with three submarginal cells; marginal cell of forewing narrowly or not separated from anterior margin of wing; apical free section of radial vein less than half width of pterostigma or absent; mesoscutum with a lateral longitudinal carina opposite tegula, or with a projecting lobe at posterolateral apex; middle femur usually with a trochanter, sometimes without..**2**

2. Mesoscutum with parategula (lobe projecting backward from posterolateral apex); tarsal claws bifid (cleft at tip), rarely simple and smooth ventrally; solitary or subsocial (Eumeninae *s.l.*) ....·**3**

— Mesoscutum without parategula; tarsal claws simple and smooth ventrally; eusocial ....·**5**

3. Forewing with the first recurrent vein received by the second submarginal cell, and the second recurrent vein received by third submarginal cell..**Raphiglossinae**

— Forewing with both recurrent veins received by second submarginal cell..**4**

4. Mid-tibia with only a single spur at distal end..**Eumeninae** ***s.str*.**

— Mid-tibia with two spurs at distal end..**Zethinae**

5. Hind coxa with dorsal longitudinal carina; hindwing without jugal lobe; metasoma sessile, first tergum truncate in dorsal view and having abrupt declivity anteriorly in lateral view ....**Vespinae**

— Hind coxa smooth dorsally; hindwing usually with jugal lobe, sometimes without; metasoma petiolate or subsessile, first tergum not truncate in dorsal view, with smooth declivity anteriorly in lateral view..·**Polistinae**

THE MIDDLE EAST

The Middle East (7,207,575 km^2) is a geopolitical region comprising Arabia, Asia Minor (Asian part of Turkey except Hatay province), East Tharace (European part of Turkey), Egypt, Iran, the Levant (including Ash-Sham and Cyprus), Iraq, and the Socotra Archipelago (a part of Yemen), or in other words, it represents the lands around the southeastern and eastern shores of the Mediterranean Sea. The central part of this general area formerly was called the Near East, so the term Middle East has become a replacement of the formerly used term "Near East" in the early 20th century. Most of the Middle East countries (13 out of 16) are part of the Arab World; Egypt, Iran, and Turkey are the most populous countries in the region, whereas Saudi Arabia is the largest in terms of area. The Middle East region generally is characterized by its hot, arid climate especially in Arabia and Egypt, with several major rivers providing irrigation to support agriculture. The great dependence on the fossil fuel industry in the region has a very negative impact on the fragile environments directly due to the extraction, transport, and refining of oil, causing serious pollution and degradation of natural habitats. Combustion of fossil fuels results in the emission of greenhouse gases, contributing to global warming. The following countries traditionally are included in the Middle East region (Figure 1.6): Bahrain, Cyprus, Egypt, Iraq, Iran, Israel-Palestine,[1] Jordan, Kuwait, Lebanon, Oman, Qatar, Saudi Arabia, Syria, Turkey, United Arab Emirates, and Yemen.

FIGURE 1.6 Map of Iran in relation to Middle East and adjacent countries.

DATA SOURCES

In this book, we summarize the available faunistic knowledge of Vespidae in the Middle East, incorporating both species reported in the literature and those confirmed through examination of specimens in entomological collections. Much of our current understanding of vespids in Iran is due to the efforts of a small number of dedicated collectors and taxonomists. Most reliable sources are taxonomic reviews and revisions. Reports from ecological or biological control studies were included only if co-authored by a taxonomic specialist or if a specialist was explicitly acknowledged for the identifications. Additionally, several specimens from private collections, particularly those of Iranian vespids recorded in recent years, were re-examined by taxonomists. Consequently, misidentified records have been excluded from this catalog. Due to numerous remaining taxonomic challenges within the group, including a lack of reliable identification keys for many groups, we applied stringent criteria. When accurate data on local distribution in Iran are lacking in a cited reference, the phrase "Iran (no specific locality cited)" is used. Locality data reported as "Persia" or "Persian" were retained as traditional synonyms for Iran.

This book is organized into chapters corresponding to the four subfamilies confirmed to occur in the Middle East. An identification key to these subfamilies is provided, along with keys to genera for selected, better-known groups. Each chapter includes comprehensive and up-to-date literature references covering taxonomy, distribution, economic importance, bioecology, and, where relevant, phylogenetic remarks.

This book thus represents the first comprehensive, modern appraisal of the true diversity and global distribution of Vespidae in the Middle East. We hope it will serve as a foundation for future research on vespid fauna in the Middle East, the Palaearctic Region, and beyond. It is also intended as an educational resource for early career entomologists, particularly hymenopterists. We further hope that this work will stimulate more extensive studies in Iran and other Middle Eastern countries, which hold significant potential for future discoveries.

NOTE

1 We treat Israel and Palestine as a single region, referred to as 'Israel-Palestine', because several taxa recorded from this area may originate from locations formerly designated as Palestine, or from areas still within present-day Palestine. To avoid ambiguity regarding the exact locality of such records, the use of 'Israel-Palestine' is most appropriate.

BIBLIOGRAPHY

Aguiar, A.P., Deans, A.R., Engel, M.S., Forshage, M., Huber, J.T., Jennings, J.T., Johnson, N.F., Lelej, A.S., Longino, J.T., Lohrmann, V., Mikó, I., Ohl, M., Rasmussen, C., Taeger, A. and Yu, D.S.K. 2013. Order Hymenoptera Linnaeus, 1758, pp. 51–62. In: Zhang, Z.-Q. (ed.), Animal biodiversity: An outline of higher-level classification and survey of taxonomic richness (addenda 2013). *Zootaxa* 3703(1): 1–82.

Akre, R.D. 1982. Social wasps, pp. 1–105. In: Hermann, H.R. (ed.), *Social insects*. Vol. 4. Academic Press, New York, xiii + 385 pp.

Akre, R.D. and Davis, H.G. 1978. Biology and pest status of venomous wasps. *Annual Review of Entomology* 23: 215–238.

Al-Mahdawi, Q.H.A. and Al-Kihani, A. 2011. Economical damage of the red wasp *Vespa orientalis* and yellow wasp *Polistes olivaceus* on grapes. *Diyala Agricultural Sciences Journal* 3(2): 216–222.

Antropov, A.V. and Fateryga, A.V. 2017. Family Vespidae, pp. 175–196. In: Belokobylskij, S.A. and Lelej, A.S. (eds), Annotated catalogue of the Hymenoptera of Russia. Volume I. Symphyta and Apocrita: Aculeata. *Proceedings of the Zoological Institute of the Russian Academy of Sciences*, Supplement No. 6, Zoological Institute of the Russian Academy of Sciences: Saint Petersburg, Russia, 475 pp.

Archer, M.E. 1999. Taxonomy, distribution and nesting biology of *Vespa binghami*, *V. basalis*, *V. variabilis*, *V. fervida*, *V. luctuosa*, *V. multimaculata* and *V. bellicosa* (Hym., Vespinae). *Entomologist's Monthly Magazine* 135: 43–50.

Archer, M.E. 2006. Taxonomy, distribution and nesting biology of species of the genus *Dolichovespula* (Hymenoptera, Vespidae). *Entomological Science* 9(3): 281–293.

Archer, M.E. 2008. Taxonomy, distribution and nesting biology of species of the genera *Provespa* Ashmead and *Vespa* Linnaeus (Hymenoptera, Vespidae). *Entomologist's Monthly Magazine* 144: 69–101.

Ashmead, W.H. 1902. Classification of the fossorial, predaceous and parasitic wasps, or the superfamily Vespoidea. *Canadian Entomologist* 34(9): 219–231.

Austin, A.D. and Dowton, M. 2000. *Hymenoptera: Evolution, biodiversity and biological control.* CSIRO Publishing, Collingwood, 468 pp.

Bale, J.S., van Lenteren, J.C. and Bigler, F. 2008. Biological control and sustainable food production. *Philosophical Transactions of the Royal Society B: Biological Sciences* 363: 761–776.

Banks, L.M., Kuper, H. and Polack, S. 2017. Poverty and disability in low- and middle-income countries. A sytematic review. *PLOS One* 13(9): e0204881.

Benda, D., Pohl, H., Beutel, R. and Straka, J. 2024. Solitary folded-winged wasps of the genus *Zethus* Fabricius (Vespidae, Zethinae) parasitized by two new species of Strepsiptera on different continents. *Journal of Hymenoptera Research* 97: 721–739.

Barthélémy, C. 2021. *Identification guide to the social wasps of Hong Kong (Hymenoptera; Vespidae).* Self-published; Hong Kong, 63 pp.

Beukeboom, L.W. and Perrin, N. 2014. *The evolution of sex determination.* Oxford University Press, New York, xvi + 222 pp.

Blaimer, B.B., Santos, B.F., Cruaud, A., Gates, M.W., Kula, R.R., Mikó, I., Rasplus, J.-Y., Smith, D.R., Talamas, E.J., Brady, S.G. and Buffington, M.L. 2023. Key innovations and diversification of Hymenoptera. *Nature Communications* 14(1212).

Bohart, R.M. 1989. A review of the genus *Euparagia* (Hymenoptera, Masaridae). *Journal of the Kansas Entomological Society* 62(4): 462–467.

Boomsma, J.J. and Gawne, R. 2018. Superorganismality and caste differentiation as points of no return: how the major evolutionary transitions were lost in translation. *Biological Reviews of the Cambridge Philosophical Society* 93(1): 28–54.

Borror, D.J., Triplehorn, C.A. and Johnson, N.F. 1989. *An Introduction to the study of insects.* Saunders College Publishing, Philadelphia, 875 pp.

Breed, M.D. and Moore, J. 2022. Chapter 14. Comparative social behavior, pp. 498–530. In: Animal behavior. Third edition. Academic Press, London, 600 pp.

Brock, R.E., Cini, A. and Sumner, S. 2021. Ecosystem services provided by aculeate wasps. *Biological Reviews* 96(4): 1645–1675.

Brodeur, J. and Boivin, G. 2004. Functional ecology of immature parasitoids. *Annual Review of Entomology* 49: 27–49.

Brothers, D.J. and Carpenter, J.M. 1993. Phylogeny of aculeata: Chrysidoidea and vespoidea. *Journal of Hymenoptera Research* 2: 227–302.

Brothers, D.J. and Finnamore, A.T. 1993. Chapter 8. Superfamily Vespoidea, pp. 161–278. In: Goulet, H. and Huber, J.T. (eds.), Hymenoptera of the world: An identification guide to families. Research Branch, Agriculture Canada, Ottawa, 668 pp.

Budriene, A., Budrys, E. and Nevronyte, Z. 2004. Solitary hymenoptera aculeata inhabiting trap-nests in Lithuania: nesting cavity choice and niche overlap. *Latvijas Entomologs* 41: 19–31.

Buschini, M.L.T. and Buss, C.E. 2010. Biological aspects of different species of *Pachodynerus* (Hymenoptera; Vespidae; Eumeninae). *Brazilian Journal of Biology* 70(3): 623–629.

Carpenter, J.M. 1982. The Phylogenetic relationships and natural classification of the Vespoidea (Hymenoptera). *Systematic Entomology* 7: 11– 38.

Carpenter, J.M. 1988. The phylogenetic system of the Stenogastrinae (Hymenoptera: Vespidae). *Journal of the New York Entomological Society* 96(2): 140–175.

Carpenter, J.M. 1989. Testing scenarios: Wasp social behavior. *Cladistics* 5(2): 131–144.

Carpenter, J.M. 1991. Phylogenetic relationships and origin of social behavior in the Vespidae, pp. 7–32. In: Ross, K.G. and Matthews, R.W. (eds.), *The social biology of wasps.* Cornell University Press, Ithaca, 678 pp.

Carpenter, J.M. and Gadallah, N.S. 2020. Biodiversity of the aculeate wasps (Hymenoptera: Aculeata) of the Arabian Peninsula: Vespoidea, Vespidae. *Zootaxa* 4754(1): 191–216.

Carpenter, J.M. and Kimsey, L.S. 2009. The genus *Euparagia* Cresson (Hymenoptera, Vespidae, Euparagiinae). *American Museum Novitates* 3643: 1–11.

Carpenter, J.M. and Kojima, J. 1996. Checklist of the subfamily Stenogastrinae (Hymenoptera: Vespidae). *Journal of the New York Entomological Society* 104: 21–36.

Carpenter, J.M. and Kojima, J. 1997. Checklist of the species in the subfamily Vespinae (Insecta: Hymenoptera: Vespidae). *Natural History Bulletin of Ibaraki University* 1: 51–92.

Carpenter, J.M. and Wenzel, J.W. 1999. The relative abundance of swarm-founding social wasps in the Congo Basin (Insecta: Hymenoptera: Vespidae: Polistinae). *Natural History Bulletin of Ibaraki University* 3: 9–14.

Cervo, R. 2006. *Polistes* wasps and their social parasites: an overview. *Annales Entomologici Fennici* 43: 531–549.

Chen, L., Chen, W., Yang, H. and Lai, R. 2010. A novel bioactive peptide from wasp venom. *Journal of Venom Research* 1: 43–47.

Choi, M.B. and Kwon, O. 2015. Occurrence of hymenoptera (wasps and bees) and their foraging in the Southwestern part of Jirisan National Park, South Korea. *Journal of Ecology and Environment* 38(3): 367–374.

Cowan, D.P. 1991. The solitary and presocial Vespidae, pp. 33–73. In: Ross, K.G. and Matthews, R.W. (eds.), *The social biology of wasps.* Cornell University Press, Ithaca, 678 pp.

Cranshaw, W., Larsen, H.J. and Zimmerman, R.J. 2011. Notes on fruit damage by the European paper wasp, *Polistes dominula* (Christ) (Hymenoptera: Vespidae). *Southwestern Entomologist* 36(1): 103–105.

Das, B.P. and Gupta, V.K. 1989. The social wasps of India and the adjacent countries (Hymenoptera: Vespidae). *Oriental Insects Monograph* 11: 176–189.

Davis, R.B., Baldauf, S.L. and Mayhew, P.J. 2010. The origins of species richness in the Hymenoptera: Insights from a family-level supertree. *BMC Evolutionary Biology* 10:109.

De Jong, D. 1979. Social wasps, enemies of honeybees. *American Bee Journal* 119(7): 505–507; 529.

De Somviele, B., Lyytikainen-Saarenmaa, P. and Niemela, P. 2004. Sawfly (Hym., Diprionidae) outbreaks on Scots pine: effect of stand structure, site quality and relative tree position on defoliation intensity. *Forest Ecology and Management* 194: 305–317.

Detoni, M., Feás, X., Jeanne, R.L., Loope, K.J., O'Donnell, S., Santoro, D. and Sumner, S. 2021. Evolutionary and ecological pressures shaping social wasps collective defenses. *Annals of the Entomological Society of America* 114(5): 581–595.

Dongol, Y., Dhananjaya, B.L., Shrestha, R.K. and Aryal, G. 2014. Pharmacological and immunological properties of wasp venom. *Pharmacology and Therapeutics* 5(8): 783–792.

Donovan, B.J. 2003. Potential manageable exploitation of social wasps, *Vespula* spp. (Hymenoptera: Vespidae), as generalist predators of insect pests. *International Journal of Pest Management* 49(4): 281–285.

Douwes, P., Abenius, J., Cederberg, B. and Wahlstedt, U. 2012. *Nationalnyckeln till Sveriges flora och fauna*, Steklar. ArtDatabanken, SLU, Uppsala, 382 pp.

Dvořák, L. and Straka, J. 2007. Vespoidea: Vespidae (vosovití). *Acta Entomologica Musei Nationalis Pragae*, Supplementum 11, pp. 171–189.

Ebrahimi, E. and Carpenter, J.M. 2008. Catalog of the vespid wasps of Iran (Hymenoptera, Vespidae). *Zootaxa* 1785, pp. 1–42.

Edery, H., Ishay, J., Gitter, S. and Joshua, H. 1978. Venoms of Vespidae, pp. 691–771. In: Bettini, S. (ed.), Arthropod venoms. Springer, Berlin Heidelberg New York, 977 pp.

Edwards, R. 1980. *Social wasps*. Their Biology and Control. Rentokil, East Grinstead, UK, 398 pp.

Evans, H.E. 1977. Observations on the nests and prey of eumenid wasps (Hymenoptera, Eumenidae). *Psyche* 83: 255–259.

Evans, H.E. and West-Eberhard, M.J. 1970. The wasps. University of Michigan Press, Ann Arbor, Michigan, vi + 265 pp.

Faraz, A., Jamali, M.M., Kumar, G. and Mir, S.J. 2023. Chapter 8. Importance of Hymenopteran insects in pollination, pp. 113–125. In: Ghoneim, K. (ed.), *Advances in agricultural entomology*. AkiNik Publications, 140 pp.

Fateryga, A.V. 2009. Trophic relations between vespid wasps (Hymenoptera: Vespidae) and flowering plants in the Crimea. *Entomological Review* 90(6): 698–705.

Fateryga, A.V., Proshchalykin, M.Y., Maharramov, M.M. and Astafurova, Y.V. 2021. New records of solitary vespid wasps (Hymenoptera: Vespidae: Masarinae and Eumeninae *s.l.*) from the Nokhchivan Autonomous Republic of Azerbaijan. *Zootaxa* 5027(1): 36–60.

Fateryga, A.V. and Proshchalykin, M.Y. 2024. New records of the solitary Vespidae (Hymenoptera: Vespidae: Masarinae and Eumeninae s. l.) from Central Asia, with new synonyms and description of a new species of *Eumenes* Latreille. *Zootaxa*, 5529(3): 551–569.

Field, J., Shreeves, G., Sumner, S. and Casiraghi, M. 2000. Insurance-based advantage to helpers in a tropical hover wasp. *Nature* 404: 869–871.

Field, J.P. 2021. Hover wasps (Stenogastrinae). In: Starr, C.K. (ed.), *Encyclopedia of Social Insects*. Springer, Cham, pp. 500–504.

Fitzgerald, K.T. and Flood, A.A. 2006. Hymenoptera stings. *Clinical Techniques in Small Animal Practice* 21: 194–204.

Flabbee, J., Petit, N., Jay, N., Guenard, L., Codreanu, F., Mazeyrat, R., Kanny, G. and Moneret-Vautrin, D.A. 2008. The economic costs of severe anaphylaxis in France: an inquiry carried out by the allergy Vigilance Network. *Allergy* 63: 360–365.

Forbes, A.A., Bagley, R.K., Beer, M.A., Hippee, A.C. and Widmayer, H.A. 2018. Quantifying the unquantifiable: why Hymenoptera, not Coleoptera, is the most speciose animal order. *BMC Ecology* 18:21. https://doi.org/10.1186/s12898-018-0176-x

Gadallah, N.S. and Brothers, D.J. 2020. Biodiversity of the aculeate wasps (Hymenoptera: Aculeata) of the Arabian Peninsula: Overview. *Zootaxa* 4754(1): 8–16.

Gallai, N., Salles, J.M., Settele, J. and Vaissière, B.E. 2009. Economic valuation of the vulnerability of world agriculture confronted with pollinator decline. *Ecological Economics* 68: 810–821.

Gaston, K.J. 1991. The magnitude of global insect species richness. *Conservation Biology* 5: 283–296.

Gauld, I.D. and Bolton, B. 1988. *The Hymenoptera*. British Museum (Natural History), Oxford University Press, London, 344 pp.

Gauld, I.D. and Hanson, P.E. 1995. The evolution, classification and identification of the Hymenoptera, pp. 138–156. In: Hanson, P.E. and Gauld, I.D. (eds.), *The hymenoptera of costa rica*. Oxford University Press, London, 893 pp.

Gess, S.K. 1996. *The pollen wasps: Ecology and natural history of the Masarinae*. Harvard University Press, Cambridge, MA, 340 pp.

Gess, S.K. and Roosenschoon, P.A. 2016. A preliminary survey of flower visiting by aculeate wasps and bees in the Dubai Desert Conservation Reserve, UAE. *Journal of Hymenoptera Research* 52: 81–141.

Ghahari, H. 2017. Pollinator insects and honey bee. Jame Elmi Karbordi University, Karaj, 312 pp. [in Persian]

Gogala, A. 2022. Vespid wasps of Slovenia (Hymenoptera: Vespoidea: Vespidae). *Scopolia* 102: 1–79.

Gong, J., Yuan, H., Gao, Z. and Hu, F. 2019. Wasp venom and acute kidney injury: The mechanisms and therapeutic role of renal replacement therapy. *Toxicon* 63: 1–7.

Gould, W.P. and Jeanne, R.L. 1984. *Polistes* wasps (Hymenoptera: Vespidae) as control agents for lepidopterous cabbage pest. *Environmental Entomology* 13(1): 150–156.

Goulet, H. and Huber, J.T. 1993. *Hymenoptera of the world: An identification guide to families*. Agriculture Canada Publication, Ottawa, 668 pp.

Grimaldi, D. and Engel, M.S. 2005. *Evolution of insects*. Cambridge University Press, New York, 771 pp.

Grissell, E. 2010. *Bees, wasps, and ants: The indispensable role of Hymenoptera in gardens*. Timber Press, Portland, Oregon, 336 pp.

Hanson, P.E. 1995. Economic importance of Hymenoptera, pp. 89–101. In: Hanson, P.E. and Gauld, I.D. (eds.), *The hymenoptera of costa rica*. Oxford University Press, Oxford, 893 pp.

Hanson, P.E. and Gauld, I.D. 1995. The biology of Hymenoptera, pp. 20–88. In: Hanson, P.E. and Gauld, I.D. (eds.), *The hymenoptera of costa rica*. Oxford University Press, Oxford, 893 pp.

Heimpel, G.E. and de Boer, J.G. 2008. Sex determination in the Hymenoptera. *Annual Review of Entomology* 53: 209–230.

Hermann, H.R. (ed.) 1982. *Social insects. Volume 3*. Academic Press, New York, xiii+ 459 pp.

Hermes, M.G., Melo, G.A.R. and Carpenter, J.M. 2014. The higher-level phylogenetic relationships of the Eumeninae (Insecta, Hymenoptera, Vespidae), with emphasis *on Eumenes sensu lato*. *Cladistics* 30(5): 453–484.

Hölldobler, B. and Wilson, E.O. 1990. *The ants*. Harvard University Press, Cambridge, MA, 752 pp.

Holway, D.A., Lach, L., Suarez, A.V., Tsutsui, N.D. and Case, T.J. 2002. The causes and consequences of ant invasions. *Annual Review of Ecology and Systematics* 33: 181–233.

Huber, J.T. 2009. Biodiversity of Hymenoptera, pp. 303–323. In: Foottit, R.G. and Adler, P.H. (eds.), *Insect biodiversity: Science and society*. Wiley-Blackwell, West Sussex, UK, 632 pp.

Hunt, J.H. 2007. *The evolution of social wasps*. Oxford University Press, Oxford, xxi + 280 pp.

Hunt, J.H. and Toth, A.L. 2017. Sociality in wasps, pp. 84–123. In: Rubenstein, D.R. and Abbot, P. (eds), *Comparative social evolution*. Cambridge University Press, Cambridge, UK, 476 pp.

Johnson, A.D., Rozenberg, T. and Segoli, M. 2023. Notes on the parasitoids found within the nests of *Delta dimidiatipenne* (Hymenoptera, Vespidae). *Journal of Hymenoptera Research* 96: 925–936.

Junior, P.A.S., Gonring, A.H.R., Picanço, M.C., Ramos, R.S., Martins, J.C. and Ferreira, D.O. 2012. Natural biological control of *Diaphania* spp. (Lepidoptera: Crambidae) by social wasps. *Sociobiology* 59(2): 561–572.

Karsai, I. and Hunt, J.H. 2002. Food quantity affects traits of offspring in the paper wasp *Polistes metricus* (Hymenoptera: Vespidae). *Environmental Entomology* 31(1): 99–106.

Kelstrup, H., West-Eberhard, M., Nascimento, F., Riddiford, L.M. and Hartfelder, K. 2023. Behavior, ovarian status, and juvenile hormone titer in the emblematic social wasp *Zethus miniatus* (Vespidae, Eumeninae). *Behavioral Ecology and Sociobiology* 77: 58.

Khuat, D.L., Nguyen, T.P.L. and Pham T.N. 2004. A survey of beneficial hymenopteran bees and wasps and their use of value in the buffer zone of Cuc Phuong National Park, North Vietnam. *Sylvatrop* 14(1–2): 67–94.

Klotz, J.H., Klotz, S.A. and Pinnas, J.L. 2009. Animal bites and stings with anaphylactic potential. *Journal of Emergency Medicine* 36: 148–156.

Kojima, J. 1993. Relative abundance of swarm-founding Ropalidia in the eusocial wasps (Hymenoptera: Vespidae) in tropical Australia. *New Entomologist* 4: 4–7.

Kolarich, D., Loos, A., Leonard, R., Mach, L., Marzban, G., Hemmer, W. and Altmann, F. 2007. A proteomic study of the major allergens from yellow jacket venoms. *Proteomics* 7: 1615–1623.

Krombein, K.V., Hurd, P.D., Smith, D.R. and Burks, B.D. 1979. *Catalog of Hymenoptera in America north of Mexico*. Vol. 2. Smithsonian Institution Press, Washington, pp. 1199–2209.

Kukalova-Peck, J. 1991. Fossil history and the evolution of hexapod structures, pp. 141–179. In: Naumann, I.D. (ed.), *The insects of Australia*. Volume 1. Cornell University Press, 1137 pp.

Kurzenko, N.V. 1995. Family Vespidae – vespid wasps, pp. 295–324. In: Lelej, A.S. (ed.), *Keys to the insects of Russian Far East*. Vol. IV, pt 1. Dalnauka (Vladivostok), Nauka, St. Petersburg, 606 pp. [in Russian]

Kurzenko, N.V. 2012. Family Vespidae – vespid wasps, pp. 415–423 pp. In: Lelej, A.S. (ed.), *Annotated catalogue of the insects of Russian Far East*. Vol. 1. Hymenoptera. Dalnauka, Vladivostok, 635 pp. [in Russian]

Labandeira, C.C. and Sepkoski, J.J. 1993. Insect diversity in the fossil record. *Science* 261: 310–315.

Lach, L. and Thomas, M.L. 2008. Invasive ants in Australia: documented and potential ecological consequences. *Australian Journal of Entomology* 47: 275–288.

LaSalle, J. and Gauld, I.D. 1993. Hymenoptera: their diversity, and their impact on the diversity of other organisms, pp. 1–26. In: LaSalle, J. and Gauld, I.D. (eds.), *Hymenoptera and biodiversity*. CAB International, Wallington, UK, 348 pp.

Lee, S.H., Baek, J.H. and Yoon, K.A. 2016. Differential properties of venom peptides and proteins in solitary vs. social hunting wasps. *Toxins* 8(2): 32. https://doi.org/10.3390/toxins8020032

Lester, P.J. 2018. *The vulgar wasp: The story of a ruthless invader and ingenious predator*. Victoria University Press, Wellington, New Zealand, 200 pp.

Linksvayer, T.A. and Wade, M.J. 2005. The evolutionary origin and elaboration of sociality in the aculeate Hymenoptera: maternal effects, sib-social effects, and heterochrony. *Quarterly Review of Biology* 80(3): 317–336.

Lohrmann, V. and Engel, M.S. 2017. The wasp larva's last supper: 100 million years of evolutionary stasis in the larval development of rhopalosomatid wasps (Hymenoptera: Rhopalosomatidae). *Fossil Record* 20(2): 239–244.

Luo, L., Kamau, P.M. and Lai, R. 2022. Bioactive peptides and proteins from wasp venoms. *Biomolecules* 12, 527. https://doi.org/10.3390/biom12040527

Mahmood, K., Ullah, M., Aziz, A., Hasan, S.A. and Inayatullah, M. 2012. To the knowledge of Vespidae (Hymenoptera) of Pakistan. *Zootaxa* 3318: 26–50.

Matsuura, M., 1988. Ecological study on vespine wasps (Hymenoptera: Vespidae) attacking honeybee colonies. I. Seasonal changes in the frequency of visits to apiaries by vespine wasps and damage inflicted, especially in the absence of artificial protection. *Applied Entomology and Zoology* 23(4): 428–440.

Matsuura, M. and Sakagami, S.F. 1973. A bionomic sketch of the giant hornet, *Vespa mandarinia*, a serious pest for Japanese apiculture. *Journal of the Faculty of Science, Hokkaido University. Series VI, Zoology* 19: 125–162.

Matsuura, M. and Yamane, S. 1990. *Biology of the vespine wasps*. Springer, Berlin, 323 pp.

Matthews, J.R. and Matthews, R.W. 2018. Nesting biology of an Australian potter wasp, *Delta latreillei* (Saussure) (Hymenoptera: Vespidae: Eumeninae). *Australian Entomologist* 45(1): 93–104.

Matthews, R.W. and Matthews, J.R. 2010. Insect behavior. 2nd edition. Springer Science and Business Media, 514 pp.

Mauss, V. 2007. Evolution verschiedener Lebensformtypen innerhalb basaler Teilgruppen der Faltenwespen (Hymenoptera, Vespidae). *Denisia* 20: 701–722.

Mauss, V., Kuba, K. and Krenn, H.W. 2019. Evolution of the multifunctional mouthparts of adult Vespidae. In: Krenn, H.W. (ed.), *Insect mouthparts. Zoological Monographs 5*. Springer, Cham, pp. 443–478.

Mead, F., Habersetzer, C., Gabouriaut, D. and Gervet, J. 1994. Dynamics of colony development in the paper wasp *Polistes dominulus* Christ (Hymenoptera, Vespidae): The influence of prey availability. *Journal of Ethology* 12: 43–51.

Michener, C.D. 1958. The evolution of social behavior in bees. *Proceedings of the Tenth International Congress of Entomology*, Montreal, 17–25 August 1956, vol. 2, pp. 441–448.

Michener, C.D. 1969. Comparative social behavior of bees. *Annual Review of Entomology* 14: 299–342.

Michener, C.D. 1974. *The social behavior of bees. A comparative study*. Harvard University Press, Cambridge, xii + 404 pp.

Michener, C.D. 2000. *The bees of the world*. Johns Hopkins University Press, Baltimore and London, 913 pp.

Mitra, A., Saha, P., Chaoulideer, M.E., Bhadra, A. and Gadagkar, R. 2011. Chemical communication in Ropalidia marginata: Dufour's gland contains queen signal that is perceived across colonies and does not contain colony signal. *Journal of Insect Physiology* 57(2): 280–284.

Monceau, K., Bonnard, O. and Thiéry, D. 2014. *Vespa velutina*: a new invasive predator of honeybees in Europe. *Journal of Pest Science* 87: 1–16.

Noll, F. B., da Silva, M., Soleman, R.A., Lopes, R.B., Grandinete, Y.C., Almeida, E.A.B., Wenzel, J.W. and Carpenter, J.M. 2021. Marimbondos: systematics, biogeography, and evolution of social behaviour of Neotropical swarm-founding wasps (Hymenoptera: Vespidae: Epiponini). *Cladistics* 37(4): 423–441.

Noyes, J.S. 1989. The diversity of Hymenoptera in the tropics with special reference to Parasitica in Sulawesi. *Ecological Entomology* 14: 197–207.

Nyman, T., Farrell, B.D., Zinovjev, A.G. and Vikberg, V. 2006. Larval habits, host-plant associations, and speciation in nematine sawflies (Hymenoptera: Tenthredinidae). *Evolution* 60: 1622–1637.

O'Neill, K.M. 2001. *Solitary wasps: Behavior and natural history*. Cornell University Press, Ithaca, 424 pp.

O'Toole, C. and Raw, A. 1991. *Bees of the world*. Blandford Press, New York, 192 pp.

Oi, C.A., Brown, R.L., da Silva, R.C. and Wenseleers, T. 2020. Reproduction and signals regulating worker policing under identical hormonal control in social wasps. *Scientific Reports* 10(1): 18971.

Oi, C.A., Oliveira, R.C., van Zweden, J.S., Mateus, S., Millar, J.G., Nascimento, F.S. and Wenselleers, T. 2019. Do primitively eusocial wasps use queen pheromones to regulate reproduction? A case study of the paper wasp *Polistes satan*. *Frontier Ecology and Evolution* 7: 199.

Oi, C.A., van Oystaeyen, A., Oliveira, R.C., Millar, J.G., Verstrepen, K.J., van Zweden, J.S. and Wenseleers, T. 2015. Dual effect of wasp queen pheromone in regulating insect sociality. *Current Biology* 25(12): 1638–1640.

Oster, G.F. and Wilson, E.O. 1978. *Caste and ecology in the social insects*. Princeton University Press, Princeton, xv + 352 pp.

Otis, G.W., Taylor, B.A. and Mattila, H.R. 2023. Invasion potential of hornets (Hymenoptera: Vespidae: *Vespa* spp.). *Frontiers in Insect Science* 3:1145158. https://doi.org/10.3389/finsc.2023.1145158

Peeters, T.M.J., Verheyde, F., van Achterberg, C., De Ketelaere, A., Dekoninck, W., Hoekstra, P.H., Klein, W., Kos, M., Libert, P.-N., Mol, A.W.M., Nieuwenhuijsen, H., Noordijk, J., Ravoet, J., Rhebergen, F., Sijstermans, L.E.N., Smit, J., Soethof, R., Ulenberg, S., Vertommen, W., Vlug, H. and Zwakhals, K. 2024. Familie-indeling en diversiteit van wespen (Hymenoptera) in de Lage Landen. *Entomologische Berichten* 84 (5): 142–153.

Perrard, A., Grimaldi, D. and Carpenter, J.M. 2017. Early lineages of Vespidae (Hymenoptera) in Cretaceous amber. *Systematic Entomology* 42: 379–386.

Peters, R.S., Krogmann, L., Mayer, C., Donath, A., Gunkel, S., Meusemann, K., Kozlov, A., Podsiadlowski, L., Petersen, M., Lanfear, R., Diez, P.A., Heraty, J., Kjer, K.M., Klopfstein, S., Meier, R., Polidori, C., Schmitt, T., Liu, S., Zhou, X., Wappler, T., Rust, J., Misof, B. and Niehuis, O. 2017. Evolutionary history of the Hymenoptera. *Current Biology* 27(7): 1013–1018.

Pham, P.H. 2014. A checklist of Ropalidiini wasps (Hymenoptera: Vespidae: Polistinae) in Indochina. *Archives of Biological Sciences*, Belgrade 66(3): 1061–1074.

Pickett, K.M. and Carpenter, J.M. 2010. Simultaneous analysis and the origin of eusociality in the Vespidae (Insecta: Hymenoptera). *Arthropod Systematics & Phylogeny* 68(1): 3–33.

Piek, T. 1986. *Venoms of the hymenoptera*. Academic Press, London, 570 pp.

Piekarski, P.K., Carpenter, J.M., Lemmon, A.R., Lemmon, E.M. and Sharanowski, B.J. 2018. Phylogenomic evidence overturns current conceptions of social evolution in wasps (Vespidae). *Molecular Biology and Evolution* 35(9): 2097–2109.

Piekarski P.K., Longair, R. and Rogers, S. 2014. Monophyly of eusocial wasps (Hymenoptera: Vespidae): molecules and morphology tell opposing histories. *Journal of Undergraduate Research in Alberta* 4: 11–14.

Pilgrim, E.M., von Dohlen, C.D. and Pitts, J.P. 2008. Molecular phylogenetics of Vespoidea indicate paraphyly of the superfamily and novel relationships of its component families and subfamilies. *Zoologica Scripta* 37: 539–560.

Prezoto, F., Tagliati Maciel, T., Detoni, M., Mayorquin, A.Z. and Barbosa, B.C. 2019. Pest control potential of social wasps in small farms and urban gardens. *Insects* 10: 192.

Quicke, D.L.J. 1997. *Parasitic wasps*. Chapman and Hall, London, 488 pp.

Rasnitsyn, A.P. 1969. The origin and evolution of the lower Hymenoptera. *Trudy Paleontologicheskogo Instituta Rossijskoj Akademii Nauk SSSR* 123: 1–196.

Rasnitsyn, A.P. 1980. Origin and evolution of Hymenoptera. *Transactions of the Paleontological Institute of the Academy of Sciences of the USSR* 174: 1–192.

Rasnitsyn, A.P. 1988. An outline of the evolution of hymenopterous insects (Order Vespida). *Oriental Insects* 22: 115–145.

Rasnitsyn, A.P. and Quicke, D.L.J. 2002. *History of insects*. Kluwer Academic, Dordrecht, 517 pp.

Rasplus, J.-Y., Villemant, C., Paiva, M.R., Delvare, G. and Roques, A. 2010. Hymenoptera Chapter 12. *BioRisk* 4(2): 669–776.

Reeve, H.K. 1991. *Polistes*, pp. 99–148. In: Ross, K.G. and Matthews, R.W. (eds.), *The social biology of wasps*. Cornell University Press, Ithaca, 678 pp.

Rehan, S.M. and Toth, A.L. 2015. Climbing the social ladder: the molecular evolution of sociality. *Trends in Ecology and Evolution* 30(7): 426–433.

Richards, O.W. 1962. *A revisional study of the Masarid wasps (Hymenoptera: Vespoidea)*. British Museum (Natural History), London, vii + 294 pp.

Richards, O.W. 1971. The biology of the social wasps (Hymenoptera, Vespidae). *Biological Reviews of the Cambridge Philosophical Society* 46(4): 483–528.

Richards, O.W. 1982. A revision of the genus *Belonogaster* de Saussure (Hymenoptera: Vespidae). *Bulletin of the British Museum Natural History, Entomology* 44(2): 31–114.

Richards, O.W. 1985 [1984]. Insects of Saudi Arabia. Hymenoptera: Fam. Masaridae (The Arabian species). *Fauna of Saudi Arabia* 6: 413–422.

Richards, O.W. and Richards, M.J. 1951. Observations on the social wasps of South America (Hymenoptera Vespidae). *Transactions of the Royal Entomological Society of London* 102: 1–170.

Richter, M.R. 2000. Social wasp (Hymenoptera, Vespidae) foraging behavior. *Annual Review of Entomology* 45: 121–150.

Saito, F. and Kojima, J.I. 2007. A taxonomic revision of the hover wasp genus *Eustenogaster* van der Vecht (Insecta: Hymenoptera; Vespidae, Stenogastrinae). *Zootaxa* 1556: 1–30.

Schmid-Egger, C. and Schmidt, S. 2021. Unexpected diversity in Central European Vespoidea (Hymenoptera, Mutillidae, Myrmosidae, Sapygidae, Scoliidae, Tiphiidae, Thynnidae, Vespidae), with description of two species of *Smicromyrme* Thomson, 1870. *ZooKeys* 1062: 49–72.

Schmidt, J.O. 1982. Biochemistry of insect venoms. *Annual Review of Entomology* 27: 339–368.

Sharkey, M. 2007. Phylogeny and classification of Hymenoptera. *Zootaxa* 1668: 521–548.

Sharkey, M.J., Carpenter, J.M., Vilhelmsen, L., Heraty, J., Liljeblad, J., Dowling, A.P., Schulmeister, S., Murray, D., Deans, A.R., Ronquist, F., Krogmann, L. and Wheeler, W.C. 2012. Phylogenetic relationships among superfamilies of Hymenoptera. *Cladistics* 28: 80–112.

Sharma, S. and Upadhyay, R.K. 2021. Hymenoptera toxins: biological activity, pharmaceutical and therapeutic uses. *International Journal of Pharmacy and Pharmaceutical Sciences* 13(4): 8–20.

Soares, J.J. and Lara, F.M. 1994. Predation of *Anthonomus grandis* Boh. by *Brachygastra lecheguana* (Latreille) (Hymenoptera: Vespidae). *Anais da Sociedade Entomológica do Brasil* 23(1): 135–136.

Spradbery, J.P. 1973. *Wasps: an account of the biology and natural history of solitary and social wasps*. University of Washington Press, Seattle, 408 pp.

Sumner, S., Law, G. and Cini, A. 2018. Why we love bees and hate wasps. *Ecological Entomology* 43(6): 836–845.

Tscharntke, T., Bommarco, R., Clough, Y., Crist, T.O., Kleijn, D., Rand, T.A., Tylianakis, J.M., van Nouhuys, S. and Vidal, S. 2007. Conservation biological control and enemy diversity on a landscape scale. *Biological Control* 43: 294–309.

Tscharntke, T., Gathmann, A. and Steffan-Dewenter, I. 1998. Bioindication using trap nesting bees and wasps and their natural enemies – community structure and interactions. *Journal of Applied Ecology* 35: 708–719.

Turillazzi, S. 1991. The Stenogastrinae, pp. 74–98. In: Ross, K.G. and Matthews, R.W. (eds.), *The social biology of wasps*. Cornell University Press, Ithaca, 678 pp.

Turillazzi, S. 2012. *The Biology of Hover Wasps*. Springer, Berlin, 284 pp.

Udayakumar, A., Aravindaram, K. and Shivalingaswamy, T.M. 2022. Nesting and predatory behaviour of potter wasp, *Rhynchium brunneum brunneum* (Eumeninae: Vespidae: Hymenoptera) in an urban farm landscape. *Biocontrol Science and Technology* 32(7): 794–810.

Varghese, T. and Girish Kumar, P. 2023. A review of the taxonomy, biology and distribution of the social wasps (Hymenoptera: Vespidae) of the Indian subcontinent. *Journal of the Indian Institute of Science* 103: 1019–1047.

Vilhelmsen, L. and Turrisi, G.F. 2011. Per arborem ad astra: Morphological adaptations to exploiting the woody habitat in the early evolution of Hymenoptera. *Arthropod Structure and Development* 40: 2–20.

Wenzel, J. 1991. Evolution of nest architecture, pp. 480–519. In: Ross, K.G. and Matthews, R.W. (eds.), *The social biology of wasps*. Cornell University Press, Ithaca, 678 pp.

Wenzel, J.W. 1998. A generic key to the nests of hornets, yellowjackets, and paper wasps worldwide (Vespidae: Vespinae, Polistinae). *American Museum Novitates* 3224: 1–39.

West-Eberhard, M.J. 2005. Behavior of the primitively social wasp *Montezumia cortesioides* Willink (Vespidae, Eumeninae) and the origins of vespid sociality. *Ethology Ecology & Evolution* 17: 201–215.

Westerfelt, P., Widenfalk, O., Lindelöw, A., Gustafsson, L. and Weslien, J. 2015. Nesting of solitary wasps and bees in natural and artificial holes in dead wood in young boreal forest stands. *Insect Conservation and Diversity* 8(6): 493–504.

Whitfield, J.B. 1998. Phylogeny and evolution of host-parasitoid interactions in Hymenoptera. *Annual Review of Entomology* 43: 129–151.

Wilson, E.O. 1971. *The insect societies*. Harvard University Press, Cambridge, 562 pp.

Brachyodynerus magnificus (Morawitz, 1867) (Eumeninae). (Photo courtesy of Alexander Fateryga)

Paragymnomerus signaticollis (Morawitz, 1888) (Eumeninae). (Photo courtesy of Alexander Fateryga)

Paravespa rex (von Schulthess, 1924) (Eumeninae). (Photo courtesy of Alexander Fateryga)

Tropidodynerus interruptus (Brullé, 1832) (Eumeninae). (Photo courtesy of Alexander Fateryga)

Katamenes dimidiatus (Brullé, 1832) (Eumeninae). (Photo courtesy of Alexander Fateryga)

Katamenes dimidiatus (Brullé, 1832) (Eumeninae). (Photo courtesy of Alexander Fateryga)

2 Subfamily Eumeninae Leach, 1815

Neveen S. Gadallah, Hassan Ghahari, Kees van Achterberg, and Patrick K. Piekarski

INTRODUCTION

The cosmopolitan subfamily Eumeninae (commonly known as potter wasps) is the largest subfamily within the family Vespidae, comprising more than 3900 valid species of solitary (or rarely subsocial) wasps in 205 genera (Tan *et al.*, 2015, 2018; Pannure *et al.*, 2016; Kumar *et al.*, 2019a; Li *et al.*, 2019; Luo *et al.*, 2022). Despite their diversity, the phylogeny and taxonomy of Eumeninae remain poorly resolved, with approximately 65% of all vespid species belonging to this subfamily (Carpenter and Cumming, 1985; Hermes *et al.*, 2014; Pickett and Carpenter, 2010; Vernier, 1997). Recently, the subfamily has been restricted to include species within the tribes Eumenini and Odynerini *sensu* Hermes *et al.* (2014); however, these tribes have also been shown to be paraphyletic. Together, they comprise approximately 3407 described species (Bank *et al.*, 2017; Piekarski *et al.*, 2018).

Eumenine species are morphologically highly diverse (Tan *et al.*, 2018), with considerable variation in adult body size. All known eumenines are predators, most of which are solitary mass provisioners. However, 17 species are known to exhibit parental care by progressively provisioning larvae, and a few species even display nest-sharing behavior (West-Eberhard, 2005; Srinivasan and Kumar, 2010). Nest construction habits can be categorized into three types: **burrowers**, which excavate nests in soil; **modifiers**, which use pre-existing cavities such as hollow stems, abandoned burrows, or nest holes; and **builders**, which construct nests using mud or plant material (Spradbery, 1973; Cowan, 1991). Nests are provisioned with the larvae of Lepidoptera (e.g., Geometridae, Tortricidae), Coleoptera (e.g., Chrysomelidae and Curculionidae) (Jennings and Houseweart, 1984; Pannure *et al.*, 2016), or Symphyta (Goulet and Huber, 1993). While developing larvae require lipids and proteins obtained from prey, adult wasps primarily require carbohydrates which they acquire from floral nectaries. Typically, females search for a suitable nest site and establish a nest of a single or few cells, provisioning each cell with multiple prey items. One egg is laid per cell before provisioning starts, and adults build and provision one cell at a time, completing and sealing it before constructing the next cell (Spradbery, 1973). While most eumenine wasps seal cells with mud or clay, the use of plant material in nest construction appears to be common among zethine and raphiglossine wasps, with many genera such as *Discoelius*, *Ischnocoelia*, *Protodiscoelius*, *Zethus*, and *Elimus* utilizing masticated vegetable matter for cell partitioning or closure in nest burrows, *Psiliglossa* and *Raphiglossa* using pith, and *Calligaster* and *Zethus* (subgenus *Zethoides*) building aerial nests primarily from plant material (Bohart and Stange, 1965).

In potter wasps, males typically emerge a few days earlier than females and mate with females soon after emergence from their cell. Males often engage in nuptial flights near emergence sites, waiting for receptive females to eclose. Upon encountering a female, a male mounts her, gripping tightly with legs and mandibles while attempting copulation through coordinated movements of the abdomen and genitalia (Spradbery, 1973).

DOI: 10.1201/9781003740438-2

This family represents a primary lineage within Vespidae and plays a significant role in the biological control of agricultural pests (Cowan, 1991; Dai *et al.*, 2023). Their behaviour, including nest construction, mating strategies, and evolution of exaggerated secondary sexual traits such as male weaponry (Longair, 2004), has attracted the attention of evolutionary biologists (Smith and Alcock, 1980; Matthews and Matthews, 2010; Carpenter and Brown, 2021). Due to their relatively close relationship with the highly social polistine and vespine wasps, eumenine species exhibiting subsocial and/or communal behavior in particular have been central to studies exploring the origins of reproductive division of labor and cooperative breeding (West-Eberhard, 1987; Kelstrup *et al.*, 2018, 2023).

The generic classification of Eumeninae is problematic and has a complex taxonomic history (Hermes *et al.*, 2014). Richards (1962) proposed a classification for the "Eumenidae", recognizing three subfamilies: Discoeliinae (= Zethinae), Eumeninae, and Raphiglossinae. In 1982, Carpenter revised this system, recognizing only one family, Vespidae, and treating the former subfamilies as a part of broader Eumeninae. He did not provide a tribal classification and rejected the use of the names Raphiglossinae and Zethinae. The lack of comprehensive identification keys for sboth genera and species illustrates that this group of vespid wasps is understudied (Pannure *et al.*, 2016).

Although most existing keys are country- or region-specific, several are particularly valuable: Guichard (1986) for the Arabian Peninsula; Yamane (1990) for the Japanese fauna; Carpenter and Garcete-Barrett (2002) for the Neotropical Eumeninae; Pannure *et al.* (2016) and Kumar *et al.* (2019b) for the Indian genera; Tan *et al.* (2018) for the Chinese Eumeninae; and Carpenter and Brown (2021) for Australian genera.

According to modern taxonomic treatments (Bank *et al.*, 2017; Piekarski *et al.*, 2018; Mauss *et al.*, 2019; Luo *et al.*, 2022; Benda *et al.*, 2024), eumenine wasps should be subdivided into three subfamilies: Eumeninae *s.s.*, Raphiglossinae, and Zethinae (Fateryga and Proshchalykin, 2024). However, due to poor sampling of raphiglossines in molecular studies to date, both their phylogenetic position and appropriate taxonomic rank remain uncertain.

Checklists of Iranian Eumeninae s.l.: Ebrahimi and Carpenter (2008) listed 149 species and subspecies in 37 genera. Modarres Awal (2012) represented 37 species and subspecies in 20 genera. Recently, Rahmani *et al.* (2020) listed 184 species and subspecies in 45 genera. The present catalogue comprises 605 species and subspecies in 57 genera for the fauna of the Middle East, and 211 species and subspecies in 45 genera for Iran.

In this chapter, we follow recent phylogenetic studies (Bank *et al.*, 2017; Piekarski *et al.*, 2018; Luo *et al.*, 2022; Benda *et al.*, 2024), which divides the subfamily Eumeninae *sensu* Carpenter, 1982 into three subfamilies: Eumeninae *s.s.*, Raphiglossinae, and Zethinae.

This chapter is dedicated to the late Josef Gusenleitner [Austria; 17 September 1929 – 6 July 2023; see Gusenleitner and Ockermüller (2024)], who collaborated extensively with H. Ghahari and many other Iranian entomologists in identifying numerous specimens of Iranian Vespidae and authored several papers on the Vespidae of the Middle East.

Key to the genera of the subfamily Eumeninae *s.l.* of the Middle East (modified after Gusenleitner, 2013a and including Eumeninae, Raphiglossinae, and Zethinae)

1. Forewing with the first recurrent vein received by the second submarginal cell, and the second recurrent vein received by third submarginal cell; mouth parts prolonged and reaching level of hind coxae (subfamily Raphiglossinae) 2

— Forewing with both recurrent veins received by second submarginal cell; mouth parts remain distinctly removed from level of hind coxae (subfamilies Eumeninae *s.str.* and Zethinae) 3

2. First metasomal tergum sessile (bell-shaped), not longer than wide and not much narrower than second tergum ....*Psiloglossa* S.S. Saunders

— First metasomal tergum petiolate, much longer than its apical width, and narrower than second tergum..*Raphiglossa* S.S. Saunders

3. Middle tibia with two spurs at distal end (subfamily Zethinae) 4

— Middle tibia with only a single spur at distal end (subfamily Eumeninae *s.str.*)................................6

4. Second metasomal tergum sessile, broadly attached to first tergum .... *Discoelius* Latreille

— Second metasomal tergum petiolate..5

5. Basal peduncle of second tergum 1–2 times as long as wide.................................. *Zethus* Fabricius

— Basal peduncle of second tergum more than five times as long as wide .... *Paramischocyttarus* Magretti

6. Second submarginal cell of forewing petiolate...*Alastor* Lepeletier

— Second submarginal cell of forewing not petiolate 7

7. First metasomal tergum significantly narrower than second tergum in dorsal view.............................8

— First metasomal tergum about as wide as second tergum in dorsal view..16

8. First metasomal sternum with regular transverse striae................................*Pareumenes* de Saussure

— First metasomal sternum without transverse striae.. 9

9. Mesopleuron with a sharp longitudinal scar posteriorly; apical margin of clypeus rounded; second tergum with weakly indented terminal lamella.. *Afreumenes* Bequaert

— Mesopleuron rounded posteriorly, without scar..10

10. Second metasomal tergum depressed in a stepped manner distally, with lamella clearly separated from remainder of tergum 11

— Distal end of second metasomal tergum without depressed lamella distally and lamella not separated...13

11. First metasomal tergum about three times as long as its distal width; mesosoma without well-developed pubescence..*Micreumenes* Ashmead

— First metasomal tergum at most twice as long as its distal width; mesosomal pubescence variable, (pilosity in large species long)..12

12. Large species, body longer than 10 mm; depressed apical lamella of second metasomal tergum of uniform colour..*Eumenes* Latreille

— Small species, body shorter than 8 mm; depressed apical lamella of second metasomal tergum with coarse, elongated pits basally and spaces between pits opaquely pigmented.................*Cyrtolabulus* van der Vecht

13. Apical margin of clypeus obtusely angled (♀) or flattened convex (♂)..14

— Apical margin of clypeus rounded...15

14. Mesosoma with erect pilosity, except in *sichelii tauriae* Giordani Soika; last metasomal tergum of male rounded distally..*Katamenes* Meade-Waldo

— Mesosoma densely covered with microscopic pilosity; last tergum of male wing-like widened distally ....*Delta* de Saussure

15. Small species, body length approximately 15 mm; first metasomal tergum often longer than mesosoma, and parallel-sided or with slightly concave margins on 2/3 of its distal part in dorsal view.............. ....*Ischnogasteroides* Magretti

— Robust species, body length more than 20 mm; first metasomal tergum pear-shaped and swollen at distal half, slightly longer than mesosoma; [section of first tergum behind spiracles shorter than section before spiracles; male terminal sternum with a longitudinal groove]................................*Delta* de Saussure

16. Tegula rounded posteriorly; 4–5 apical antennal segments rolled up when at rest..........................17

— Tegula pointed posteriorly; apical antennal segments not rolled up...25

17. Labial palp 3-segmented (at least in male) more elongated, with long bristles on both sides 18

— Labial palp 4-segmented, not elongated, with no or short bristles, and with sparse setae 21

18. Body length less than 9.0 mm; maxillary palp 5- or 6-segmented; mesoscutum without parapsidal furrows; apex of clypeus very wide, about 2/3 wider than distance between antennal sockets, and truncate in female, barely notched in male; second segment of labial palp in female thick basally, second and third labial palp segments straight, fringed with dense yellow setae along its entire length.......................... ....*Pterocheilus* Klug

— Body length more than 10 mm (if smaller than 10 mm, then head, mesosoma and first and second metasomal sterna with long, black erect setae); maxillary palp 6-segmented; clypeal apex with more or less distinct rim in female and deep rim in male (if anterior rim of clypeus slightly convex in female and straight, then mesoscutum in front of scutellum with clearly impressed parapsidal furrows); labial palp segments without yellow fringe of setae 19

19. Third labial palp segment narrower than the second, pointed at end, bent apically like a needle and bare; in male setae not longer than diameter of second segment; clypeal apex with sharp corners ....*Hemipterochilus* Ferton

— Third labial palp segment broad, elliptical, curved or almost cylindrical, if pointed distally, then covered with long setae to its tip, and head, mesosoma, first and second sterna with erect black setae; clypeal apex rounded laterally in male...20

20. Parastigma of forewing significantly longer than pterostigma; female without parietal pits on vertex; second and third labial palp segments of female almost cylindrical, and third segment parallel-sided; clypeal apex straight in male and slightly convex in female; sterna of male without dense setae; tarsal claws simple ....*Cephalochilus* Blüthgen

— Parastigma of forewing significantly shorter than pterostigma; female with two parietal pits; second and third labial palp segments of female slender, flat and curved; clypeal apex of male emarginate (excised); sterna 2–5 (or 2–6) of male with dense setae, if not then tarsal claws tridentate.......................... ....*Onychopterocheilus* Blüthgen

21. Parastigma of forewing significantly longer than pterostigma; large species over 15 mm long; mesoscutum almost glabrous, with only microscopic pubescence...................................*Paravespa* Radoszkowski

— Parastigma of forewing significantly shorter than pterostigma; species are smaller than 14 mm, if larger than 15 mm, then mesoscutum with long, wavy pilosity..22

22. Mesosoma and temple with very fine pubescence; female without cephalic foveae; slope of vertex weakly longitudinally rugose posteriorly; in male, apex of posterior slope with an often indistinct median line, and terga 3–7 with short (slightly shorter than the base diameter of third tarsomere of hind tarsus) dense and wooly pilosity; male sterna 2–6 with conspicuous fringe of setae ....*Tropidodynerus* Blüthgen

— Mesosoma and temple with long pilosity; female with cephalic foveae and the male without a longitudinal line; sterna 3–5 of male with indistinct setae or only yellow fringe of setae, only sternum 7 may have longer setae, but they are not wooly...23

23. Female clypeus deeply emarginate; female mandible with a deep indentation between 3rd and 4th teeth; vertex with very small point-like close fossae; male coxae and femora normal; male clypeus margined with black apically..*Gymnomerus* Blüthgen

— Female clypeus with flattened edge, or truncate, almost straight; mandible of female without indentation, and vertex fossae are apart from each with interspaces as that present between posterior ocelli; male coxae particularly prominent (middle coxae obtusely widened in basal third, with sharp edges below, or with deep incisions, or thickened and distally widened at obtuse angle), or femora and cheeks with special modifications; yellow clypeus of male not edged with black apically.................................... 24

24. Large species, 13–17 mm; with distinct parapsidal furrow on mesoscutum in front of scutellum; in female, crown cavity is larger than an ocellus and is set transversely; middle femur of male thickened and widened to an obtuse angle towards the distal end; middle and hind femora normal ....
....*Paragymnomerus* Blüthgen

— Smaller species, 0.8–13 mm; mesoscutum without parapsidal furrows; in female cephalic fovea and vertex not rugose; in male, sterna without fringe of setae, middle coxa with deep cleavages, or middle femur (sometimes hind) and cheeks have distinctive features.................................*Odynerus* Latreille

25. First metasomal tergum with one or two transverse edges, without lamella at base 26

— First metasomal tergum without transverse edge..35

26. Horizontal surface of first metasomal tergum with more or less distinctly developed longitudinal furrow in middle extending over whole surface; with one exception (*S. murarius*, a species which is very large and has long wavy pilosity on mesosoma), a transverse ridge is formed at the transition from front vertical to the horizontal surface; last antennal flagellomeres of male are not folded back like a hook..*Symmorphus* Wesmael

— First metasomal tergum without a longitudinal furrow, with only a pit or a short longitudinal indentation could be seen in front of distal end; transition between horizontal and vertical surface is evenly rounded or appears angled to edge in profile; last antennal flagellomeres of male hook-like...........................27

27. Transition from the front vertical to the horizontal surface of first metasomal tergum not evenly rounded, but with one or two transverse edges, if transverse edge is difficult to see, then tegula enlarged and translucently punctate...28

— Transition from front vertical to horizontal surface of first metasomal tergum evenly rounded............35

28. Mesosoma (as head) with dense, long and erect setae; [tegula narrow, much longer than wide; first metasomal tergum more or less uniformly punctate behind transverse edge on its horizontal surface, vertical surface is almost impunctate; female vertex with pit]........................*Ancistrocerus* Wesmael

— Mesosoma (as metasoma) with extremely short, microscopic setae 29

29. First metasomal tergum with two transverse ridges...............*Pseudonortonia* Giordani Soika (in part)

— First metasomal tergum with one transverse ridge..30

30. Tegula with dense strong, sieve-like translucent punctures...31

— Tegula with inconspicuously finer and superficial punctures...32

31. First metasomal tergum short, wider than long; the transverse edge on first tergum approximately in middle of tergite...*Eustenancitrocerus* (*Eustenancistrocerus*) Blüthgen

— First metasomal tergum only slightly wider than long or longer than wide; transverse edge of first tergum lower than its horizontal plane and closer to base of tergum....................... *Jucancistrocerus* Blüthgen

32. Scutellum serrated above and falls steeply to propodeum 33

— Scutellum not serrated above and propodeum is more or less elongated backwards behind scutellum......34

33. Tegula very broad, almost as wide as its length, inconspicuously, finely, and superficially punctate, usually with only two slightly larger, glassy translucent spots; second metasomal sternum clearly elevated above basal fissure; male clypeus broad, flattened apically.. ....*Eustenancistrocerus* (*Parancistrocerus*) Blüthgen

— Tegula broad, with only a translucent glassy spot in middle; second metasomal sternum slightly or not elevated above basal fissure (except in *S. liliput*); male clypeus deeply emarginates apically; second tergum with light transverse band or lateral spots at base......................*Stenancistrocerus* de Saussure

34. A smooth, slopping surface without or with only few coarse dots present above transverse edge of first metasomal tergum; third metasomal tergum much more coarsely punctate than base of second tergum distally; clypeal apex never deep or wide (wider than distance between antennal sockets) .... *Tachyancistrocerus* Giordani Soika

— Transverse edge of first metasomal tergum mostly raised or formed as a lamella on the slopping surface, in front of base or at transition in front of horizontal surface; first tergum usually somewhat narrower than second tergum; third tergum very often with very coarse spots that are much larger than those on second and fourth tergum ....*Pseudonortonia* Giordani Soika (in part)

35. Second metasomal tergum with a glassy, translucent lamella behind its transverse band, which is flat in the plane of tergum or indented (then it can also be black) and clearly distinguished from remainder of the tergum; first tergum without translucent light border behind light terminal band; small species, 8 mm or less (if larger than 8 mm, then tegula is larger than half area of scutellum or setaeare longer than distal length of antennal flagellum)..36

— Second metasomal tergum without translucent lamella, at most a very narrow one that does not stand out from tergum (if in some species somewhat larger, then first tergum behind the light terminal band also with broad light translucent lamella); mostly larger than 10 mm, rarely smaller 38

36. Face and mesoscutum much longer than wide, very shiny; shoulders form a right angle; second metasomal tergum with an indented, translucent terminal margin with row of dots at base.................... .... *Leptodynerus* Blüthgen

— With the exception of some *Microdynerus* species, face and mesoscutum are broader and less shiny; row of dots at base of bright terminal lamella absent, except in *Leptochilus* 37

37. Mesoscutum shorter than wide; terminal lamella on second metasomal tergum in most cases noticeably indented, usually with a row of dots at base, in some cases spaces between which opaquely pigmented in the color of the terminal band; end of antenna pointed in male; mandible and clypeus in female without special features, tongue short (if tongue extends to middle of coxae see *Labochilus* Blüthgen)............... .... *Leptochilus* de Saussure

— Mesoscutum longer than wide; terminal lamella on second metasomal tergum at least in the middle, not or hardly noticeably depressed, never punctate; antennal tip pointed or broad and flattened; mandible of female without special features, or with angularly projected outer edge in one subgenus and clypeus appears tridentate in other...*Microdynerus* Thomson

38. First metasomal tergum with a well-developed translucent border of variable width behind the terminal band, which is usually demarcated by a fine, dotted line; propodeum designated like a strip or edge behind scutellum or rounded..39

— First metasomal tergum without such border behind the terminal band; propodeum never with developed edges behind scutellum 45

39. Propodeum without edges or lamellae behind scutellum..40

— Propodeum with clearly raised edges or lamellae behind scutellum, which are closer to scutellum than to apex of propodeum...41

40. Propodeum with a semicircular edge between horizontal and slopping area in dorsal view ....*Antodynerus* de Saussure

— Edge formed between horizontal surface and slopeing area of propodeum straight, not curved in dorsal view..*Euodynerus* (*Euodynerus*) Dalla Torre

41. Parategula not visible in dorsal view; tegula very narrow and slightly incurved at end, mostly punctate; anterior part of pronotum densely punctate laterally...................................*Knemodynerus* Blüthgen

— Parategula visible in dorsal view; tegula with different shapes, its end is not inwardly curved; anterior area of pronotum differently punctate...42

42. Pronotum without sharp glassy translucent edge; epicnemial ridge of mesoscutum barely discernable in general sculpture; upper side of metasoma covered with short, sloping, even set of bristles; in male, middle femur emarginate on the underside over 2/3 of its length, and hind tibia suddenly greatly widened at distal end...*Chlorodynerus* Blüthgen

— Pronotum with clearly developed transparent, moderately wide lamella, which stands out from the basic sculpture (missing in *Xanthodynerus jordanicus*); male middle femur not emarginate, hind tibia not suddenly widened at distal end...43

43. Propodeum in lateral view with horizontal upper surface and vertical surface separated by an angled edge, this edge is evenly serrated; [upper lateral boundary of the concavity of propodeum directed upwards or backwards in a tooth-like manner behind scutellum]... ....*Euodynerus* (*Pareuodynerus* Blüthgen)

— Propodeum rounded in profile, or with indistinct edge...44

44. Propodeum rounded in profile, not serrated; lamellar raised upper side borders of propodeal concavity partly bent inwards like a hook, their distance above is less than the width of scutellum ....*Syneuodynerus* Blüthgen

— Propodeum with a tooth-like structure on both sides below scutellum; [scutellum serrated above; temple edge linear]...*Xanthodynerus* Blüthgen

45. Upper half of propodeal concavity with a transverse ridge (sometimes difficult to see); tegula broadly rounded on the outside...46

— Propodeal cavity without transverse ridge on upper half; tegula of different shape 47

46. Tegula without sieve-like, glassy, translucent punctures; in many cases, male with a deep tooth gap in mandible above last tooth...*Pseudepipona* de Saussure

— Tegula with sieve-like, glassy, translucent punctures; mandible of male without tooth gap .... *Brachypipona* Gusenleitner

47. Propodeum with an erect tooth on both sides, which is clearly raised than its level when seen from behind or from the front, the tooth could also be flat..48

— Propodeum without such raised teeth...50

48. Tegula very large, almost as large as scutellum, broadly rounded posteriorly; [scutellum edged medially forming lateral humps, without upright lamella; second metasomal tergum with three humps in almost all specimens]...*Cyphodynerus* van der Vecht

— Tegula small, not reaching half side of scutellum...49

49. Clypeus with two sharp edges laterally and a sharp longitudinal carina between antennal sockets continuing to upper part of clypeus; second metasomal tergum with a longitudinal carina medially; tegula broadly rounded at outer side and shorter........................*Pseudodontodynerus* Giordani Soika

— Clypeus without sharp edges; second metasomal tergum rarely with a longitudinal sharp carina medially (only occurs in few Afrotropical species); in most cases, propodeum with a horizontal surface; tegula significantly longer than wide..*Antepipona* de Saussure

50. Scutellum with three humps or teeth, if indistinct, then mesoscutum with two longitudinal carinae; [propodeum without side edges] ....*Tricarinodynerus* Giordani Soika

— Scutellum not as above; mesoscutum without such carinae..51

51. Axillary fossae in dorsal view much narrower than long, often slit-like......................................52

— Axillary fossae in dorsal view not slit-like, at least as wide as long, oval....................................53

52. Mesoscutum and scutellum entirely punctate; propodeum deeply eroded; maxillary palp 6-segmented; body over 20 mm in length; mesosoma entirely black...*Synagris* Latreille

— Mesoscutum and scutellum smooth and shiny posteriorly, very sparsely and finely punctate; propodeum only depressed medially; body less than 20 mm in length; mesosoma without black areas ....*Rhynchium* Spinola

53. First metasomal tergum with a narrow longitudinal indentation medially in front of distal edge which often appears as a line; occipital foramen present in female...54

— First metasomal tergum evenly punctated on horizontal surface, without any longitudinal depression or pit in front of distal edge; except in *Allodynerus*, female without occipital foramen 57

54. Clypeus very broad, truncate, with a weak edge only in male; underside of head and upper side of mesosoma with a long (sometimes longer than the diameter of antennal flagellum), wavy and dark setae ....*Acanthodynerus* Gusenleitner

— Clypeus deep and shallower, but more narrowly margined than above; underside of head and upper side of mesosoma with only short, erect, light setae..55

55. First metasomal tergum significantly longer than wide..........................*Eumicrodynerus* Gusenleitner

— First metasomal tergum wider than long or as wide as long 56

56. Clypeal edge narrow and noticeably flat; propodeal concavity deep; last antennal segment of male very short (about as long as wide)...*Pseudosymmorphus* Blüthgen

— Clypeal edge deeper and slightly wider; propodeal concavity flat and slightly indented; last antennal segment of male longer than wide; [with a light marking on third metasomal tergum ....*Pseudoleptochilus* Blüthgen

57. Enlarged tegula with sieve-like, glassy, translucent punctures; [clypeus punctate without longitudinal striations; last antennal segment of male very short and rounded or pointed]............................. ....*Brachyodynerus* Blüthgen

— Tegula of normal size; without sieve-like, translucent punctures...58

58. Propodeum with blunt, tooth-like structures on the sides; [last antennal segment of male long, slightly curved and rounded at tip] ....*Parodontodynerus* Blüthgen

— Propodeum without teeth like structures on sides; [clypeus not striated lengthwise].......................59

59. Head and mesosoma with very short fine pubescence; tegula short, rounded on the outside, hardly surpassing parategula; anterior face of pronotum with two small, deeply impressed pits in the middle; light posterior bands could be seen on third and following terga in some species, while in others terga 3–6 (7) entirely black .... *Stenodynerus* de Saussure

— Head and mesosoma with light setae of median length; tegula narrower and longer, tapered at end, clearly surpassing parategula posteriorly; anterior face of pronotum smooth and shiny, without impressed pits or foveae; more than two light tergal bands always present .... *Allodynerus* Blüthgen

List of Species of the Subfamily Eumeninae in the Middle East

Subfamily Eumeninae Leach, 1815

Eumeninae *s.str.* comprises wasps formerly classified within the tribes Eumenini and Odynerini (Hermes *et al.*, 2014). The monophyly of both tribes is questionable, as phylogenomic data indicate that Eumenini and Odynerini are paraphyletic (Piekarski *et al.*, 2018). However, the "Eumenini" clade *sensu* Hermes *et al.* (2014) may represent a monophyletic group if *Hypodynerus* is included (Piekarski, 2017).

Genus *Acanthodynerus* Gusenleitner, 1969

Acanthodynerus Gusenleitner, 1969: 13. Type species: *Acanthodynerus giordanii* Gusenleitner, 1969, by original designation.

Acanthodynerus giordanii Gusenleitner, 1969

Acanthodynerus giordanii Gusenleitner, 1969 (1966): 14, ♂, ♀; Turkey.
Distribution in the Middle East: Known only from Turkey (Gusenleitner, 1969, 1988, 2013a; van der Vecht and Fischer, 1972; Yildirim and Kojima, 1999).
Extralimital distribution: None.

Acanthodynerus multimaculatus Gusenleitner, 2000c

Acanthodynerus multimaculatus Gusenleitner, 2000c: 932, ♀; Turkey.
Catalogues with Iranian records: Rahmani *et al.* (2020).
Distribution in Iran: Isfahan (Gusenleitner, 2011a).
Distribution in the Middle East: Iran (Gusenleitner, 2011a, 2013a; Selis *et al.*, 2025), Iraq (Selis *et al.*, 2025), Israel-Palestine, Turkey (Gusenleitner, 2000c, 2013a; Selis *et al.*, 2025).
Extralimital distribution: None.

Genus *Afreumenes* Bequaert, 1926

Afreumenes Bequaert, 1926: 86. Type species: *Eumenes melanosoma* de Saussure, 1852, by monotypy.

Afreumenes melanosoma yemenensis Giordani Soika, 1996

Afreumenes melanosoma yemenensis Giordani Soika, 1996: 34, ♀; Yemen: Taizz.
Distribution in the Middle East: Known only from Yemen (Giordani Soika, 1996; Carpenter and Gadallah, 2020).
Extralimital distribution: None.

Afreumenes nigrorufus Giordani Soika, 1968

Afreumenes nigrorufus Giordani Soika, 1968: 15 (key), 82, ♀, ♂; Kenya, Malawi, Zambia.

Distribution in the Middle East: Saudi Arabia (Guichard, 1986; Carpenter and Gadallah, 2020), Yemen (Giordani Soika, 1968; Gusenleitner, 2007, 2013a; Carpenter and Gadallah, 2020).
Extralimital distribution: Kenya, Malawi, Tanzania, Zambia (Carpenter *et al.*, 2010; Carpenter and Gadallah, 2020).

Genus *Alastor* Lepeletier, 1841

Alastor Lepeletier, 1841: 668. Type species: *Alastor atropos* Lepeletier, 1841; designated by Ashmead, 1902.

***Alastor* (*Megalastor*) *aegyptiacus* Blüthgen, 1955b**
Alastor aegyptiacus Blüthgen, 1955b: 65, 68, 75, ♀, ♂; Egypt.
Distribution in the Middle East: Egypt (Giordani Soika, 1935 as *A. savignyi*; Blüthgen, 1955b, 1956; Gusenleitner, 2013a), Israel-Palestine (Giordani Soika, 1952 as *A. specularis savignyi*; Bytinski-Salz and Gusenleitner, 1971).
Extralimital distribution: Libya (Gusenleitner, 2013a).

***Alastor* (*Alastor*) *arabicus* Giordani Soika, 1979**
Alastor arabicus Giordani Soika, 1979: 272, ♀; Dhofar.
Distribution in the Middle East: Oman (Giordani Soika, 1979; Guichard, 1986; Gusenleitner, 2013a; Carpenter and Gadallah, 2020), Yemen (Gusenleitner, 2013a; Carpenter and Gadallah, 2020).
Extralimital distribution: None.

***Alastor* (*Alastor*) *ardens* Kostylev, 1935**
Alastor ardens Kostylev, 1935: 144, ♀; Turkmenistan.
Distribution in the Middle East: Turkey (Gusenleitner, 1988; Yildirim and Kojima, 1999; Fateryga *et al.*, 2023).
Extralimital distribution: Turkmenistan (Kostylev, 1935 as *A. problematicus*; Fateryga *et al.*, 2023), Uzbekistan (Fateryga *et al.*, 2023).

***Alastor* (*Megalastor*) *asiaticus* Morawitz, 1895**
Alastor asiaticus Morawitz, 1895: 492, ♂; Transcaspia: Pul-i-Chatun.
Distribution in the Middle East: Israel-Palestine (van der Vecht and Fischer, 1972; Gusenleitner, 2013a), Turkey (Blüthgen, 1951a; van der Vecht and Fischer, 1972; Gusenleitner, 2013a).
Extralimital distribution: Turkmenistan (van der Vecht and Fischer, 1972).

***Alastor* (*Alastor*) *atropos* Lepeletier, 1841**
Alastor atropos Lepeletier, 1841: 669, ♀, ♂; France.
Distribution in the Middle East: Syria (Giordani Soika, 1970), Turkey (Tüzün, 1986; Yildirim and Kojima, 1999).
Extralimital distribution: Central, South and Southwest Europe (Gusenleitner, 2013a), Italy (Selis, 2023).

***Alastor* (*Alastor*) *dalyi* Giordani Soika, 1979**
Alastor dalyi Giordani Soika, 1979: 272, ♀, ♂; Oman.
Distribution in the Middle East: Oman (Giordani Soika, 1979; Gusenleitner, 2013a), Saudi Arabia (Guichard, 1986; Gusenleitner, 2013a), United Arab Emirates (Gusenleitner, 2010b, 2012, 2013a).
Extralimital distribution: None.

***Alastor* (*Alastor*) *darius* Gusenleitner, 1986b**
Alastor (*Alastor*) *darius* Gusenleitner, 1986b: 30, ♀; Iran.
Catalogues with Iranian records: Ebrahimi and Carpenter (2008), Rahmani *et al.* (2020).
Distribution in Iran: Fars, Hormozgan (Gusenleitner, 1986b), Iran (no specific locality cited) (Gusenleitner, 2013a).
Distribution in the Middle East: Known only from Iran (Gusenleitner, 1986b, 2013a).
Extralimital distribution: None.

Alastor (Alastor) elisaei von Schulthess, 1925

Alastor elisaei von Schulthess, 1925: 61, 64, 201, ♀; Palestina, Jericho, and Fontem Elisae.

Distribution in the Middle East: Egypt (Giordani Soika, 1952; Gusenleitner, 2013a), Israel-Palestine (von Schulthess, 1925; Gusenleitner, 2013a).

Extralimital distribution: None.

Alastor (Megalastor) esfandiarii Giordani Soika, 1970

Alastor (Megalastor) esfandiarii Giordani Soika, 1970: 58, ♂; Iran: Khash.

Distribution in Iran: Sistan & Baluchestan (Giordani Soika, 1970).

Distribution in the Middle East: Iran (Giordani Soika, 1970; Gusenleitner, 2013a), Turkey (Yildirim and Kojima, 1999; Yildirim and Özbek, 1999).

Extralimital distribution: None.

Alastor (Alastor) iranus Blüthgen, 1956

Alastor (Alastor) iranus Blüthgen, 1956: 125, ♀; Iran: Belutschistan.

Catalogues with Iranian records: Ebrahimi and Carpenter (2008), Rahmani *et al.* (2020).

Distribution in Iran: Kerman (Schmid-Egger *et al.*, 2021), Sistan & Baluchestan (Blüthgen, 1956; Blüthgen and Gusenleitner, 1970).

Distribution in the Middle East: Iran (see references above, and Gusenleitner, 2013a), Israel-Palestine (van der Vecht and Fischer, 1972).

Extralimital distribution: None.

Alastor (Megalastor) mediomaculatus mediomaculatus Giordani Soika, 1952

Alastor mediomaculatus Giordani Soika, 1952: 61, ♀; Kotaik, Armenia.

Alastor iconius Blüthgen, 1955b: 67, 71, 74.

Catalogues with Iranian records: None.

Distribution in Iran: Fars (Schmid-Egger *et al.*, 2021).

Distribution in the Middle East: Iran (Schmid-Egger *et al.*, 2021; Selis *et al.*, 2025), Iraq (Giordani Soika, 1970 as *A. Iconius*; Yildirim and Kojima, 1999; Selis *et al.*, 2025), Israel-Palestine (Giordani Soika, 1952; Bytinski-Salz and Gusenleitner, 1971 as both *A. mediomaculatus* and *A. iconius*; Yildirim and Kojima, 1999; Gusenleitner, 2013a; Selis *et al.*, 2025), Jordan (Gusenleitner, 2013a; Selis *et al.*, 2025), Syria (Giordani Soika, 1970; Yildirim and Kojima, 1999; Gusenleitner, 2013a, 2013b; Selis *et al.*, 2025), Turkey (Blüthgen, 1955b, 1956 both as *A. (Megalastor) iconius*; Gusenleitner, 1966 both as *A. (Megalastor) iconius*, 1997; Giordani Soika, 1970 as *A. (Megalator) iconius*; Yildirim and Özbek, 1996a as *A. iconius*, 1996b as *A. (Megalastor) iconius*; Yildirim and Kojima, 1999; Selis *et al.*, 2025).

Extralimital distribution: Armenia (Giordani Soika, 1970; Selis *et al.*, 2025).

Alastor (Alastor) mocsaryi (André, 1884)

Odynerus mocsaryi André, 1884: 741, ♀; Russia.

Catalogues with Iranian records: This species is a new record for the fauna of Iran.

Distribution in Iran: East Azarbaijan province, Arasbaran forest, 4♀♀, August 2009, leg. P. Farhadi, det. J. Gusenleitner (Figure 2.1).

Distribution in the Middle East: Iran (new record), Lebanon, Syria (Giordani Soika, 1970 as *A. biegelebeni*; Fateryga, 2017b, 2018), Turkey (Giordani Soika, 1952, 1970 both as *Alastor biegelebeni*; Giordani Soika, 1960 as *A. (Alastor) kuehlhorni*; Gusenleitner, 1967 as *A. (A.) biegelebeni*; Yildirim and Özbek, 1996a as *A. biegelebeni*, 1996b, 1999 both as *A. (A.) biegelebeni*; Yildirim and Kojima, 1999 as *A. (A.) biegelebeni*; Yildirim and Gusenleitner, 2001, 2012; Fateryga, 2017b, 2018; Yildirim and Tezcan, 2018 as *A. moczaryi*).

Extralimital distribution: Caucasus, Europe, Russia (Fateryga, 2017b, 2018; Fateryga *et al.*, 2019), Armenia, Azerbaijan (Fateryga *et al.*, 2019), Italy (Selis, 2023), Slovenia (Gogala, 2022).

FIGURE 2.1 *Alastor* (*Alastor*) *mocsaryi* (André, 1884).

(Photo courtesy of Alexander Fateryga)

***Alastor* (*Megalastor*) *nitens* Gusenleitner, 2000c**

Alastor (*Megalastor*) *nitens* Gusenleitner, 2000c: 927, 929, ♂; Syria.

Distribution in the Middle East: Known only from Syria (Gusenleitner, 2000c, 2013a).

Extralimital distribution: None.

***Alastor* (*Alastor*) *pentheri* Kohl, 1905**

Alastor pentheri Kohl, 1905: 231, ♂; Turkey.

Distribution in the Middle East: Turkey (Kohl, 1905; Giordani Soika, 1960 as *A.* (*A.*) *korbi*, 1970; Yildirim and Özbek, 1996a, 1999; Yildirim, 1998; Gusenleitner, 1973, 2007; Yildirim and Kojima, 1999; Yildirim and Gusenleitner, 2007, 2012; Gusenleitner, 2013a).

Extralimital distribution: Armenia (Gusenleitner, 2013a).

***Alastor* (*Alastor*) *ruficornis* Gusenleitner, 2001a**

Alastor ruficornis Gusenleitner, 2001a: 663, ♂, ♀; Turkey.

Distribution in the Middle East: Iraq (Selis *et al.*, 2025), Turkey (Gusenleitner, 2001a, 2013a; Selis *et al.*, 2025).

Extralimital distribution: None.

***Alastor* (*Alastor*) *sanctus* Blüthgen, 1956**

Alastor sanctus Blüthgen, 1956: 124, 139, 140, ♀; Israel: Jerusalem.

Distribution in the Middle East: Israel-Palestine (Blüthgen, 1956; Gusenleitner, 2013a), Jordan (Gusenleitner, 2013a).

Extralimital distribution: None.

***Alastor* (*Megalastor*) *savignyi* de Saussure, 1853**
Alastor savignyi de Saussure, 1853: 260; sex not stated; Egypt.
Distribution in the Middle East: Israel-Palestine (von Schulthess, 1925; Blüthgen, 1955b; Bytinski-Salz and Gusenleitner, 1971; Gusenleitner, 2013a), Saudi Arabia (Gusenleitner, 2013a).
Extralimital distribution: None.

***Alastor* (*Megalastor*) *schwarzi* Gusenleitner, 1967**
Alastor (*Megalastor*) *schwarzi* Gusenleitner, 1967: 667, ♀, ♂; Turkey.
Distribution in the Middle East: "Arabia" (Yildirim and Kojima, 1999), Saudi Arabia (Guichard, 1986; Carpenter and Gadallah, 2020), Turkey (Gusenleitner, 1967, 1988, 2013a; Giordani Soika, 1970; Yildirim and Kojima, 1999; Carpenter and Gadallah, 2020).
Extralimital distribution: None.

***Alastor* (*Alastor*) *seidenstueckeri* Blüthgen, 1956**
Alastor (*Alastor*) *seidenstueckeri* Blüthgen, 1956: 130, 139, ♂; Turkey.
Distribution in the Middle East: Iraq (Selis *et al.*, 2025), Turkey (Blüthgen, 1956; Yildirim and Özbek, 1996b; Yildirim and Gusenleitner, 2012; Gusenleitner, 2013a; Selis *et al.*, 2025).
Extralimital distribution: Austria, Greece (Gusenleitner, 2008; Selis *et al.*, 2025), Bulgaria (Gusenleitner, 2013a; Selis *et al.*, 2025).

***Alastor* (*Alastor*) *thymbrinus* Blüthgen, 1956**
Alastor thymbrinus Blüthgen, 1956: 133, 139, ♂; Turkey.
Alastor rubripes Gusenleitner, 1967: 666.
Distribution in the Middle East: Known only from Turkey (Blüthgen, 1956; Gusenleitner, 1966, 1967 as *A.* (*Alastor*) *rubripes*; Giordani Soika, 1970 as *A.* (*A.*) *rubripes*; Yildirim and Özbek, 1996a; Yildirim and Kojima, 1999; Yildirim and Gusenleitner, 2004, 2012; Gusenleitner, 2013a).
Extralimital distribution: None.

***Alastor* (*Alastor*) *xerxes* Gusenleitner, 1986b**
Alastor (*Alastor*) *xerxes* Gusenleitner, 1986b: 29, 33, ♀; Iran.
Catalogues with Iranian records: Ebrahimi and Carpenter (2008), Rahmani *et al.* (2020).
Distribution in Iran: Hormozgan (Gusenleitner, 1986b).
Distribution in the Middle East: Known only from Iran (Gusenleitner, 1986b; Gusenleitner, 2013a).
Extralimital distribution: None.

***Alastor* (*Alastor*) *zoroaster* Gusenleitner, 1986b**
Alastor (*Alastor*) *zoroaster* Gusenleitner, 1986b: 29, 35, ♀; Iran.
Catalogues with Iranian records: Ebrahimi and Carpenter (2008), Rahmani *et al.* (2020).
Distribution in Iran: Fars (Schmid-Egger *et al.*, 2021), Hormozgan (Gusenleitner, 1986b; Schmid-Egger *et al.*, 2021).
Distribution in the Middle East: Known only from Iran (see references above, and Gusenleitner, 2013a).
Extralimital distribution: None.

Genus *Allodynerus* Blüthgen, 1938

Allodynerus Blüthgen, 1938: 280. Type species: "*Lionotus floricola* Sauss, 1852" [= *Odynerus floricola* de Saussure, 1853].

***Allodynerus delphinalis delphinalis* (Giraud, 1866)**
Odynerus (*Lionotus*) *delphinalis* Giraud, 1866: 464, ♀; Grenoble [Paris].
Catalogues with Iranian records: Ebrahimi and Carpenter (2008), Rahmani *et al.* (2020).
Distribution in Iran: East Azarbaijan (Ebrahimi and Carpenter, 2008; Modarres Awal, 2012).
Distribution in the Middle East: Egypt (Giordani Soika, 1935 as *Odynerus delphinalis*; Fateryga, 2018), Iran (see references above, and Fateryga, 2017b, 2018; Yalniz and Tüzün, 2019;

Kochetkov, 2020; Fateryga and Proshchalykin, 2024; Fateryga *et al.*, 2025), Lebanon (Giordani Soika, 1970; Gusenleitner, 2013a; Fateryga, 2017b, 2018; Kochetkov, 2020; Fateryga and Proshchalykin, 2024; Fateryga *et al.*, 2025), Turkey (Giordani Soika, 1970; Yildirim and Özbek, 1996a, 1996b, 1999; Yildirim and Kojima, 1999; Yildirim and Gusenleitner, 2009; Yildirim, 2012; Gusenleitner, 2013a; Fateryga, 2017b, 2018; Yalniz and Tüzün, 2019; Kochetkov, 2020; Fateryga and Proshchalykin, 2024; Fateryga *et al.*, 2025).

Extralimital distribution: Albania, Armenia, Austria, Azerbaijan, Belgium, Bulgaria, China, France, Germany, Italy, Japan, Kazakhstan, Korea, Kyrgyzstan, Portugal, Russia, Spain, Turkmenistan (Giordani Soika, 1970; Gusenleitner, 2013a; Fateryga, 2018), Slovenia (Gogala, 2022), Uzbekistan (Fateryga and Proshchalykin, 2024; Fateryga *et al.*, 2025).

***Allodynerus dignotus* (Morawitz, 1895)**

Odynerus (*Lionotus*) *dignotus* Morawitz, 1895: 457, ♀, ♂; «Transcaucasia: Tschemachlinskaja» [Azerbaijan].

Catalogues with Iranian records: Ebrahimi and Carpenter (2008), Rahmani *et al.* (2020).

Distribution in Iran: Razavi Khorasan (Salehi Fard *et al.*, 2013), Sistan & Baluchestan (Blüthgen and Gusenleitner, 1970).

Distribution in the Middle East: Iran (see references above, and Fateryga *et al.*, 2021, 2023; Selis *et al.*, 2025), Iraq (Rafi *et al.*, 2017; Selis *et al.*, 2025), Israel-Palestine (Giordani Soika, 1952 as *Pseudepipona dignotus*; Fateryga *et al.*, 2021, 2023; Selis *et al.*, 2025), Jordan (Selis *et al.*, 2025), Oman (Guichard, 1986; Gusenleitner, 2013a; Fateryga *et al.*, 2021, 2023; Selis *et al.*, 2025), Syria (Giordani Soika, 1970; Gusenleitner, 2013a; Fateryga *et al.*, 2021, 2023; Selis *et al.*, 2025), Turkey (Gusenleitner, 1966 as *Allodynerus dignotus dignotus*, 2013a; Giordani Soika, 1970 as *A. dignotus* and as *A. dignotus ankarensis*; Yildirim and Kojima, 1999 as *A. d. ankarensis* and as *A. d. dignotus*; Fateryga *et al.*, 2021, 2023; Selis *et al.*, 2025), United Arab Emirates (Guichard, 1986; Gillett and Howarth, 2004; Gusenleitner, 2012, 2013a; Fateryga *et al.*, 2021, 2023; Selis *et al.*, 2025).

Extralimital distribution: Azerbaijan (Fateryga *et al.*, 2021, 2023), China (Selis *et al.*, 2025), Greece, Kazakhstan, Pakistan, Turkmenistan (Gusenleitner, 2013a; Selis *et al.*, 2025), Uzbekistan (Fateryga *et al.*, 2023).

Comments: *Andrenosoma serratum* Hermann (Diptera: Asilidae) has been recorded by Ghahari *et al.* (2007) as being a predator of *Allodynerus dignotus*.

***Allodynerus floricola floricola* (de Saussure, 1853)**

Odynerus floricola de Saussure, 1953: 155, ♀; Algeria.

Catalogues with Iranian records: None.

Distribution in Iran: Semnan (Ghahari *et al.*, 2007).

Distribution in the Middle East: Cyprus (Fateryga, 2017b; Fateryga, 2018; Yalniz and Tüzün, 2019; Selis *et al.*, 2025), Iran (Ghahari *et al.*, 2007; Selis *et al.*, 2025), Iraq (Selis *et al.*, 2025), Jordan (Gusenleitner, 2013a; Fateryga, 2017b; Fateryga, 2018; Yalniz and Tüzün, 2019; Selis *et al.*, 2025), Turkey (Blüthgen, 1951a; Yildirim and Özbek, 1996a, 1996b, 1999; Yildirim and Kojima, 1999; Gusenleitner, 2007, 2013a; Yildirim, 2012; Yildirim and Gusenleitner, 2012; Fateryga, 2017b, 2018; Yildirim and Tezcan, 2018; Yalniz and Tüzün, 2019; Selis *et al.*, 2025).

Extralimital distribution: Albania, Algeria, Armenia, Austria, Croatia, France, Germany, Italy, Portugal, Spain, Switzerland, Tunisia (Gusenleitner, 2013a; Fateryga, 2017b, 2018; Selis *et al.*, 2025), Russia (Fateryga, 2018; Selis *et al.*, 2025), Slovenia (Gogala, 2022; Selis *et al.*, 2025).

***Allodynerus floricola inaequalis* Giordani Soika, 1970**

Allodynerus floricola inaequalis Giordani Soika, 1970: 142, ♀; Turchia: Büyükada, Istambul.

Distribution in the Middle East: Known only from Turkey (Giordani Soika, 1970; van der Vecht and Fischer, 1972; Gusenleitner, 1999; Yildirim and Kojima, 1999; Yildirim and Gusenleitner, 2007).

Extralimital distribution: None.

***Allodynerus nutatus* Gusenleitner, 1991**

Allodynerus nutatus Gusenleitner, 1991: 297, 303, ♂; Jordanien, Dehbeen.

Distribution in the Middle East: Known only from Jordan (Gusenleitner, 1991, 2013a).

Extralimital distribution: None.

***Allodynerus rossii* (Lepeletier, 1841)**

Odynerus rossii Lepeletier, 1841: 633, ♀, ♂; Environs de Paris. (?destroyed).

Catalogues with Iranian records: Ebrahimi and Carpenter (2008), Rahmani *et al.* (2020).

Distribution in Iran: Alborz (Giordani Soika, 1970), East Azarbaijan (Ebrahimi and Carpenter, 2008; Modarres Awal, 2012).

Distribution in the Middle East: Iran (see references above, and Gusenleitner, 2013a; Fateryga, 2017b, 2018), Turkey (Gusenleitner, 1966, 1967, 2013a; Giordani Soika, 1970; Yildirim and Özbek, 1996a, 1996b, 1999; Yildirim and Kojima, 1999; Yildirim and Gusenleitner, 2012; Fateryga, 2017b, 2018; Yildirim and Tezcan, 2018).

Extralimital distribution: Caucasus, South Europe, Central Europe to Germany, Northwest Africa, Russia, Ukraine (Gusenleitner, 2013a; Fateryga, 2018), Italy (Selis, 2023), Turkmenistan (Fateryga, 2017b, 2018), Slovenia (Gogala, 2022).

***Allodynerus vinciguerrae pallidulus* Giordani Soika, 1970**

Allodynerus vinciguerrae pallidulus Giordani Soika, 1970: 151, ♀, ♂; Eritrea, Sudan.

Distribution in the Middle East: Egypt (Gusenleitner, 2013a).

Extralimital distribution: Chad (Giordani Soika, 1974), Eritrea, Ethiopia, Suda (Carpenter *et al.*, 2009).

***Allodynerus vinciguerrae vinciguerrae* (Guiglia, 1929)**

Odynerus vinciguerrae Guiglia, 1929: 402, ♀, ♂; Giarabub (Cyrenaica orientalis).

Distribution in the Middle East: Egypt (Giordani Soika, 1935 as *Odynerus delphinalis*; Blüthgen, 1942), Israel-Palestine (Giordani Soika, 1952 as *Pseudepipona vinciguerrae*; Gusenleitner, 2013a), Oman (Guichard, 1986; Gusenleitner, 2013a; Carpenter and Gadallah, 2020), Saudi Arabia (Gadallah, 2003; Gusenleitner, 2013a; Carpenter and Gadallah, 2020), United Arab Emirates (Guichard, 1986; Howarth and Gillett, 2008; Gusenleitner, 2012, 2013a; Carpenter and Gadallah, 2020).

Extralimital distribution: Algeria, Libya (Gusenleitner, 2013a).

Genus *Ancistrocerus* Wesmael, 1836

Ancistrocerus Wesmael, 1836: 45. Type species: *Vespa parietum* L., 1758; designated by Girard, 1879.

***Ancistrocerus adenensis* Giordani Soika, 1952**

Ancistrocerus adenensis Giordani Soika, 1952: 31, ♂; S. Arabia: Prottetorato di Aden, Gebel Jihaf.

Distribution in the Middle East: Known only from Yemen (Giordani Soika, 1952; Guichard, 1986; Gusenleitner, 2013a; Carpenter and Gadallah, 2020).

Extralimital distribution: None.

***Ancistrocerus antilope* (Panzer, 1798)**

Vespa antílope Panzer, 1798: 9, sex not stated; "English" (destroyed).

Distribution in the Middle East: Turkey (Giordani Soika, 1970; Gusenleitner, 1988, 2013a; Yildirim and Özbek, 1996a; Yildirim, 1998; Yildirim and Kojima, 1999; Fateryga, 2017b, 2018; Kochetkov, 2020; Fateryga and Proshchalykin, 2024).

Extralimital distribution: Afghanistan, Caucasus, China, Europe, India, Japan, Kazakhstan, Kyrgyzstan, Mongolia, North America (Fateryga, 2017b, 2018), Tajikistan (Fateryga and Proshchalykin, 2024).

***Ancistrocerus auctus* (Fabricius, 1793)**

Vespa aucta Fabricius, 1793: 272, ♀; Kiliae [København] [Denmark].

Catalogues with Iranian records: Ebrahimi and Carpenter (2008), Rahmani *et al.* (2020).

Distribution in Iran: East Azarbaijan (Ebrahimi and Carpenter, 2008; Modarres Awal, 2012), Isfahan (Hayat *et al.*, 2008), Zanjan (Abbasi *et al.*, 2008a, 2008b; Sakenin *et al.*, 2008; Pashaei Rad *et al.*, 2010).

Distribution in the Middle East: Iran (see references above, and Fateryga, 2017b, 2018; Yalniz and Tüzün, 2019; Fateryga *et al.*, 2021; Selis *et al.*, 2025), Iraq (Giordani Soika, 1970; Gusenleitner, 2001b, 2013a; Fateryga, 2017b, 2018; Yalniz and Tüzün, 2019; Fateryga *et al.*, 2021; Selis *et al.*, 2025), Israel-Palestine (Gusenleitner, 2013a; Fateryga, 2017b, 2018; Fateryga *et al.*, 2021; Selis *et al.*, 2025), Jordan (Gusenleitner, 2001b; Fateryga, 2017b, 2018; Fateryga *et al.*, 2021; Selis *et al.*, 2025), Syria (Fateryga, 2017b, 2018; Yalniz and Tüzün, 2019; Fateryga *et al.*, 2021; Selis *et al.*, 2025), Turkey (Gusenleitner, 1967, 2013a; Giordani Soika, 1970; Yildirim and Özbek, 1996a, 1996b, 1999; Yildirim and Kojima, 1999; Yildirim and Gusenleitner, 2004, 2012; Yildirim, 2012; Fateryga, 2017b, 2018; Yildirim and Tezcan, 2018; Yalniz and Tüzün, 2019; Fateryga *et al.*, 2021; Selis *et al.*, 2025).

Extralimital distribution: Albania, Armenia, Austria, Czech Republic, France, Greece, Italy, Kazakhstan, North Macedonia, Russia, Slovakia, Spain, Switzerland, Turkmenistan (Gusenleitner, 2013a; Fateryga, 2018; Selis *et al.*, 2025), Slovenia (Gogala, 2022; Selis *et al.*, 2025).

Comments: *Dioctria linearis* (Fabricius) and *Machiremisca decipiens* (Wiedemann in Meigen) (Diptera: Asilidae) are the predators of *Ancistrocerus auctus* (Hayat *et al.*, 2008).

***Ancistrocerus biphaleratus biphaleratus* (de Saussure, 1852)**

Odynerus (*Ancistrocerus*) *biphaleratus* de Saussure, 1852: 134, ♀, ♂; Egypt.

Distribution in the Middle East: Egypt (de Saussure, 1852 as *Odynerus* (*Ancestrocerus*) *biphaleratus*; Giodani Soika, 1935, 1936; Gusenleitner, 2013a), Saudi Arabia (Zalat *et al.*, 1993), Southern Arabia (Bingham, 1898 as *O. biphaleratus*).

Extralimital distribution: Ethiopia (de Saussure, 1863 as *O. impunctatus*; Giordani Soika, 1941), Malta (Cassar *et al.*, 2022), Mauritania, Spain (Castro, 1986).

***Ancistrocerus biphaleratus palaestinicus* Giordani Soika, 1952**

Ancistrocerus palaestinicus Giordani Soika, 1952: 22, 30, ♂; Palestina: Wadi el Kelt.

Distribution in the Middle East: Egypt (Guichard, 1986; Gusenleitner, 2001b), Israel-Palestine (Giordani Soika, 1952, 1973 as *A. palaestinicus*; Gusenleitner, 2001b, 2013a), Jordan (Gusenleitner, 2001b), Saudi Arabia (Bingham, 1898; Guichard, 1986; Gusenleitner, 2001b, 2013a; Carpenter and Gadallah, 2020), Yemen (Gusenleitner, 2001b).

Extralimital distribution: None.

***Ancistrocerus biphaleratus triphaleratus* (de Saussure, 1855)**

Odynerus triphaleratus de Saussure, 1855: 198, 200, ♀, ♂; 200.

Distribution in Iran: Hormozgan (Gusenleitner, 2010a).

Distribution in the Middle East: Iran (Gusenleitner, 2010a, 2013a), Turkey (Blüthgen, 1943 as *Ancistrocerus triphaleratus*; Yildirim and Kojima, 1999; Gusenleitner, 2013a).

Extralimital distribution: Algeria, Italy, Morocco, Portugal, Spain (Gusenleitner, 2013a).

***Ancistrocerus claripennis claripennis* Thomson, 1874**

Ancistrocerus claripennis Thomson, 1874: 76, ♀, ♂; Sweden.

Catalogues with Iranian records: Ebrahimi and Carpenter (2008), Rahmani *et al.* (2020).

Distribution in Iran: Ardabil (Ebrahimi and Carpenter, 2008; Modarres Awal, 2012), Mazandaran (Gusenleitner, 2011a).

Distribution in the Middle East: Iran (see references above, and Gusenleitner, 2013a; Fateryga, 2017b), Turkey (Gusenleitner, 1988; Yildirim and Özbek, 1996a, 1996b; Yildirim and Kojima, 1999; Fateryga, 2017b).

Extralimital distribution: Albania, Andorra, Austria, Belarus, Bulgaria, England, France, Germany, Greece, Italy, Spain, Sweden, Switzerland (Gusenleitner, 2013a), China, Russia, Tajikistan (Fateryga, 2017b), Slovenia (Gogala, 2022).

***Ancistrocerus claripennis ponticus* Giordani Soika, 1970**

Ancistrocerus claripennis ponticus Giordani Soika, 1970: 134, ♀; Turchia: Erzurum, Koptag Gecidi.

Distribution in the Middle East: Known only from Turkey (Giordani Soika, 1970; Yildirim and Özbek, 1996a, 1996b; Yildirim and Kojima, 1999; Yildirim and Gusenleitner, 2001; Gusenleitner, 2013a).

Extralimital distribution: None.

***Ancistrocerus dusmetiolus* (Strand, 1914)**

Odynerus dusmetiolus Strand, 1914: 163, ♂, ♀; Madrid [replacement name for *O. sociabilis* Dusmet].

Distribution in the Middle East: Turkey (Yildirim and Özbek, 1996a, 1996b; Yildirim and Kojima, 1999; Yildirim and Gusenleitner, 2004; Fateryga, 2017a, 2017b).

Extralimital distribution: Afghanistan, Austria, Belgium, Central Asia, Europe (WE, SE, EE), France, Germany, Kazakhstan, Kyrgyzstan, Portugal, Russia, Spain, Switzerland, Tajikistan (Gusenleitner, 2013a; Fateryga, 2017a, 2017b).

***Ancistrocerus gazella* (Panzer, 1798)**

Vespa gazella Panzer, 1798: 10, ♂; Austria (destroyed).

Catalogues with Iranian records: Ebrahimi and Carpenter (2008), Rahmani *et al.* (2020).

Distribution in Iran: Iran (no specific locality cited) (Gusenleitner, 1995b, 2013a; Fateryga, 2017b, 2018).

Distribution in the Middle East: Cyprus (Gusenleitner, 2013a; Fateryga, 2017b, 2018; Fateryga *et al.*, 2019), Iran (see references above), Israel-Palestine (Gusenleitner, 2013a; Fateryga, 2017b, 2018; Fateryga *et al.*, 2019), Turkey (Fateryga, 2017b, 2018; Kemal and Koçak, 2018).

Extralimital distribution: Afghanistan, Albania, Austria, Belarus, Belgium, Bulgaria, eastern and North America (introduced), England, France, Germany, Greece, Italy, Morocco, New Zealand (introduced), Pakistan, Portugal, Russia, Spain, Sweden (Gusenleitner, 2013a; Fateryga, 2017b, 2018), Slovenia (Gogala, 2022).

***Ancistrocerus ichneumonideus ichneumonideus* (Ratzeburg, 1844)**

Vespa (Odynerus) ichneumonidea Ratzeburg, 1844: 52, ♂; Germany, bred from resin gall of *Tortrix resinana*.

Distribution in the Middle East: Turkey (Yildirim and Özbek, 1996b; Yildirim and Kojima, 1999; Fateryga, 2017b).

Extralimital distribution: Europe, Kazakhstan, Mongolia, Russia (Fateryga, 2017b), Italy (Selis, 2023).

***Ancistrocerus impunctatus impunctatus* (Spinola, 1838)**

Odynerus impunctatus Spinola, 1838: 503, ♂; "Egypt".

Distribution in the Middle East: Known only from Egypt (Spinola, 1838; de Saussure, 1854; Smith, 1857; André, 1884; Dalla Torre, 1894 all as *Odynerus impunctatus*; de Saussure, 1854–1856; Bequaert, 1918; Giordani Soika, 1966 as both *O. impunctatus* and *A. impunctatus*; van der Vecht and Fischer, 1972; Gusenleitner, 2013a).

Extralimital distribution: None.

***Ancistrocerus longispinosus longispinosus* (de Saussure, 1855)**

Odynerus longispinosus de Saussure, 1855: 213, ♂; L'Algérie.

Distribution in the Middle East: Cyprus (Blüthgen, 1957) as *A. gazelloides hellenicus*), Egypt (Blüthgen, 1957 as *A. g. hellenicus*; Gusenleitner, 2013a), Lebanon (Gusenleitner, 2013a), Turkey (Blüthgen, 1957 as *A. g. hellenicus*; Gusenleitner, 1966 as *A. g. hellenicus*, 1995b as

A. longispinosus; Giordani Soika, 1970 as *A. gazelloides hellenicus*; Yildirim and Özbek, 1996a as *A. g. hellenicus*, 1996b; Yildirim and Kojima, 1999; Yildirim, 2012; Yalniz and Tüzün, 2019 as *A. longispinosus*).

Extralimital distribution: Italy (Selis, 2023), Malta (Cassar *et al.*, 2022), Northwest Africa, South Europe (Gusenleitner, 2013a), Slovenia (Gogala, 2022).

***Ancistrocerus nigricornis* (Curtis, 1826)**

Odynerus nigricornis Curtis, 1826: 137, sex not stated; "Great Britain".

Catalogues with Iranian records: Ebrahimi and Carpenter (2008), Rahmani *et al.* (2020).

Distribution in Iran: Iran (no specific locality cited) (Gusenleitner, 1995b, 2013a; Fateryga, 2017b, 2018; Fateryga *et al.*, 2020).

Distribution in the Middle East: Iran (see references above, and Fateryga and Proshchalykin, 2024; Fateryga *et al.*, 2025), Turkey (Giordani Soika, 1970 as *Allodynerus nigricornis*; Gusenleitner, 1988; Yildirim and Özbek, 1996a as *A. nigricornis*; Yildirim and Kojima, 1999 as *A. nigricornis*; Fateryga, 2017b, 2018; Fateryga *et al.*, 2020; Fateryga and Proshchalykin, 2024; Fateryga *et al.*, 2025).

Extralimital distribution: Azerbaijan (Fateryga *et al.*, 2025), Europe, China, India, Japan, Kazakhstan, Mongolia, North Africa, Russia (Gusenleitner, 2013a; Fateryga, 2017b, 2018; Fateryga *et al.*, 2020; Fateryga and Proshchalykin, 2024), Italy (Selis, 2023), Slovenia (Gogala, 2022), Tajikistan, Turkmenistan (Fateryga and Proshchalykin, 2024; Fateryga *et al.*, 2025).

***Ancistrocerus oviventris caucasicus* Gusenleitner, 1994**

Ancistrocerus oviventris caucasicus Gusenleitner, 1994: 307, 318, ♀, ♂; Rußland, Kaukasus, Umgebung Teberda; Amanus-Tal.

Distribution in the Middle East: Turkey (Gusenleitner, 1994; Yildirim and Özbek, 1996b; Yildirim and Kojima, 1999).

Extralimital distribution: China, Europe, North Africa, Central Asia, Kazakhstan, Korea, Mongolia, Japan, Russia (Fateryga, 2017b as *A. oviventris*).

***Ancistrocerus oviventris oviventris* (Wesmael, 1836)**

Odynerus oviventris Wesmael, 1836: 45, ♀, ♂; "environs de Bruxelles".

Distribution in the Middle East: Turkey (Giordani Soika, 1970; Yildirim and Özbek, 1996a, 1996b, 1999; Madl, 1997; Yildirim and Kojima, 1999; Yildirim and Gusenleitner, 2001, 2012; Fateryga, 2017b, 2018; Kochetkov, 2020).

Extralimital distribution: Caucasus, Central Asia, Russia (Fateryga, 2017b, 2018), China (NE), Japan, Kazakhstan, Korea, Mongolia, N Africa (Fateryga, 2017b), Europe to East Asia (Gusenleitner, 2013a), Italy (Selis, 2023), Slovenia (Gogala, 2022).

***Ancistrocerus oviventris siculus* Blüthgen, 1955b**

Ancistrocerus oviventris siculus Blüthgen, 1955b: 155, ♀; "Taormina" (Berlin).

Distribution in the Middle East: Turkey (Gusenleitner, 1995b; Yildirim and Kojima, 1999; Yildirim and Gusenleitner, 2012).

Extralimital distribution: Italy, Spain (Gusenleitner, 1995b).

***Ancistrocerus parietinus* (Linnaeus, 1761)**

Vespa parietina Linnaeus, 1761: 418, sex not stated. Sweden.

Catalogues with Iranian records: Ebrahimi and Carpenter (2008), Rahmani *et al.* (2020).

Distribution in Iran: Iran (no specific locality cited) (Gusenleitner, 1995b; Yildirim and Kojima, 1999; Fateryga, 2017b).

Distribution in the Middle East: Iran (see references above), Turkey (Giordani Soika, 1970; Yildirim and Kojima, 1999; Yildirim and Özbek, 1999; Gusenleitner, 2013a; Fateryga, 2017b).

Extralimital distribution: Europe (Gusenleitner, 2013a; Fateryga, 2017b), Italy (Selis, 2023), Japan, Kazakhstan, Korea, Mongolia, North Africa, Russia (Fateryga, 2017b), Slovenia (Gogala, 2022).

***Ancistrocerus parietum* (Linnaeus, 1758)**

Vespa parietum Linnaeus, 1758: 572, ♀; Sweden.

Catalogues with Iranian records: Rahmani *et al.* (2020).

Distribution in Iran: Qazvin (Morice, 1921 as *Odynerus parietum* (Linnaeus, 1758)).

Distribution in the Middle East: Iran (Morice, 1921, and Kurzenko, 1978; Gusenleitner, 2013a; Fateryga, 2017b, 2018; Yalniz and Tüzün, 2019; Kochetkov, 2020; Fateryga *et al.*, 2023), Turkey (Giordani Soika, 1952; Tüzün and Tanyolag, 1988 as *O.* (*Ancistrocerus*) *parietum*; Yildirim and Özbek, 1996a, 1996b as *O.* (*A.*) *parietum*; Yildirim and Kojima, 1999; Yildirim and Gusenleitner, 2009; Yildirim, 2012; Gusenleitner, 2013a; Fateryga, 2017b, 2018; Yalniz and Tüzün, 2019; Kochetkov, 2020; Fateryga *et al.*, 2023).

Extralimital distribution: China, Kazakhstan, Korea, Mongolia, North Africa, North America (introduced), Russia (Fateryga, 2017b, 2018; Kochetkov, 2020), Europe (Gusenleitner, 2013a; Fateryga, 2017b, 2018), Italy (Selis, 2023), Korean Peninsula, Uzbekistan (Fateryga *et al.*, 2023).

***Ancistrocerus renimacula* (Lepeletier, 1841)**

Odynerus renimacula Lepeletier, 1841: 654, ♀, ♂; Algeria "Oran".

Distribution in the Middle East: Israel-Palestine (Giordani Soika, 1952), Turkey (Gusenleitner, 1966).

Extralimital distribution: Middle Europe, Northwest Africa, Southwest Europe (Gusenleitner, 2013a), Italy (Selis, 2023), Slovenia (Gogala, 2022).

***Ancistrocerus scoticus scoticus* (Curtis, 1826)**

Odynerus scoticus Curtis, 1826: 137, sex not stated; "Great Britain and Ireland".

Catalogues with Iranian records: Ebrahimi and Carpenter (2008), Rahmani *et al.* (2020).

Distribution in Iran: Iran (no specific locality cited) (Ebrahimi and Carpenter, 2008; Fateryga, 2017b; Fateryga *et al.*, 2021).

Distribution in the Middle East: Iran (see references above), Turkey (Giordani Soika, 1970; Gusenleitner, 1988, 2013a; Yildirim and Özbek, 1996a, 1996b; Yildirim and Kojima, 1999; Yildirim and Gusenleitner, 2001, 2004, 2012; Fateryga, 2017b; Fateryga *et al.*, 2021).

Extralimital distribution: Azerbaijan (Fateryga *et al.*, 2021), Europe to East Asia (Gusenleitner, 2013a), China, Kazakhstan, Korean Peninsula, Mongolia, North Africa (Fateryga, 2017b; Fateryga *et al.*, 2021), Italy (Selis, 2023), Slovenia (Gogala, 2022).

***Ancistrocerus tenebrosus* Gusenleitner, 2010a**

Ancistrocerus tenebrosus Gusenleitner, 2010a: 1349, ♂, ♀, Holotype ♂; Iran.

Catalogues with Iranian records: Rahmani *et al.* (2020).

Distribution in Iran: Mazandaran (Gusenleitner, 2010a).

Distribution in the Middle East: Known only from Iran (Gusenleitner, 2010a, 2013a; Selis *et al.*, 2025), Iraq (Selis *et al.*, 2025).

Extralimital distribution: None.

***Ancistrocerus trifasciatus trifasciatus* (Müller, 1776)**

Vespa trifasciata Müller, 1776: 163 sex not stated; Dania et Norvegia.

Distribution in the Middle East: Turkey (Giordani Soika, 1970; Yildirim and Kojima, 1999; Gusenleitner, 2013a; Fateryga, 2017b, 2018).

Extralimital distribution: Caucasus, China, Europe, Japan, Kazakhstan, Korea, Kyrgyzstan, Mongolia (Fateryga, 2017b, 2018), Italy (Selis, 2023), Slovenia (Gogala, 2022).

Genus *Antepipona* de Saussure, 1855

Antepipona Saussure, 1855: 244. Type species: *Odynerus silaos* Saussure, 1853; designated by ICZN, 1970.

***Antepipona albosignata* Gusenleitner, 1986c**

Antepipona albosignata Gusenleitner, 1986c: 361, ♀, ♂; "Türkei: Ürgüp".

Distribution in the Middle East: Known only from Turkey (Gusenleitner, 1986c, 2007, 2013a; Yildirim, 1998; Yildirim and Özbek, 1999; Yildirim and Kojima, 1999; Yildirim and Gusenleitner, 2004, 2007).

Extralimital distribution: None.

***Antepipona arabica* Giordani Soika, 1979**

Antepipona arabica Giordani Soika, 1979: 281, ♀, ♂; "Oman: Rostaq, 350 m".

Catalogues with Iranian records: None.

Distribution in Iran: Fars, Kerman (Schmid-Egger *et al.*, 2021).

Distribution in the Middle East: Iran (Schmid-Egger *et al.*, 2021), Oman (Giordani Soika, 1979; Guichard, 1986; Gusenleitner, 2013a; Carpenter and Gadallah, 2020), United Arab Emirates (Gusenleitner, 2010a, 2012, 2013a; Carpenter and Gadallah, 2020).

Extralimital distribution: None.

***Antepipona barrei barrei* (Radoszkovski, 1893)**

Odynerus barrei Radoszkovski, 1893: 76, ♀; Turkmenistan.

Catalogues with Iranian records: Ebrahimi and Carpenter (2008) as *Antepipona barrei vastifica* (Morawitz, 1895), Rahmani *et al.* (2020).

Distribution in Iran: Kerman (Schmid-Egger *et al.*, 2021), Tehran (Giordani Soika, 1970 as *Antepipona barrei vastifica* (Morawitz, 1895)).

Distribution in the Middle East: Iran (see references above, and Castro and Dvořák, 2009).

Extralimital distribution: Afghanistan, Armenia, Turkmenistan, Uzbekistan (Gusenleitner, 2013a as *A. barrei barrei*).

***Antepipona barrei vastifica* (Morawitz, 1895)**

Odynerus (*Lionotus*) *vastificus* Morawitz, 1895: 467, ♂; "Transcaspia: Tschuli".

Distribution in Iran: Iran (no specific locality cited) (Giordani Soika, 1970; Gusenleitner, 2013a).

Distribution in the Middle East: Iran (Giordani Soika, 1970; Gusenleitner, 2013a).

Extralimital distribution: Turkmenistan (Gusenleitner, 2013a).

***Antepipona cingulifera* (Walker, 1871)**

Odynerus cingulifer Walker, 1871: 33, ♀; "Harkeko".

Distribution in the Middle East: Egypt (Walker, 1871 as *Odynerus cingulifer*; Radoszkowski, 1876 as *Oplopus eversmanni*; Giordani Soika, 1935 as *Odynerus* (*Rhynchium*) *priesneri*; Gusenleitner, 2013a), Saudi Arabia (Walker, 1871 as *Odynerus cingulifer*; Guichard, 1986; Gadallah, 2003; Madl, 2018; Carpenter and Gadallah, 2020), Yemen (Giordani Soika, 1957 as *Odontodynerus eversmanni*; Gusenleitner, 2007, 2013a; Zalat *et al.*, 1993; Madl, 2018; Carpenter and Gadallah, 2020).

Extralimital distribution: Cameroon, Djibouti, Ghana, Mali, Niger, Nigeria (Carpenter *et al.*, 2009), Eritrea, Ethiopia, Sudan, West Africa (Gusenleitner, 2013a).

***Antepipona cribrata* (Morawitz, 1885)**

Lionotus cribratus Morawitz, 1885: 171, ♀; "Transcaucasia; Doroschitscach".

Distribution in the Middle East: Turkey (Gusenleitner, 2002, 2013a; Fateryga, 2017b).

Extralimital distribution: Afghanistan, Armenia, Azerbaijan, Russia (Fateryga, 2017b), Turkmenistan (Gusenleitner, 2013a).

***Antepipona curialis* (Morawitz, 1895)**

Odynerus (*Lionotus*) *curialis* Morawitz, 1895: 476, ♀; Uzbekistan.

Distribution in the Middle East: Iraq (Gusenleitner, 2013a; Selis *et al.*, 2025).

Extralimital distribution: China, Turkmenistan, Uzbekistan (Gusenleitner, 2013a; Qin *et al.*, 2024; Selis *et al.*, 2025).

***Antepipona deflenda* (Saunders, 1853)**

Ancistrocerus deflendus S. S. Saunders, 1853: 141, ♀, ♂; Albania.

Catalogues with Iranian records: Ebrahimi and Carpenter (2008), Rahmani *et al.* (2020).

Distribution in Iran: East Azarbaijan, Northern Khorasan, Tehran (Ebrahimi and Carpenter, 2008; Modarres Awal, 2012), Kerman (Pashaei Rad *et al.*, 2010), Razavi Khorasan (Giordani Soika, 1970), Sistan & Baluchestan (Hamzavi *et al.*, 2019), Zanjan (Abbasi *et al.*, 2008a, 2008b; Pashaei Rad *et al.*, 2010).

Distribution in the Middle East: Cyprus (Fateryga, 2017a, 2017b, 2018; Fateryga *et al.*, 2019; Selis *et al.*, 2025), Iran (see references above, and Fateryga, 2017a, 2017b, 2018; Fateryga *et al.*, 2019; Selis *et al.*, 2025), Iraq, Lebanon (Giordani Soika, 1970; Yildirim and Kojima, 1999; Gusenleitner, 2013a; Fateryga, 2017a, 2017b, 2018; Fateryga *et al.*, 2019; Selis *et al.*, 2025), Israel-Palestine (Gusenleitner, 2013a; Fateryga, 2017a, 2017b, 2018; Fateryga *et al.*, 2019; Selis *et al.*, 2025), Jordan (Giordani Soika, 1970; Gusenleitner, 2013a; Fateryga, 2017b, 2018; Fateryga *et al.*, 2019; Selis *et al.*, 2025), Turkey (Giordani Soika, 1952 as *Odontodynerus deflendus*, 1970; Gusenleitner, 1966, 1967 both as *O. deflendus*, 2007, 2013a; Yildirim and Özbek, 1996a, 1996b; Yildirim and Kojima, 1999; Yildirim and Gusenleitner, 2007, 2012; Fateryga, 2017a, 2017b, 2018; Yildirim and Tezcan, 2018; Fateryga *et al.*, 2019; Selis *et al.*, 2025).

Extralimital distribution: Albania, Algeria, Armenia, Austria, Azerbaijan, Croatia, France, Georgia, Germany, Kazakhstan, Montenegro, Morocco, Pakistan, Portugal, Spain, Switzerland, Ukraine (Selis *et al.*, 2025), China, Russia, Tajikistan, Turkmenistan, Uzbekistan (Fateryga, 2017a, 2017b, 2018; Selis *et al.*, 2025), Italy (Selis, 2023; Selis *et al.*, 2025), Malta (Cassar *et al.*, 2022), Slovenia (Gogala, 2022; Selis *et al.*, 2025).

***Antepipona dentella* Gusenleitner, 1991**

Antepipona dentella Gusenleitner, 1991: 297, 300, ♀, ♂; "Jordanien, Zekatal".

Distribution in the Middle East: Jordan (Gusenleitner, 2013a), Syria (Gusenleitner, 1991), Turkey (Yildirim and Gusenleitner, 2012; Gusenleitner, 2013a).

Extralimital distribution: None.

***Antepipona iconia* (Blüthgen, 1951)**

Odontodynerus iconius Blüthgen, 1951a: 169, 182, Holotype ♀; Turkey.

Distribution in Iran: Mazandaran (Gusenleitner, 2012), Rahmani *et al.* (2020).

Distribution in the Middle East: Iran (see references above, and Gusenleitner, 2013a; Selis *et al.*, 2025), Iraq (Selis *et al.*, 2025), Israel-Palestine (Giordani Soika, 1952 as *Odontodynerus deflendus tiberiacus*; Gusenleitner, 2013a; Selis *et al.*, 2025), Syria (Giordani Soika, 1970; Yildirim and Kojima, 1999; Gusenleitner, 2013a; Selis *et al.*, 2025), Turkey (Blüthgen, 1951a as *Odontodynerus iconius*; Gusenleitner, 1967 as *O. iconius*; Giordani Soika, 1970; Yildirim and Özbek, 1996a, 1996b; Yildirim and Kojima, 1999; Yildirim and Gusenleitner, 2009; Gusenleitner, 2013a; Selis *et al.*, 2025).

Extralimital distribution: None.

***Antepipona insana clava* Gusenleitner, 1988**

Antepipona insana clava Gusenleitner, 1988: 713, 728, ♀; "Türkei (Van)".

Distribution in the Middle East: Iraq (Selis *et al.*, 2025), Turkey (Blüthgen and Gusenleitner, 1970 as *Odontodynerus quadriguttatus*; Yildirim and Özbek, 1996b as both *Antepiponai insana clava* and *A. insana quadriguttata*; Yildirim and Kojima, 1999; Gusenleitner, 1988, 2013a; Selis *et al.*, 2025).

Extralimital distribution: None.

***Antepipona insana insana* (Giordani Soika, 1943)**

Odynerus (*Rhynchium*) *insanus* Giordani Soika, 1943: 10, ♀; Rhodos.

Distribution in the Middle East: Turkey (Yildirim and Özbek, 1996b, 1999; Yildirim and Kojima, 1999; Yildirim and Gusenleitner, 2004, 2012; Gusenleitner, 2013a).

Extralimital distribution: South Europe (Gusenleitner, 2013a).

***Antepipona irakensis* Giordani Soika, 1970**

Antepipona irakensis Giordani Soika, 1970: 121, ♂; "Irak: Bagdad".

Distribution in the Middle East: Known only from Iraq (Giordani Soika, 1970; van der Vecht and Fischer, 1972; Gusenleitner, 2013a; Selis *et al.*, 2025).

Extralimital distribution: None.

***Antepipona jordanica* Gusenleitner, 2004**

Antepipona jordanica Gusenleitner, 2004: 1083, ♂; "Jordania: Wadi al Haldam".
Distribution in the Middle East: Known only from Jordan (Gusenleitner, 2004, 2013a).
Extralimital distribution: None.

***Antepipona karadgensis* Giordani Soika, 1970**

Antepipona karadgensis Giordani Soika, 1970: 113, ♀; Iran.
Catalogues with Iranian records: Ebrahimi and Carpenter (2008), Rahmani *et al.* (2020).
Distribution in Iran: Alborz (Giordani Soika, 1970).
Distribution in the Middle East: Known only from Iran (Giordani Soika, 1970; Gusenleitner, 2013a).
Extralimital distribution: None.

***Antepipona kassalensis* (Giordani Soika, 1939)**

Odynerus kassalensis Giordani Soika, 1939: 359, ♀; "Sudan, Kassala".
Distribution in the Middle East: Egypt (Gusenleitner, 2004, 2013a), Oman (Giordani Soika, 1979; Gusenleitner, 2004, 2013a; Carpenter and Gadallah, 2020), Saudi Arabia (Guichard, 1986; Gusenleitner, 2013a; Carpenter and Gadallah, 2020), United Arab Emirates (Gusenleitner, 2010a; Carpenter and Gadallah, 2020), Yemen (Gusenleitner, 2007, 2013a; Carpenter and Gadallah, 2020).
Extralimital distribution: Eritrea, Mauritania, Nigeria (Carpenter *et al.*, 2009), Ethiopia, Sudan, West Africa (Gusenleitner, 2013a).

***Antepipona laevigata* (Blüthgen, 1951)**

Odontodynerus laevigatus Blüthgen, 1951a: 169, 180, ♂; Armenia.
Catalogues with Iranian records: Ebrahimi and Carpenter (2008), Rahmani *et al.* (2020).
Distribution in Iran: Alborz, Mazandaran, Semnan (Gusenleitner, 1986a).
Distribution in the Middle East: Iran (Gusenleitner, 1986a), Iraq (Gusenleitner, 2013a), Lebanon (Giordani Soika, 1970 as *A. varentzowi*; Gusenleitner, 2013a; Yildirim and Kojima, 1999), Turkey (Giordani Soika, 1970 as *A. varentzowi*, 2007; Gusenleitner, 1986b, 2013a; Yildirim and Özbek, 1996b; Yildirim and Kojima, 1999; Yildirim and Gusenleitner, 2007).
Extralimital distribution: Armenia, Pakistan (Gusenleitner, 2013a).

***Antepipona levis* Gusenleitner, 2021**

Antepipona levis Gusenleitner, 2021: 61, ♂; Iran.
Catalogues with Iranian records: None.
Distribution in Iran: Khuzestan (Gusenleitner, 2021).
Distribution in the Middle East: Known only from Iran (Gusenleitner, 2021).
Extralimital distribution: None.

***Antepipona nigricornis nigricornis* (Morawitz, 1885)**

Lionotus nigricornis Morawitz, 1885: 160, ♀, ♂; "Tauria: Balaklava".
Distribution in the Middle East: Turkey (Blüthgen, 1951a as *Allodynerus binominatus*; Giordani Soika, 1970 as *Allodynerus nigricornis*; Gusenleitner, 2013a; Fateryga, 2017b, 2018).
Extralimital distribution: South Europe (Gusenleitner, 2013a; Fateryga, 2017b), Ukraine (Gusenleitner, 2013a), West Europe, Russia, Turkmenistan (Fateryga, 2017b, 2018).

***Antepipona omanensis* Giordani Soika, 1979**

Antepipona omanensis Giordani Soika, 1979: 282, ♀, ♂; "Oman: Wadi Quryat, Ag Stn., 500 m".
Catalogues with Iranian records: None.
Distribution in Iran: Kerman (Schmid-Egger *et al.*, 2021).
Distribution in the Middle East: Iran (Schmid-Egger *et al.*, 2021), Oman (Giordani Soika, 1979; Gusenleitner, 2004, 2005, 2013a, 2013b; Carpenter and Gadallah, 2020), Saudi Arabia (Giordani Soika, 1979; Guichard, 1986; Carpenter and Gadallah, 2020), United Arab Emirates (Gusenleitner, 2010a; Carpenter and Gadallah, 2020).
Extralimital distribution: None.

Antepipona orbitalis ballioni (Morawitz, 1867)
Odynerus ballioni Morawitz, 1867: 129, ♀, ♂; Russia.
Distribution in the Middle East: Turkey (Giordani Soika, 1970; Yildirim and Özbek, 1996a, 1996b, 1999; Yildirim and Kojima, 1999; Gusenleitner, 2007, 2013a; Yildirim and Gusenleitner, 2007, 2012; Yildirim, 2012; Fateryga, 2017a, 2017b, 2018; Yalniz and Tüzün, 2019).
Extralimital distribution: Caucasus, Kazakhstan, Mongolia, Russia (Fateryga, 2017a, 2017b, 2018), Southeastern Europe (Gusenleitner, 2013a).

Antepipona orbitalis orbitalis (Herrich-Schäffer, 1839)
Odynerus orbitalis Herrich-Schäffer, 1839: 10, 15, ♂; "Aus der Wiener Gegend" (destroyed).
Distribution in the Middle East: Turkey (Giordani Soika, 1952, 1966 as both *Odontodynerus orbitalis*, 1970 and *A. orbitalis*; Yildirim and Özbek, 1999; Yildirim and Kojima, 1999; Yildirim and Gusenleitner, 2001; Gusenleitner, 2007, 2013a; Yildirim and Gusenleitner, 2007; Fateryga, 2017b; Fateryga *et al.*, 2020, 2021).
Extralimital distribution: Azerbaijan (Fateryga *et al.*, 2020, 2021), Europe, Kazakhstan, Mongolia, North Africa, Russia (Fateryga, 2017b; Fateryga *et al.*, 2020, 2021), Italy (Selis, 2023).

Antepipona osmania Gusenleitner, 1986c
Antepipona osmania Gusenleitner, 1986c: 361, ♀; Turkey.
Catalogues with Iranian records: Rahmani *et al.* (2020).
Distribution in Iran: East Azarbaijan (Gusenleitner, 2012).
Distribution in the Middle East: Iran (Gusenleitner, 2012), Turkey (Gusenleitner, 1986a, 2012, 2013a; Yildirim and Özbek, 1996a; Madl, 1997; Yildirim and Kojima, 1999).
Extralimital distribution: None.

Antepipona praeclara Giordani Soika, 1970
Antepipona praeclara Giordani Soika, 1970: 117, ♀; Iran.
Catalogues with Iranian records: Ebrahimi and Carpenter (2008), Rahmani *et al.* (2020).
Distribution in Iran: Khuzestan (Giordani Soika, 1970).
Distribution in the Middle East: Iran (Giordani Soika, 1970; Gusenleitner, 2013a).
Extralimital distribution: Pakistan (Gusenleitner, 2013a).

Antepipona prompta Giordani Soika, 1970
Antepipona prompta Giordani Soika, 1970: 123, ♀; Iran.
Catalogues with Iranian records: Ebrahimi and Carpenter (2008), Rahmani *et al.* (2020).
Distribution in Iran: Khuzestan (Giordani Soika, 1970).
Distribution in the Middle East: Known only from Iran (Giordani Soika, 1970; Gusenleitner, 2013a).
Extralimital distribution: None.

Antepipona rotunda Gusenleitner, 2014
Antepipona rotunda Gusenleitner, 2014: 533, ♂; "Oman, Ash Sharqiyah, 25 km SW Sur, Wadi Fulayi".
Distribution in the Middle East: Known only from Oman (Gusenleitner, 2014).
Extralimital distribution: None.

Antepipona specifica (Morawitz, 1895)
Odynerus (*Lionotus*) *specificus* Morawitz, 1895: 464, ♂, ♀; Turkmenistan.
Catalogues with Iranian records: Ebrahimi and Carpenter (2008), Rahmani *et al.* (2020).
Distribution in Iran: Markazi (Ebrahimi and Carpenter, 2008).
Distribution in the Middle East: Iran (Ebrahimi and Carpenter, 2008), Iraq (Giordani Soika, 1970 as *A. curialis* - misidentification; Gusenleitner, 2013a).
Extralimital distribution: Turkmenistan (Gusenleitner, 2013a), Kyrgyzstan, Uzbekistan (Castro and Dvořák, 2009).
Comments: Due to misidentification, this species was listed by Giordani Soika (1970) as *Antepipona curialis* for the fauna of Iraq (Gusenleitner, 2013a).

Antepipona tenuis **Gusenleitner, 1988**
Antepipona tenuis Gusenleitner, 1988: 713, 726, ♀; "Tuz Golu, 900 m".
Distribution in the Middle East: Known only from Turkey (Gusenleitner, 1988, 2004, 2013a).
Extralimital distribution: None.

Antepipona vagabunda **(Dalla Torre, 1889)**
Odynerus vagabundus Dalla Torre, 1889: 125, ♀, ♂; Turkmenistan.
Catalogues with Iranian records: Ebrahimi and Carpenter (2008), Rahmani *et al.* (2020).
Distribution in Iran: Kerman (Pashaei Rad *et al.*, 2010; Schmid-Egger *et al.*, 2021), Tehran (Giordani Soika, 1970), Zanjan (Abbasi *et al.*, 2006a, 2008a, 2008b; Pashaei Rad *et al.*, 2010).
Distribution in the Middle East: Iran (see references above, and Gusenleitner, 2013a).
Extralimital distribution: Afghanistan, Turkmenistan (Gusenleitner, 2013a).

Antepipona varentzowi **(Morawitz, 1895)**
Odynerus varentzowi Morawitz, 1895: 474, ♀; "Transcaspia; Cheirabad" [Uzbekistan].
Catalogues with Iranian records: Rahmani *et al.* (2020).
Distribution in Iran: North of Iran (Morawitz, 1895 as *Odynerus varentzowi* Morawitz, 1895).
Distribution in the Middle East: Iran (Morawitz, 1895; Fateryga *et al.*, 2021; Selis *et al.*, 2025), Iraq (Giordani Soika, 1970; Selis *et al.*, 2025), Lebanon (Yildirim and Kojima, 1999; Fateryga *et al.*, 2021; Selis *et al.*, 2025), Turkey (Giordani Soika, 1970; Yildirim and Kojima, 1999; Fateryga *et al.*, 2021; Selis *et al.*, 2025).
Extralimital distribution: Armenia, Azerbaijan, Pakistan (Fateryga *et al.*, 2021; Selis *et al.*, 2025), China, Turkmenistan (Selis *et al.*, 2025), Kazakhstan, Uzbekistan (Gusenleitner, 1986a; Fateryga *et al.*, 2021).

Antepipona vescovilis **Giordani Soika, 1970**
Antepipona vescovilis Giordani Soika, 1970: 116, ♀; Iran.
Catalogues with Iranian records: Ebrahimi and Carpenter (2008), Rahmani *et al.* (2020).
Distribution in Iran: Alborz, Razavi Khorasan (Giordani Soika, 1970).
Distribution in the Middle East: Iran (Giordani Soika, 1970; Gusenleitner, 2013a).
Extralimital distribution: Pakistan (Gusenleitner, 2006c).

Antepipona yemenensis **(Giordani Soika, 1957)**
Odontodynerus yemenensis Giordani Soika, 1957: 480, ♀; Yemen.
Distribution in the Middle East: Saudi Arabia (Guichard, 1986; Gusenleitner, 2013a; Carpenter and Gadallah, 2020), Yemen (Giordani Soika, 1957 as *Odontodynerus yemenensis*; Gusenleitner, 2013a; Carpenter and Gadallah, 2020).
Extralimital distribution: None.

Genus *Antodynerus* de Saussure, 1855

Antodynerus Saussure, 1855: 242, 287. Type species: *Vespa flavescens* Fabricius, 1775 ("*Odynerus punctum* (Fabricius) " *sensu* Saussure, 1853). Desginated by van der Vecht, 1959.

Antodynerus ignaruris **(Kohl, 1906)**
Odynerus ignaruris Kohl, 1906: 256, ♀; Südarbien: Ras Faerták.
Distribution in the Middle East: Oman (Guichard, 1986; Gusenleitner, 2013a), Saudi Arabia (Kohl, 1906 as *Odynerus ignaruris*).
Extralimital distribution: None.

Antodynerus igneus **Gusenleitner, 2004**
Antodynerus igneus Gusenleitner, 2004: 151, 154, ♀; "Wadi Quitbit".
Distribution in the Middle East: Known only from Oman (Gusenleitner, 2004, 2007, 2013a).
Extralimital distribution: None.

***Antodynerus ignoratus* Gusenleitner, 2007**

Antodynerus ignoratus Gusenleitner, 2007: 119, 123, ♀, ♂; "Jemen-W, Jabal Bura NNE Al Hudayda".

Distribution in the Middle East: Known only from Yemen (Gusenleitner, 2007, 2013a).

Extralimital distribution: None.

***Antodynerus incomparabilis* Giordani Soika, 1970**

Antodynerus incomparabilis Giordani Soika, 1970: 138, ♀; Iran.

Catalogues with Iranian records: Ebrahimi and Carpenter (2008), Rahmani *et al.* (2020).

Distribution in Iran: Fars (Giordani Soika, 1970).

Distribution in the Middle East: Iran (Giordani Soika, 1970; Yildirim and Kojima, 1999; Gusenleitner, 2013a), Turkey (Gusenleitner, 1988, 2013a; Yildirim and Kojima, 1999).

Extralimital distribution: None.

***Antodynerus multicolor dhufariensis* (Giordani Soika, 1957)**

Pseudepipona (*Parepipona*) *multicolor dhufariensis* Giordani Soika, 1957: 477, ♀; "Oman: Dhufar, foothills".

Distribution in the Middle East: Oman (Giordani Soika, 1957, 1973 both as *Pseudepipona multicolor dhufariensis*; Gusenleitner, 2013a), Saudi Arabia (Guichard, 1986 as *P. m. dhufarien sis*; Gusenleitner, 2013a), Yemen (Gusenleitner, 2007, 2013a).

Extralimital distribution: None.

Genus *Brachyodynerus* Blüthgen, 1938

Brachyodynerus Blüthgen, 1938: 450, 459. Type species: *Odynerus magnificus* Morawitz, 1867; by original designation.

***Brachyodynerus binominatus* Blüthgen, 1939**

Brachyodynerus binominatus Blüthgen, 1939: 244, [replacement name], ♀, ♂; Algeria "Biskra".

Distribution in the Middle East: Israel-Palestine (Gusenleitner, 2013a; Fateryga *et al.*, 2017).

Extralimital distribution: Algeria (Gusenleitner, 2013a).

***Brachyodynerus chloroticus* Gusenleitner, 2004**

Brachyodynerus chloroticus Gusenleitner, 2004: 1093, 1100, ♀; "Türkei-Ost, Kagizman/Kars, 1100 m".

Distribution in the Middle East: Known only from Turkey (Gusenleitner, 2004; Fateryga *et al.*, 2017).

Extralimital distribution: None.

***Brachyodynerus djebaili* Giordani Soika, 1983**

Brachyodynerus djebaili Giordani Soika, 1983 (1982): 200, ♂, ♀; Algeria "Biskra, Chegga".

Distribution in the Middle East: Jordan (Gusenleitner, 2004; Fateryga *et al.*, 2017).

Extralimital distribution: Algeria (Gusenleitner, 2013a), Tunisia (Fateryga *et al.*, 2017).

***Brachyodynerus kopetdagicus* (Kostylev, 1940)**

Odynerus (*Brachyodynerus*) *kopetdagicus* Kostylev, 1940: 27, ♀; Koptdag, Tschuli (Turkomani).

Distribution in the Middle East: Turkey (Yildirim and Özbek, 1999; Yildirim and Kojima, 1999; Gusenleitner, 2011b, 2013a; Fateryga *et al.*, 2017).

Extralimital distribution: Turkmenistan (Gusenleitner, 2013a).

***Brachyodynerus kusdasi* Gusenleitner, 1967**

Brachyodynerus kusdasi Gusenleitner, 1967: 670, ♀; "Mut" [Turkey].

Distribution in the Middle East: Turkey (Gusenleitner, 1967, 2013a; Giordani Soika, 1970; Yildirim, 1998; Yildirim and Özbek, 1999; Yildirim and Kojima, 1999; Yildirim and Gusenleitner, 2001; Fateryga *et al.*, 2017, 2021).

Extralimital distribution: Azerbaijan (Fateryga *et al.*, 2021).

***Brachyodynerus magnificus magnificus* (Morawitz, 1867)**
Odynerus (Leionotus) magnificus Morawitz, 1867: 119, ♀; Russia.
Distribution in the Middle East: Turkey (Giordani Soika, 1970; Yildirim and Kojima, 1999; Gusenleitner, 2013a; Fateryga, 2017b, 2018; Fateryga *et al.*, 2017).
Extralimital distribution: Armenia, Azerbaijan, Kazakhstan, Romania, Russia, Turkmenistan, Ukraine (Fateryga, 2017b, 2018).

***Brachyodynerus quadrimaculatus* (André, 1884)**
Odynerus quadrimaculatus André, 1884: 699, ♀; Sarepta (Russia).
Distribution in the Middle East: Turkey (Gusenleitner, 1967, 2013a; Giordani Soika, 1970; Yildirim and Özbek, 1996b; Yildirim and Kojima, 1999; Yildirim and Gusenleitner, 2012; Fateryga, 2017b, 2018; Fateryga *et al.*, 2017).
Extralimital distribution: Azerbaijan, Greece, Russia (Fateryga, 2017b, 2018).

***Brachyodynerus zhelochovtzevi* (Kostylev, 1929)**
Odynerus (Lionotus) zhelochvtzevi Kostylev, 1929: 112, ♂; Turkestan.
Catalogues with Iranian records: Rahmani *et al.* (2020).
Distribution in Iran: Golestan (Gusenleitner, 2011b).
Distribution in the Middle East: Iran (Gusenleitner, 2011b, 2013a; Fateryga *et al.*, 2017, 2023; Rahmani *et al.*, 2020).
Extralimital distribution: Kazakhstan, Uzbekistan (Fateryga *et al.*, 2023), Turkmenistan (Gusenleitner, 2013a).

Genus *Brachypipona* Gusenleitner, 1967

Brachypipona Gusenleitner, 1967: 671. Type species: *Pseudepipona schmidti* Gusenleitner, 1967, by original designation.

***Brachypipona irana* Gusenleitner and Schmid-Egger in Schmid-Egger *et al.*, 2021**
Brachypipona irana Gusenleitner and Schmid-Egger in Schmid-Egger *et al.*, 2021: 1108, ♀; Iran.
Catalogues with Iranian records: None.
Distribution in Iran: Fars (Schmid-Egger *et al.*, 2021).
Distribution in the Middle East: Known only from Iran (Schmid-Egger *et al.*, 2021).
Extralimital distribution: None.

***Brachypipona montana* Gusenleitner and Schmid-Egger in Schmid-Egger *et al.*, 2021**
Brachypipona montana Gusenleitner and Schmid-Egger in Schmid-Egger *et al.*, 2021: 1108, ♂; Iran.
Catalogues with Iranian records: None.
Distribution in Iran: Fars, Kerman (Schmid-Egger *et al.*, 2021).
Distribution in the Middle East: Known only from Iran (Schmid-Egger *et al.*, 2021).
Extralimital distribution: None.

***Brachypipona orientalis* Gusenleitner, 2004**
Brachypipona orientalis Gusenleitner, 2004: 1102, ♀; "Türkie, Gürün" [Turkey].
Catalogues with Iranian records: Ebrahimi and Carpenter (2008), Rahmani *et al.* (2020).
Distribution in Iran: Hamadan (Gusenleitner, 2004).
Distribution in the Middle East: Iran (Gusenleitner, 2004; Fateryga *et al.*, 2021), Turkey (Gusenleitner, 2004, 2013a; Fateryga *et al.*, 2021).
Extralimital distribution: Azerbaijan (Fateryga *et al.*, 2021).

***Brachypipona schlaeflei* Gusenleitner, 2001b**
Brachypipona schlaeflei Gusenleitner, 2001b: 233, 235, ♀; Jordanien: Petra.
Distribution in the Middle East: Israel-Palestine (Gusenleitner, 2004), Jordan (Gusenleitner, 2001b, 2004, 2013a).
Extralimital distribution: None.

***Brachypipona schmidti* (Gusenleitner, 1967)**

Pseudepipona schmidti Gusenleitner, 1967: 671, ♂, ♀; Turkey.

Distribution in the Middle East: Iraq (Selis *et al*., 2025), Turkey (Gusenleitner, 1967 as *Pseudepipona* (*Brachypipona*) *schmidti*, 1988, 2013a; Yildirim and Kojima, 1999; Selis *et al*., 2025).

Extralimital distribution: None.

Genus *Cephalochilus* Blüthgen, 1939

Cephalochilus Blüthgen, 1939: 13. Type species: *Pterochilus grandis* Lepeletier, 1841 (= *Vespa labiata* Fabricius 1798), by original designation and monotypy.

***Cephalochilus draco* Giordani Soika, 1970**

Cephalochilus draco Giordani Soika, 1970: 40, ♀, ♂; Turkey.

Catalogues with Iranian records: Ebrahimi and Carpenter (2008), Rahmani *et al.* (2020).

Distribution in Iran: Markazi (Ebrahimi and Carpenter, 2008; Modarres Awal, 2012).

Distribution in the Middle East: Iran (see references above), Syria (Giordani Soika, 1970; Gusenleitner, 2013a), Turkey (Giordani Soika, 1970; Gusenleitner, 1988, 2013a; Yildirim and Özbek, 1996a; Madl, 1997; Yildirim and Kojima, 1999; Yildirim and Gusenleitner, 2004).

Extralimital distribution: None.

Genus *Chlorodynerus* Blüthgen, 1951

Chlorodynerus Blüthgen, 1951b: 67, 75 (as subgenus of *Euodynerus* Blüthgen). Type species: *Odynerus chloroticus* Spinola, 1838, by original designation.

***Chlorodynerus araxanus* (Blüthgen, 1956)**

Euodynerus (*Chlorodynerus*) *araxanus* Blüthgen, 1956: 281, ♀; "Caucasus: Araxes-Tal".

Catalogues with Iranian records: Ebrahimi and Carpenter (2008), Rahmani *et al.* (2020).

Distribution in Iran: Tehran (Giordani Soika, 1970).

Distribution in the Middle East: Known only from Iran (Giordani Soika, 1970; Gusenleitner, 2013a).

Extralimital distribution: None.

***Chlorodynerus chloroticus* (Spinola, 1838)**

Odynerus chloroticus Spinola, 1838: 500, ♀; Egypt.

Distribution in the Middle East: Egypt (Spinola, 1838 as *Odynerus chloroticus*; de Saussure, 1853 as *Odynerus testaceus*; Giordani Soika, 1935 as *O. testaceus*), Israel-Palestine (Gusenleitner, 2013a), Oman (Gusenleitner, 2004, 2013a; Carpenter and Gadallah, 2020), Saudi Arabia (Guichard, 1986; Gadallah, 2003; Gusenleitner, 2013a; Carpenter and Gadallah, 2020), United Arab Emirates (Roche, 1981; Guichard, 1986; Gillett and Gillett, 2005 as *Chlorodynerus chloriticus*; Howarth and Gillett, 2008; Gusenleitner, 2013a; Carpenter and Gadallah, 2020).

Extralimital distribution: Chad, Eretria, North Africa, Sudan (Gusenleitner, 2013a).

Comments: Morice (1921) recorded this species from Iraq (close to the Iranian border). *Chlorodynerus chloroticus* is a mainly North African species, reaching the Levant and the Arabian Peninsula in the east, without further findings in the areas near the locations cited by Morice (1921). Many species of *Chlorodynerus* were described after 1921, and several also in Iran, so it is impossible to ascertain the identity of the specimens examined by Morice (Selis *et al*., 2025).

***Chlorodynerus diglaensis* (Blüthgen, 1954)**

Euodynerus (*Chlorodynerus*) *diglaensis* Blüthgen, 1954: 237, 247, 254, sex not stated; Wadi Digla (S.O. Kairo).

Distribution in the Middle East: Known only from Egypt (Blüthgen, 1954 as *Euodynerus* (*Chlorodynerus*) *diglaensis*; van der Vecht and Fischer, 1972; Gusenleitner, 2013a).

Extralimital distribution: None.

***Chlorodynerus flavoferrugineus* Giordani Soika, 1970**

Chlorodynerus flavoferrugineus Giordani Soika, 1970: 170, ♀, ♂; Iran.

Catalogues with Iranian records: Ebrahimi and Carpenter (2008), Rahmani *et al.* (2020).

Distribution in Iran: Khuzestan (Giordani Soika, 1970).
Distribution in the Middle East: Known only from Iran (Giordani Soika, 1970; Gusenleitner, 2013a).
Extralimital distribution: None.

Chlorodynerus gratus **(Kostylev, 1940)**
Odynerus (*Pareuodynerus*) *gratus* Kostylev, 1940: 31, ♂; Iran.
Catalogues with Iranian records: Ebrahimi and Carpenter (2008), Rahmani *et al.* (2020).
Distribution in Iran: Kerman (Kostylev, 1940 as *Odynerus* (*Pareuodynerus*) *gratus*).
Distribution in the Middle East: Known only from Iran (Kostylev, 1940; Gusenleitner, 2013a).
Extralimital distribution: None.

Chlorodynerus incisipes **(Kostylev, 1935)**
Odynerus incisipes Kostylev, 1935 (1934): 128, ♂; Turkmenistan.
Catalogues with Iranian records: None.
Distribution in Iran: Kerman (Schmid-Egger *et al.*, 2021).
Distribution in the Middle East: Iran (Schmid-Egger *et al.*, 2021; Fateryga *et al.*, 2023).
Extralimital distribution: Turkmenistan (Kostylev, 1935 as *Odynerus incisipes*; Fateryga *et al.*, 2023), Uzbekistan (Fateryga *et al.*, 2023).

Chlorodynerus infrenis **Giordani Soika, 1970**
Chlorodynerus infrenis Giordani Soika, 1970: 172, ♀, ♂; Iran.
Catalogues with Iranian records: Ebrahimi and Carpenter (2008), Rahmani *et al.* (2020).
Distribution in Iran: Khuzestan (Giordani Soika, 1970), Sistan & Baluchestan (Ebrahimi and Carpenter, 2008; Modarres Awal, 2012).
Distribution in the Middle East: Known only from Iran (see references above, and Gusenleitner, 2013a).
Extralimital distribution: None.

Chlorodynerus intricatus **Giordani Soika, 1958**
Chlorodynerus intricatus Giordani Soika, 1958: 140, 141, ♀; "Egitto: Via di Suez".
Distribution in the Middle East: Known only from Egypt (Giordani Soika, 1958, 1973; van der Vecht and Fischer, 1972; Borsato and Ratti, 1999; Gusenleitner, 2013a).
Extralimital distribution: None.

Chlorodynerus kelidopterus **(Kohl, 1907)**
Rhynchium kelidopterum Kohl, 1907: 252, ♂; "Südarabien (Aden)".
Distribution in the Middle East: Israel-Palestine (Gusenleitner, 2013a), Oman (Gusenleitner, 2004, 2005, 2013a; Madl, 2018; Carpenter and Gadallah, 2020), Saudi Arabia (Guichard, 1986; Gadallah, 2003; Gusenleitner, 2013a; Madl, 2018; Carpenter and Gadallah, 2020), United Arab Emirates (Guichard, 1986; Howarth and Gillett, 2008 as *Euodynerus kelidopterus*; Carpenter and Gadallah, 2020), Yemen (Kohl, 1906 as *Rhynchium kelidopterum*; Blüthgen, 1956 as *E. kelidopterus*; Gusenleitner, 2013a; Madl, 2018; Carpenter and Gadallah, 2020).
Extralimital distribution: Algeria, Libya, Morocco (Gusenleitner, 2013a), Burkina Faso, Chad, Djibouti, Mali, Sudan (Gusenleitner, 1992, 2005).

Chlorodynerus loeffleri **Giordani Soika, 1958**
Chlorodynerus loeffleri Giordani Soika, 1958: 144, ♀; Iran.
Catalogues with Iranian records: Ebrahimi and Carpenter (2008), Rahmani *et al.* (2020).
Distribution in Iran: Fars (Giordani Soika, 1958).
Distribution in the Middle East: Iran (Giordani Soika, 1958; Gusenleitner, 2013a).
Extralimital distribution: India (Giordani Soika, 1970), Pakistan (Giordani Soika, 1970; Gusenleitner, 2013a).

***Chlorodynerus mochii* Giordani Soika, 1958**

Chlorodynerus mochii Giordani Soika, 1958 (1957): 140, 146, ♀; "Egitto: Salloum".

Distribution in the Middle East: Known only from Egypt (Giordani Soika, 1958, 1973; van der Vecht and Fischer, 1972; Borsato and Ratti, 1999; Gusenleitner, 2013a).

Extralimital distribution: None.

***Chlorodynerus sanctus* (Blüthgen, 1954)**

Euodynerus (*Chlorodynerus*) *sanctus* Blüthgen, 1954: 242, 257, 264, ♀; "Wadi Ain Faran bei Jerusalem".

Distribution in the Middle East: Egypt (Giordani Soika, 1958; Gusenleitner, 2013a), Israel-Palestine (Blüthgen, 1954 as *Euodynerus* (*Chlorodynerus*) *sanctus*; Giordani Soika, 1958, 1973; Gusenleitner, 2013a).

Extralimital distribution: None.

***Chlorodynerus schulthessianus* (Kostylev, 1937)**

Odynerus schulthessianus Kostylev, 1937: 226, ♀, ♂; Armenia.

Catalogues with Iranian records: Ebrahimi and Carpenter (2008), Rahmani *et al.* (2020).

Distribution in Iran: Khuzestan, Tehran (Giordani Soika, 1970).

Distribution in the Middle East: Iran (Giordani Soika, 1970; Gusenleitner, 2013a), Israel-Palestine (Blüthgen, 1956 as *Euodynerus schulthessianusi*; Gusenleitner, 2013a).

Extralimital distribution: Armenia (Gusenleitner, 2013a).

***Chlorodynerus ypsilon rhodius* (Blüthgen, 1954)**

Euodynerus (*Chlorodynerus*) *ypsilon rhodius* Blüthgen, 1954: 231, 244, 259, ♀, ♂; "Rhodus".

Distribution in the Middle East: Turkey (Gusenleitner, 1966, 1967 both as *Euodynerus* (*Chlorodynerus*) *ypsilon*; Giordani Soika, 1970 as *Chlorodynerus ypsilon*; Gusenleitner, 1997c as *Chlorodynerus ypsilon ypsilon*; Gusenleitner, 2013a; Yildirim and Kojima, 1999 as *Euodynerus* (*Chlorodynerus*) *ypsilon rhodius*).

Extralimital distribution: Greece (Gusenleitner, 2013a).

***Chlorodynerus ypsilon ypsilon* (Kostylev, 1929)**

Odynerus ypsilon Kostylev, 1929: 79, ♀; Armenia.

Distribution in the Middle East: Cyprus (Giordani Soika, 1958; Gusenleitner, 2013a; Fateryga, 2017a, 2017b), Israel-Palestine (Giordani Soika, 1952 as *Pseudepipona ypsilon*; Blüthgen, 1956 as *Euodynerus ypsilon ypsilon*; Gusenleitner, 2013a; Fateryga, 2017a, 2017b), Syria (Giordani Soika, 1970; Gusenleitner, 2013a; Fateryga, 2017a, 2017b), Turkey (Giordani Soika, 1970; Gusenleitner, 2013a; Fateryga, 2017a, 2917b).

Extralimital distribution: Caucasus, Europe (SE), Russia (Fateryga, 2017a, 2017b), Italy (Selis, 2023).

Genus *Cyphodynerus* van der Vecht, 1971

Cyphodynerus van der Vecht, 1971: 127. Type species: *Odynerus dimidiatus* Spinola, 1838, by original designation.

***Cyphodynerus canaliculatus* (de Saussure, 1855)**

Odynerus canaliculatus de Saussure, 1855: 260, ♀; Egypte.

Distribution in the Middle East: Arabia (without specific locality, de Saussure, 1855 as *Odynerus* (*Odynerus*) *canaliculatus*; Carpenter and Gadallah, 2020), Egypt (Spinola, 1838 as *Odynerus dimidiatus*; Giordani Soika, 1935 as *Odynerus canaliculatus*; Gusenleitner, 2013a), Saudi Arabia (Zalat *et al.*, 1993 as *Cyphodynerus dimidiatus*; Carpenter and Gadallah, 2020).

Extralimital distribution: Malta (Carpenter and Gadallah, 2020).

***Cyphodynerus sculpturatus* (Dover, 1925)**

Odynerus sculpturatus Dover, 1925: 300, ♀; Karachi [Pakistan].

Catalogues with Iranian records: Ebrahimi and Carpenter (2008), Rahmani *et al.* (2020).

Distribution in Iran: Kerman (Kostylev, 1937 as *Odynerus canaliculatus nigroflammeus* Kostylev, 1937).
Distribution in the Middle East: Iran (Kostylev, 1937).
Extralimital distribution: Pakistan (Dover, 1925).

Genus *Cyrtolabulus* van der Vecht, 1969

Cyrtolabus van der Vecht, 1963: 11. Type species: *Cyrtolabus suavis* van der Vecht, 1963, by original designation. Replaced by *Cyrtolabulus* van der Vecht, 1969.

***Cyrtolabulus anatolicus* Gusenleitner, 2006d**
Cyrtolabulus anatolicus Gusenleitner, 2006d: 663, 665, ♂; Turkey.
Distribution in the Middle East: Known only from Turkey (Gusenleitner, 2006d, 2013a).
Extralimital distribution: None.

***Cyrtolabulus caputabnormis* Gusenleitner, 2000b**
Cyrtolabulus caputabnormis Gusenleitner, 2000b: 19, 22, ♀; Mali, 60 km SW Segou.
Distribution in the Middle East: Yemen (Gusenleitner, 2007, 2013a).
Extralimital distribution: Mali (Gusenleitner, 2013a; Carpenter and Gadallah, 2020).

***Cyrtolabulus chudeaui* (du Buysson, 1908)**
Labus chudeaui du Buysson, 1908: 132, ♀; Ahaggar.
Distribution in the Middle East: Egypt (Giordani Soika, 1939 as *Leptomenes* (*Eumenidiopsis*) *exiguous*).
Extralimital distribution: Algeria (du Buysson, 1908), Sudan (Giordani Soika, 1939 as *Leptomenes* (*Eumenidiopsis*) *exiguus*; Giordani Soika, 1954 as *Leptomenes chudeaui*; Carpenter *et al.*, 2009).

***Cyrtolabulus elbanus* Giordani Soika, 1977**
Cyrtolabulus elbanus Giordani Soika, 1977 (1976): 164, ♀; "Egitto: Gebel Elba, Wadi Cansisob".
Distribution in the Middle East: Known only from Egypt (Giordani Soika, 1977; Gusenleitner, 2013a).
Extralimital distribution: None.

***Cyrtolabulus eremicus* (Giordani Soika, 1952)**
Labus eremicus Giordani Soika, 1952: 15, 16, ♀, ♂; "Plaestina: Gerico".
Distribution in the Middle East: "Arabia", Egypt (Gusenleitner, 2013a; Carpenter and Gadallah, 2020), Israel-Palestine (Giordani Soika, 1952, 1973; Gusenleitner, 2013a; Carpenter and Gadallah, 2020).
Extralimital distribution: None.

***Cyrtolabulus exiguus* (de Saussure, 1853)**
Leptochilus exiguus de Saussure, 1853: 237, for Savigny, 1812, Descr. Égypte pl. 8, fig. 11, ♀; Egypt.
Distribution in the Middle East: Egypt (de Saussure, 1853 as *Leptochilus exiguous*; Gusenleitner, 2013a), Saudi Arabia (Guichard, 1986; Carpenter and Gadallah, 2020), Syria (Gusenleitner, 2013a), Yemen (Gusenleitner, 2007, 2013a; Carpenter and Gadallah, 2020).
Extralimital distribution: Algeria, Libya, Sudan (Gusenleitner, 2013a; Carpenter and Gadallah, 2020).

***Cyrtolabulus gracilis* (Kohl, 1906)**
Labus gracilis Kohl, 1906: 227, 242, ♂; "Halbinsel Sinai (Tor ...)".
Catalogues with Iranian records: None.
Distribution in Iran: Kerman (Schmid-Egger *et al.*, 2021).
Distribution in the Middle East: Egypt (Kohl, 1906; Gusenleitner, 2013a), Iran (Schmid-Egger *et al.*, 2021), Oman (Hamer, 1985; Gusenleitner, 2004, 2005, 2012, 2013a; Carpenter and Gadallah, 2020), Saudi Arabia (Guichard, 1986; Gadallah, 2003; Gusenleitner, 2013a; Carpenter and

Gadallah, 2020), United Arab Emirates (Hamer, 1985; Guichard, 1986; Gillett and Gillett, 2005; Howarth and Gillett, 2008; Gusenleitner, 2010a, 2012, 2013a; Carpenter and Gadallah, 2020).

Extralimital distribution: Algeria (Giordani Soika, 1954 as *Labus swalei*; Capenter, 2020), Sudan (Carpenter and Gadallah, 2020).

***Cyrtolabulus grossepunctatus* (Kirby, 1900)**

Icaria grossepunctata Kirby, 1900: 249, ♀; Sokotra.

Distribution in the Middle East: Yemen (Socotra) (Kirby, 1900, 1903 as *Icaria grossepunctata*; Kohl, 1906 as both *Labus grossepunctatus* and *Labus fraterculus*; Giordani Soika, 1934 as *L. fraterculus*, 1974; Carpenter and Gadallah, 2020).

Extralimital distribution: Palaearctic region (Carpenter and Gadallah, 2020).

***Cyrtolabulus iranus* (Giordani Soika, 1968)**

Cyrtolabus iranus Giordani Soika, 1968: 112, ♀; Iran.

Catalogues with Iranian records: Ebrahimi and Carpenter (2008), Rahmani *et al.* (2020).

Distribution in Iran: Fars (Giordani Soika, 1968 as *Cyrtolabus iranus*).

Distribution in the Middle East: Iran (see references above, and Yildirim and Kojima, 1999; Gusenleitner, 2013a), Turkey (Gusenleitner, 1988; Yildirim and Kojima, 1999).

Extralimital distribution: None.

***Cyrtolabulus karachiensis* Gusenleitner, 2006a**

Cyrtolabulus karachiensis Gusenleitner, 2006a: 1298, ♀; Pakistan.

Catalogues with Iranian records: Rahmani *et al.* (2020).

Distribution in Iran: Sistan & Baluchestan (Hamzavi *et al.*, 2019).

Distribution in the Middle East: Iran (Hamzavi *et al.*, 2019).

Extralimital distribution: Pakistan (Gusenleitner, 2006a).

***Cyrtolabulus mutinensis* Gusenleitner, 2006a**

Cyrtolabulus mutinensis Gusenleitner, 2006a: 155, ♀; Turkey.

Distribution in the Middle East: Known only from Turkey (Gusenleitner, 2006a, 2013a).

Extralimital distribution: None.

***Cyrtolabulus omanicus* Gusenleitner, 2014**

Cyrtolabulus omanicus Gusenleitner, 2014: 531, ♂; Oman: Dhofar prov.

Distribution in the Middle East: Known only from Oman (Gusenleitner, 2014; Carpenter and Gadallah, 2020).

Extralimital distribution: None.

***Cyrtolabulus sexspinosus* Giordani Soika, 1977**

Cyrtolabulus sexspinosus Giordani Soika, 1977 (1976): 166, ♂; "Egitto: Gebel Elba, Aideb".

Distribution in the Middle East: Known only from Egypt (Giordani Soika, 1977; Gusenleitner, 2013a).

Extralimital distribution: None.

***Cyrtolabulus socotrae* (Kohl, 1906)**

Labus socotrae Kohl, 1906: 226, ♀; Sokotra.

Distribution in the Middle East: Known only from Yemen (Socotra) (Kohl, 1906 as *Labus socotrae*; Giordani Soika, 1944 as *Leptomenes socotrae*; Giordani Soika, 1974; Carpenter and Gadallah, 2020).

Extralimital distribution: None.

***Cyrtolabulus spinithorax* (Giordani Soika, 1968)**

Cyrtolabus spinithorax Giordani Soika, 1968: 110, 115, ♀; "Sahara centrale: Hoggar, Timesdelssine".

Distribution in the Middle East: Saudi Arabia (Guichard, 1986; Gusenleitner, 2013a).
Extralimital distribution: Algeria (Giordani Soika, 1968; Carpenter and Gadallah, 2020).

Cyrtolabulus syriacus **(Giordani Soika, 1968)**
Cyrtolabus syriacus Giordani Soika, 1968: 120, ♀, ♂; "Siria: Wadi Raggad".
Catalogues with Iranian records: Rahmani *et al.* (2020).
Distribution in Iran: Sistan & Baluchestan (Hamzavi *et al.*, 2019).
Distribution in the Middle East: Iran (Hamzavi *et al.*, 2019), Israel-Palestine (Gusenleitner, 2006a, 2013a), Syria (Giordani Soika, 1968 as *Cyrtolabus syriacus*; Gusenleitner, 2013a).
Extralimital distribution: Pakistan (Gusenleitner, 2006a).

Cyrtolabulus yemenensis **Gusenleitner, 2007**
Cyrtolabulus yemenensis Gusenleitner, 2007:102, 119, ♀; "Jemen-E, Jabal Falk, Damau".
Distribution in the Middle East: Known only from Yemen (Gusenleitner, 2007, 2013a; Carpenter and Gadallah, 2020).
Extralimital distribution: None.

Cyrtolabulus yildirimi **Gusenleitner, 2006d**
Cyrtolabulus yildirimi Gusenleitner, 2006d: 663, 667, ♀; "Turkey: Perr Hakkari, Habur Deresi-Tal S Beytisebap, 110 m".
Distribution in the Middle East: Known only from Turkey (Gusenleitner, 2006a, 2013a).
Extralimital distribution: None.

Cyrtolabulus zarudnyi **(Kostylev, 1939)**
Labus zarudnyi Kostylev, 1939: 166, 168, ♀; [Kirman, Iran].
Catalogues with Iranian records: Ebrahimi and Carpenter (2008), Rahmani *et al.* (2020).
Distribution in Iran: Hormozgan (Schmid-Egger *et al.*, 2021), Kerman (Kostylev, 1939).
Distribution in the Middle East: Iran (see references above, and Yildirim and Kojima, 1999; Gusenleitner, 2013a; Fateryga *et al.*, 2021), Turkey (Gusenleitner, 1988, 2006a, 2013a; Yildirim and Kojima, 1999; Fateryga *et al.*, 2021).
Extralimital distribution: Azerbaijan (Fateryga *et al.*, 2021).

Genus *Delta* de Saussure, 1855

Delta Saussure, 1855: 130, 132, 143. Type species: *Vespa maxillosa* DeGeer, 1775 (= *Vespa emarginata* Linnaeus, 1758); by subsequent designation of Bequaert, 1925.

Delta anomala **(Zavattari, 1909)**
Eumenes anomalus Zavattari, 1909: 4, ♂; Kashmir.
Catalogues with Iranian records: Ebrahimi and Carpenter (2008), Rahmani *et al.* (2020).
Distribution in Iran: Sistan & Baluchestan (Blüthgen and Gusenleitner, 1970 as *Alfieria anomala* (Zavattari, 1909)).
Distribution in the Middle East: Egypt (Zavattari, 1909 as *Eumenes anomalus* (?); Giordani Soika, 1934 as both *E. anomalus* and *Alfieria anomalia*; Giordani Soika, 1935 as *A. anomalia*; Gusenleitner, 2013a as *A. anomala*; Carpenter and Gadallah, 2020), Iran (Blüthgen and Gusenleitner, 1970; Carpenter and Gadallah, 2020), Saudi Arabia (Giordani Soika, 1934; Zalat *et al.*, 1993 as *Alfieria anomala*; Carpenter and Gadallah, 2020).
Extralimital distribution: None.

Delta asina mixtum **(Giordani Soika, 1944)**
Eumenes (*Delta*) *mixtus* Giordani Soika, 1944: 166, ♂; "Africa orientale: Baidole".
Distribution in the Middle East: Arabia (Zalat *et al.*, 1993), Israel-Palestine (Giordani Soika, 1952 as *Eumenes* (*Delta*) *asinus maxitus*; van der vecht and Fischer, 1972; Zalat *et al.*, 1993; Gusenleitner, 2013a), Jordan (Giordani Soika, 1970; Zalat *et al.*, 1993; van der Vecht and Fischer, 1972; Gusenleitner, 2013a), Oman (Giordani Soika, 1957 as *Eumenes maxirus*; Guichard, 1986;

Gusenleitner, 2013a; Carpenter and Gadallah, 2020), Saudi Arabia (Giordani Soika, 1970; Guichard, 1986; Gusenleitner, 2013a; Carpenter and Gadallah, 2020).

Extralimital distribution: Chad, Eritrea, Ethiopia, Niger, Somalia, Sudan (Carpenter and Gadallah, 2020).

***Delta campaniforme chloroticum* Giordani Soika, 1979**

Delta campaniforme chloroticum Giordani Soika, 1979: 263, ♀; Pakistan: Deesa.

Catalogues with Iranian records: Ebrahimi and Carpenter (2008), Rahmani *et al.* (2020).

Distribution in Iran: Fars (Giordani Soika, 1970), Hormozgan (Khoobdel *et al.*, 2014), Khuzestan (Giordani Soika, 1970; Boyeri Shemi and Pashaei Rad, 2017), Sistan & Baluchestan (Blüthgen and Gusenleitner, 1970; Giordani Soika, 1979).

Distribution in the Middle East: Iran (see references above, and Gusenleitner, 2013a).

Extralimital distribution: India (Giordani Soika, 1979; Gusenleitner, 2013a).

Comments: Giordani Soika (1979) and Gusenleitner (2013a) erroneously recorded this subspecies from Pakistan, however, the collecting locality was Gujairat, which occurs in India not Pakistan.

***Delta dimidiatipenne* (de Saussure, 1852)**

Eumenes dimidiatipennis de Saussure, 1852: 51, ♀, ♂; Djidda (Arabie), les Indes Orientales, «l'Egypte».

Catalogues with Iranian records: Ebrahimi and Carpenter (2008), Rahmani *et al.* (2020).

Distribution in Iran: Alborz, Fars (Giordani Soika, 1970), Ardabil, Kuhgiloyeh & Boyerahmad, Kordestan, Lorestan, Markazi (Ebrahimi and Carpenter, 2008; Modarres Awal, 2012), East Azarbaijan (Sakenin *et al.*, 2008), Golestan (Schmid-Egger *et al.*, 2021), Hormozgan, Tehran (Blüthgen and Gusenleitner, 1970 as *Delta emarginatus dimidiatipennis* (de Saussure, 1852); Ebrahimi and Carpenter, 2008; Modarres Awal, 2012), Kerman (Blüthgen and Gusenleitner, 1970 as *D. emarginatus dimidiatipennis*; Pashaei Rad *et al.*, 2010), Kermanshah (Hayat *et al.*, 2008), Khuzestan (Blüthgen and Gusenleitner, 1970 as *D. emarginatus dimidiatipennis*; Ebrahimi and Carpenter, 2008; Modarres Awal, 2012), Sistan & Baluchestan (Giordani Soika, 1970; Ghahari *et al.*, 2007; Ebrahimi and Carpenter, 2008; Modarres Awal, 2012; Hamzavi *et al.*, 2019), Southern Khorasan (Blüthgen and Gusenleitner, 1970 as *D. emarginatus dimidiatipennis*), Iran (no specific locality cited) (Esmaili and Rastegar, 1974 as *Eumenes maxillosa dimidiatipennis* de Saussure; Ebrahimi, 2014; Fateryga *et al.*, 2021).

New material examined: Isfahan province, Najaf-Abad, 2♀, 1♂, August 2016; Mazandaran province, Qaemshahr, Ahangarkola, 3♀, July 2018.

Distribution in the Middle East: Egypt (Walker, 1871 as *D. dimidiatipennis*; Giordani Soika, 1935 as *D. maxillosus* var. *dimidiatipennis*; Zalat *et al.*, 1993, Fateryga *et al.*, 2021, Selis *et al.*, 2025), Iran (see references above, and Fateryga *et al.*, 2023; Selis *et al.*, 2025), Iraq (Derwesh, 1965; Augul, 2017; Selis *et al.*, 2025), Israel-Palestine (Yildirim and Kojima, 1999), Jordan (Gusenleitner, 2001b; Fateryga *et al.*, 2021, 2023; Selis *et al.*, 2025), Kuwait (Walker and Pittaway, 1987; Al-Houty, 1989, 1997; Carpenter and Gadallah, 2020), Oman (Walker and Pittaway, 1987; Yildirim and Kojima, 1999; Gusenleitner, 2004, 2005, 2013a; Castro and Dvořák, 2009; Monks *et al.*, 2019; Carpenter and Gadallah, 2020; Fateryga *et al.*, 2021, 2023; Selis *et al.*, 2025), Qatar (Guichard, 1986; Carpenter and Gadallah, 2020; Fateryga *et al.*, 2021, 2023; Selis *et al.*, 2025), Saudi Arabia (de Saussure, 1852 as *Eumenes dimidiatipennis*; Giordani Soika, 1970; Guichard, 1986; Walker and Pittaway, 1987; Yildirim and Kojima, 1999; Gadallah, 2003; El-Hawagry *et al.*, 2013 as *E. dimidiatipennis*; Carpenter and Gadallah, 2020; Bosly, 2021; Fateryga *et al.*, 2021, 2023; Selis *et al.*, 2025), Syria (Blüthgen, 1952 as *D. maxillosus* var. *dimidiatipennis*; Abo Kaf and Allewa, 2016; Selis *et al.*, 2025), Turkey (Yildirim, 1998; Yildirim and Kojima, 1999; Gusenleitner, 2001a; Yildirim and Gusenleitner, 2012; Kemal and Koçak, 2018; Yildirim and Tezcan, 2018; Fateryga *et al.*, 2021, 2023; Selis *et al.*, 2025), United Arab Emirates (Roche, 1981; Guichard, 1986; Gillett and Howarth, 2004; Gillett and Gillett, 2005 as *Delta diminuatipenne*; Howarth and Gillett, 2008; Gusenleitner, 2010a; Carpenter and Gadallah, 2020; Fateryga *et al.*, 2021, 2023; Selis *et al.*, 2025), Yemen (including Socotra) (Kirby, 1903 as *E. dimidiatipennis*; Kohl, 1906 as *E. dimidiatipennis*; Giordani Soika, 1974; Yildirim and Kojima, 1999; Wahis, 2000; Gusenleitner, 2007; Castro and Dvořák, 2010; Gusenleitner, 2013a; Madl, 2018; Carpenter and Gadallah, 2020; Fateryga *et al.*, 2021; Selis *et al.*, 2025).

Extralimital distribution: Azerbaijan (Fateryga *et al.*, 2021; Selis *et al.*, 2025), Afghanistan, Algeria, Chad, Djibouti, Eritrea, Ethiopia, India, Mauritania, Morocco, Nepal, Niger, Pakistan, Somalia, South Africa, Spain, Sudan, Tajikistan, Turkmenistan, Uganda (Carpenter and Gadallah, 2020; Fateryga *et al.*, 2021; Selis *et al.*, 2025), China (Li and Chen, 2023; Selis *et al.*, 2025), Uzbekistan (Fateryga *et al.*, 2023).

Comments: *Choerades gilva* (Linnaeus) (Diptera: Asilidae) is the predator of *Delta dimidiatipenne* (Hayat *et al.*, 2008).

Two parasitoid species were found within *D. dimidiatipenne* nests: *Cotesia vanessae* (Reinhard) (Hymenoptera: Braconidae) [Golestan province, Kordkoy, Salikandeh, 2♀♀, June 2017]; and *Netelia fuscicornis* (Holmgren) (Hymenoptera: Ichneumonidae) [Mazandaran province, Qaemshahr, Ahangarkola, 3♀♀, July 2018].

***Delta emarginatum* (Linnaeus, 1758)**

Vespa emarginata Linnaeus, 1758: 574, sex not stated; "in America" [error].

Distribution in the Middle East: Egypt (Guérin-Méneville, 1844 as *Eumenes savignyi*; Walker, 1871 as *E. savignyi* and as *E. tinctor*; Spinola, 1838 as *E. savignyi*; Giordani Soika, 1935 as *E. maxillosus* and as *E. maxillosus* var. *savignyi*; Zalat *et al.*, as both *Delta emarginatum emarginatum* and *D. e. savignyi*), Saudi Arabia (Giordani Soika, 1957 as *E. maxillosus savignyi*; Guichard, 1986; Gadallah, 2003; Carpenter and Gadallah, 2020), Yemen (Giordani Soika, 1957 as *E. m. savignyi*; Gusenleitner, 2007; Carpenter and Gadallah, 2020).

Extralimital distribution: Algeria, Angola, Botswana, Burkina Faso, Burundi, Cape Verde Islands, Central African Republic, Comoros, Congo, Equatorial Guinea, Eritrea, Ethiopia, Gabon, Gambia, Ivory Coast, Kenya, Liberia, Libya, Malawi, Mali, Madagascar, Mauritania, Mozambique, Namibia, Niger, Nigeria, Rwanda, Seychelles, Sierra Leone, Somalia, South Africa, Sudan, Tanzania, Togo, Zambia (Carpenter *et al.*, 2010; Carpenter and Gadallah, 2020; Selis, 2025c).

***Delta esuriens esuriens* (Fabricius, 1787)**

Vespa esuriens Fabricius, 1787: 393, sex not stated; India.

Catalogues with Iranian records: Ebrahimi and Carpenter (2008), Rahmani *et al.* (2020).

Distribution in Iran: Hormozgan (Blüthgen and Gusenleitner, 1970 as *Delta companiforme esuriens* (Fabricius, 1787)), Kerman (Pashaei Rad *et al.*, 2010; Schmid-Egger *et al.*, 2021), Khuzestan (Blüthgen and Gusenleitner, 1970 as *D. companiforme esuriens*).

Distribution in the Middle East: Egypt (Walker, 1871 as *Eumenes esuriens*; Selis *et al.*, 2025), Iran (see references above, and Carpenter, 2023; Selis *et al.*, 2025), Iraq (Morice, 1921 as *E. esuriens*; Derwesh, 1965; Giordani Soika, 1970 as *D. campaniforme esuriens*: van der Vecht and Fischer, 1972; Abdul-Rassoul, 1976; Augul, 2017 as *D. Esuriens*; Al-Saffar and Augul, 2021; Carpenter, 2023; Selis *et al.*, 2025), Israel-Palestine (Castro and Dvořák, 2009; Selis *et al.*, 2025), Kuwait (Walker and Pittaway, 1987 as *D. campaniforme*; Al-Houty, 1989, 1997 both as *D. campaniforme*; Carpenter and Gadallah, 2020; Carpenter, 2023), Oman (Giordani Soika, 1957, 1970 both as *D. campaniforme esuriens*; Walker and Pittaway, 1987 as *D. campaniforme*; Castro and Dvořák, 2009; Gusenleitner, 2004, 2005, 2013a; Carpenter and Gadallah, 2020; Carpenter, 2023; Selis *et al.*, 2025), Qatar (Giordani Soika, 1957, 1970 both as *D. campaniforme esuriens*; Carpenter and Gadallah, 2020; Carpenter, 2023; Selis *et al.*, 2025), Saudi Arabia (Walker and Pittaway, 1987 as *D. campaniforme*; Gadallah, 2003; Castro and Dvořák, 2009; Gusenleitner, 2013a; Carpenter and Gadallah, 2020; Bosly, 2021; Carpenter, 2023; Selis *et al.*, 2025), Syria (Abo Kaf and Allewa, 2016 as *D. esuriens*), United Arab Emirates (Gillett and Howarth, 2004 as *D. campaniforme*; Gillett and Gillett, 2005 as *D. campaniforme*; Howarth and Gillett, 2008 as *D. campaniforme*; Gusenleitner, 2010a; Gess and Roosenschoon, 2016; Carpenter and Gadallah, 2020; Carpenter, 2023; Selis *et al.*, 2025).

Extralimital distribution: Bhutan (Carpenter, 2023; Selis *et al.*, 2025), China (Li and Chen, 2023; Selis *et al.*, 2025), India, Laos, Malaysia, Mauritius, Myanmar, New Caledonia (adventive), Pakistan, Sri Lanka, Thailand, Vietnam (Carpenter and Gadallah, 2020; Carpenter, 2023; Selis *et al.*, 2025).

***Delta esuriens gracile* (de Saussure, 1852)**

Eumenes gracilis de Saussure, 1852: 57, ♂, ♀; "L'Egypte, le Senegal".

Catalogues with Iranian records: Ebrahimi and Carpenter (2008), Rahmani *et al.* (2020).

Distribution in Iran: Markazi (Ebrahimi and Carpenter, 2008; Modarres Awal, 2012), Sistan & Baluchestan (Ebrahimi and Carpenter, 2008; Modarres Awal, 2012; Hamzavi *et al.*, 2019).

Distribution in the Middle East: Egypt (de Saussure, 1852 as *Eumenes gracilis*; Radoszkowski, 1876 as *Eumenes wagae* (doubtful); Giordani Soika, 1935 as *E. campaniformis campaniformis*, 1970 as *Delta campaniforme gracile*; Zalat *et al.*, 1993 as *D. c. gracile*; Gusenleitner, 2013a; Carpenter and Gadallah, 2020), Iran (see references above, and Carpenter and Gadallah, 2020), Oman, United Arab Emirates (Roche, 1981 as *D. c. gracile*; Gillett and Gillett, 2005 as *D. c. gracile*; Howarth and Gillett, 2008 as *D. c. gracile*; Castro and Dvořák, 2009; Gusenleitner, 2013a; Carpenter and Gadallah, 2020), Qatar, Yemen (Castro and Dvořák, 2009), Turkey (Carpenter and Gadallah, 2020).

Extralimital distribution: Algeria, edge of South Sahara to West Africa, Sychelles (Bingham, 1897 as *E. gracilis*; Cameron, 1905 as *E. gracilis*; Carpenter and Gadallah, 2020).

***Delta fenestrale* (de Saussure, 1852)**

Eumenes fenestralis de Saussure, 1852: 53, ♀; "le Congo".

Distribution in the Middle East: Saudi Arabia (Guichard, 1986; Gusenleitner, 2013a; Carpenter and Gadallah, 2020), Yemen (Wahis, 2000; Gusenleitner, 2007, 2010a, 2013a; Carpenter and Gadallah, 2020).

Extralimital distribution: Tropical Africa (Carpenter *et al.*, 2010; Carpenter and Gadallah, 2020).

***Delta hottentottum elegans* (de Saussure, 1852)**

Eumenes elegans de Saussure, 1852: 58, ♀; "Djidda en Arabie. Les Indes Orientales".

Catalogues with Iranian records: Ebrahimi and Carpenter (2008), Rahmani *et al.* (2020).

Distribution in Iran: East Azarbaijan, Fars, Khuzestan (Ebrahimi and Carpenter, 2008; Modarres Awal, 2012), Kerman (Pashaei Rad *et al.*, 2010), Sistan & Baluchestan, Southern Khorasan (Blüthgen and Gusenleitner, 1970).

Distribution in the Middle East: Arabia (Zalat *et al.*, 1993; Gusenleitner, 2001b), Egypt (Giordani Soika, 1935 as *Eumenes hottentotus* var. *elegans*; Zalat *et al.*, 1993 as *D. hottentotum elegans*; Gusenleitner, 2001b, 2013a), Iran (see references above, and Zalat *et al.*, 1993 as *D. hottentotum elegansi*; Gusenleitner, 2001b), Israel-Palestine (Giordani Soika, 1952 as *E. hottentotus* var. *elegans*; Zalat *et al.*, 1993; Gusenleitner, 2001b, 2013a), Jordan (Giordani Soika, 1970; Gusenleitner, 2001b, 2013a; Haddad *et al.*, 2007 as *D. hottentottus*), Oman (Gusenleitner, 2004, 2013a; Monks *et al.*, 2019), Saudi Arabia (de Saussure, 1852 as *Eumenes elegans*; Guichard, 1986; Gadallah, 2003; Gusenleitner, 2013a; Carpenter and Gadallah, 2020), United Arab Emirates (Roche, 1981 as *D. h. elegans*; Guichard, 1986; Gillett and Gillett, 2005; Howarth and Gillett, 2008; Gusenleitner, 2010a; Carpenter and Gadallah, 2020), Yemen (Giordani Soika, 1952; Wahis, 2000; Gusenleitner, 2007, 2010a, 2013a; Dvořák and Carpenter, 2010; Carpenter and Gadallah, 2020).

Extralimital distribution: Burkina Faso, Cameroon, Central African Republic, Democratic Republic of Congo, Ethiopia, Gabon, Gambia, Ivory Coast, Kenya, Mozambique, Namibia, Nigeria, Senegal, South Africa, Tanzania (Carpenter and Gadallah, 2020).

***Delta lepeleterii lepeleterii* (de Saussure, 1852)**

Eumenes lepeleterii de Saussure, 1852: 45, ♀; "Le Senegal".

Distribution in the Middle East: Cyprus (Carpenter and Gadallah, 2020), Egypt (Giordani Soika, 1935 as *Eumenes campaniformis* var. *lepeleterii*; van der Vecht and Fischer, 1972; Zalat *et al.*,

1993), Jordan (Haddad *et al.*, 2007), Oman (Gusenleitner, 2005; Monks *et al.*, 2019; Carpenter and Gadallah, 2020), Saudi Arabia (Guichard, 1986), Yemen (including Socotra) (Kirby, 1900, 1903 both as *Eumenes granti*; Kohl, 1906 as *E. granti*; Giordani Soika, 1974; Zalat *et al.*, 1993; Gusenleitner, 2007, 2013a; Madl, 2018; Carpenter and Gadallah, 2020).

Extralimital distribution: From SW Asia to several parts of Africa (Carpenter *et al.*, 2010; Carpenter and Gadallah, 2020).

***Delta lepeleterii pilosellum* Giordani Soika, 1982**

Delta lepeleterii pilosellum Giordani Soika, 1982 (1981): 212, ♂, ♀; Eritrea, Ethiopia, Kenya.

Distribution in the Middle East: Yemen (Giordani Soika, 1982a; Guichard, 1986; Carpenter and Gadallah, 2020).

Extralimital distribution: Eritrea, Ethiopia, Kenya (Carpenter *et al.*, 2010; Carpenter and Gadallah, 2020).

***Delta unguiculatum libanicum* (Giordani Soika, 1941)**

Eumenes unguiculatus var. *libanicus* Giordani Soika, 1941: 106, ♂; "Libano: Djezin".

Distribution in the Middle East: Israel-Palestine (Giordani Soika, 1941 as *Eumenes unguiculatus* var. *libanicus*), Lebanon (Giordani Soika, 1941 as *Eumenes unguiculatus* var. *libanicus*; Yildirim and Kojima, 1999), Syria (Giordani Soika, 1970 as *Delta unguiculatum libanense*), Turkey (Yildirim and Özbek, 1996a; Yildirim and Kojima, 1999).

Extralimital distribution: None.

***Delta unguiculatum sporadense* (von Schulthess, 1929)**

Eumenes unguiculatus var. *sporadensis* von Schulthess, 1929: 137, ♀; "Rodi".

Distribution in the Middle East: Cyprus (Fateryga, 2017b as *D. unguiculatum*), Egypt (Zalat *et al.*, 1993).

Extralimital distribution: Italy, Greece.

***Delta unguiculatum unguiculatum* (Villers, 1789)**

Vespa unguiculata Villers, 1789: 282, ♂; Galliae Aust., floribes caepae frequens.

Catalogues with Iranian records: Rahmani *et al.* (2020).

Distribution in Iran: Fars (Lehr *et al.*, 2007; Saghaei *et al.*, 2010), Guilan, Mazandaran (Giordani Soika, 1970), Semnan (Schmid-Egger *et al.*, 2021), Sistan & Baluchestan (Giordani Soika, 1970; Ebrahimi and Carpenter, 2008; Modarres Awal, 2012), Tehran (Blüthgen and Gusenleitner, 1970; Giordani Soika, 1970).

Distribution in the Middle East: Cyprus (Fateryga, 2018; Yalniz and Tüzün, 2019; Fateryga 2017b; Fateryga *et al.*, 2021; Selis *et al.*, 2025), Iran (see references above, and Gusenleitner, 2013a; Fateryga, 2017b, 2018; Yalniz and Tüzün, 2019; Fateryga *et al.*, 2021; Selis *et al.*, 2025), Iraq (Khalaf and Al-Omar, 1974; Selis *et al.*, 2025), Israel-Palestine, Lebanon (Gusenleitner, 2013a; Fateryga, 2017b, 2018; Yalniz and Tüzün, 2019; Fateryga *et al.*, 2021; Selis *et al.*, 2025), Saudi Arabia (Guichard, 1986; Gusenleitner, 2013a; Fateryga 2017b, 2018; Yalniz and Tüzün, 2019; Carpenter and Gadallah, 2020; Fateryga *et al.*, 2021; Selis *et al.*, 2025), Syria (Fateryga, 2017b, 2018; Yalniz and Tüzün, 2019; Fateryga *et al.*, 2021; Selis *et al.*, 2025), Turkey (Giordani Soika, 1952, 1970; Gusenleitner, 1966; Tüzün and Tanyolag, 1988 as *E. anguiculus*; Yildirim and Özbek, 1996a, 1996b, 1999; Yildirim and Kojima, 1999; Yildirim, 2012; Yildirim and Gusenleitner, 2012; Fateryga 2017b, 2018; Kemal and Koçak, 2018; Yildirim and Tezcan, 2018; Yalniz and Tüzün, 2019; Fateryga *et al.*, 2021; Selis *et al.*, 2025).

Extralimital distribution: Albania, Austria, Bulgaria, France, Germany, Italy, Russia, Spain, Switzerland (Fateryga, 2018; Carpenter and Gadallah, 2020; Selis *et al.*, 2025), Armenia, Azerbaijan, Kyrgyzstan, Turkmenistan, Uzbekistan (Fateryga, 2018; Fateryga *et al.*, 2021; Selis *et al.*, 2025), Greece, Morocco, Portugal (Selis *et al.*, 2025), Slovenia (Gogala, 2022; Selis *et al.*, 2025).

Comments: *Antipalus truncatus* (Loew) (Diptera: Asilidae) (Lehr *et al.*, 2007), *Antiphrisson trifarius* (Loew), and *Machimus rusticus* (Meigen) (Diptera: Asilidae) (Saghaei *et al.*, 2010) are the predators of *D. unguiculatum*.

A parasitoid species, *Ophion similis* (Szépligeti) (Hymenoptera: Ichneumonidae) [Semnan province, Garmsar, 2♀♀, August 2019] was collected within *D. dimidiatipenne* nest.

***Delta viatrix* (Nurse, 1903)**

Eumenes viatrix Nurse, 1903: 531, ♂; "Murree-Kashmir Road, about 4000 ft".

Distribution in the Middle East: Turkey (Gusenleitner, 1988; Yildirim and Kojima, 1999).

Extralimital distribution: India (Dover, 1925).

Genus *Eumenes* Latreille, 1802

Eumenes Latreille, 1802: 360. Type species: "*Eumenes coarctata*, Fab." [= *Vespa coarctata* Linnaeus, 1758], by subsequent designation of Latreille, 1810: 438.

***Eumenes (Eumenes) coarctatus coarctatus* (Linnaeus, 1758)**

Vespa coarctata Linnaeus, 1758: 573, ♀; «in Europa».

Catalogues with Iranian records: Ebrahimi and Carpenter (2008), Rahmani *et al.* (2020).

Distribution in Iran: Alborz, Khorasan (Gusenleitner, 1972), Golestan (Schmid-Egger *et al.*, 2021), Kermanshah (Morice, 1921), Kordestan (Ebrahimi and Carpenter, 2008 as *E. coarctatus ordubadensis* Blüthgen), Tehran (Gusenleitner, 1972).

Distribution in the Middle East: Cyprus (Yildirim and Kojima, 1999; Fateryga, 2018; Fateryga *et al.*, 2020; Kochetkov, 2020; Selis *et al.*, 2025), Iran (see references above, and Fateryga *et al.*, 2020; Kochetkov, 2020; Selis *et al.*, 2025), Iraq (Selis *et al.*, 2025), Israel-Palestine (Yildirim and Kojima, 1999; Fateryga, 2018; Fateryga *et al.*, 2020; Kochetkov, 2020; Selis *et al.*, 2025), Jordan (Fateryga, 2018; Fateryga *et al.*, 2020; Kochetkov, 2020; Selis *et al.*, 2025), Syria (Fateryga, 2018; Fateryga *et al.*, 2020; Selis *et al.*, 2025), Turkey (Gusenleitner, 1972; Yildirim and Özbek, 1994, 1996b, 1999; Yildirim, 1998; Yildirim and Kojima, 1999; Yildirim and Gusenleitner, 2001; Fateryga, 2018; Kemal and Koçak, 2018 as *E. coarctatus*; Fateryga *et al.*, 2020; Kochetkov, 2020; Selis *et al.*, 2025).

Extralimital distribution: Armenia, Azerbaijan (Fateryga *et al.*, 2020; Selis *et al.*, 2025), Belarus, Bulgaria, China, Finland, France, Greece, Italy, Japan, Kazakhstan, Kyrgyzstan, Mongolia, Netherlands, North Africa, Portugal, Romania, Russia, Serbia, Spain, Sweden, Switzerland, Tajikistan, Turkmenistan, United Kingdom (Fateryga, 2018; Fateryga *et al.*, 2019, 2020; van Achterberg *et al.*, 2023; Selis *et al.*, 2025), Austria, Czech Republic, Germany, Hungary, Libya, Morocco, Poland, Slovakia, Uzbekistan (Selis *et al.*, 2025), Malta (Cassar *et al.*, 2022), Slovenia (Gogala, 2022; Selis *et al.*, 2025).

Comments: The taxonomy of *Eumenes coarctatus* is quite complex and needs to be resolved completely. In the past it has been divided into several species and subspecies (Gusenleitner, 1999). Recently they were grouped under two subspecies: *E. c. coarctatus* and *E. c. lunulatus* Fabricius (van Achterberg *et al.*, 2023). Pending a revision that clarifies the taxonomy of its numerous forms, a broad concept of *Eumenes coarctatus* is maintained (Selis *et al.*, 2025).

***Eumenes (Eumenes) coarctatus lunulatus* Fabricius, 1804**

Eumenes lunulata Fabricius, 1804: 290, ♀; Austria.

Eumenes pomiformis limissicus Blüthgen, 1938: 493, ♀, ♂; «Limassol (Cypern)».

Distribution in Iran: East Azarbaijan (Ebrahimi and Carpenter, 2008), Tehran (Ebrahimi and Carpenter, 2008).

Distribution in the Middle East: Cyprus (Blüthgen, 1938 as *E. pomiformis limissicus*; Gusenleitner, 1972 as *E. lunulatus limissicus*; Gusenleitner, 1998d, 2013a; Fateryga, 2018; Yalniz and Tüzün, 2019), Iran (see references above, and Gusenleitner, 2013a as *E.* (*E.*) *coarctatus ordubadensis* Blüthgen; Fateryga, 2018; Yalniz and Tüzün, 2019; van Achterberg *et al.*, 2023), Israel-Palestine (Yildirim and Kojima, 1999 as *E. c. ordubadensis*; Gusenleitner, 2013a; Fateryga, 2018; Yalniz and Tüzün, 2019), Jordan (Yildirim and Kojima, 1999 as *E. c. ordubadensis*; Gusenleitner, 2013a; Fateryga, 2018; Yalniz and Tüzün, 2019), Syria (Gusenleitner, 2013a as *E.* (*E.*) *coarctatus ordubadensis* Blüthgen; Fateryga, 2018; Yalniz and Tüzün, 2019), Turkey (Gusenleitner, 1972, 2013a as both *E. c. lunulatus* and *E. l. ordubadensis*; Yildirim and Özbek, 1994 as *E. l. ordubadensis*;

Madl, 1997; Gusenleitner, 1998d as *E. coarctatus ordubadensis*; Yildirim and Kojima, 1999 as *E. c. ordubadensis*; Yildirim, 2012; Yildirim and Gusenleitner, 2009, 2012; Yildirim and Tezcan, 2018; Yalniz and Tüzün, 2019; van Achterberg *et al.*, 2023).

Extralimital distribution: Armenia, China, Tajikistan (Gusenleitner, 2013a as *E.* (*E.*) *coarctatus ordubadensis* Blüthgen), Austria, Bulgaria, Czech Republic, Greece, Hungary, Italy, Slovakia (van Achterberg *et al.*, 2023), Slovenia (Gogala, 2022).

***Eumenes* (*Eumenes*) *comberi* Dover, 1925**

Eumenes comberi Dover, 1925: 293, ♀; "Karachi; ....Shapali, Bombay Pres."

Catalogues with Iranian records: Ebrahimi and Carpenter (2008), Rahmani *et al.* (2020).

Distribution in Iran: Kerman (Pashaei Rad *et al.*, 2010), Khuzestan (Gusenleitner, 1972 as *Eumenes mirandus* Kostylev, 1940), Markazi (Ebrahimi and Carpenter, 2008).

Distribution in the Middle East: Iran (see reference above, and Gusenleitner, 2013a as *E. mirandus*; Fateryga *et al.*, 2023).

Extralimital distribution: Pakistan, Tajikistan, India (Kumar *et al.*, 2017; Fateryga *et al.*, 2023), Uzbekistan (Fateryga *et al.*, 2023).

***Eumenes* (*Eumenes*) *coronatus* (Panzer, 1799)**

Vespa coronatus Panzer, 1799: 12, ♂; "Habitat Norimbergae in floribus" [destroyed].

Catalogues with Iranian records: Ebrahimi and Carpenter (2008 as *Eumenes coronatus detonsus* Blüthgen, 1943), Rahmani *et al.* (2020).

Distribution in Iran: Alborz (Blüthgen, 1943; Gusenleitner, 1972 both as *Eumenes coronatus detonsus* Blüthgen, 1943), Mazandaran (Schmid-Egger *et al.*, 2021).

Distribution in the Middle East: Iran (see references above, and Kurzenko, 1977; Yildirim and Kojima, 1999 as *E. c. detonsus*; Gusenleitner, 2013a as *E. c. detonsus*; Fateryga, 2017b, 2018; Fateryga *et al.*, 2020), Israel-Palestine (Yildirim and Kojima, 1999 as *E. c. coreutus*; Gusenleitner, 2013a as *E. c. coreutus*; Fateryga, 2017b, 2018; Fateryga *et al.*, 2019; Fateryga *et al.*, 2020; van Achterberg *et al.*, 2023), Turkey (Gusenleitner, 1972 as both *E. coronatus coronatus* and *E. c. dentosus*; Gusenleitner, 2013a as both *E. c. coreutes* and *E. c. detonsus*; Yildirim and Özbek, 1994, 1996b as *E. c. detonsus*; Madl, 1997 as *E. c. detonsus*; Yildirim, 1998 as *E. c. detonsus*; Yildirim and Kojima, 1999 as *E. c. detonsus*; Yıldırım and Gusenleitner, 2001, 2007 both as *E. c. detonus*; Gusenleitner, 2013a as *E. c. coreutus*; Yildirim and Kojima, 1999 as *E. c. coreutus*; Fateryga, 2017b, 2018; Fateryga *et al.*, 2020; van Achterberg *et al.*, 2023).

Extralimital distribution: Azerbaijan, Caucasus, Central Asia, China, Europe, Japan, Kazakhstan, Korean Peninsula, Mongolia, Russia (Fateryga, 2017b, 2018; Fateryga *et al.*, 2019; Fateryga *et al.*, 2020; van Achterberg *et al.*, 2023), Pakistan, Turkmenistan (Fateryga *et al.*, 2020), Slovenia (Gogala, 2022).

***Eumenes* (*Eumenes*) *crimensis* Blüthgen, 1938**

Eumenes dubius crimensis Blüthgen, 1938: 463, ♀, ♂; Jalta [Crimea].

Catalogues with Iranian records: Ebrahimi and Carpenter (2008) as both *Eumenes dubius crimensis* (Blüthgen) and *E. sareptanus scabrosus* Gusenleitner), Rahmani *et al.* (2020).

Distribution in Iran: Tehran (Gusenleitner, 1972 as *Eumenes sareptanus scabrosus* Gusenleitner, 1972), Zanjan (Abbasi *et al.*, 2006a, 2008a, 2008b; Ebrahimi and Carpenter, 2008; Pashaei Rad *et al.*, 2010 all as *E. dubius crimensis* Blüthgen, 1938).

Distribution in the Middle East: Iran (see references above, and Gusenleitner, 2013a as *E. crimensis*; Fateryga, 2017b, 2018).

Extralimital distribution: Kazakhstan, Russia, Turkmenistan, Uzbekistan (Fateryga, 2017b, 2018).

***Eumenes* (*Eumenes*) *cyrenaicus* Blüthgen, 1938**

Eumenes dubius cyrenaicus Blüthgen, 1938: 464, 468, ♀, ♂; Libya.

Distribution in the Middle East: Turkey (van Achterberg *et al.*, 2023).

Extralimital distribution: Bulgaria, Greece, Italy, North Africa, North Macedonia, Portugal, Spain (van Achterberg *et al.*, 2023).

***Eumenes* (*Eumenes*) *dubius dubius* de Saussure, 1852**

Eumenes dubia de Saussure, 1852: 32, ♀, ♂; France.

Catalogues with Iranian records: Ebrahimi and Carpenter (2008), Rahmani *et al.* (2020).

Distribution in Iran: Kordestan, Markazi (Ebrahimi and Carpenter, 2008).

Distribution in the Middle East: Cyprus (Yildirim and Kojima, 1999; Fateryga, 2017a, 2017b, 2018; Fateryga *et al.*, 2019; Yalniz and Tüzün, 2019 as *E. d. dubius*; Yalniz and Tüzün, 2019; Fateryga *et al.*, 2023; Fateryga and Proshchalykin, 2024), Iran (Ebrahimi and Carpenter, 2008; Yildirim and Kojima, 1999; Ebrahimi, 2014; Fateryga, 2017a, 2017b, 2018; van Achterberg *et al.*, 2023; Fateryga *et al.*, 2023; Fateryga and Proshchalykin, 2024), Iraq, Israel-Palestine (Yildirim and Kojima, 1999; Fateryga, 2017a, 2017b, 2018; van Achterberg *et al.*, 2023; Fateryga *et al.*, 2023; Fateryga and Proshchalykin, 2024), Jordan, Lebanon, Syria (Yildirim and Kojima, 1999; Fateryga, 2017a, 2017b, 2018; Fateryga *et al.*, 2019; Fateryga *et al.*, 2023; van Achterberg *et al.*, 2023), Turkey (Gusenleitner, 1966, 1972, 2007, 2013a; Yildirim and Özbek, 1994, 1996b, 1999; Yildirim and Kojima, 1999; Yıldırım and Gusenleitner, 2007, 2012; Yildirim, 2012; Fateryga, 2017a, 2017b, 2018; Fateryga *et al.*, 2019, 2023; Yalniz and Tüzün, 2019; van Achterberg *et al.*, 2023; Fateryga and Proshchalykin, 2024).

Extralimital distribution: Azerbaijan, Europe, Kazakhstan, North Africa, Russia, South America (introduced), Turkmenistan, Tajikistan (Gusenleitner, 1999; Fateryga, 2017a, 2017b, 2018; Fateryga *et al.*, 2019; van Achterberg *et al.*, 2023), Kyrgyzstan (Fateryga and Proshchalykin, 2024), Malta (Schembri, 1980), Italy (Selis, 2023), Uzbekistan (Fateryga *et al.*, 2023).

***Eumenes* (*Eumenes*) *dubius palaestinensis* Blüthgen, 1938**

Eumenes dubius dubius var. *palaestinensis* Blüthgen, 1938: 466, 467, ♀, ♂; "ohne Fundort, sicher von Jericho stammend".

Catalogues with Iranian records: Ebrahimi and Carpenter (2008), Rahmani *et al.* (2020).

Distribution in Iran: Iran (no specific locality cited) (Gusenleitner, 1972, 2013a).

Distribution in the Middle East: Iran (Gusenleitner, 2013a; Selis *et al.*, 2025), Iraq (Yildirim and Kojima, 1999; Gusenleitner, 1972, 2013a; Selis *et al.*, 2025), Israel-Palestine (Blüthgen, 1938 as *E. dubius dubius* var. *palaestinensis*; Giordani Soika, 1952; Gusenleitner, 2013a; Selis *et al.*, 2025), Lebanon, Syria (Yildirim and Kojima, 1999; Gusenleitner, 2013a; Selis *et al.*, 2025), Turkey (Giordani Soika, 1952; Gusenleitner, 1972, 2013a; Yildirim, 1998; Yildirim and Kojima, 1999; Selis *et al.*, 2025).

Extralimital distribution: Russia (Selis *et al.*, 2025).

***Eumenes* (*Eumenes*) *interpositus* Gusenleitner, 1972**

Eumenes interpositus Gusenleitner, 1972: 103, ♀, ♂; Iran.

Catalogues with Iranian records: Ebrahimi and Carpenter (2008), Rahmani *et al.* (2020).

Distribution in Iran: Alborz (Gusenleitner, 1972).

Distribution in the Middle East: Iran (Gusenleitner, 1972; Gusenleitner, 2013a).

Extralimital distribution: Turkmenistan (Gusenleitner, 2013a).

***Eumenes* (*Eumenes*) *jarkandensis* Blüthgen, 1938**

Eumenes mediterraneus jarkandensis Blüthgen, 1938: 487, ♀, ♂; Uss-Lusch (1600 m) bei *Jarkand* (*Ostturkestan*) *gesammelt* [Turkestan].

Catalogues with Iranian records: Ebrahimi and Carpenter (2008), Rahmani *et al.* (2020).

Distribution in Iran: Alborz, Khorasan (Gusenleitner, 1972), East Azarbaijan, Kordestan (Ebrahimi and Carpenter, 2008).

Distribution in the Middle East: Iran (see references above, and Yildirim and Kojima, 1999 as *E. jarkandensis*; Ebrahimi, 2014; Fateryga *et al.*, 2023), Lebanon (Yildirim and Kojima, 1999;

Fateryga *et al.*, 2019, 2023), Turkey (Gusenleitner, 1972, 2013a; Yildirim and Özbek, 1994, 1996b, 1999; Yildirim, 1998; Yildirim and Kojima, 1999; Yildirim and Gusenleitner, 2012; Fateryga *et al.*, 2019, 2023).

Extralimital distribution: Azerbaijan, Kyrgyzstan, Mongolia, Tajikistan, Turkmenistan (Fateryga *et al.*, 2019, 2023), Uzbekistan (Fateryga *et al.*, 2023).

***Eumenes* (*Eumenes*) *mediterraneus cypricus* Blüthgen, 1938**

Eumenes mediterraneus cypricus Blüthgen, 1938: 488, ♀, ♂; Cyprus.

Distribution in the Middle East: Known only from Cyprus (Blüthgen, 1938; Gusenleitner, 1972, 2013a).

Extralimital distribution: None.

***Eumenes* (*Eumenes*) *mediterraneus mediterraneus* Kriechbaumer, 1879**

Eumenis (!) *mediterranea* Kriechbaumer, 1879: 85, ♀, ♂; lectotype ♂ from "Dalmatia".

Eumenes mediterraneus var. *opacus* Gusenleitner, 1972: 92, ♂, ♀.

Catalogues with Iranian records: Ebrahimi and Carpenter (2008), Rahmani *et al.* (2020).

Distribution in Iran: Alborz (Esmaili and Rastegar, 1974), East Azarbaijan, Guilan, Kordestan, Markazi, Mazandaran, Qom, Razavi Khorasan (Ebrahimi and Carpenter, 2008), Fars, Golestan (Schmid-Egger *et al.*, 2021), Isfahan (Blüthgen and Gusenleitner, 1970; Ebrahimi and Carpenter, 2008), Kerman (Ebrahimi and Carpenter, 2008; Pashaei Rad *et al.*, 2010; Schmid-Egger *et al.*, 2021), Sistan & Baluchestan, Southern Khorasan (Blüthgen and Gusenleitner, 1970), Tehran (Esmaili and Rastegar, 1974; Ebrahimi and Carpenter, 2008), Zanjan (Abbasi *et al.*, 2008a, 2008b; Pashaei Rad *et al.*, 2010).

Distribution in the Middle East: Cyprus (Fateryga, 2018; Fateryga *et al.*, 2019; Yalniz and Tüzün, 2019), Egypt (Giordani Soika, 1935 as *Labus superbus*; Krombein, 1969; Zalat *et al.*, 1993 as *E. m. opacus*; Fateryga, 2018; Carpenter, 2023; Selis *et al.*, 2025), Iran (see references above, and Zalat *et al.*, 1993 as *Eumenes mediterraneus opacus*; Gusenleitner, 2013a; Ebrahimi, 2014; Carpenter, 2023; van Achterberg *et al.*, 2023; Kaplan and Efil, 2025; Selis *et al.*, 2025), Iraq (Gusenleitner, 1972; Augul, 2017 as *Eumenes pomiformis*; Fateryga, 2018; Selis *et al.*, 2025), Israel-Palestine (Fateryga, 2018; Yalniz and Tüzün, 2019; Zalat *et al.*, 1993 as *E. m. opacus*; Carpenter, 2023; Kaplan and Efil, 2025; Selis *et al.*, 2025), Jordan (Zalat *et al.*, 1993 as *E. m. opacus*; Haddad *et al.*, 2007; Fateryga, 2018; Yalniz and Tüzün, 2019; Carpenter, 2023; Selis *et al.*, 2025), Lebanon (Zalat *et al.*, 1993 as *E. m. opacus*; Fateryga, 2018; Yalniz and Tüzün, 2019; Carpenter, 2023; Kaplan and Efil, 2025; Selis *et al.*, 2025), Oman (Kriechbaumer, 1879; Giordani Soika, 1952 as *E. houskai*; Fateryga, 2018; Yalniz and Tüzün, 2019; Carpenter, 2023), Saudi Arabia (Guichard, 1986; Gusenleitner, 2013a; Fateryga, 2018; Fateryga *et al.*, 2019; Yalniz and Tüzün, 2019; Carpenter, 2023; van Achterberg *et al.*, 2023; Kaplan and Efil, 2025; Selis *et al.*, 2025), Syria (Kriechbaumer, 1879; Fateryga, 2018; Fateryga *et al.*, 2019; Yalniz and Tüzün, 2019; Selis *et al.*, 2025), Turkey (Giordani Soika, 1952; Gusenleitner, 1966, 1970, 2007; Gusenleitner, 1972 as *E. m. opacus*; Zalat *et al.*, 1993 as *E. m. opacus*; Yildirim and Özbek, 1994, 1996b, 1999; Madl, 1997; Yildirim and Kojima, 1999; Yildirim and Gusenleitner, 2004, 2009, 2007, 2012; Gusenleitner, 2013a; Yildirim, 2012; Yildirim and Tezcan, 2018; Fateryga *et al.*, 2019; Yalniz and Tüzün, 2019; van Achterberg *et al.*, 2023; Carpenter, 2023; Kaplan and Efil, 2025; Selis *et al.*, 2025), United Arab Emirates (Gusenleitner, 2010b; Fateryga, 2018; Fateryga *et al.*, 2019; Yalniz and Tüzün, 2019; Carpenter, 2023; Selis *et al.*, 2025), Yemen (Fateryga, 2018; Fateryga *et al.*, 2019; Yalniz and Tüzün, 2019; Carpenter, 2023; Kaplan and Efil, 2025; Selis *et al.*, 2025).

Extralimital distribution: Afghanistan, Armenia, Azerbaijan, China, Kazakhstan, Korea, Mongolia, Pakistan, Russia (Fateryga, 2018; Fateryga *et al.*, 2019; Carpenter and Gadallah, 2020; Carpenter, 2023; Selis *et al.*, 2025), Albania, Austria, Bulgaria, France, Germany, Greece, Kyrgyzstan, Morocco, Portugal, Serbia, Spain, Sudan, Switzerland, Tunisia (Carpenter, 2023; Selis *et al.*, 2025), India (van Achterberg *et al.*, 2023 as *E. mediterraneus*), Italy (Selis, 2023; Selis *et al.*, 2025), Malta (Cassar *et al.*, 2022), Tahiti (introduced) (Fateryga, 2018; Selis *et al.*, 2025).

Comments: Augul (2017) recorded this species from Iraq, misidentified as *Eumenes pomiformis* (Selis *et al.*, 2025).

***Eumenes (Eumenes) mediterraneus quettaensis* Cameron, 1907**

Eumenes affinissima quettaensis Cameron, 1907: 132, sex not stated; Quetta.

Distribution in Iran: Alborz (Gusenleitner, 1972), Fars (Gusenleitner, 1972), Khorasan (Gusenleitner, 1972), Tehran (Gusenleitner, 1972).

Distribution in the Middle East: Iran (see references above, and Gusenleitner, 1972, 2013a).

Extralimital distribution: Afghanistan, Kazakhstan, Pakistan, Tajikistan, Turkmenistan, Uzbekistan (Gusenleitner, 2013a).

***Eumenes (Eumenes) modestus* Gusenleitner, 2006b**

Eumenes modestus Gusenleitner, 2006b: 1361, ♂; Iran.

Catalogues with Iranian records: Ebrahimi and Carpenter (2008), Rahmani *et al.* (2020).

Distribution in Iran: Kerman (Schmid-Egger *et al.*, 2021), Tehran (Gusenleitner, 2006b).

Distribution in the Middle East: Iran (see references above, and Gusenleitner, 2013a), Turkey (Yildirim and Gusenleitner, 2009).

Extralimital distribution: None.

***Eumenes (Eumenes) papillarius papillarius* (Christ, 1791)**

Sphex papillaria Christ, 1791: 325, sex not stated; locality not mentioned (destroyed).

Catalogues with Iranian records: Ebrahimi and Carpenter (2008), Rahmani *et al.* (2020).

Distribution in Iran: East Azarbaijan (Ebrahimi and Carpenter, 2008), Zanjan (Abbasi *et al.*, 2008a, 2008b; Pashaei Rad *et al.*, 2010).

Distribution in the Middle East: Iran (see references above, and Gusenleitner, 2013a as *Eumenes papillarius papillarius*; Fateryga, 2017b, 2018; Fateryga *et al.*, 2023; Fateryga and Proshchalykin, 2024), Israel-Palestine, Lebanon, Syria (Gusenleitner, 1972, 2013a as *Eumenes papillarius rubricornis*; Fateryga, 2017b, 2018; Fateryga *et al.*, 2019, 2023; Fateryga and Proshchalykin, 2024), Turkey (Gusenleitner, 1972, 2013a; Madl, 1997; Yildirim and Özbek, 1994, 1996b, 1999; Yildirim and Kojima, 1999; Fateryga, 2017b, 2018; Fateryga *et al.*, 2019, 2023).

Extralimital distribution: Azerbaijan, Europe, Kazakhstan, Mongolia, Pakistan, Russia (Fateryga, 2017b, 2018; Fateryga *et al.*, 2019; van Achterberg *et al.*, 2023), Italy (Selis, 2023), Kyrgyzstan (Fateryga and Proshchalykin, 2024), Slovenia (Gogala, 2022), Uzbekistan (Fateryga *et al.*, 2023).

***Eumenes (Eumenes) pedunculatus pedunculatus* (Panzer, 1799)**

Vespa pedunculata Panzer, 1799: 8, ♀; "Habitat et apud nos in hortis, minus frequens."

Catalogues with Iranian records: None.

Distribution in Iran: Golestan (Schmid-Egger *et al.*, 2021).

Distribution in the Middle East: Iran (Schmid-Egger *et al.*, 2021), Turkey (Blüthgen, 1959; Gusenleitner, 1967; Yildirim and Özbek, 1994, 1996b, 1999; Yildirim and Kojima, 1999; Yildirim and Gusenleitner, 2009; Fateryga, 2017b; Kochetkov, 2020; Kaplan and Efil, 2025).

Extralimital distribution: Austria, China, Czech Republic, Finland, France, Italy, Mongolia, Netherlands, Poland, Russia, Serbia, Spain, Sweden, Switzerland (Fateryga, 2017b), Japan, Korea (van Achterberg *et al.*, 2023; Kaplan and Efil, 2025), Kazakhstan (Fateryga, 2017b), Slovenia (Gogala, 2022).

***Eumenes (Eumenes) pomiformis* (Fabricius, 1781)**

Eumenes pomiformis Fabricius, 1781: 467, ♀; "Italia".

Catalogues with Iranian records: This species is a new record for the fauna of Iran.

Distribution in Iran: Golestan province, Golestan National Park, 1♀, 1♂, June 2007, leg. H. Ghahari, det. J. Gusenleitner (Figure 2.2).

Distribution in the Middle East: Egypt (Giordani Soika, 1935), Iran (new record), Lebanon (Yildirim and Kojima, 1999; Gusenleitner, 2013a; Fateryga, 2017a, 2017b, 2018; Fateryga *et al.*, 2019; Yalniz and Tüzün, 2019; van Achterberg *et al.*, 2023), Turkey (Giordani Soika,

FIGURE 2.2 *Eumenes* (*Eumenes*) *pomiformis* (Fabricius, 1781).

(Photo courtesy of Yuri Palamarchuk)

1952; Tüzün and Tanyolag, 1988; Madl, 1997; Gusenleitner, 1972 as both *E. p. pomiformis* and *E. p. turcicus*, 1997c, 2007; Yildirim and Özbek, 1994, 1996b; Madl, 1997; Yildirim and Özbek, 1999; Yildirim and Kojima, 1999; Yildirim and Gusenleitner, 2009, 2007, 2012; Gusenleitner, 2013a; Yildirim, 2012; Fateryga, 2018; Fateryga *et al.*, 2019; Yalniz and Tüzün, 2019; van Achterberg *et al.*, 2023), United Arab Emirates (Gusenleitner, 2010a; Fateryga, 2017b, 2018; Fateryga *et al.*, 2019; Yalniz and Tüzün, 2019; Carpenter and Gadallah, 2020).

Extralimital distribution: Armenia, Azerbaijan, Caucasus, Central Asia, China, Europe, India (doubtful), Kazakhstan, Korea, Pakistan, Russia (Fateryga, 2017a, 2017b, 2018; Fateryga *et al.*, 2019), Italy (Selis, 2023), Malta (Cassar *et al.*, 2022), Tunisia (van Achterberg *et al.*, 2023), Slovenia (Gogala, 2022).

Comments: *Eumenes pomiformis* was excluded from the fauna of Iraq by Selis *et al.* (2025). It was listed as the fauna of Iraq by Khalaf (1958), Derwesh (1965), Augul (2017).

***Eumenes* (*Eumenes*) *pseudubius* Gusenleitner, 1972**

Eumenes pseudubius Gusenleitner, 1972: 69, 110, ♀, ♂; "Baghdad, Irak".

Distribution in the Middle East: Iraq (Gusenleitner, 1972, 2013a; Castro and Dvořák, 2010; Selis *et al.*, 2025), Jordan (Castro and Dvořák, 2010; Selis *et al.*, 2025).

Extralimital distribution: None.

***Eumenes* (*Eumenes*) *punctaticlypeus* Giordani Soika, 1943**

Eumenes (*Eumenes*) *punctaticlypeus* Giordani Soika, 1943: 29, ♀, ♂; Viterbo [Italy].

Catalogues with Iranian records: Rahmani *et al.* (2020).

Distribution in Iran: Northern Iran (Kurzenko, 1977 as *Eumenes punctaticlypeus kostylevi* Kurzenko).

Distribution in the Middle East: Iran (Kurzenko, 1977; Fateryga, 2017b, 2018), Turkey (Giordani Soika, 1943; Yildirim and Özbek, 1994 as *E. kostylevi*; Yildirim and Özbek, 1999 as *E. kostylevi punctaticlypeus*; Yildirim and Özbek, 1999 as *E. p. kostylevi*; Gusenleitner, 1972, 1979 as *E. k. punctaticlypeus*; 2007; Yildirim and Kojima, 1999 as *E. p. kostylevi*; Yildirim and Gusenleitner, 2007; Fateryga, 2017b, 2018; van Achterberg *et al.*, 2023).

Extralimital distribution: Central Asia, Europe, Kazakhstan, Russia (Fateryga, 2017b, 2018), Italy (Selis, 2023), Ukraine (van Achterberg *et al.*, 2023).

***Eumenes* (*Eumenes*) *sareptanus* André, 1884**

Eumenes sareptanus sareptanus André, 1884: 638, ♀; Russia.

Distribution in the Middle East: Turkey (Yildirim and Özbek, 1994 as both *Eumenes sareptanus isolatus* and *E. s. sareptanus*; Yildirim and Özbek, 1999; Yildirim and Kojima, 1999; Gusenleitner, 2007, 2013a; Yildirim and Gusenleitner, 2007, 2009, 2012; Fateryga, 2017a, 2017b, 2018; Fateryga *et al.*, 2023).

Extralimital distribution: Austria, Bulgaria, Caucasus, Czech Republic, France, Germany, Hungary, Italy, Kazakhstan, Malta, Poland, Russia, Slovakia, Switzerland, Turkmenistan (Fateryga, 2017a, 2017b, 2018), Uzbekistan (Fateryga *et al.*, 2023).

***Eumenes* (*Eumenes*) *subpomiformis* Blüthgen, 1938**

Eumenes subpomiformis Blüthgen, 1938: 480, 494, ♀, ♂; Austria.

Distribution in the Middle East: Iraq (Selis *et al.*, 2025; Fateryga *et al.*, 2025), Israel-Palestine (Giordani Soika, 1952; Gusenleitner, 2013a; Fateryga, 2017b; van Achterberg *et al.*, 2023; Selis *et al.*, 2025; Fateryga *et al.*, 2025), Lebanon (Yildirim and Kojima, 1999; Gusenleitner, 2013a; Fateryga, 2017b; van Achterberg *et al.*, 2023; Selis *et al.*, 2025; Fateryga *et al.*, 2025), Turkey (Gusenleitner, 1972, 2013a; Yildirim and Özbek, 1994, 1996b, 1999; Yildirim, 1998; Yildirim and Kojima, 1999; Yildirim and Gusenleitner, 2004, 2012; Fateryga, 2017b; van Achterberg *et al.*, 2023; Selis *et al.*, 2025; Fateryga *et al.*, 2025).

Extralimital distribution: Albania, Algeria, Austria, Belgium, Bulgaria, Czech Republic, France, Germany, Greece, Italy, Morocco, Portugal, Serbia, Spain, Switzerland (van Achterberg *et al.*, 2023; Selis *et al.*, 2025), Azerbaijan (Fateryga *et al.*, 2025), Kazakhstan, Russia (Fateryga, 2017b).

***Eumenes* (*Eumenes*) *tripunctatus* (Christ, 1791)**

Sphex tripunctata Christ, 1791: 317, sex not stated; locality not stated (destroyed).

Catalogues with Iranian records: Ebrahimi and Carpenter (2008), Rahmani *et al.* (2020).

Distribution in Iran: Markazi (Ebrahimi and Carpenter, 2008).

Distribution in the Middle East: Iran (Kurzenko, 1977; Ebrahimi and Carpenter, 2008; Fateryga, 2017b, 2018; Fateryga *et al.*, 2020).

Extralimital distribution: Afghanistan (Fateryga *et al.*, 2020), China, Kazakhstan, Mongolia, Russia, Turkmenistan, Ukraine (Fateryga, 2018; Fateryga *et al.*, 2020; van Achterberg *et al.*, 2023), Europe (Fateryga, 2017b; Fateryga *et al.*, 2020).

***Eumenes* spp.**

Distribution in Iran: Hormozgan (Khoobdel *et al.*, 2014), Isfahan (Ebadi, 1995).

Comments: *Eumenes* sp. was introduced as a pollinator insect by Ebadi (1995).

Genus *Eumicrodynerus* Gusenleitner, 1972

Eumicrodynerus Gusenleitner, 1972: 74, subgenus of *Microdynerus* Thomson. Type species: *Leptomenes europaeus* Giordani Soika, 1942, by original designation and monotypy.

***Eumicrodynerus europaeus* (Giordani Soika, 1942)**

Leptomenes (*Leptomenes*) *europaeus* Giordani Soika, 1942: 74, ♀; "Grecia: Parnasso".

Distribution in the Middle East: Iraq (Selis *et al.*, 2025), Turkey (Gusenleitner, 1972 as *Microdynerus* (*Eumicrodynerus*) *europaeus*, 2013a; Yildirim and Gusenleitner, 2012; Selis *et al.*, 2025).

Extralimital distribution: Bulgaria (Castro and Dvořák, 2009; Selis *et al.*, 2025), Greece (Gusenleitner, 2013a; Selis *et al.*, 2025).

Genus *Euodynerus* Dalla Torre, 1904

Euodynerus Dalla Torre, 1904: 38. Type species: *Vespa dantici* Rossi, 1790; designated by Blüthgen, 1938 [1937]: 277.

***Euodynerus* (*Euodynerus*) *amplus* Gusenleitner, 2011a**
Euodynerus (*Euodynerus*) *amplus* Gusenleitner, 2011a: 324, ♂; Iran.
Catalogues with Iranian records: Rahmani *et al.* (2020).
Distribution in Iran: Kerman (Gusenleitner, 2011a).
Distribution in the Middle East: Known only from Iran (Gusenleitner, 2011a, 2013a).
Extralimital distribution: None.

***Euodynerus* (*Euodynerus*) *annae* (Kostylev, 1937)**
Odynerus annae Kostylev, 1937: 223, ♀; Iran.
Catalogues with Iranian records: Ebrahimi and Carpenter (2008), Rahmani *et al.* (2020).
Distribution in Iran: Kerman (Kostylev, 1937).
Distribution in the Middle East: Iran (Kostylev, 1937; Gusenleitner, 2013a; Selis *et al.*, 2025), Iraq (Selis *et al.*, 2025).
Extralimital distribution: None.

***Euodynerus* (*Euodynerus*) *caspicus caspius* (Morawitz, 1873)**
Leionotus caspicus Morawitz, 1873: 295, ♀, ♂; Krasnowodsk [Turkmenistan].
Catalogues with Iranian records: Rahmani *et al.* (2020).
Distribution in Iran: Iran (no specific locality cited) (Fateryga, 2017b; Fateryga *et al.*, 2019).
Distribution in the Middle East: Iran (see references above), Turkey (Fateryga *et al.*, 2019).
Extralimital distribution: Armenia, Azerbaijan, China, Europe, Georgia, Kazakhstan, Kyrgyzstan, Mongolia, Russia, Turkmenistan (Fateryga, 2017b; Fateryga *et al.*, 2019).

***Euodynerus* (*Euodynerus*) *cherkensis* (Giordani Soika, 1942)**
Odynerus (*Rhynchium*) *cherkensis* Giordani Soika, 1942: 61, "♀" [♂]; "Cipro: Cherkes".
Distribution in the Middle East: Cyprus (Giordani Soika, 1942 as *Odynerus* (*Rhynchium*) *cherkensis*; Yildirim and Kojima, 1999; Gusenleitner, 2013a), Turkey (Giordani Soika, 1970; Yildirim and Kojima, 1999; Gusenleitner, 2013a).
Extralimital distribution: None.

***Euodynerus* (*Euodynerus*) *clatratus* Blüthgen, 1951**
Euodynerus (*Euodynerus*) *pseudocaspius clatratus* Blüthgen, 1951a: 172, 199, ♀; "Kulp (Armenien)".
Distribution in the Middle East: Syria (Giordani Soika, 1970 as *E.* (*E.*) *paracaspicus*; Gusenleitner, 2013a), Turkey (Giordani Soika, 1970 as *E.* (*E.*) *paracaspicus*; Gusenleitner, 1979, 1988, 2013a; Yildirim and Kojima, 1999).
Extralimital distribution: Armenia (Gusenleitner, 2013a).

***Euodynerus* (*Euodynerus*) *curictensis curictensis* Blüthgen, 1940**
Euodynerus curictensis Blüthgen, 1940: 210, ♀, ♂; "Insel KrK in Jugoslavien" [Croatia].
Catalogues with Iranian records: Ebrahimi and Carpenter (2008), Rahmani *et al.* (2020).
Distribution in Iran: Alborz (Giordani Soika, 1970), Sistan & Baluchestan (Blüthgen and Gusenleitner, 1970).
Distribution in the Middle East: Iran (see references above, and Yildirim and Kojima, 1999; Gusenleitner, 2001a, 2013a; Fateryga *et al.*, 2021; Selis *et al.*, 2025), Iraq (Selis *et al.*, 2024, 2025), Jordan (Gusenleitner, 2001b; Fateryga and Mokrousov, 2019; Fateryga *et al.*, 2021; Selis *et al.*, 2025), Lebanon, Syria (Yildirim and Kojima, 1999; Gusenleitner, 2013a; Fateryga *et al.*, 2021; Selis *et al.*,

2025), Turkey (Gusenleitner, 1966, 1997c, 2013a; Giordani Soika, 1970; Yildirim and Özbek, 1996a, 1996b; Madl, 1997; Yildirim and Özbek, 1999; Yildirim and Kojima, 1999; Yildirim and Gusenleitner, 2009; Fateryga and Mokrousov, 2019; Fateryga *et al.*, 2021; Selis *et al.*, 2025).

Extralimital distribution: Albania, Austria, Croatia, France, Greece, Morocco, Spain, Switzerland (Fateryga *et al.*, 2021; Selis *et al.*, 2024, 2025), Armenia, Kazakhstan, Mongolia, Russia, (Fateryga and Mokrousov, 2019; Selis *et al.*, 2024, 2025), Azerbaijan (Fateryga *et al.*, 2021), Italy (Selis, 2023; Selis *et al.*, 2025), Slovenia (Gogala, 2022; Selis *et al.*, 2025).

***Euodynerus* (*Euodynerus*) *dantici dantici* (Rossi, 1790)**

Vespa dantici Rossi, 1790: 89, ♀; Italia.

Catalogues with Iranian records: Rahmani *et al.* (2020).

Distribution in Iran: Golestan (Schmid-Egger *et al.*, 2021), Kerman (Pashaei Rad *et al.*, 2010).

Distribution in the Middle East: Egypt, Iran (Pashaei Rad *et al.*, 2010; Fateryga, 2018; Schmid-Egger *et al.*, 2021), Israel-Palestine, Jordan (Castro and Dvořák, 2010; Fateryga, 2017b, 2018), Lebanon, Syria (Giordani Soika, 1970; Fateryga, 2017b, 2018; Fateryga *et al.*, 2019), Southern Arabia, Turkey (Gusenleitner, 1966, 1997c; Giordani Soika, 1970; Yildirim and Özbek, 1996a, 1996b; Yildirim and Gusenleitner, 2009, 2012; Yildirim, 2012; Fateryga, 2017b, 2018; Fateryga *et al.*, 2019; Yalniz and Tüzün, 2019).

Extralimital distribution: Afghanistan, Armenia, Azerbaijan, Central Asia, China, Europe, Georgia, Japan, Kazakhstan, Korea, Mongolia, North Africa, Russia, South East of Asia, (Fateryga, 2017b, 2018; Fateryga *et al.*, 2019), Malta (Cassar *et al.*, 2022), Italy (Selis, 2023), Slovenia (Gogala, 2022).

***Euodynerus* (*Euodynerus*) *dantici tinctus* (Walker, 1871)**

Odynerus tinctus Walker, 1871: 33, 34, ♀; "Wadi Gennèh".

Distribution in the Middle East: Known only from Egypt (Walker, 1871 as *Odynerus tinctus*; Giordani Soika, 1941 as *O. tinctus*; van der Vecht and Fischer, 1972; Gusenleitner, 2013a).

Extralimital distribution: None.

***Euodynerus* (*Euodynerus*) *disconotatus disconotatus* (Lichtenstein, 1884)**

Odynerus disconotatus Lichtenstein, 1884: L. sex not stated. "Montpellier" [France] (? destroyed).

Catalogues with Iranian records: Ebrahimi and Carpenter (2008), Rahmani *et al.* (2020) as both *Euodynerus* (*Euodynerus*) *disconotatus disconotatus* and *Euodynerus* (*Euodynerus*) *disconotatus sulfuripes* (Morawitz, 1885).

Distribution in Iran: Fars, Razavi Khorasan, Tehran (Giordani Soika, 1970 as *Euodynerus sulfuripes*), Hormozgan, Southern Khorasan (Blüthgen and Gusenleitner, 1970 as *E. sulfuripes*), Isfahan (Lehr *et al.*, 2007), Kerman (Blüthgen and Gusenleitner, 1970 as *E. sulfuripes*; Schmid-Egger *et al.*, 2021), Zanjan (Abbasi *et al.*, 2008a; Pashaei Rad *et al.*, 2010 both as *E. disconotatus sulfuripes*).

Distribution in the Middle East: Cyprus (Fateryga, 2018; Fateryga *et al.*, 2021; Selis *et al.*, 2025), Iran (see references above, and Blüthgen and Gusenleitner, 1970; Giordani Soika, 1970 both as *Euodynerus* (*E.*) *sulfuripes* (Morawitz, 1885); Gusenleitner, 2013a; Yildirim and Kojima, 1999; Abbasi *et al.*, 2008a, 2008b all as *E.* (*E.*) *sulfuripes*; Fateryga, 2018; Carpenter and Gadallah, 2020 as *E.* (*E.*) *sulfuripes*; Fateryga *et al.*, 2021; Selis *et al.*, 2025), Iraq (Giordani Soika, 1970; Yildirim and Kojima, 1999 both as *E. sulfuripes*; Castro and Dvořák, 2010; Fateryga, 2018; Carpenter and Gadallah, 2020 as *E. sulfuripes*; Fateryga *et al.*, 2021; Selis *et al.*, 2025), Jordan (Castro and Dvořák, 2010; Fateryga, 2018; Fateryga *et al.*, 2021; Selis *et al.*, 2025), Saudi Arabia, Turkey (Gusenleitner, 1966, 1967, 1997c, 2007, 2013a; Gusenleitner, 1979 as *E. sulfuripes*; Yildirim and Özbek, 1996a, 1996b, 1999; Yildirim and Kojima, 1999; Madl, 1997; Yildirim and Gusenleitner, 2007, 2009, 2012; Fateryga, 2018; Yildirim and Tezcan, 2018; Carpenter and Gadallah, 2020 as *E. sulfuripes*; Fateryga *et al.*, 2021; Selis *et al.*, 2025).

Extralimital distribution: Afghanistan, Pakistan, Russia (Fateryga, 2018; Fateryga *et al.*, 2021; Selis *et al.*, 2025), Albania, Austria, Croatia, France, Georgia, Germany, Greece, Kazakhstan, Spain,

Serbia, Tajikistan, Turkmenistan, Uzbekistan (Fateryga *et al.*, 2021; Selis *et al.*, 2024, 2025), Armenia, Azerbaijan (Fateryga *et al.*, 2021; Selis *et al.*, 2025), India (Gusenleitner, 2007 as *Euodynerus sulfuripes*), Italy (Selis, 2023; Selis *et al.*, 2025).

Comments: This species was recorded from Iraq by Giordani Soika (1970) under the name *Euodynerus sulfuripes* (Morawitz), now synonymized with *E. disconotatus* by Selis *et al.* (2024, 2025), based on morphological and DNA barcoding data.

Dioctria abdominalis (Becker) (Diptera: Asilidae) has been recorded by Lehr *et al.* (2007) as a predator of *Euodynerus disconotatus*.

***Euodynerus* (*Euodynerus*) *disconotatus laniensis* Giordani Soika, 1979**

Euodynerus disconotatus laniensis Giordani Soika, 1979: 253, ♂, ♀; "Cipro: Lania".

Distribution in the Middle East: Known only from Cyprus (Giordani Soika, 1979; Gusenleitner, 2013a).

Extralimital distribution: None.

***Euodynerus* (*Euodynerus*) *diversus* (Walker, 1871)**

Odynerus diversus Walker, 1871: 33, 34, ♂; "Wadi Gennèh".

Distribution in the Middle East: Egypt (Walker, 1871 as *Odynerus diversus*; Giordani Soika, 1935 as *O. enslini*; Bytinski-Salz and Gusenleitner, 1972; Gusenleitner, 2001b as *Euodynerus diversus*, 2013a), Israel-Palestine (von Schulthess, 1928 as *O. enslini*; Bytinski-Salz and Gusenleitner, 1972; Gusenleitner, 2001b, 2013a), Jordan (Gusenleitner, 2001b), Saudi Arabia, Syria (Gusenleitner, 2001b).

Extralimital distribution: None.

***Euodynerus* (*Euodynerus*) *effrenatus* Giordani Soika, 1970**

Euodynerus (*Euodynerus*) *effrenatus* Giordani Soika, 1970: 157, ♀; Iran.

Catalogues with Iranian records: Ebrahimi and Carpenter (2008), Rahmani *et al.* (2020).

Distribution in Iran: Fars (Giordani Soika, 1970).

Distribution in the Middle East: Known only from Iran (Giordani Soika, 1970; Gusenleitner, 2013a).

Extralimital distribution: None.

***Euodynerus* (*Euodynerus*) *fastidiosus* (de Saussure, 1853)**

Odynerus (*Leionotus*) *fastidiosus* de Saussure, 1853: 154, 189, ♀; Algeria.

Catalogues with Iranian records: Ebrahimi and Carpenter (2008) Rahmani *et al.* (2020).

Distribution in Iran: Alborz, Fars, Razavi Khorasan (Giordani Soika, 1970), Zanjan (Abbasi *et al.*, 2008a, 2008b; Pashaei Rad *et al.*, 2010).

Distribution in the Middle East: Iran (see references above, and Gusenleitner, 2013a; Fateryga, 2017a, 2017b, 2018; Fateryga *et al.*, 2025), Lebanon, Syria (Giordani Soika, 1970; Yildirim and Kojima, 1999; Fateryga, 2017a, 2017b, 2018; Fateryga *et al.*, 2025), Oman (Gusenleitner, 2012, 2013a; Fateryga, 2017a, 2017b, 2018; Carpenter and Gadallah, 2020; Fateryga *et al.*, 2025), Turkey (Giordani Soika, 1952 as *Pseudepipona* (*Euodynerus*) *fastidiosus*; Gusenleitner, 1966, 1997c; Yildirim and Özbek, 1996a, 1996b; Yildirim, 1998; Yildirim and Kojima, 1999; Fateryga, 2017a, 2017b, 2018; Fateryga *et al.*, 2025).

Extralimital distribution: Afghanistan, Algeria, China, Kazakhstan, Morocco, Russia, Southern continental Europe, Transcaucasica, Ukraine, Uzbekistan (Castro and Dvořák, 2010; Fateryga, 2018; Fateryga *et al.*, 2025), Azerbaijan (Fateryga *et al.*, 2025), Pakistan (Fateryga, 2017a, 2017b, 2018; Fateryga *et al.*, 2025), Italy (Sclis, 2023).

***Euodynerus* (*Pareuodynerus*) *garaensis* Selis, 2025a**

Euodynerus (*Pareuodynerus*) *garaensis* Selis, 2025a [in Selis *et al.*, 2025]: 954, ♂; Kurdistan: Iraq.

Distribution in the Middle East: Known only from Iraq (Selis *et al.*, 2025).

Extralimital distribution: None.

***Euodynerus* (*Euodynerus*) *hellenicus* Blüthgen, 1942**

Euodynerus dantici hellenicus Blüthgen, 1942: 301, ♀, ♂; "Amorgos (Kykladen)".

Distribution in the Middle East: Turkey (Yildirim and Özbek, 1996a; Gusenleitner, 1997c; Yildirim and Kojima, 1999; Yildirim and Gusenleitner, 2001, 2009; Yildirim and Tezcan, 2018).
Extralimital distribution: Afghanistan (Blüthgen, 1961), Greece, Spain (Gusenleitner, 2013a).

***Euodynerus* (*Euodynerus*) *khuzestanicus* Giordani Soika, 1970**
Euodynerus (*Euodynerus*) *khuzestanicus* Giordani Soika, 1970: 155, ♀; Iran.
Catalogues with Iranian records: Ebrahimi and Carpenter (2008), Rahmani *et al.* (2020).
Distribution in Iran: Khuzestan (Giordani Soika, 1970).
Distribution in the Middle East: Known only from Iran (Giordani Soika, 1970; Gusenleitner, 2013a).
Extralimital distribution: None.

***Euodynerus* (*Euodynerus*) *maximilianus* Gusenleitner, 1998c**
Euodynerus (*Euodynerus*) *maximilianus* Gusenleitner, 1998c: 503, 506, ♂; "Türkei, Sirt, 20 km E Batman, 700 m".
Distribution in the Middle East: Known only from Turkey (Gusenleitner, 1998a, 2013a; Yildirim and Kojima, 1999).
Extralimital distribution: None.

***Euodynerus* (*Euodynerus*) *niloticus ebneri* (von Schulthess, 1921)**
Odynerus (*Lionotus*) *ebneri* von Schulthess, 1921 (1920): 286, ♀, ♂; "Kairo, Ägypten; Tuti-Insel bei Khartum; ägyptischer Sudan ..., Manora, Charachi".
Distribution in the Middle East: Egypt (von Schulthess, 1922 as *Lionotus ebneri*; Giordani Soika, 1935 as *O. ebneri*; Gusenleitner, 2013a; Carpenter and Gadallah, 2020), Palestina (von Schulthess, 1928 as *O.* (*L.*) *meyeri* var. *palaestinensis*), Saudi Arabia (Guichard, 1986; Gusenleitner, 2013a; Carpenter and Gadallah, 2020), United Arab Emirates (Guichard, 1986; Howarth and Gillett, 2008; Carpenter and Gadallah, 2020), Yemen (including Socotra) (Kohl, 1906 as *Rhynchium niloticum*; Giordani Soika, 1934 as *Odynerus* (*Rhynchium*) *niloticus*, 1974).
Extralimital distribution: Chad, Eritrea, Ethiopia, Pakistan, Sudan (Carpenter *et al.*, 2010; Carpenter and Gadallah, 2020).

***Euodynerus* (*Euodynerus*) *niloticus niloticus* (de Sausssure, 1855)**
Rhynchium niloticum de Saussure, 1855: 181, ♀; "l'Abyssinie".
Distribution in the Middle East: Egypt (Giordani Soika, 1935 as *Odynerus niloticus*; Gusenleitner, 2013a; Carpenter and Gadallah, 2020; Selis, 2024), Israel-Palestine, Jordan, Oman (Selis, 2024), Saudi Arabia (Zalat *et al.*, 1993; Gusenleitner, 2013a; Carpenter and Gadallah, 2020; Selis, 2024), United Arab Emirates (Roche, 1981; Gillett and Gillett, 2005; Carpenter and Gadallah, 2020; Selis, 2024), Yemen (Giordani Soika, 1935 as *Odynerus niloticus*; Carpenter and Gadallah, 2020; Selis, 2024).
Extralimital distribution: Chad, Eritrea, Ethiopia, Sudan (Carpenter *et al.*, 2010; Carpenter and Gadallah, 2020; Selis, 2024), Pakistan (Selis, 2024).

***Euodynerus* (*Euodynerus*) *niloticus sauditus* (Giordani Soika, 1957)**
Pseudepipona (*Euodynerus*) *niloticus saudica* Giordani Soika, 1957: 478, ♀, ♂; Aden.
Distribution in the Middle East: Saudi Arabia (Giordani Soika, 1970 as *Pseudepipona* (*Euodynerus*) *niloticus saudita*; Gusenleitner, 2013a; Carpenter and Gadallah, 2020), United Arab Emirates (Gusenleitner, 2013a; Carpenter and Gadallah, 2020), Yemen (Giordadni Soika, 1970 as *Pseudepipona* (*Euodynerus*) *niloticus saudita*; Carpenter and Gadallah, 2020).
Extralimital distribution: None.

***Euodynerus* (*Euodynerus*) *rufinus confictus* Giordani Soika, 1970**
Euodynerus (*Euodynerus*) *rufinus confictus* Giordani Soika, 1970: 156, ♂; "Iran: Haft Tapeh, 300 km. N di Abadan, Rive fiume Dez".
Catalogues with Iranian records: None.
Distribution in Iran: Iran (no specific locality cited) (Giordani Soika, 1970; Gusenleitner, 2013a).

Distribution in the Middle East: Known only from Iran (Giordani Soika, 1970; Gusenleitner, 2013a).
Extralimital distribution: None.

***Euodynerus* (*Euodynerus*) *rufinus rufinus* Blüthgen, 1942**
Euodynerus (*Euodynerus*) *rufinus* Blüthgen, 1942: 303, ♀; Transcaspien [Turkmenistan].
Catalogues with Iranian records: Ebrahimi and Carpenter (2008), Rahmani *et al.* (2020).
Distribution in Iran: Khuzestan (Giordani Soika, 1970 as *Euodynerus rufinus confictus*), Tehran (Ebrahimi and Carpenter, 2008).
Distribution in the Middle East: Egypt (Carpenter and Gadallah, 2020), Iran (see references above, and Gusenleitner, 2013a; Carpenter and Gadallah, 2020), Israel-Palestine (Gusenleitner, 2013a; Carpenter and Gadallah, 2020), United Arab Emirates (Gusenleitner, 2010b; Fateryga and Mokrousov, 2019; Carpenter and Gadallah, 2020).
Extralimital distribution: Armenia, China, Kazakhstan, Mongolia, Turkmenistan, Uzbekistan (Castro and Dvořák, 2010; Carpenter and Gadallah, 2020).

***Euodynerus* (*Euodynerus*) *salzi* (Giordani Soika, 1952)**
Pseudepipona (*Euodynerus*) *salzi* Giordani Soika, 1952: 42, ♂, ♀; "Palestina: Gerusalemme".
Distribution in the Middle East: Egypt (Selis, 2024), Israel-Palestine, Jordan (Giordani Soika, 1952 as *Pseudepipona* (*Euodynerus*) *salzi*; Gusenleitner, 2013a; Carpenter and Gadallah, 2020; Selis, 2024), Saudi Arabia (Guichard, 1986; Gusenleitner, 2013a; Carpenter and Gadallah, 2020; Selis, 2024), United Arab Emirates (Gusenleitner, 2010a; Carpenter and Gadallah, 2020; Selis, 2024).
Extralimital distribution: Chad (Giordani Soika, 1954; Gusenleitner, 2013a; Carpenter and Gadallah, 2020).

***Euodynerus* (*Euodynerus*) *semidantici semidantici* (Giordani Soika, 1952)**
Pseudepipona (*Euodynerus*) *semidantici* Giordani Soika, 1952: 47, ♂; Palestina.
Catalogues with Iranian records: Rahmani *et al.* (2020).
Distribution in Iran: Iran (no specific locality cited) (Gusenleitner, 2011a, 2013a).
Distribution in the Middle East: Iran (Gusenleitner, 2011a, 2013a; Selis *et al.*, 2025), Iraq (Selis *et al.*, 2025), Israel-Palestine (Giordani Soika, 1952 as *Pseudepipona* (*Euodynerus*) *semidantici*; Bytinski-Salz and Gusenleitner, 1971 as *E. semidantici*; Yildirim and Kojima, 1999; Selis *et al.*, 2025), Turkey (Bytinski-Salz and Gusenleitner, 1971; Gusenleitner, 1972; Yildirim and Kojima, 1999; Selis *et al.*, 2025).
Extralimital distribution: Armenia, China, Crimea, Greece, Kazakhstan, Pakistan, Tajikistan, Turkmenistan, Uzbekistan (Castro and Dvořák, 2010).

***Euodynerus* (*Euodynerus*) *semisaecularis macedonicus* Blüthgen, 1951**
Euodynerus (*Euodynerus*) *macedonicus* Blüthgen, 1951a: 172, 195, ♀, ♂; Greece.
Catalogues with Iranian records: Ebrahimi and Carpenter (2008), Rahmani *et al.* (2020).
Distribution in Iran: Alborz, Razavi Khorasan (Giordani Soika, 1970 as *Euodynerus macedonicus* (Blüthgen)).
Distribution in the Middle East: Iran (see references above, and Yildirim and Kojima, 1999), Lebanon, Syria (Yildirim and Kojima, 1999), Turkey (Giordani Soika, 1970 as *E.* (*E.*) *macedonicus*; Gusenleitner, 1972; Yildirim and Kojima, 1999).
Extralimital distribution: Armenia (Giordani Soika, 1970), China (Kurzenko, 1977), Greece (Blüthgen, 1951a).

***Euodynerus* (*Euodynerus*) *semisaecularis semisaecularis* (Dalla Torre, 1889)**
Odynerus humeralis André, 1884: 701, ♀; "Turkestan (Tashkend)" [Uzbekistan].
Odynerus semisaecularis Dalla Torre, 1889: 125, replacement name for *Odynerus humeralis* André, 1884, nec *O. humeralis* Haliday.
Catalogues with Iranian records: Rahmani *et al.* (2020).

Distribution in Iran: Golestan, Guilan, Semnan (Schmid-Egger *et al.*, 2021).

Distribution in the Middle East: Iran (see references above, and Gusenleitner, 2013a; Carpenter and Gadallah, 2020; Fateryga *et al.*, 2021; Fateryga and Proshchalykin, 2024), Lebanon, Syria (Fateryga *et al.*, 2021; Fateryga and Proshchalykin, 2024), Turkey (Yildirim and Özbek, 1996a, 1996b, 1999; Yildirim and Kojima, 1999; Yildirim and Gusenleitner, 2001; Gusentleitner, 2013a; Carpenter and Gadallah, 2020; Fateryga *et al.*, 2021; Fateryga and Proshchalykin, 2024), United Arab Emirates (Gusenleitner, 2010a, 2013a; Carpenter and Gadallah, 2020; Fateryga *et al.*, 2021; Fateryga and Proshchalykin, 2024).

Extralimital distribution: Armenia, China, Crimea, Greece, Kazakhstan, Pakistan, Tajikistan, Turkmenistan, Uzbekistan (Castro and Dvořák, 2009; Carpenter and Gadallah, 2020; Fateryga *et al.*, 2021), Azerbaijan (Fateryga *et al.*, 2021), Kyrgyzstan (Fateryga and Proshchalykin, 2024).

***Euodynerus* (*Euodynerus*) *setosus rubicundus* Gusenleitner, 1994**

Euodynerus setosus rubicundus Gusenleitner, 1994: 307, 315, ♂; "UAE, Oman, W Agran".

Distribution in the Middle East: Known only from United Arab Emirates (Gusenleitner, 1994, 2013a; Howarth and Gillett, 2008; Carpenter and Gadallah, 2020).

Extralimital distribution: None.

***Euodynerus* (*Euodynerus*) *setosus setosus* Gusenleitner in Blüthgen and Gusenleitner, 1970**

Euodynerus (*Euodynerus*) *setosus* Gusenleitner in Blüthgen and Gusenleitner, 1970: 11, ♀, ♂; Iran.

Catalogues with Iranian records: Ebrahimi and Carpenter (2008), Rahmani *et al.* (2020).

Distribution in Iran: Southern Khorasan (Blüthgen and Gusenleitner, 1970).

Distribution in the Middle East: Iran (Blüthgen and Gusenleitner, 1970; Gusenleitner, 2013a; Fateryga and Proshchalykin, 2024).

Extralimital distribution: Afghanistan (in Blüthgen and Gusenleitner, 1970; Gusenleitner, 1973), Turkmenistan, Uzbekistan (Fateryga and Proshchalykin, 2024).

***Euodynerus* (*Euodynerus*) *shirazensis* Giordani Soika, 1970**

Euodynerus shirazensis Giordani Soika, 1970: 162, ♀; Iran.

Catalogues with Iranian records: Ebrahimi and Carpenter (2008), Rahmani *et al.* (2020).

Distribution in Iran: Fars (Giordani Soika, 1970).

Distribution in the Middle East: Known only from Iran (Giordani Soika, 1970; Gusenleitner, 2013a).

Extralimital distribution: None.

***Euodynerus* (*Euodynerus*) *socotrae* (Kohl, 1906)**

Odynerus (*Leiodynerus*) *socotrae* Kohl, 1906: 257, ♂, ♀; Sokotra.

Distribution in the Middle East: Known only from Yemen (including Socotra) (Kohl, 1906 as *Odynerus* (*Leiodynerus*) *socotrae*; Bequaert, 1918 as *O.* (*Rhynchium*) *socotrae*; von Schulthess, 1928 as *O. socotrae*; Giordani Soika, 1934, 1940 as *O.* (*Rhynchium*) *socotrae*, 1974; Carpenter and Gadallah, 2020; Selis, 2024).

Extralimital distribution: None.

***Euodynerus* (*Euodynerus*) *soikai* Guichard, 1986**

Euodynerus soikai Guichard, 1986 [1985]: 202, 228, ♂; "Saudi Arabia: Wadi Maraba below Abha 300 m".

Distribution in the Middle East: Known only from Saudi Arabia (Guichard, 1986; Gusenleitner, 2013a; Carpenter and Gadallah, 2020).

Extralimital distribution: None.

***Euodynerus* (*Euodynerus*) *variegatus kruegeri* (von Schulthess, 1928)**

Odynerus crenatus var. *krügeri* von Schulthess, 1928: 71, 77, ♀, ♂; "Cyrenaica: Bengasi".

Distribution in the Middle East: Egypt (Giordani Soika, 1935 as *Odynerus crenatus* var. *krügeri*; Gusenleitner, 2013a), Israel-Palestine (Gusenleitner, 2013a).

Extralimital distribution: China, Libya (Gusenleitner, 2013a).

***Euodynerus* (*Pareuodynerus*) *breviventris* (Giordani Soika, 1952)**

Pseudepipona (*Euodynerus*) *breviventris* Giordani Soika, 1952: 383, ♂; Turkey.

Catalogues with Iranian records: Ebrahimi and Carpenter (2008), Rahmani *et al.* (2020).

Distribution in Iran: Fars (Giordani Soika, 1970).

Distribution in the Middle East: Iran (Giordani Soika, 1970; Yildirim and Kojima, 1999; Gusenleitner, 2013a), Turkey (Giordani Soika, 1952 as *Pseudepipona* (*Euodynerus*) *breviventris*; Yildirim and Kojima, 1999; Gusenleitner, 2013a).

Extralimital distribution: None.

***Euodynerus* (*Pareuodynerus*) *enodatus* (Kostylev, 1940)**

Odynerus (*Pareuodynerus*) *enodatus* Kostylev, 1940: 32, ♀; Beloutchistan.

Catalogues with Iranian records: Ebrahimi and Carpenter (2008), Rahmani *et al.* (2020).

Distribution in Iran: Sistan & Baluchestan (Hamzavi *et al.*, 2019).

Distribution in the Middle East: Iran (Hamzavi *et al.*, 2019; Kostylev, 1940; Gusenleitner, 2013a).

Extralimital distribution: Afghanistan, Pakistan (Gusenleitner, 2013a).

***Euodynerus* (*Pareuodynerus*) *notatus notatus* (Jurine, 1807)**

Vespa notata Jurine, 1807: 170, ♀; Geneve [Switzerland].

Catalogues with Iranian records: Rahmani *et al.* (2020).

Distribution in Iran: Mazandaran (Castro and Dvořák, 2009 as *Euodynerus notatus notatus*).

Distribution in the Middle East: Iran (Castro and Dvořák, 2009; Fateryga, 2018; Fateryga *et al.*, 2020), Turkey (Gusenleitner, 1966 as *E.* (*Pareuodynerus*) *notatus nigripes*; Giordani Soika, 1970 as *E.* (*P.*) *notatus nigripes*; Yildirim and Özbek, 1996b, 1999; Gusenleitner, 1997c, 2013a; Yildirim, 1998 as *E.* (*P.*) *notatus nigripes*; Yildirim and Gusenleitner, 2004; Fateryga, 2018; Fateryga *et al.*, 2020).

Extralimital distribution: Caucasus, China, Europe, Kazakhstan, Korea, Mongolia, North Africa, Russia (Fateryga, 2017b, 2018; Fateryga *et al.*, 2020), Italy (Selis, 2023), Slovenia (Gogala, 2022).

***Euodynerus* (*Pareuodynerus*) *posticus* (Herrich-Schäffer, 1841)**

Odynerus posticus Herrich-Schäffer, 1841: 176, ♀; type locality not stated.

Catalogues with Iranian records: Ebrahimi and Carpenter (2008), Rahmani *et al.* (2020).

Distribution in Iran: Alborz, Lorestan, Sistan & Baluchestan (Ebrahimi and Carpenter, 2008; Modarres Awal, 2012), Guilan (Blüthgen and Gusenleitner, 1970 as *Euodynerus posticus punctatissimus* (Giordani Soika, 1952); Ebrahimi and Carpenter, 2008; Modarres Awal, 2012), Mazandaran (Ebrahimi and Carpenter, 2008; Modarres Awal, 2012; Schmid-Egger *et al.*, 2021), Zanjan (Abbasi *et al.*, 2008a, 2008b; Pashaei Rad *et al.*, 2010).

Distribution in the Middle East: Iran (see references above, and Fateryga, 2017b, 2018), Israel-Palestine (Gusenleitner, 2013a; Fateryga, 2017b, 2018), Turkey (Giordani Soika, 1952, 1970 both as *Euodynerus* (*Pareuodynerus*) *postica punctatissima*; Gusenleitner, 1966, 1967 both as *E.* (*P.*) *p. punctatissima*, 1997c as *E. p. punctatissimus*, 2013a; Tüzün and Tanyolag, 1988 as *Odynerus* (*Lionotus*) *innumirabilis*; Yildirim and Özbek, 1996b as *E.* (*P.*) *p. punctatissima*, 1996a as *E. p. punctatissimus*; Yildirim, 1998 as *O.* (*L.*) *innumirabilis*; Yildirim and Özbek, 1999; Yildirim and Kojima, 1999; Gusenleitner, 2007 as *Euodynerus posticus*; Blüthgen and Gusenleitner, 1970 as *E. posticus punctatissima*; Yildirim and Gusenleitner, 2007, 2012; Gusenleitner, 2013a; Fateryga, 2017b, 2018; Yildirim and Tezcan, 2018 as *E. posticus*; Fateryga *et al.*, 2019).

Extralimital distribution: Armenia, Azerbaijan (Fateryga *et al.*, 2019), Caucasus, China, Europe, Georgia, Japan, Kazakhstan, Korea, Mongolia, North Africa, Russia (Fateryga, 2017b, 2018; Fateryga *et al.*, 2019), Italy (Selis, 2023).

***Euodynerus* (*Pareuodynerus*) *quadrifasciatus quadrifasciatus* (Fabricius, 1793)**

Vespa quadrifasciata Fabricius, 1793: 266, ♂; "Dania".

Distribution in the Middle East: Turkey (Giordani Soika, 1970; Yildirim and Özbek, 1996b; Gusenleitner, 1997c, 2013a; Yildirim and Gusenleitner, 2004; Fateryga, 2017b; Fateryga *et al.*, 2020).

Extralimital distribution: China, Europe, Kazakhstan, Korea, Mongolia, North Africa, Russia (Fateryga, 2017b; Fateryga *et al.*, 2020), Japan (Fateryga *et al.*, 2020), Slovenia (Gogala, 2022).

***Euodynerus (Pareuodynerus) quadrifasciatus rufipes* Gusenleitner, 1984**

Euodynerus (Pareuodynerus) quadrifasciatus rufipes Gusenleitner, 1984: 165, 168, ♂, ♀; "Türkei, Horasan, Arastal".

Distribution in the Middle East: Known only from Turkey (Gusenleitner, 1984, 2013a; Yildirim and Özbek, 1996a, 1999, 1996b; Yildirim and Kojima, 1999; Yildirim and Gusenleitner, 2004).

Extralimital distribution: None.

***Euodynerus (Euodynerus) velutinus* Blüthgen, 1951**

Euodynerus velutinus Blüthgen, 1951a: 172, ♀, ♂; "Süd-Morea; Peloponnes; Attika".

Distribution in the Middle East: Cyprus, Iraq (Selis *et al.*, 2025), Lebanon, Syria (Giordani Soika, 1970; van der Vecht and Fischer, 1972; Gusenleitner, 2013a, all as *Euodynerus mavromoustakisi* – synonymy; Selis *et al.*, 2025), Turkey (Giordani Soika, 1970; Gusenleitner, 1997c, 2013a; Yildirim and Özbek, 1999; Yildirim and Kojima, 1999; Gusenleitner, 2013a, as *E. mavromoustakisi*; Fateryga, 2018; Fateryga *et al.*, 2020; Fateryga *et al.*, 2021; Selis *et al.*, 2025).

Extralimital distribution: Armenia, Greece, North Macedonia (Giordani Soika, 1970; Selis *et al.*, 2024, 2025), Azerbaijan (Fateryga *et al.*, 2021; Selis *et al.*, 2025), Russia, Southern Europe (Fateryga, 2018; Fateryga *et al.*, 2020; Fateryga *et al.*, 2021; Selis *et al.*, 2025).

Comments: Recently, *Euodynerus (Euodynerus) mavromoustakisi* Giordani Soika, 1970 synonymised with *E. velutinus* by Selis *et al.* (2025).

Genus *Eustenancistrocerus* Blüthgen, 1938

Eustenancistrocerus Blüthgen, 1938: 443, 460. Type species: *Odynerus blanchardianus* Saussure, 1855, by original designation.

***Eustenancistrocerus (Eustenancistrocerus) askhabadensis askhabadensis* (Radoszkowski, 1886)**

Odynerus (Lionotus) askhabadensis Radoszkowski, 1886: 47, ♀, ♂; Askhabad.

Catalogues with Iranian records: Ebrahimi and Carpenter (2008), Rahmani *et al.* (2020).

Distribution in Iran: Alborz (Giordani Soika, 1970; Ebrahimi and Carpenter, 2008; Modarres Awal, 2012), Markazi (Ebrahimi and Carpenter, 2008), Golestan, Sistan & Baluchestan (Ebrahimi and Carpenter, 2008; Modarres Awal, 2012).

Distribution in the Middle East: Iran (see references above, and Gusenleitner, 2013a; Fateryga *et al.*, 2023), Israel-Palestine, Jordan, Syria, Turkey (Fateryga *et al.*, 2023).

Extralimital distribution: Azerbaijan, China, Kazakhstan, Mongolia (Fateryga *et al.*, 2023), Pakistan, Turkmenistan (Gusenleitner, 2013a), Uzbekistan (Giordani Soika, 1943; Fateryga *et al.*, 2023).

***Eustenancistrocerus (Eustenancistrocerus) inconstans* (de Saussure, 1863)**

Odynerus (Stenancistrocerus) inconstans de Saussure, 1863: 217, ♀; Abyssinie.

Catalogues with Iranian records: Ebrahimi and Carpenter (2008), Rahmani *et al.* (2020).

Distribution in Iran: Mazandaran (Ebrahimi and Carpenter, 2008), Tehran (Giordani Soika, 1970).

Distribution in the Middle East: Egypt (Walker, 1871 as *Odynerus flavus*; Giordani Soika, 1935 as *Ancistrocerus inconstans*; Zalat *et al.*, 1993), Iran (see references above, and Zalat *et al.*, 1993 as *Eustenancistrocerus inconstans*; Gusenleitner, 2013a), Israel-Palestine (Gusenleitner, 2006a, 2013a), Jordan (Giordani Soika, 1970; Gusenleitner, 2006a, 2013a), Oman (Gusenleitner, 2004, 2013a; Carpenter and Gadallah, 2020), Saudi Arabia (Guichard, 1986; Gadallah, 2003; Gusenleitner, 2013a; Carpenter and Gadallah, 2020), United Arab Emirates (Gusenleitner, 2010a, 2012, 2013a; Carpenter and Gadallah, 2020), Yemen (Gusenleitner, 2007, 2013a; Carpenter and Gadallah, 2020).

Extralimital distribution: Central Asia, Ethiopia, North Africa (Gusenleitner, 2006a, 2013a).

***Eustenancistrocerus (Eustenancistrocerus) jerichoensis iconius* Blüthgen, 1957**

Eustenancistrocerus jerichoensis iconius Blüthgen, 1957: 164, ♀; Turkey.

Distribution in the Middle East: Known only from Turkey (Blüthgen, 1957, 1961; Yildirim and Özbek, 1996a, 1996b as *E. j. iconius*; Yildirim and Kojima, 1999; Gusenleitner, 2013a).
Extralimital distribution: None.

***Eustenancistrocerus* (*Eustenancistrocerus*) *jerichoensis jerichoensis* (von Schulthess, 1928)**
Odynerus jerichoensis von Schulthess, 1928: 73–74, ♀, ♂; Jericho [Israel].
Catalogues with Iranian records: Ebrahimi and Carpenter (2008), Rahmani *et al.* (2020).
Distribution in Iran: Alborz (Giordani Soika, 1970), Kerman (Schmid-Egger *et al.*, 2021), Sistan & Baluchestan (Hamzavi *et al.*, 2019).
Distribution in the Middle East: Iran (see references above, and Gusenleitner, 2013a), Israel-Palestine (von Schulthess, 1928 as *Odynerus jerichoensis*; Fateryga and Mokrousov, 2019), Jordan (Castro and Dvořák, 2010; Fateryga and Mokrousov, 2019), Syria (Giordani Soika, 1970; Gusenleitner, 2013a; Fateryga and Mokrousov, 2019), Turkey (Gusenleitner, 1966; Yildirim and Özbek, 1996b as *E. jerichoensis*; Yildirim and Kojima, 1999; Fateryga and Mokrousov, 2019; Kaplan and Efil, 2025).
Extralimital distribution: Azerbaijan, Russia (Fateryga and Mokrousov, 2019).

***Eustenancistrocerus* (*Eustenancistrocerus*) *malus* Giordani Soika, 1970**
Eustenancistrocerus (*Eustenancistrocerus*) *malus* Giordani Soika, 1970: 96, ♂; Iran.
Catalogues with Iranian records: Ebrahimi and Carpenter (2008), Rahmani *et al.* (2020).
Distribution in Iran: Fars (Giordani Soika, 1970; Ebrahimi and Carpenter; 2008; Modarres Awal, 2012), Tehran (Giordani Soika, 1970).
Distribution in the Middle East: Known only from Iran (see references above, and Gusenleitner, 2013a).
Extralimital distribution: None.

***Eustenancistrocerus* (*Eustenancistrocerus*) *spinosissimus* Gusenleitner, 2000c**
Eustenancistrocerus (*Eustenancistrocerus*) *spinosissimus* Gusenleitner, 2000c: 934, ♂; Iran.
Catalogues with Iranian records: Ebrahimi and Carpenter (2008), Rahmani *et al.* (2020).
Distribution in Iran: Yazd (Gusenleitner, 2000c; Modarres Awal, 2012).
Distribution in the Middle East: Known only from Iran (see references above, and Gusenleitner, 2013a).
Extralimital distribution: None.

***Eustenancistrocerus* (*Eustenancistricerus*) *sulcithorax* Giordani Soika, 1970**
Eustenancistrocerus (*Eustenancistrocerus*) *sulcithorax* Giordani Soika, 1970: 93, ♀, ♂; Iran.
Catalogues with Iranian records: Ebrahimi and Carpenter (2008), Rahmani *et al.* (2020).
Distribution in Iran: Fars, Khuzestan, Tehran (Giordani Soika, 1970).
Distribution in the Middle East: Known only from Iran (Giordani Soika, 1970; Gusenleitner, 2013a).
Extralimital distribution: None.

***Eustenancistrocerus* (*Eustenancistrocerus*) *tegularis* (Morawitz, 1885)**
Lionotus tegularis Morawitz, 1885: 165, ♀, ♂; Trauscaucasia [Azerbaijan].
Catalogues with Iranian records: Ebrahimi and Carpenter (2008) as *Eustenancistrocerus israelensis* Giordani Soika, 1952), Rahmani *et al.* (2020).
Distribution in Iran: Alborz (Giordani Soika, 1970; Pashaei Rad *et al.*, 2006 both as *Eustenancistrocerus israelensis* Giordani Soika, 1952), Golestan (Ebrahimi and Carpenter, 2008), Khorasan (Kohl and Handlirsch, 1889 as *Lionotus tegularis* Morawitz, 1885), Mazandaran (Pashaei Rad *et al.*, 2006 as *E. israelensis*), Northern Khorasan, Tehran (Pashaei Rad *et al.*, 2006), Razavi Khorasan (Giordani Soika, 1970; Pashaei Rad *et al.*, 2006 both as *E. israelensis*), Zanjan (Pashaei Rad *et al.*, 2006; Abbasi *et al.*, 2008a, 2008b all as *E. israelensis*).
Distribution in the Middle East: Iran (see references above; Gusenleitner, 2013a as *E. israelensis*; Fateryga *et al.*, 2023; Fateryga and Proshchalykin, 2024), Israel-Palestine (Giordani Soika, 1952

as *E. israelensis*; Bytinski-Salz and Gusenleitner, 1971 as *E. israelensis*; Gusenleitner, 2013a; Fateryga *et al.*, 2019, 2023; Fateryga and Proshchalykin, 2024), Syria (Giordani Soika, 1970 as *E. israelensis*; Gusenleitner, 2013a; Fateryga *et al.*, 2019, 2023; Fateryga and Proshchalykin, 2024), Turkey (Gusenleitner, 1966, 1967; Giordani Soika, 1970 as *E. israelensis*; Yildirim and Özbek, 1996a, 1996b, 1999; Yildirim and Kojima, 1999; Yildirim and Gusenleitner, 2004, 2009 both as *E. israelensisi*; Fateryga *et al.*, 2019, 2023; Fateryga and Proshchalykin, 2024).

Extralimital distribution: Armenia, Azerbaijan, Europe; Kazakhstan (Fateryga *et al.*, 2019), Kyrgyzstan (Fateryga and Proshchalykin, 2024), Bulgaria, Greece, Uzbekistan (Fateryga *et al.*, 2023).

***Eustenancistrocerus* (*Hemistenancistrocerus*) *pharao* (de Saussure, 1863)**

Odynerus (*Ancistrocerus*) *pharao* de Saussure, 1863: 212, ♀, ♂; Egypt.

Distribution in the Middle East: Egypt (de Saussure, 1863 as *Odynerus* (*Ancistrocerus*) *pharao*; Giordani Soika, 1935 as *Ancistrocerus pharaoh*; Zalat *et al.*, 1993 as *Eustenancistrocerus pharo*; Gusenleitner, 2013a as both *Pseudonortonia parvula* and *E.* (*H.*) *pharao*).

Extralimital distribution: Algeria, Burkina Faso, Eritrea, Ethiopia, Niger, Senegal (Gusenleitner, 2013a).

***Eustenancistrocerus* (*Parastenancistrocerus*) *amadanensis amadanensis* (de Saussure, 1855)**

Odynerus (*Ancistrocerus*) *amadanensis* de Saussure, 1855: 214, ♂; Amadan, en Perse [Iran].

Catalogues with Iranian records: Ebrahimi and Carpenter (2008), Rahmani *et al.* (2020).

Distribution in Iran: Alborz (Giordani Soika, 1970 as both *Odynerus amadanensis* de Saussure, 1855 and *Eustenancistrocerus transitorius* (Morawitz, 1867)), Bushehr (Giordani Soika, 1943 as *Ancistrocerus iranensis* Giordani Soika, 1943), East Azarbaijan, Golestan, Sistan & Baluchestan (Ebrahimi and Carpenter, 2008; Modarres Awal, 2012), Fars (Giordani Soika, 1970 as both *O. amadanensis* and *E. transitorius*; Ebrahimi and Carpenter, 2008; Modarres Awal, 2012), Kerman (Schmid-Egger *et al.*, 2021), Khuzestan (de Saussure, 1855; Giordani Soika, 1970 as both *O. amadanensis* and *E. transitorius*; Gusenleitner, 2000b; Ebrahimi and Carpenter, 2008; Modarres Awal, 2012), Markazi (Ebrahimi and Carpenter, 2008), Razavi Khorasan (Giordani Soika, 1970 as both *O. amadanensis* and *E. transitorius*; Ebrahimi and Carpenter, 2008; Modarres Awal, 2012), Tehran (Giordani Soika, 1970 as both *O. amadanensis* and *E. transitorius*; Ebrahimi and Carpenter, 2008; Modarres Awal, 2012), Zanjan (Abbasi *et al.*, 2008a, 2008b; Pashaei Rad *et al.*, 2010).

Distribution in the Middle East: Cyprus (Fateryga and Proshchalykin, 2024; Selis *et al.*, 2025), Iran (see references above, and Gusenleitner, 2013a; Fateryga, 2018; Fateryga and Proshchalykin, 2024; Selis *et al.*, 2025), Iraq (de Saussure, 1863 as *Odynerus* (*Ancistrocerus*) *turca*; Morice, 1921 as *Odynerus transitorius*; Giordani Soika, 1970; Gusenleitner, 2013a; Fateryga, 2018; Carpenter and Gadallah, 2020; Fateryga and Proshchalykin, 2024; Selis *et al.*, 2025), Israel-Palestine (Giordani Soika, 1952 as *Eustenancistrocerus lobatus flavellus*; Gusenleitner, 2013a; Fateryga, 2018; Carpenter and Gadallah, 2020; Fateryga and Proshchalykin, 2024; Selis *et al.*, 2025), Saudi Arabia (Guichard, 1986; Yildirim and Kojima, 1999; Fateryga, 2018; Carpenter and Gadallah, 2020; Fateryga and Proshchalykin, 2024; Selis *et al.*, 2025), Syria (Giordani Soika, 1970; Fateryga, 2018; Carpenter and Gadallah, 2020; Fateryga and Proshchalykin, 2024; Selis *et al.*, 2025), Turkey (de Saussure, 1863 as *O.* (*Ancistrocerus*) *turca*; Giordani Soika, 1952 as *Eustenancistrocerus transitorius hoberlandti*, 1970 as *E. transitorius*; Gusenleitner, 1966, 1967 both as *Parastenancistrocerus amadanensis*; Yildirim and Özbek, 1996a, 1996b as *E. amadanensis amadanensis*; Yildirim, 1998; Yildirim and Kojima, 1999; Yildirim and Gusenleitner, 2001, 2004, 2007, 2012; Gusenleitner, 2006a, 2007; Fateryga, 2018; Yildirim and Tezcan, 2018; Carpenter and Gadallah, 2020; Fateryga and Proshchalykin, 2024; Selis *et al.*, 2025).

Extralimital distribution: Azerbaijan, China, Kazakhstan, Pakistan, Russia (Fateryga, 2018; Fateryga *et al.*, 2019; Selis *et al.*, 2025), Greece, Hungary, Italy, Kyrgyzstan, Portugal, Spain, Tajikistan (Fateryga and Proshchalykin, 2024), Uzbekistan (Fateryga and Proshchalykin, 2024; Selis *et al.*, 2025).

***Eustenancistrocerus* (*Parastenancistrocerus*) *iranicus* Gusenleitner, 2013a**

Eustenancistrocerus (*Parastenancistrocerus*) *iranicus* Gusenleitner, 2013a [in: Gusenleitner *et al.*, 2013]: 110, ♂; Iran.

Catalogues with Iranian records: Rahmani *et al.* (2020).

Distribution in Iran: Fars (Gusenleitner *et al.*, 2013; Fateryga *et al.*, 2025).

Distribution in the Middle East: Azerbaijan (Fateryga *et al.*, 2025), Iran (Gusenleitner, 2013a; Gusenleitner *et al.*, 2013; Fateryga *et al.*, 2025).

Extralimital distribution: None.

***Eustenancistrocerus* (*Parancistrocerus*) *khuzestanicus* Giordani Soika, 1970**

Eustenancistrocerus khuzestanicus Giordani Soika, 1970: 97, ♀, ♂; Iran.

Catalogues with Iranian records: Ebrahimi and Carpenter (2008), Rahmani *et al.* (2020).

Distribution in Iran: Kerman (Schmid-Egger *et al.*, 2021), Khuzestan (Giordani Soika, 1970; Ebrahimi and Carpenter, 2008; Modarres Awal, 2012).

Distribution in the Middle East: Iran (see references above, and Gusenleitner, 2013a).

Extralimital distribution: Pakistan (Gusenleitner, 2006c).

***Eustenancistrocerus* (*Parancistrocerus*) *nilensis* (Giordani Soika, 1935)**

Ancistrocerus nilensis Giordani Soika, 1935: 170, 174, ♂; Egypt.

Distribution in the Middle East: Known only from Egypt (Giordani Soika, 1935, 1973 both as *Ancistrocerus nilensisi*; Zalat *et al.*, 1993; Gusenleitner, 2013a).

Extralimital distribution: None.

***Eustenancistrocerus* sp.**

Distribution in Iran: Southern Khorasan (Blüthgen and Gusenleitner, 1970).

Genus *Extraepipona* Gusenleitner, 2014

Extraepipona Gusenleitner, 2014: 537. Type species: *Extraepipona occulta* Gusenleitner, 2014, by original designation.

***Extraepipona occulta* Gusenleitner, 2014**

Extraepipona occulta Gusenleitner, 2014: 537, ♀; Iran.

Catalogues with Iranian records: Rahmani *et al.* (2020).

Distribution in Iran: Fars (Gusenleitner, 2014).

Distribution in the Middle East: Known only from Iran (Gusenleitner, 2014).

Extralimital distribution: None.

Genus *Gymnomerus* Blüthgen, 1938

Gymnomerus Blüthgen, 1938 [1937]: 286. Type species: *Odynerus laevipes* Shuckard, 1837, by original designation.

***Gymnomerus laevipes laevipes* (Shuckard, 1837)**

Odynerus laevipes Shuckard, 1837: 495, ♀, ♂; "vicinity of Blackwater, upon the borders of Hampshire" [Great Britain].

Catalogues with Iranian records: Ebrahimi and Carpenter (2008), Rahmani *et al.* (2020).

Distribution in Iran: Alborz (Ebrahimi and Carpenter, 2008).

Distribution in the Middle East: Iran (Ebrahimi and Carpenter, 2008; Fateryga, 2018; Fateryga *et al.*, 2021), Turkey (Gusenleitner, 1966, 1967, 2013a; Giordani Soika, 1970; Yildirim and Özbek, 1996a, 1999, 1996b; Madl, 1997; Yildirim and Kojima, 1999; Yildirim and Gusenleitner, 2009, 2012; Fateryga, 2018; Kemal and Koçak, 2018; Yildirim and Tezcan, 2018; Fateryga *et al.*, 2021; Kaplan and Efil, 2025).

Extralimital distribution: Armenia, Azerbaijan, Georgia, Caucasus, Europe, Kazakhstan, Mongolia, North Africa, Russia, Ukraine (Fateryga, 2018; Fateryga *et al.*, 2021), Italy (Selis, 2023), Slovenia (Gogala, 2022).

Genus *Hemipterochilus* Ferton, 1909

Hemipterochilus Ferton, 1909 [1908]: 572. Type species: *Odynerus terricola* Mocsáry, 1883 (= *Hemipterochilus bembeciformis terricola*), by monotypy.

***Hemipterochilus aberrans* (Morawitz, 1885)**

Pterochilus aberrans Morawitz, 1885: 145, ♀; Azerbaijan.

Distribution in the Middle East: Lebanon (Yildirim and Kojima, 1999; Gusenleitner, 2013a; Fateryga *et al*., 2025), Syria (Giordani Soika, 1970; Yildirim and Kojima, 1999; Gusenleitner, 2013a; Fateryga *et al*., 2025), Turkey (Blüthgen, 1951a as *Pterochilus* (*Pseudopterochilus*) *aberrans*; Gusenleitner, 1966, 2007, 2013a; Yildirim and Özbek, 1996a, 1996b, 1999; Yildirim, 1998; Yildirim and Kojima, 1999; Yildirim and Gusenleitner, 2007; Fateryga *et al*., 2025).

Extralimital distribution: Armenia (Kostylev, 1929 as *Pterochilus aberrans*; Fateryga *et al*., 2025), Azerbaijan (Morawitz, 1885 as *P. aberrans*; Fateryga *et al*., 2025), Greece (Gusenleitner, 2013a; Fateryga *et al*., 2025).

***Hemipterochilus bembeciformis bembeciformis* (Morawitz, 1867)**

Odynerus (*Pterochilus*) *bembeciformis* Morawitz, 1867: 139, ♀, ♂; Russia.

Distribution in the Middle East: Turkey (Giordani Soika, 1970; Yildirim and Özbek, 1996a, 1996b, 1999; Yildirim and Kojima, 1999; Gusenleitner, 2013a; Fateryga, 2017a, 2017b; Fateryga, 2018; Yildirim and Tezcan, 2018; Fateryga *et al*., 2021).

Extralimital distribution: Armenia, France, Greece, Russia, Turkmenistan, Ukraine, Uzbekistan (Gusenleitner, 2013a), Caucasus, Kazakhstan, Russia (Fateryga, 2017a, 2017b, 2018; Fateryga *et al*., 2021), Azerbaijan (Fateryga *et al*., 2021), Slovenia (Gogala, 2022).

***Hemipterochilus bembeciformis terricola* (Mocsáry, 1883)**

Odynerus (*Epipona*) *terricola* Mocsáry, 1883: 52, ♀, ♂; "in Hungaria centrali ad Budapestinum et in Slavonia ad Daljam".

Distribution in the Middle East: Turkey (Giordani Soika, 1970; Yildirim and Kojima, 1999).

Extralimital distribution: Austria (Gusenleitner, 2008), Hungary (Mocsáry, 1883 as *Odynerus* (*Epipona*) *terricola*), Greece, Italy (Costa, 1888 as *Odynerus sinuatofasciatus*).

***Hemipterochilus bicoloricornis* (Giordani Soika, 1952)**

Pterocheilus bicoloricornis Giordani Soika, 1952: 59, ♂, ♀; "Palestina; Gerusalemme".

Distribution in the Middle East: Israel-Palestine (Giordani Soika, 1952 as *Pterocheilus bicoloricornis*; Bytinski-Salz and Gusenleitner, 1972; Gusenleitner, 2001b, 2013a), Jordan (Gusenleitner, 2001b), Turkey (Castro and Dvořák, 2009).

Extralimital distribution: None.

***Hemipterochilus hohlbecki* (Kostylev, 1935)**

Pterochilus hohlbecki Kostylev, 1935: 120, ♀; Transkaspien [Turkmenistan].

Catalogues with Iranian records: Ebrahimi and Carpenter (2008), Rahmani *et al*. (2020).

Distribution in Iran: Alborz (Giordani Soika, 1970).

Distribution in the Middle East: Iran (Giordani Soika, 1970; Gusenleitner, 2013a).

Extralimital distribution: Afghanistan, Turkmenistan (Gusenleitner, 2013a).

***Hemipterochilus punctiventris punctiventris* (Morawitz, 1885)**

Pterochilus punctiventris Morawitz, 1885: 143, ♀; Iran.

Catalogues with Iranian records: Ebrahimi and Carpenter (2008), Rahmani *et al*. (2020).

Distribution in Iran: Semnan (Morawitz, 1885), Sistan & Baluchestan (Blüthgen and Gusenleitner, 1970 as *Pterochilus punctiventris* Morawitz, 1885).

Distribution in the Middle East: Known only from Iran (see references above, and Gusenleitner, 2013a).

Extralimital distribution: None.

***Hemipterochilus punctiventris subquadricolor* Gusenleitner in Blüthgen and Gusenleitner, 1970**

Hemipterochilus punctiventris subquadricolor Gusenleitner, 1970 [in Blüthgen and Gusenleitner, 1970]: 3, ♀; Iran.

Distribution in Iran: Sistan & Baluchestan (Blüthgen and Gusenleitner, 1970).

Distribution in the Middle East: Known only from Iran (Blüthgen and Gusenleitner, 1970; Gusenleitner, 2013a).

Extralimital distribution: None.

***Hemipterochilus rubrosignatus* (André, 1884)**

Odynerus rubrosignatus André, 1884: 742, ♀; "Turkesta (Tachkend)".

Catalogues with Iranian records: None.

Distribution in Iran: Fars (Schmid-Egger *et al.*, 2021).

Distribution in the Middle East: Iran (Schmid-Egger *et al.*, 2021).

Extralimital distribution: Turkmenistan (André, 1884 as *Odynerus rubrosignatus*).

***Hemipterochilus simplex* Gusenleitner, 2000c**

Hemipterochilus simplex Gusenleitner, 2000c: 927, ♀, ♂; Turkey.

Catalogues with Iranian records: Ebrahimi and Carpenter (2008), Rahmani *et al.* (2020).

Distribution in Iran: Hamadan (Gusenleitner, 2000c).

Distribution in the Middle East: Iran (Gusenleitner, 2000c, 2013a), Turkey (Gusenleitner, 2000c, 2013a; Kemal and Koçak, 2018).

Extralimital distribution: None.

Genus *Ischnogasteroides* Magretti, 1884

Ischnogasteroides Magretti, 1884 [1883]: 603. Type species: *Ischnogasteroides flavus* Magretti, 1884 (= *Ischnogasteroides leptogaster flavus*), by monotypy.

***Ischnogasteroides annae* (Kostylev, 1939)**

Eumenes (*Ischnogasteroides*) *annae* Kostylev, 1939: 163, 168, sex not stated; "Arabie du Sud".

Distribution in the Middle East: Egypt (Zalat *et al.*, 1993; Carpenter and Gadallah, 2020), Saudi Arabia (Zalat *et al.*, 1993; Gusenleitner, 2013a), Yemen (Kostylev, 1939 as *Eumenes* (*Ischnogasteroides*) *annae*; Zalat *et al.*, 1993; Carpenter and Gadallah, 2020).

Extralimital distribution: None.

Comments: The records for Saudi Arabia may be erroneous, as from the French summary atating «Arabie du Sud», despite the holotype being given in the Russian text as from Sanaa (Yemen) (Carpenter and Gadallah, 2020).

***Ischnogasteroides carinatus* (Kostylev, 1939)**

Eumenes (*Ischnogasteroides*) *carinatus* Kostylev, 1939: 165, 168, ♂; "Arabie du Sud".

Distribution in the Middle East: Saudi Arabia (Gusenleitner, 2013a; Carpenter and Gadallah, 2020), Yemen (Kostylev, 1939 as *Eumenes* (*Ischnogasteroides*) *carinatus*; Carpenter and Gadallah, 2020).

Extralimital distribution: None.

Remarks: The record for Saudi Arabia may be incorrect, resulting from the French summary stating «Arabie du Sud», despite the holotype being given in Russian text as from Sanna, Yemen (Carpenter and Gadallah, 2020).

***Ischnogasteroides leptogaster* (Walker, 1871)**

Eumenes leptogaster Walker, 1871: 30, ♂; "Wâdy Ferran".

Distribution in the Middle East: Egypt (Walker, 1871 as *Eumenes leptogaster*; Giordani Soika, 1935 as *E. leptogaster*; Zalat *et al.*, 1993 both as *Ischnogasteroides leptogaster* and as *I. l. flavus*; Gusenleitner, 2013a), Israel-Palestine (Giordani Soika, 1952 as *E. flavus leptogaster*; Gusenleitner, 2013a), Oman (Gusenleitner, 2004, 2004, 2012, 2013a; Monks *et al.*, 2019; Carpenter and Gadallah, 2020), Saudi Arabia (Guichard, 1986; Gusenleitner, 2013a; Carpenter and Gadallah,

2020), United Arab Emirates (Gusenleitner, 2010a; Carpenter and Gadallah, 2020), Yemen (Gusenleitner, 2007; Carpenter and Gadallah, 2020).
Extralimital distribution: None.

***Ischnogasteroides picteti picteti* (de Saussure, 1852)**
Eumenes picteti de Saussure, 1852: 67, ♀, ♂; "Le midi de la France. Montpellier".
Distribution in the Middle East: Tukey (Tüzün and Tanyolag, 1988 as *Eumenes picteti*; Yildirim and Kojima, 1999).
Extralimital distribution: France, Italy, Portugal, Spain (van der Vecht and Fischer, 1972).

***Ischnogasteroides picteti tenuis* (Morawitz, 1888)**
Eumenes (?Ischnogasteroides) tenuis Morawitz, 1888: 261, ♂; "In territorio transcaspico (bei Kukulàbi)".
Catalogues with Iranian records: Ebrahimi and Carpenter (2008), Rahmani *et al.* (2020).
Distribution in Iran: Mazandaran (Lehr *et al.*, 2007 as *Ischnogasteroides picteti picteti* – misidentification), Semnan (Ghahari *et al.*, 2007 as *I. p. picteti* – misidentification), Sistan & Baluchestan (Ebrahimi and Carpenter, 2008).
Distribution in the Middle East: Iran (see references above, and Gusenleitner, 2013a; Fateryga *et al.*, 2025), Turkey (Gusenleitner, 1988, 2013a; Yildirim and Özbek, 1999; Yildirim and Kojima, 1999; Fateryga *et al.*, 2025).
Extralimital distribution: Azerbaijan, Mongolia (Fateryga *et al.*, 2025), Kazakhstan, Turkmenistan (van der Vecht and Fischer, 1972; Fateryga *et al.*, 2025), Kyrgyzstan (Gusenleitner, 2011b; Fateryga *et al.*, 2025).
Comments: *Pogonosoma maroccanum* (Fabricius) and *Pycnopogon leucostomus* Engel (Diptera: Asilidae) were recorded as the predators of *Ischnogasteroides picteti* by Ghahari *et al.* (2007) and Lehr *et al.* (2007), respectively.

***Ischnogasteroides tenuissimus* (Giordani Soika, 1941)**
Eumenes (Ischnogasteroides) gracillimus Giordani Soika, 1941: 97, 98, ♂; [Homonym: *Ischnogasteroides tenuissimus* Giordani Soika is a replacement name for *Eumenes gracillimus* Giordani Soika] "Alto Egitto: Gabal Elba Wadi Aideb".
Distribution in the Middle East: Egypt (Giordani Soika, 1935 as *Eumenes leptogasteri*, 1941 as *E. (Ischnogasteroides) gracillimus*; Zalat *et al.*, 1993; Gusenleitner, 2013a; Carpenter and Gadallah, 2020), Israel-Palestine (Gusenleitner, 2013a; Carpenter and Gadallah, 2020), Oman (Gusenleitner, 2004, 2013a; Carpenter and Gadallah, 2020), Saudi Arabia (Guichard, 1986; Gusenleitner, 2013a; Carpenter and Gadallah, 2020), Yemen (Gusenleitner, 2007, 2013a; Carpenter and Gadallah, 2020).
Extralimital distribution: None.

***Ischnogasteroides zarudnyi* (Kostylev, 1939)**
Labus zarudnyi Kostylev, 1939: 166, ♀; Iran.
Catalogues with Iranian records: Ebrahimi and Carpenter (2008), Rahmani *et al.* (2020).
Distribution in Iran: Kerman (Kostylev 1939 as *Eumenes zarudnyi*), Markazi (Ebrahimi and Carpenter, 2008; Modarres Awal, 2012), Southern Khorasan (Blüthgen and Gusenleitner, 1970).
Distribution in the Middle East: Iran (see references above, and Yildirim and Kojima, 1999; Gusenleitner, 2013a), Turkey (Gusenleitner, 1988, 2013a; Yildirim and Kojima, 1999).
Extralimital distribution: Afghanistan (Gusenleitner, 2013a).

***Ischnogasteroides* sp.**
Distribution in Iran: Sistan & Baluchestan (Blüthgen and Gusenleitner, 1970).

Genus *Jucancistrocerus* Blüthgen, 1938
Jucancistrocerus Blüthgen, 1938: 442, 460. Type species: *Odynerus jucundus* Mocsáry, 1883, by monotypy.

***Jucancistrocerus* (*Eremodynerus*) *atrofasciatus* (Morawitz, 1885)**
Lionotus atrofasciatus Morawitz, 1885: 162, ♀, ♂; China montibus Alaschanensibus.
Catalogues with Iranian records: Rahmani *et al.* (2020).
Distribution in Iran: Kerman (Schmid-Egger *et al.*, 2021).
Distribution in the Middle East: Iran (Gusenleitner, 2013a; Schmid-Egger *et al.*, 2021).
Extralimital distribution: Kyrgyzstan, Mongolia (Gusenleitner, 2013a), China (Tan *et al.*, 2018).

***Jucancistrocerus* (*Jucancistrocerus*) *caspicus* Giordani Soika, 1970**
Jucancistrocerus caspicus Giordani Soika, 1970: 102, ♀, ♂; Iran.
Catalogues with Iranian records: Ebrahimi and Carpenter (2008), Rahmani *et al.* (2020).
Distribution in Iran: Alborz (Giordani Soika, 1970), Kerman (Schmid-Egger *et al.*, 2021).
Distribution in the Middle East: Iran (see references above, and Gusenleitner, 2013a; Fateryga, 2017a, 2017b), Turkey (Yildirim and Gusenleitner, 2001; Fateryga, 2017a, 2017b).
Extralimital distribution: Armenia, Azerbaijan, Georgia, Kazakhstan, Russia, Ukraine (Fateryga, 2010, 2017a).

***Jucancistrocerus* (*Jucancistrocerus*) *citreodecoratus* Giordani Soika, 1970**
Jucancistrocerus citreodecoratus Giordani Soika, 1970: 104, ♀; Iran.
Catalogues with Iranian records: Ebrahimi and Carpenter (2008), Rahmani *et al.* (2020).
Distribution in Iran: Alborz (Giordani Soika, 1970; Gusenleitner, 2010a), Kordestan (Ebrahimi and Carpenter, 2008; Modarres Awal, 2012), Markazi (Ebrahimi and Carpenter, 2008), Sistan & Baluchestan (Hamzavi *et al.*, 2019).
Distribution in the Middle East: Iran (see references above, and Gusenleitner, 2013a), Turkey (Gusenleitner, 1988; Yildirim and Özbek, 1996a; Yildirim and Kojima, 1999; Yildirim and Gusenleitner, 2012).
Extralimital distribution: Serbia (Gusenleitner, 2013a).

***Jucancistrocerus* (*Eremodynerus*) *consimilis* (Morawitz, 1895)**
Odynerus (*Lionotus*) *consimilis* Morawitz, 1895: 449, ♀; Turkmenistan.
Catalogues with Iranian records: Rahmani *et al.* (2020).
Distribution in Iran: Kerman (Schmid-Egger *et al.*, 2021).
Distribution in the Middle East: Iran (Schmid-Egger *et al.*, 2021; Fateryga and Proshchalykin, 2024).
Extralimital distribution: Tajikistan (Fateryga and Proshchalykin, 2024), Turkmenistan, Uzbekistan (Morawitz, 1895 as *Odynerus* (*Lionotus*) *consimilis*).

***Jucancistrocerus* (*Jucancistrocerus*) *jucundus jucundus* (Mocsáry, 1883)**
Odyneus (*Ancistrocerus*) *jucundus* Mocsáry, 1883: 49, ♀, ♂; Hungária centralis (Budapestinum) et Asia minor (Brussa).
Catalogues with Iranian records: Ebrahimi and Carpenter (2008), Rahmani *et al.* (2020).
Distribution in Iran: Iran (no specific locality cited) (Ebrahimi and Carpenter, 2008 as *Jucancistrocerus jucundus deceptrix* (Morice, 1921)).
Distribution in the Middle East: Iran (Ebrahimi and Carpenter, 2008; Selis *et al.*, 2025), Iraq (Morice, 1921 as *Nortonia deceptrix*; Derwesh, 1965; Giordani Soika, 1970 as *Jucancistrocerus jucundus deceptrix*; Gusenleitner, 2013a; Selis *et al.*, 2025), Turkey (Mocsáry, 1883 as *Odynerus* (*Ancistrocerus*) *jucundus*; Blüthgen, 1951a as *Iucancistrocerus jucundus*; Giordani Soika, 1971; Yildirim and Özbek, 1996a, 1996b both as *J. jucundus*; Yildirim and Kojima, 1999; Gusenleitner, 2007, 2013a; Yildirim and Gusenleitner, 2007, 2008; Yildirim and Tezcan, 2018; Kaplan and Efil, 2025; Selis *et al.*, 2025).
Extralimital distribution: Afghanistan, Austria, Bulgaria, Greece, Hungary, Turkmenistan (Gusenleitner, 2013a; Selis *et al.*, 2025), Slovenia (Gogala, 2022; Selis *et al.*, 2025).

***Jucancistrocerus* (*Jucancistrocerus*) *pareumeniformis* Giordani Soika, 1973**
Jucancistrocerus pareumeniformis Giordani Soika, 1973: 102, ♂; Iran.

Catalogues with Iranian records: Ebrahimi and Carpenter (2008), Rahmani *et al.* (2020).
Distribution in Iran: Guilan (Giordani Soika, 1973).
Distribution in the Middle East: Known only from Iran (Giordani Soika, 1973).
Extralimital distribution: None.

***Jucancistrocerus (Eremodynerus) saharensis* (Giordani Soika, 1934)**
Odynerus (Rhynchium) saharensis Giordani Soika, 1934: 42, sex not stated; [replacement name].
Distribution in the Middle East: Egypt, Israel-Palestine (Gusenleitner, 2013a).
Extralimital distribution: Libya (Gusenleitner, 2013a).

Genus *Katamenes* Meade-Waldo, 1910
Katamenes Meade-Waldo, 1910: 46. Type species: *Katamenes watsoni* Meade-Waldo, 1910, by monotypy.

***Katamenes dimidiativentris dimidiativentris* (Giordani Soika, 1941)**
Eumenes niger var. *dimidiativentris* Giordani Soika, 1941: 104, ♀; "Palestina: fra Gerico e Gerusalemme".
Distribution in the Middle East: Egypt, Jordan (Gusenleitner, 2013a), Israel-Palestine (Giordani Soika, 1941, 1949, 1960 all as *Eumenes niger* var. *dimidiativentris*; Gusenleitner, 2013a).
Extralimital distribution: None.

***Katamenes dimidiativentris francisi* Zalat in Zalat et al., 1993**
Katamenes dimidiativentris francisi Zalat in Zalat *et al.*, 1993: 106, ♀, ♂; Egypt.
Distribution in the Middle East: Known only from Egypt (Zalat *et al.*, 1993).
Extralimital distribution: None.

***Katamenes dimidiatus dimidiatus* (Brullé, 1832)**
Eumenes dimidiatus Brullé, 1832: 361, ♀; Plaine de Modon [Greece].
Catalogues with Iranian records: Ebrahimi and Carpenter (2008), Rahmani *et al.* (2020).
Distribution in Iran: East Azarbaijan, Fars, Markazi (Ebrahimi and Carpenter, 2008; Modarres Awal, 2012), Kerman (Schmid-Egger *et al.*, 2021), Tehran (Giordani Soika, 1970 as *Katamenes dimidiatus montanus* (Nurse, 1904)), Zanjan (Abbasi *et al.*, 2008a, 2008b; Pashaei Rad *et al.*, 2010).
Distribution in the Middle East: Cyprus (Yildirim and Kojima, 1999; Gusenleitner, 2013a; Fateryga, 2017b, 2018), Iran (see references above and Fateryga, 2017b, 2018), Israel-Palestine (Yildirim and Kojima, 1999; Gusenleitner, 2013a; Fateryga, 2017b, 2018), Syria (Fateryga, 2017b, 2018), Turkey (Blüthgen, 1951a as *Eumenes (Katamenes) sesquisinctus dimidiatus*; Gusenleitner, 1967, 2013a; Giordani Soika, 1970; Yildirim and Özbek, 1996a, 1996b; Yildirim and Kojima, 1999; Fateryga, 2017b, 2018).
Extralimital distribution: Afghanistan, Caucasus, Central Asia, China, Europe, Kazakhstan, Mongolia, Pakistan, Russia (Fateryga, 2017b, 2018).

***Katamenes dimidiatus arrectus* Giordani Soika, 1970**
Katamenes dimidiatus Giordani Soika, 1970: 179, ♀, ♂; "Turchia: Kars, Ararat, Serdarbulak, 5000 ft".
Distribution in the Middle East: Known only from Turkey (Giordani Soika, 1970; Van der Vecht and Fischer, 1972; Yildirim and Özbek, 1996a; Yildirim and Kojima, 1999; Gusenleitner, 2013a).
Extralimital distribution: None.

***Katamenes dimidiatus montanus* (Nurse, 1904)**
Eumenes montana Nurse, 1904: 26, ♀; Pakistan.
Distribution in Iran: Iran (no specific locality cited) (Giordani Soika, 1970; Yildirim and Kojima, 1999; Gusenleitner, 2013a).
Distribution in the Middle East: Iran (see references above), Turkey (Gusenleitner, 2013a).
Extralimital distribution: Afghanistan, Bulgaria, North Macedonia, Pakistan (Gusenleitner, 2013a).

***Katamenes flavigularis* (Blüthgen, 1951)**

Eumenes (*Katamenes*) *arbustorum flavigularis* Blüthgen, 1951a: 176, ♀, ♂; Turkey.

Catalogues with Iranian records: Ebrahimi and Carpenter (2008) Rahmani *et al.* (2020).

Distribution in Iran: Golestan (Schmid-Egger *et al.*, 2021), Qazvin (Ebrahimi and Carpenter, 2008), Tehran (Ebrahimi and Carpenter, 2008; Modarres Awal, 2012 as *Katamenes arbustorum* Panzer, 1799).

Distribution in the Middle East: Iran (see references above, and Fateryga, 2017b, 2018; Fateryga *et al.*, 2021), Israel-Palestine (Fateryga *et al.*, 2021), Lebanon, Syria (Giordani Soika, 1960 as *Eumenes arbustorum flavigularis*; Fateryga, 2017b, 2018; Fateryga *et al.*, 2021), Turkey (Gusenleitner, 1966 as *Katamenes arbustorum flavigularis*; Giordani Soika, 1970; Yildirim and Özbek, 1996a, 1996b, 1999; Yildirim and Kojima, 1999; Yildirim and Gusenleitner, 2004, 2007, 2009, 2012; Gusenleitner, 2007; Fateryga, 2017b, 2018; Yildirim and Tezcan, 2018; Fateryga *et al.*, 2021).

Extralimital distribution: Afghanistan, Caucasus, Europe, Russia, Tajikistan (Fateryga, 2017b, 2018), Armenia, Azerbaijan, Georgia, India (Fateryga *et al.*, 2021).

***Katamenes jenjouristei jenjouristei* (Kostylev, 1939)**

Eumenes jenjouristei Kostylev, 1939: 165, 168, ♀, ♂; "Arabie du Sud".

Distribution in the Middle East: Israel-Palestine (Carpenter and Gadallah, 2020), Oman (Giordani Soika, 1949 as *Eumenes* (*Katamenes*) *rubroniger* var. *pseudoconicus*; Gusenleitner, 2013a; Carpenter and Gadallah, 2020), Saudi Arabia (Guichard, 1986; Gusenleitner, 2013a; Carpenter and Gadallah, 2020), Yemen (Kostylev, 1939 as both *Eumenes jenjouristei* and *E. zeniouristei*; Gusenleitner, 2013a; Carpenter and Gadallah, 2020).

Extralimital distribution: None.

***Katamenes jenjouristei rubroniger* (Giordani Soika, 1941)**

Eumenes (*Delta*) *libycus* var. *rubroniger* Giordani Soika, 1941: 103, ♀; [Egypt] "Sinai: Wadi um Mitla".

Distribution in the Middle East: Egypt (Giordani Soika, 1941 as *Eumenes* (*Delta*) *libycus* var. *rubroniger*; Zalat *et al.*, 1993; Gusenleitner, 2013a; Carpenter and Gadallah, 2020), Jordan (Giordani Soika, 1970; Carpenter and Gadallah, 2020), Palestine (Giordani Soika, 1952 as *Eumenes rubroniger*; Zalat *et al.*, 1993; Gusenleitner, 2013a; Carpenter and Gadallah, 2020), Saudi Arabia (Guichard, 1986; Zalat *et al.*, 1993; Carpenter and Gadallah, 2020), United Arab Emirates (Gusenleitner, 2010a, 2013a; Carpenter and Gadallah, 2020).

Extralimital distribution: None.

***Katamenes lybicus lybicus* (Giordani Soika, 1941)**

Eumenes (*Delta*) *lybicus* Giordani Soika, 1941: 102, ♀, ♂; "Cirenaica: Tobruk".

Distribution in the Middle East: Egypt (Zalat *et al.*, 1993 as *Katamenes lybicus*).

Extralimital distribution: Libya (Gusenleitner, 2013a).

***Katamenes niger* (Brullé in Barker-Webb and Berthelot, 1839)**

Eumenes niger Brullé, 1839: 89, ♀, ♂; Canaries [Canary Islands].

Catalogues with Iranian records: Ebrahimi and Carpenter (2008), Rahmani *et al.* (2020).

Distribution in Iran: Iran (no specific locality cited) (Giordani Soika, 1970).

Distribution in the Middle East: Bahrain (Guichard, 1986; Carpenter and Gadallah, 2020), Egypt (Giordani Soika, 1935, 1953 both as *Eumenes nigra*; Zalat *et al.*, 1993; Gusenleitner, 2013a), Iran (Giordani Soika, 1970), Israel-Palestine, Jordan (Gusenleitner, 2013a), Saudi Arabia (Giordani Soika, 1957 as *Eumenes nigra*; Guichard, 1986; Zalat *et al.*, 1993; Gusenleitner, 2013a; Carpenter and Gadallah, 2020).

Extralimital distribution: Algeria (Saunders, 1905 as *Eumenes nigra*; Carpenter and Gadallah, 2020), Chad (Carpenter and Gadallah, 2020), Morocco (Giordani Soika, 1953 as *E. nigra*; Carpenter and Gadallah, 2020), Spain (Brullé, 1839 as *E. nigra*; Gusenleitner, 1990; Carpenter and Gadallah, 2020), Sudan (Carpenter and Gadallah, 2020).

***Katamenes priesneri* (Giordani Soika, 1941)**

Eumenes (*Delta*) *priesneri* Giordani Soika, 1941: 104, ♂; "Alto Egitto: Gabal Elba, Wadi Cansisnob".

Distribution in the Middle East: Known only from Egypt (Giordani Soika, 1941, 1949, 1973 all as *Eumenes* (*Delta*) *priesneri*; van der Vecht and Fischer, 1972; Zalat *et al.*, 1993; Borsato and Ratti, 1999; Gusenleitner, 2013a).

Extralimital distribution: None.

***Katamenes rauensis* (Giordani Soika, 1960)**

Eumenes rauensis Giordani Soika, 1960 (1958): 62, ♂; Arabia.

Distribution in the Middle East: Known only from Saudi Arabia (Giordani Soika, 1960 as *Eumenes rauensis*; Guichard, 1986; Gusenleitner, 2013a; Carpenter and Gadallah, 2020).

Extralimital distribution: None.

***Katamenes sichelii baerii* (Radoszkowski, 1865)**

Eumenes baerii Radiszkowski, 1865: 59, ♀, ♂; Azerbaijan.

Catalogues with Iranian records: Ebrahimi and Carpenter (2008), Rahmani *et al.* (2020).

Distribution in Iran: Alborz, Lorestan (Giordani Soika, 1949 as *Eumenes sichelii baeri* (Radoszkowski, 1865).

Distribution in the Middle East: Iran (Giordani Soika, 1949, 1970; Gusenleitner, 2013a), Turkey (Blüthgen, 1951a as *Eumenes* (*Katamenes*) *sichelii baerii*; Giordani Soika, 1949 as *E. sichelii* var. *baerii*; Gusenleitner, 1966 as *K. sichelii*, 2013a; Giordani Soika, 1970 as *K. sichelii*; Yildirim and Özbek, 1996a; Yildirim, 1998; Yildirim and Kojima, 1999).

Extralimital distribution: Armenia (Gusenleitner, 2013a), Azerbaijan (Radoszkowski, 1865 as *Eumenes baerii*).

Katamenes sichelii biblicus (Giordani Soika, 1935)

Eumenes sichelii var. *biblicus* Giordani Soika, 1935: 162, 167, ♀; "Sinai".

Distribution in the Middle East: Egypt (Giordani Soika, 1935, 1941 both as *Eumenes sichelii* var. *biblicus*; Zalat *et al.*, 1993; Gusenleitner, 2013a; Carpenter and Gadallah, 2020), Israel-Palestine (Giordani Soika, 1949; Zalat *et al.*, 1993; Gusenleitner, 2013a; Carpenter and Gadallah, 2020), Jordan (Giordani Soika, 1952; Zalat *et al.*, 1993; Gusenleitner, 2013a; Carpenter and Gadallah, 2020), United Arab Emirates (Gusenleitner, 2010a; Carpenter and Gadallah, 2020).

Extralimital distribution: None.

***Katamenes sichelii coranicus* Giordani Soika, 1970**

Katamenes sichelii coranicus Giordani Soika, 1970: 177, ♀, ♂; "Arabia Saudita: El Riyadh".

Distribution in the Middle East: Saudi Arabia (Giordani Soika, 1970; Guichard, 1986; Gusenleitner, 2013a; Carpenter and Gadallah, 2020).

Extralimital distribution: Asia, North Africa, South Europe (Carpenter and Gadallah, 2020).

***Katamenes sichelii fulvus* (Eversmann, 1854)**

Eumenes fulva Eversmann, 1854: 199, sex not stated; "die Steppen zwischen der Wolga und dem Uralfluss; die südlichen nackten Vorgebirge des Urals, und die südlichen Kirgisensteppen".

Catalogues with Iranian records: Ebrahimi and Carpenter (2008), Rahmani *et al.* (2020).

Distribution in Iran: Lorestan (Giordani Soika, 1949 as *Eumenes sichelii* var. *luristanensis*).

Distribution in the Middle East: Iran (Giordani Soika, 1949; Gusenleitner, 2013a).

Extralimital distribution: China (Kurzenko, 1977), Russia (Fateryga, 2017b).

***Katamenes sichelii sichelii* (de Saussure, 1852)**

Eumenes sichelii de Saussure, 1852: 36, ♀; Albania.

Catalogues with Iranian records: Ebrahimi and Carpenter (2008), Rahmani *et al.* (2020).

Distribution in Iran: Alborz (Giordani Soika, 1970), Golestan (Sakenin *et al.*, 2008), Markazi (Ebrahimi and Carpenter, 2008; Modarres Awal, 2012), Sistan & Baluchestan (Blüthgen and Gusenleitner, 1970), Tehran (Castro and Dvořák, 2009).

Distribution in the Middle East: Iran (see references above, and Fateryga, 2017b), Iraq, Israel-Palestine, Jordan, Saudi Arabia, United Arab Emirates (Fateryga, 2017b; Fateryga and Mokrousov, 2019), Turkey (Giordani Soika, 1970; Gusenleitner, 2013a; Fateryga, 2017a, 2017b).
Extralimital distribution: Afghanistan, Armenia, Azerbaijan, Caucasus, Central Asia, China, Europe (SE), India, Kazakhstan, North Africa (Fateryga, 2017b; Fateryga *et al.*, 2019), Mongolia, Russia (Fateryga, 2017a, 2017b; Fateryga *et al.*, 2019).

***Katamenes sichelii tauriae* (Giordani Soika, 1960)**
Eumenes sichelii tauriae Giordani Soika, 1960 (1958): 63, ♂; Turkey.
Catalogues with Iranian records: Ebrahimi and Carpenter (2008), Rahmani *et al.* (2020).
Distribution in Iran: East Azarbaijan (Hayat *et al.*, 2008 as *Katamenes sichelii* (Giordani Soika)), Sistan & Baluchestan (Hayat *et al.*, 2008 as *K. sichelii*).
Distribution in the Middle East: Iran (see references above, and Blüthgen and Gusenleitner, 1970), Turkey (Giordani Soika, 1960 as *Eumenes sichelli tauriae*; Gusenleitner, 1966; Yildirim and Özbek, 1996a; Madl, 1997; Yildirim and Kojima, 1999; Yildirim and Gusenleitner, 2001; Gusenleitner, 2013a; Yildirim and Tezcan, 2018).
Extralimital distribution: Afghanistan, Turkmenistan, Uzbekistan (Gusenleitner, 2013a).
Comments: *Dasypogon octonotatus* Loew, *Dioctria linearis* (Fabricius) and *Habropogon longiventris* Loew (Diptera: Asilidae) are the predators of *Katamenes sichelii* (Hayat *et al.*, 2008).

***Katamenes sichelii timanensis* Giordani Soika, 1986**
Katamenes sichelii timanensis Giordani Soika, 1986: 123, ♀; "*Medio Oriente*: Timan".
Distribution in the Middle East: Known only from Iraq (Giordani Soika, 1986; Selis *et al.*, 2025).
Extralimital distribution: None.

***Katamenes tauricus tauricus* (de Saussure, 1855)**
Eumenes tauricus de Saussure, 1855: 137, ♀; Russia.
Catalogues with Iranian records: Ebrahimi and Carpenter (2008), Rahmani *et al.* (2020).
Distribution in Iran: Alborz (Giordani Soika, 1970 as *Katamenes sesquicinctus viratus* (Giordani Soika, 1949); Ebrahimi and Carpenter, 2008; Modarres Awal, 2012), North of Iran (Gusenleitner, 1972; Kurzenko, 1977).
Distribution in the Middle East: Iran (see references above, and Giordani Soika, 1949 as *Eumenes* (*Katamenes*) *latipes viratus* Giordani Soika, 1949; Fateryga, 2017b, 2018; Fateryga *et al.*, 2020).
Extralimital distribution: Afghanistan, China, India, Kazakhstan, Kyrgyzstan, Mongolia, Russia (Fateryga, 2017b, 2018; Fateryga *et al.*, 2020).

***Katamenes tauricus viratus* (Giordani Soika, 1949)**
Eumenes (*Katamenes*) *latipes* var. *viratus* Giordani Soika, 1949: 43, ♂; "Persia".
Distribution in Iran: Elborz (Giordani Soika, 1949 as *Eumenes* (*Katamenes*) *latipes* var. *viratus*; Blüthgen, 1952 as *E.* (*K.*) *sesquicinctus* var. *caspica*).
Distribution in the Middle East: Known only from Iran (see references above, and Giordani Soika, 1970; van der Vecht and Fischer, 1972; Gusenleitner, 2013a).
Extralimital distribution: None.

Genus *Knemodynerus* Blüthgen, 1940
Knemodynerus Blüthgen, 1940: 43. Type species: *Odynerus excellens* Pérez, 1907, by original designation.

***Knemodynerus djarabubensis djarabubensis* (von Schulthess, 1928)**
Odynerus djarabubensis von Schulthess, 1928: 70, ♂, ♀; "Cyrenaica: Djarabub".
Distribution in the Middle East: Israel-Palestine (Gusenleitner, 2013a).
Extralimital distribution: North Africa (Gusenleitner, 2013a).

Knemodynerus euodyneroides **Gusenleitner, 1997c**

Knemodynerus euodyneroides Gusenleitner, 1997c: 137, 142, ♂; "Zentral-Türkei, Provinz Nevsehir, Zelve".

Distribution in the Middle East: Known only from Turkey (Gusenleitner, 1997c, 2013a; Yildirim and Kojima, 1999).

Extralimital distribution: None.

Knemodynerus excellens **(Pérez, 1907)**

Odynerus (*Lionotus*) *excellens* Pérez, 1907: 493, ♀, ♂; Dibba [Oman], ♀; Bahrein, ♂.

Catalogues with Iranian records: Ebrahimi and Carpenter (2008), Rahmani *et al.* (2020).

Distribution in Iran: Kerman (Pashaei Rad *et al.*, 2010; Schmid-Egger *et al.*, 2021), Khuzestan (Blüthgen and Gusenleitner, 1970; Giordani Soika, 1970 as *Euodynerus excellens* (Pérez, 1907)), Markazi (Ebrahimi and Carpenter, 2008; Modarres Awal, 2012), Sistan & Baluchestan (Blüthgen and Gusenleitner, 1970), Zanjan (Abbasi *et al.*, 2008a, 2008b; Pashaei Rad *et al.*, 2010).

Distribution in the Middle East: Bahrain (Pérez, 1907 as *Odynerus excellens*; Carpenter and Gadallah, 2020), Iran (see references above), Oman (Pérez, 1907 as *O. excellens*; Hamer, 1985 as *Euodynerus excellens*; Gusenleitner, 2004, 2013a; Carpenter and Gadallah, 2020), Qatar (Guichard, 1986 as *E. excellens*; Carpenter and Gadallah, 2020), Saudi Arabia (Guichard, 1986 as *E. excellens*; Walker and Pittaway, 1987 as *E. excellens*; Gusenleitner, 2013a; Carpenter and Gadallah, 2020), United Arab Emirates (Roche, 1981 as *E. excellens*; Hamer, 1985 as *E. excellens*; Walker and Pittaway, 1987 as *E. excellens*; Gillett and Gillett, 2005 as *E. excellens*; Howarth and Gillett, 2008 as *E. excellens*; Gusenleitner, 2010a, 2013a; Carpenter and Gadallah, 2020).

Extralimital distribution: India, Turkmenistan (Gusenleitner, 2013a; Carpenter and Gadallah, 2020), Pakistan (Gusenleitner, 2006c; Carpenter and Gadallah, 2020).

Knemodynerus familiaris familiaris **(Giordani Soika, 1939)**

Odynerus (*Rhynchium*) *familiaris* Giordani Soika, 1939: 3, ♂, ♀; "Egitto: Ezbet el Nakl; Marg; Abu Zabal".

Distribution in the Middle East: Egypt (Giordani Soika, 1939 as *Odynerus* (*Rhynchium*) *familiaris*; Gusenleitner, 2013a), Saudi Arabia (Guichard, 1986 as *Euodynerus familiaris*: Carpenter and Gadallah, 2020), Yemen (Gusenleitner, 2007, 2013a; Carpenter and Gadallah, 2020).

Extralimital distribution: None.

Knemodynerus familiaris judaicus **(Giordani Soika, 1952)**

Pseudepipona (*Euodynerus*) *familiaris judaica* Giordani Soika, 1952: 42, ♀; "Wadi el Kelt".

Distribution in the Middle East: Known only from Israel-Palestine (Giordani Soika, 1952 as *Pseudpipona* (*Euodynerus*) *familiaris judaica*; van der Vecht and Fischer, 1972 as *Euodynerus familiaris judaicus*; Gusenleitner, 2013a).

Extralimital distribution: None.

Knemodynerus familiaris muscatensis **(Giordani Soika, 1970)**

Euodenus familiaris muscatensis Giordani Soika, 1970: 154, ♀; "Arabia: Muscat".

Distribution in the Middle East: Known only from Oman (Giordani Soika, 1970 as *Euodynerus familiaris muscatensis*; Guichard, 1986 as *E. f. muscatensis*; Gusenleitner, 2013a; Carpenter and Gadallah, 2020).

Extralimital distribution: None.

Knemodynerus lahijensis **Gusenleitner, 2002**

Knemodynerus lahijensis Gusenleitner, 2002: 307, 315, ♀; "Yemen, Lahj".

Distribution in the Middle East: Known only from Yemen (Gusenleitner, 2002; Carpenter and Gadallah, 2020).

Extralimital distribution: None.

***Knemodynerus pseudolateralis* (Meade-Waldo, 1915)**
Odynerus pseudolateralis Meade-Waldo, 1915: 494, 509, ♀, ♂; "N. Nigeria: Minna".
Distribution in the Middle East: Egypt (Giordani Soika, 1979 as *Euodynerus pseudolateralis*), Oman (Gusenleitner, 2013a).
Extralimital distribution: Burkina Faso, Congo, Eritrea, Gambia, Ghana, Ivory Coast, Mali, Niger, Nigeria, Tanzania (Gusenleitner, 2013a).

***Knemodynerus rhynchoides inclinans* (Giordani Soika, 1935)**
Odynerus (*Rhynchium*) *tectus* var. *inclinans* Giordani Soika, 1935: 183, 187, ♂, ♀; "Aegypten".
Distribution in the Middle East: Known only from Egypt (von Schulthess, 1928 as *Odynerus crenatus* var. *aegyptiaca*; Giordani Soika, 1935 as *Odynerus* (*Rhynchium*) *tectus* var. *inclinans*; Blüthgen, 1939 as *Euodynerus tectus* var. *aegyptiaca*, 1955a as *E. rhynchoides* var. *aegyptiaca*; van der Vecht and Fischer, 1972 as *E.* (*Knemodynerus*) *rhynchoides inclinans*; Giordani Soika, 1973 as *O. tectus inclinans*; Gusenleitner, 2013a).
Extralimital distribution: None.

***Knemodynerus rhynchoides rhynchoides* (de Saussure, 1853)**
Odynerus (*Leionotus*) *rhynchoides* de Saussure, 1853: 152, 174, ♂, "Le Sénégal".
Distribution in the Middle East: Egypt (de Saussure, 1863 as *Odynerus* (*Odynerus*) *interruptus*; André, 1884 as *O. saussurei*; Giordani Soika, 1935 as *O. tectus* var. *rhynchoides*), Saudi Arabia (Guichard, 1986 as *Euodynerus rhynchoides*; Carpenter and Gadallah, 2020), Yemen (Gusenleitner, 2007, 2013a; Madl, 2018; Carpenter and Gadallah, 2020).
Extralimital distribution: Algeria, Burkina Faso, Central African Republic, Congo, Djibouti, Eritrea, Ethiopia, Gambia, Kenya, Libya, Mali, Niger, Senegal, Sudan (Carpenter *et al.*, 2010; Carpenter and Gadallah, 2020).

***Knemodynerus sauditus* (Giordani Soika, 1989)**
Trachyodenerus sauditus Giordani Soika, 1989: 19, 62, ♀; "Saudi Arabia: dintorni di Hail".
Distribution in the Middle East: Known only from Saudi Arabia (Giordani Soika, 1989 as *Trachyodenerus sauditus*; Carpenter and Gadallah, 2020).
Extralimital distribution: None.

***Knemodynerus sinaiticus* (Giordani Soika, 1939)**
Odynerus (*Rhynchium*) *tectus* var. *sinaiticus* Giordani Soika, 1939: 6, ♀; "Sinai: Wadi Nasib".
Distribution in the Middle East: Egypt (Giordani Soika, 1939 as *Odynerus* (*Rhynchium*) *tectus* var. *sinaiticus*; Gusenleitner, 2013a), Saudi Arabia (Guichard, 1986; Gusenleitner, 2013a; Carpenter and Gadallah, 2020), Yemen (Giordani Soika, 1957 as *Pseudepipona sinaitica*; Guichard, 1986 as *Euodynerus sinaiticus*; Zalat *et al.*, 1993 as *E. sinaiticus*; Gusenleitner, 2002, 2013a; Carpenter and Gadallah, 2020).
Extralimital distribution: None.

***Knemodynerus stigma* (de Saussure, 1863)**
Odynerus stigma de Saussure, 1863: 219, ♂; «Abyssinie».
Distribution in the Middle East: Egypt (de Saussure, 1863 as *Odynerus stigma*; Walker, 1871 as *Rhynchium zonatum*; Giordani Soika, 1935 as *O. patrizi*; Gusenleitner, 2013a), Oman (Giordani Soika, 1957 as *Pseudepipona stigma arabica*; Carpenter and Gadallah, 2020), Saudi Arabia (Giordani Soika, 1970 as *P. stigma*; Guichard, 1986 as *Euodynerus stigma*; Zalat *et al.*, 1993 as *E. stigma*; Madl, 2018; Carpenter and Gadallah, 2020), Yemen (Gusenleitner, 2013a; Madl, 2018; Carpenter and Gadallah, 2020).
Extralimital distribution: Algeria (Giordani Soika, 1954 as *Pseudepipona stigma*; Carpenter and Gadallah, 2020), Chad, Djibouti, Eretria, Ethiopia, Sudan (Carpenter *et al.*, 2010; Carpenter and Gadallah, 2020).

Genus *Labochilus* Blüthgen, 1939
Labochilus Blüthgen, 1939: 12. Type species: *Pterochilus linguarius* Saunders, 1905, by monotypy.

***Labochilus felix* Guichard, 1986**

Labochilus felix Guichard, 1986 (1985): 202, 207, 216, ♀; "Oman (Dhofar): km 48 Salalah-Thamarit main road, 670 m".

Distribution in the Middle East: Known only from Oman (Guichard, 1986; Gusenleitner, 2013a; Carpenter and Gadallah, 2020).

Extralimital distribution: None.

***Labochilus orientalis* Gusenleitner, 2001a**

Labochilus orientalis Gusenleitner, 2001a: 680, ♀, ♂; Iran.

Catalogues with Iranian records: Ebrahimi and Carpenter (2008), Rahmani *et al.* (2020).

Distribution in Iran: Hormozgan (Gusenleitner, 2001a).

Distribution in the Middle East: Iran (Gusenleitner, 2001a, 2013a), Jordan (Gusenleitner, 2001a).

Extralimital distribution: None.

***Labochilus pulawskyi pulawskyi* Giordani Soika, 1970**

Labochilus pulawskyi Giordani Soika, 1970: 83, ♀, ♂; "Egitto: Luxor".

Distribution in the Middle East: Known only from Egypt (Giordani Soika, 1970; van der Vecht and Fischer, 1972; Zalat *et al.*, 1993; Gusenleitner, 2013a).

Extralimital distribution: None.

***Labochilus pulawskyi soharensis* Giordani Soika, 1979**

Labochilus pulawskyi soharensis Giordani Soika, 1979: 257, ♀; "Oman: Sohar".

Distribution in the Middle East: Oman (Giordani Soika, 1979; Guichard, 1986; Gusenleitner, 2013a; Carpenter and Gadallah, 2020), United Arab Emirates (Gusenleitner, 2010a; Carpenter and Gadallah, 2020).

Extralimital distribution: None.

Genus *Leptochilus* de Saussure, 1853

Leptochilus de Saussure, 1853: 233. Type species: *Pterochilus mauritanicus* Lepeletier, 1841; designated by Ashmead, 1902.

***Leptochilus* (*Lionotulus*) *aegineticus* Gusenleitner in Blüthgen and Gusenleitner, 1970**

Leptochilus (*Lionotulus*) *aegineticus* Gusenleitner, 1970 [in Blüthgen and Gusenleitner, 1970]: 292, ♂; Greece.

Distribution in the Middle East: Turkey (Gusenleitner, 1985b; Yildirim and Kojima, 1999).

Extralimital distribution: Greece (Blüthgen and Gusenleitner, 1970).

***Leptochilus* (*Lionotulus*) *alborufulus* Gusenleitner, 1977**

Leptochilus (*Lionotulus*) *alborufulus* Gusenleitner, 1977: 174, ♂, ♀; "Tunis, Carthago".

Distribution in the Middle East: Egypt (Gusenleitner, 2013a).

Extralimital distribution: Morocco, Tunisia (Gusenleitner, 2013a).

***Leptochilus* (*Lionotulus*) *alpestris alpestris* (de Saussure, 1855)**

Odynerus (*Odynerus*) *alpestris* de Saussure, 1855: 272, ♀, ♂; "L'Allemagne. Ratisbonne".

Distribution in the Middle East: Jordan, Turkey (Giordani Soika, 1952, 1970; Yildirim and Özbek, 1999, 1996b; Yildirim and Kojima, 1999; Yildirim and Gusenleitner, 2012; Gusenleitner 2013a; Fateryga, 2017b, 2018).

Extralimital distribution: Azerbaijan, North Africa, Russia, Turkmenistan, Ukraine (Fateryga, 2017b, 2018), Central and South Europe (Gusenleitner, 2013a), Italy (Selis, 2023).

***Leptochilus* (*Sarochilus*) *alterego* Gusenleitner, 1970**

Leptochilus (*Sarochilus*) *alterego* Gusenleitner, 1970: 4, ♀; "Birecik, Turkey".

Distribution in the Middle East: Israel-Palestine (Gusenleitner, 1970; Yildirim and Kojima, 1999; Gusenleitner, 2013a), Turkey (Gusenleitner, 1970, 2013a; van der Vecht and Fischer, 1972; Yildirim and Kojima, 1999).

Extralimital distribution: None.

***Leptochilus* (*Lionotulus*) *ambitiosus* Giordani Soika, 1986**

Leptochilus (*Lionotulus*) *ambitiosus* Giordani Soika, 1986: 104, ♀; "Turchia: 25 Km SW di Hakkari, 1200 m".

Distribution in the Middle East: Iraq (Selis *et al*., 2025), Turkey (Giordani Soika, 1986; Yildirim and Kojima, 1999; Gusenleitner, 2013a; Selis *et al*., 2025).

Extralimital distribution: None.

***Leptochilus* (*Neoleptochilus*) *amos amos* Gusenleitner in Bytinski-Salz and Gusenleitner, 1971**

Leptochilus (*Neoleptochilus*) *amos* Gusenleitner, 1971 [in Bytinski-Salz and Gusenleitner, 1971]: 239, 264. 293, ♀; "Israel: Mishor Rotem (between Dimona and Dead Sea)".

Distribution in the Middle East: Known only from Israel-Palestine (Bytinski-Salz and Gusenleitner, 1971; Gusenleitner, 2013a).

Extralimital distribution: None.

***Leptochilus* (*Lionotulus*) *arabicus* Giordani Soika, 1979**

Leptochilus (*Lionotus*) *arabicus* Giordani Soika, 1979: 274, ♀; "Muscat: Seeb".

Distribution in the Middle East: Known only from Oman (Giordani Soika, 1979, 1981; Gusenleitner, 2013a; Carpenter and Gadallah, 2020).

Extralimital distribution: None.

***Leptochilus* (*Lionotulus*) *aterrimus* (Kirby, 1900)**

Icaria aterrima Kirby, 1900: 23, sex unstated; Socotra.

Distribution in the Middle East: Known only from Yemen (including Socotra) (Kirby, 1900, 1903 both as *Icaria aterrima*; Kohl, 1906 as both *Odynerus* (*Leptochilus*) *melanochrous* and *Labus aterrimus*; Bequaert, 1918 as both *I. aterrima* and *O.* (*L.*) *melanochrous*; Giordani Soika, 1974 as both *O. melanochrous* and *Leptochilus aterrimus*; Carpenter *et al*., 2010; Carpenter and Gadallah, 2020).

Extralimital distribution: None.

***Leptochilus* (*Euleptochilus*) *ayunensis* Giordani Soika, 1981**

Leptochilus (*Euleptochilus*) *ayunensis* Giordani Soika, 1981: 111, ♀; Arabia.

Distribution in the Middle East: Oman (Giordani Soika, 1981; Guichard, 1986; Carpenter and Gadallah, 2020), Yemen (Gusenleitner, 2007, 2013a; Carpenter and Gadallah, 2020).

Extralimital distribution: None.

***Leptochilus* (*Lionotulus*) *bellus* Blüthgen, 1955a**

Leptochilus (*Lionotulus*) *bellus* Blüthgen, 1955a: 25, ♀; «Jericho».

Distribution in the Middle East: Egypt (Giordani Soika, 1970; Zalat *et al*., 1993 as *Leptochilus bellus*; Gusenleitner, 2001b, 2013a), Israel-Palestine (Blüthgen, 1955a; Bytinski-Salz and Gusenleitner, 1971; Zalat *et al*., 1993; Gusenleitner, 2001b, 2013a), Jordan (Gusenleitner, 2001b).

Extralimital distribution: None.

***Leptochilus* (*Lionotulus*) *bipunctatus* Gusenleitner, 2016**

Leptochilus (*Lionotulus*) *bipunctatus* Gusenleitner, 2016: 89, 94, ♂; "Jordanien, 80 km NE Aqaba (Straße nach Amman)".

Distribution in the Middle East: Known only from Jordan (Gusenleitner, 2016).

Extralimital distribution: None.

***Leptochilus* (*Neoleptochilus*) *bytinskii* Giordani Soika, 1952**

Leptochilus bytinskii Giordani Soika, 1952: 8, 52, ♀, ♂; "Bat Jam".

Distribution in the Middle East: Israel-Palestine (Giordani Soika, 1952, 1973; van der Vecht and Fischer, 1972; Borsato and Ratti, 1999; Gusenleitner, 2006b, 2013a).

Extralimital distribution: Tunisia (Gusenleitner, 2006c, 2013a).

***Leptochilus* (*Lionotulus*) *callidus* (Kostylev, 1940)**

Odynerus (*Lionotulus*) *callidus* Kostylev, 1940: 33, ♀, ♂; Syr-Daria [Kazakhstan].

Catalogues with Iranian records: Rahmani *et al.* (2020).
Distribution in Iran: Kerman (Gusenleitner, 2012).
Distribution in the Middle East: Iran (Gusenleitner, 2012; Gusenleitner, 2013a; Fateryga *et al.*, 2023).
Extralimital distribution: Kazakhstan, Mongolia (Gusenleitner, 2013a), Uzbekistan (Fateryga *et al.*, 2023).

***Leptochilus* (*Euleptochilus*) *camurus* (Giordani Soika, 1938)**
Odynerus (*Leptochilus*) *camurus* Giordani Soika, 1938: 8, ♀, ♂; "Egitto: Abu Rawash".
Distribution in the Middle East: Known only from Egypt (Giordani Soika, 1938, 1973 both as *Odynerus* (*Leptochilus*) *camurus*; van der Vecht and Fischer, 1972; Giordani Soika, 1979; Zalat *et al.*, 1993; Borsato and Ratti, 1999; Gusenleitner, 2013a).
Extralimital distribution: None.

***Leptochilus* (*Lionotulus*) *errabundus* (Giordani Soika, 1938)**
Odynerus (*Leptochilus*) *errabundus* Giordani Soika, 1938: 11, ♀, ♂; "Egitto: Wadi Hoff".
Distribution in the Middle East: Known only from Egypt (Giordani Soika, 1938, 1973 both as *Odynerus* (*Leptochilus*) *erabundus*; Giordani Soika, 1974; van der Vecht and Fischer, 1972; Zalat *et al.*, 1993; Borsato and Ratti, 1999; Gusenleitner, 2013a).
Extralimital distribution: None.

***Leptochilus* (*Euleptochilus*) *flegias* (Giordani Soika, 1938)**
Odynerus (*Leptochilus*) *flegias* Giordani Soika, 1938: 6, ♀; Egypt.
Distribution in the Middle East: Egypt (Giordani Soika, 1938 as *Odynerus* (*Leptochilus*) *flegias*; Zalat *et al.*, 1993; Gusenleitner, 2013a), Israel-Palestine (Bytinski-Salz and Gusenleitner, 1971; Zalat *et al.*, 1993).
Extralimital distribution: Portugal (Gusenleitner, 2007).

***Leptochilus* (*Lionotulus*) *flexilis* (Giordani Soika, 1938)**
Odynerus (*Leptochilus*) *flexilis* Giordani Soika, 1938: 10, ♀, ♂; "Egitto: Wadi Hoff".
Distribution in the Middle East: Known only from Egypt (Giordani Soika, 1939, 1973 both as *Odynerus* (*Leptochilus*) *flexilis*; Giordani Soika, 1970; van der Vecht and Fischer, 1972; Zalat *et al.*, 1993; Borsato and Ratti, 1999; Gusenleitner, 2013a).
Extralimital distribution: None.

***Leptochilus* (*Lionotulus*) *fuscipes* Gusenleitner, 1985b**
Leptochilus (Lionotulus) fuscipes Gusenleitner, 1985b: 98, ♂; Iran.
Catalogues with Iranian records: Ebrahimi and Carpenter (2008), Rahmani *et al.* (2020).
Distribution in Iran: Hamadan, Qazvin (Gusenleitner, 1985b).
Distribution in the Middle East: Iran (Gusenleitner, 1985b; Yildirim and Kojima, 1999), Turkey (Gusenleitner, 1985b; Yildirim and Kojima, 1999).
Extralimital distribution: None.

***Leptochilus* (*Sarochilus*) *gemma* Giordani Soika, 1970**
Leptochilus (*Lionotulus*) *gemma* Giordani Soika, 1970: 66, ♀; "Ghiordania: Al Maghatas".
Catalogues with Iranian records: None.
Distribution in Iran: Hormozgan (Gusenleitner, 2001b).
Distribution in the Middle East: Iran (Gusenleitner, 2001b), Israel-Palestine (Gusenleitner, 1970 as *Leptochilus* (*Lionotulus*) *gemma*, 2001b; Bytinski-Salz and Gusenleitner, 1972), Jordan (Giordani Soika, 1970 as *L.* (*Lionotulus*) *gemma*; Gusenleitner, 2001b, 2013a).
Extralimital distribution: None.

***Leptochilus* (*Lionotulus*) *gusenleitneri* Yildirim and Özbek, 1995**
Leptochilus (*Lionotulus*) *gusenleitneri* Yildirim and Özbek, 1995: 241, ♂; "Turkey, Erzincan Bahçe Kültürleri Araştırma Enstitüsü, 1250 m".

Distribution in the Middle East: Known only from Turkey (Yildirim and Özbek, 1995; Yildirim and Kojima, 1999; Gusenleitner, 2013a).
Extralimital distribution: None.

***Leptochilus* (*Lionotulus*) *hermon* Gusenleitner in Bytinski-Salz and Gusenleitner, 1971**
Leptochilus (*Lionotulus*) *hermon* Gusenleitner, 1971 [in Bytinski-Salz and Gusenleitner, 1971]: 239, 263, 290, ♀, ♂; Israel.
Distribution in the Middle East: Iraq (Selis *et al.*, 2025), Israel-Palestine (Bytinski-Salz and Gusenleitner, 1971; Yildirim and Kojima, 1999; Gusenleitner, 2013a; Kaplan and Efil, 2025; Selis *et al.*, 2025), Turkey (Bytinski-Salz and Gusenleitner, 1971; Yildirim and Özbek, 1996b, 1999; Yildirim and Kojima, 1999; Yildirim and Gusenleitner, 2012; Gusenleitner, 2013a; Yildirim and Tezcan, 2018; Kaplan and Efil, 2025; Selis *et al.*, 2025).
Extralimital distribution: Greece (Gusenleitner, 2013a; Kaplan and Efil, 2025; Selis *et al.*, 2025).

***Leptochilus* (*Lionotulus*) *hethiticus* Gusenleitner, 1985b**
Leptochilus hethiticus Gusenleitner, 1985b: 81, 91, ♂, ♀; "Türkei, Mut, Sertavul".
Distribution in the Middle East: Known only from Turkey (Gusenleitner, 1985a, 2013a; Yildirim and Kojima, 1999; Yildirim and Özbek, 1999; Yildirim and Gusenleitner, 2012).
Extralimital distribution: None.

***Leptochilus* (*Sarochilus*) *iranus* Gusenleitner and Schmid-Egger in Schmid-Egger *et al.*, 2021**
Leptochilus (*Sarochilus*) *iranus* Gusenleitner and Schmid-Egger in Schmid-Egger *et al.*, 2021: 1112, ♀; Iran.
Catalogues with Iranian records: None.
Distribution in Iran: Fars, Hormozgan (Schmid-Egger *et al.*, 2021).
Distribution in the Middle East: Known only from Iran (Schmid-Egger *et al.*, 2021).
Extralimital distribution: None.

***Leptochilus* (*Lionotulus*) *jordaneus* Gusenleitner, 1994**
Leptochilus jordaneus Gusenleitner, 1994: 307, 310, ♀; "Jordanien, 80 km NE Aqaba (Straße nach Amman)".
Distribution in the Middle East: Israel-Palestine, Jordan (Gusenleitner, 1994, 2013a).
Extralimital distribution: None.

***Leptochilus* (*Lionotulus*) *josephi* Giordani Soika, 1947**
Leptochilus josephi Giordani Soika, 1947: 130, ♀, ♂; Italy.
Distribution in the Middle East: Turkey (Gusenleitner, 1993, 2013a; Yildirim and Kojima, 1999; Yildirim and Gusenleitner, 2012).
Extralimital distribution: Bulgaria, Italy, Morocco (Gusenleitner, 2013a).

***Leptochilus* (*Neoleptochilus*) *kemali* Gusenleitner, 1977**
Leptochilus (*Neoleptochilus*) *kemali* Gusenleitner, 1977: 163, ♂, ♀; "Mut, Türkei".
Distribution in the Middle East: Known only from Turkey (Gusenleitner, 1977, 2013a; Yildirim and Kojima, 1999).
Extralimital distribution: None.

***Leptochilus* (*Lionotulus*) *kurnubensis* Blüthgen, 1955a**
Leptochilus (*Lionotulus*) *kurnubensis* Blüthgen, 1955a: 25, ♀, ♂; "Kurnub (30 km ESE Beersheba, Negev)".
Distribution in the Middle East: Known only from Israel-Palestine (Blüthgen, 1955a; van der Vecht and Fischer, 1972; Gusenleitner, 2013a).
Extralimital distribution: None.

***Leptochilus* (*Euleptochilus*) *limbiferoides* (Giordani Soika, 1938)**
Odynerus (*Euleptochilus*) *limbiferoides* Giordani Soika, 1938: 9, ♀, ♂; "Egitto Helwan, W. Garrawi".

Distribution in the Middle East: Egypt (Giordani Soika, 1938 as *Odynerus* (*Euleptochilus*) *limbiferoides*; Zalat *et al.*, 1993; Gusenleitner, 2013a), Israel-Palestine (Giordani Soika, 1952; Gusenleitner, 2013a).
Extralimital distribution: Algeria (Gusenleitner, 2013a).

***Leptochilus* (*Euleptochilus*) *limbiferus anatolicus* Blüthgen, 1955a**
Leptochilus (*Euleptochilus*) *limbiferus anatolicus* Blüthgen, 1955a: 24, ♀; "Kawak (S.W. Samsun, Anatolia)".
Distribution in the Middle East: Israel-Palestine (Blüthgen, 1955a; Gusenleitner, 1966, 1993; Giordani Soika, 1970; van der Vecht and Fischer, 1972), Turkey (Blüthgen, 1955a; Gusenleitner, 1966, 2013a; Giordani Soika, 1970; Yildirim and Özbek, 1996a, 1999; Yildirim and Kojima, 1999; Yildirim and Gusenleitner, 2012).
Extralimital distribution: Armenia, Europe (SE, EE), Russia (Fateryga, 2017b).

***Leptochilus* (*Lionotulus*) *longipilis* Gusenleitner, 1988**
Leptochilus (*Lionotulus*) *longipilis* Gusenleitner, 1988: 108, sex is not stated; Jordan, Agaba.
Distribution in the Middle East: Known only from Jordan (Gusenleitner, 1988, 2013a).
Extralimital distribution: None.

***Leptochilus* (*Lionotulus*) *lorestanicus* Gusenleitner, 1995a**
Leptochilus (*Lionotulus*) *lorestanicus* Gusenleitner, 1995a: 174, ♀; Iran.
Catalogues with Iranian records: Ebrahimi and Carpenter (2008), Rahmani *et al.* (2020).
Distribution in Iran: Lorestan (Gusenleitner, 1995a).
Distribution in the Middle East: Known only from Iran (Gusenleitner, 1995a, 2013a).
Extralimital distribution: None.

***Leptochilus* (*Neoleptochilus*) *masirahensis* Giordani Soika, 1981**
Leptochilus (*Neoleptochilus*) *masirahensis* Giordani Soika, 1981: 113, ♀, ♂; Oman.
Distribution in the Middle East: Known only from Oman (Giordani Soika, 1981; Guichard, 1986; Gusenleitner, 2013a; Carpenter and Gadallah, 2020).
Extralimital distribution: None.

***Leptochilus* (*Neoleptochilus*) *medanae falkenhayni* (Dusmet y Alonso, 1917)**
Odynerus (*Lionotus*) *falkenhayni* Dusmet, 1917: 365, ♀, ♂; "Marruecos: Mogador ... Amismiz ... Marraquesh".
Distribution in the Middle East: Israel-Palestine (Carpenter and Gadallah, 2020), Oman (Gusenleitner, 1993; Carpenter and Gadallah, 2020), Saudi Arabia (Gusenleitner, 2013a; Carpenter and Gadallah, 2020), United Arab Emirates (Gusenleitner, 2010b, 2012, 2013a; Carpenter and Gadallah, 2020).
Extralimital distribution: Morocco (Dusmet, 1917 as *Odynerus* (*Lionotus*) *falkenhayni*; Carpenter and Gadallah, 2020).

***Leptochilus* (*Neoleptochilus*) *medanae medanae* (Gribodo, 1886)**
Odynerus medanae Gribodo, 1886: 874, ♀; Tunisie.
Distribution in the Middle East: Egypt (Zalat *et al.*, 1993), Israel-Palestine, Saudi Arabia (Guichard, 1986; Carpenter and Gadallah, 2020).
Extralimital distribution: Algeria, France, Italy, Spain, Tunisia (Gusenleitner, 2013a).

***Leptochilus* (*Lionotulus*) *membranaceus luxuriosus* Blüthgen, 1955b**
Leptochilus (*Lionotulus*) *membranaceus luxuriosus* Blüthgen, 1955b: 25, ♀; "Jerusalem".
Distribution in the Middle East: Israel-Palestine (Blüthgen, 1955b; van der Vecht and Fischer, 1972; Gusenleitner, 1993), Jordan, Syria (Gusenleitner, 2023).
Extralimital distribution: None.

***Leptochilus* (*Lionotulus*) *membranaceus membranaceus* (Morawitz, 1867)**
Odynerus (*Leionotus*) *membranaceus* Morawitz, 1867: 135, ♀; Gouvernement von Saratov.

Catalogues with Iranian records: Ebrahimi and Carpenter (2008), Rahmani *et al.* (2020).
Distribution in Iran: Alborz (Giordani Soika, 1970).
Distribution in the Middle East: Iran (Giordani Soika, 1970; Gusenleitner, 2013a; Fateryga, 2017b, 2018; Fateryga *et al.*, 2021; Selis *et al.*, 2025), Iraq (Selis *et al.*, 2025), Israel-Palestine (Gusenleitner, 2013a; Fateryga, 2017b, 2018; Fateryga *et al.*, 2021; Selis *et al.*, 2025), Turkey (Blüthgen, 1951a; Gusenleitner, 1966, 2013a; Giordani Soika, 1970 as *Leptochilus* (*Neoleptochilus*) *m. membranaceusi*; Yildirim and Kojima, 1999; Fateryga, 2017b, 2018; Fateryga *et al.*, 2021; Selis *et al.*, 2025).
Extralimital distribution: Armenia, Azerbaijan, Bulgaria, Greece, Ukraine (Fateryga *et al.*, 2021; Selis *et al.*, 2025), Kazakhstan, Russia, Turkmenistan, Uzbekistan (Fateryga, 2017b, 2018; Fateryga *et al.*, 2021; Selis *et al.*, 2025), Italy (Selis, 2023; Selis *et al.*, 2025).

***Leptochilus* (*Leptochilus*) *mauritanicus derufatus* Blüthgen, 1955b**
Leptochilus mauritanicus derufatus Blüthgen, 1955b: 24, ♀; Israel.
Distribution in the Middle East: Known only from Israel-Palestine (Blüthgen, 1955b; Gusenleitner, 1970, 2013a as *L. m. derufatus*; van der Vecht and Fischer, 1972; Gusenleitner, 1993).
Extralimital distribution: None.

***Leptochilus* (*Neoleptochilus*) *mochianus mochianus* Giordani Soika, 1970**
Leptochilus (*Neoleptochilus*) *mochianus* Giordani Soika, 1970: 75, ♀; "Siria, Palmira".
Distribution in the Middle East: Known only from Syria (Giordani Soika, 1970; van der Vecht and Fischer, 1972; Gusenleitner, 2013a).
Extralimital distribution: None.

***Leptochilus* (*Euleptochilus*) *modestus* de Saussure, 1853**
Leptochilus modestus de Saussure, 1853: 234, ♀; "L'Algérie".
Distribution in the Middle East: Arabian Peninsula (no specific locality cited) (Gusenleitner, 2013a), Israel-Palestine, Syria (Giordani Soika, 1970; Gusenleitner, 2013a).
Extralimital distribution: Algeria (de Saussure, 1853; Blüthgen, 1956).

***Leptochilus* (*Lionotulus*) *montivagus* Gusenleitner, 2002**
Leptochilus (*Lionotulus*) *montivagus* Gusenleitner, 2002: 335, 338, ♂; "Türkei, 7 km W Darende, 1400 m".
Distribution in the Middle East: Known only from Turkey (Gusenleitner, 2002, 2013a).
Extralimital distribution: None.

***Leptochilus* (*Euleptochilus*) *muscatensis* Giordani Soika, 1979**
Leptochilus (*Euleptochilus*) *muscatensis* Giordani Soika, 1979: 273, ♀; "Muscat: Ruwi".
Distribution in the Middle East: Oman (Giordania Soika, 1979; Guichard, 1986; Gusenleitner, 2013a; Carpenter and Gadallah, 2020), United Arab Emirates (Gusenleitner, 2010b; Carpenter and Gadallah, 2020).
Extralimital distribution: None.

***Leptochilus* (*Lionotulus*) *nabataeus* Gusenleitner, 1990**
Leptochilus (*Lionotulus*) *nabataeus* Gusenleitner, 1990: 193, 199, ♂, ♀; "Jordanien, 80 km NE Aqaba (Straße nach Amman)".
Distribution in the Middle East: Israel-Palestine, Jordan (Gusenleitner, 1990, 2013a, 2016).
Extralimital distribution: None.

***Leptochilus* (*Lionotulus*) *neutraliformis* Gusenleitner, 1977**
Leptochilus (*Lionotulus*) *neutraliformis* Gusenleitner, 1977: 177, ♂, ♀; "Birecik, Türkei".
Distribution in the Middle East: Known only from Turkey (Gusenleitner, 1977, 2013a; Yildirim and Kojima, 1999).
Extralimital distribution: None.

***Leptochilus* (*Lionotulus*) *neutralis neutralis* (Giordani Soika, 1943)**

Odynerus (*Leptochilus*) *neutralis* Giordani Soika, 1943: 12, ♀; Asia minore, Amasia [Turkey].

Catalogues with Iranian records: Ebrahimi and Carpenter (2008), Rahmani *et al.* (2020).

Distribution in Iran: Kerman (Schmid-Egger *et al.*, 2021), Khuzestan (Giordani Soika, 1961).

Distribution in the Middle East: Iran (see references above, and Fateryga *et al.*, 2025), Israel-Palestine (Giordani Soika, 1952), Syria (Giordani Soika, 1970), Turkey (Giordani Soika, 1943 as *Odynerus* (*Leptochilus*) *neutralis*; Gusenleitner, 1966, 2013a; Giordani Soika, 1970; Yildirim and Kojima, 1999; Yildirim and Gusenleitner, 2009; Gusenleitner, 2013a; Fateryga *et al.*, 2025).

Extralimital distribution: Azerbaijan (Fateryga *et al.*, 2025).

***Leptochilus* (*Lionotulus*) *neutralis rufior* Giordani Soika, 1952**

Leptochilus neutralis rufior Giordani Soika, 1952: 52, ♀; "Kiriat".

Distribution in the Middle East: Israel-Palestine (Giordani Soika, 1952; Blüthgen, 1955b), Syria (Giordani Soika, 1970; Gusenleitner, 2013a).

Extralimital distribution: None.

***Leptochilus* (*Neoleptochilus*) *nigrocitrinus* Giordani Soika, 1970**

Leptochilus (*Neoleptochilus*) *nigrocitrinus* Giordani Soika, 1970: 76, ♀; "Siria: Damasco, Mezzé".

Distribution in the Middle East: Known only from Syria (Giordani Soika, 1970; van der Vecht and Fischer, 1972; Gusenleitner, 2013a).

Extralimital distribution: None.

***Leptochilus* (*Lionotulus*) *nigroclypeus* Gusenleitner, 1985b**

Leptochilus (*Lionotulus*) *nigroclypeus* Gusenleitner, 1985b: 81, 82, ♂; "Türkei, Nevsehir".

Distribution in the Middle East: Known only from Turkey (Gusenleitner, 1985a; Yildirim and Kojima, 1999).

Extralimital distribution: None.

***Leptochilus* (*Lionotulus*) *ornatulus* Gusenleitner, 1977**

Leptochilus orantulus Gusenleitner, 1977: 169, ♀; "Urfa, Türkei".

Distribution in the Middle East: Jordan, Turkey (Gusenleitner, 1977; Yildirim and Kojima, 1999; Gusenleitner, 2013a).

Extralimital distribution: None.

***Leptochilus* (*Lionotulus*) *osiris* (Schmiedeknecht, 1900)**

Odynerus (*Microdynerus*) *osiris* Schmiedeknect, 1900: 229, ♀, ♂; "In den Wadis der arabischen Wüste in Unterägypten".

Distribution in the Middle East: Egypt (Schmiedeknecht, 1900 as *Odynerus* (*Microdynerus*) *osiris*; Giordani Soika, 1938 as *O.* (*M.*) *osiris*, 1958; van der Vecht and Fischer, 1972; Zalat *et al.*, 1993; Gusenleitner, 2013a), Israel-Palestine (Zalat *et al.*, 1993).

Extralimital distribution: None.

***Leptochilus* (*Lionotulus*) *osmanicus* Gusenleitner, 1988**

Leptochilus (*Lionotulus*) *osmanicus* Gusenleitner, 1988: 713, 721, ♀; "Türkei, Provinz Hakkari, Habur Derese-Tal, S Beylisebap, 1100 m".

Distribution in the Middle East: Known only from Turkey (Gusenleitner, 1988, 2013a; Yildirim, 1998; Yildirim and Kojima, 1999).

Extralimital distribution: None.

***Leptochilus* (*Lionotulus*) *palandokenicus* Yildirim and Özbek, 1995**

Leptochilus (*Lionotulus*) *palandokenicus* Yildirim and Özbek, 1995: 241, 242, ♂; "Turkey, Erzurum, Palandöken, 2200 m".

Distribution in the Middle East: Known only from Turkey (Yildirim and Özbek, 1995; Yildirim and Kojima, 1999).

Extralimital distribution: None.

***Leptochilus* (*Lionotulus*) *perterritus* Giordani Soika, 1970**

Leptochilus (*Lionotulus*) *perterritus* Giordani Soika, 1970: 67, ♀, ♂; "Arabia: Dhufar, bassa pianura vicino Salala".

Distribution in the Middle East: Known only from Oman (Giordani Soika, 1970; Gusenleitner, 2013a; Carpenter and Gadallah, 2020).

Extralimital distribution: None.

***Leptochilus* (*Sarochilus*) *praestans* Giordani Soika, 1970**

Leptochilus (*Lionotulus*) *praestans* Giordani Soika, 1970: 71, ♀; "Sinai: W. Mitla".

Distribution in the Middle East: Egypt (Giordani Soika, 1970 as *Leptochilus* (*Lionotulus*) *praestans*; Zalat *et al.*, 1993; Gusenleitner, 2013a), Israel-Palestine (Giordani Soika, 1970 as *L.* (*Lionotulus*) *praestans*), Oman (Gusenleitner, 2013a), Palestine (Zalat *et al.*, 1993).

Extralimital distribution: None.

***Leptochilus* (*Lionotulus*) *pseudojosephi* Giordani Soika, 1952**

Leptochilus pseudojosephi Giordani Soika, 1952: 8, 51, ♂; Israel.

Distribution in the Middle East: Cyprus (Gusenleitner, 2013a), Israel-Palestine (Giordani Soika, 1952, 1973; Bytinski-Salz and Gusenleitner, 1971; van der Vecht and Fischer, 1972; Borsato and Ratti, 1999; Gusenleitner, 2013a).

Extralimital distribution: None.

***Leptochilus* (*Neoleptochilus*) *regulus* (de Saussure, 1855)**

Odynerus (*Odynerus* div. *Parodynerus*) *regulus* de Saussure, 1855: 247, ♀, ♂; « L'Algerie » [Algeria].

Catalogues with Iranian records: Ebrahimi and Carpenter (2008), Rahmani *et al.* (2020).

Distribution in Iran: East Azarbaijan (Ebrahimi and Carpenter, 2008; Modarres Awal, 2012), Semnan (Schmid-Egger *et al.*, 2021).

Distribution in the Middle East: Cyprus (Fateryga, 2017a, 2017b, 2018; Fateryga *et al.*, 2021), Iran (see references above, and Fateryga, 2017a, 2017b, 2018; Fateryga *et al.*, 2021), Israel-Palestine (Giordani Soika, 1952), Jordan, Syria (Giordani Soika, 1970; Gusenleitner, 2013a; Fateryga, 2017a, 2017b, 2018; Fateryga *et al.*, 2021), Turkey (Giordani Soika, 1952; Gusenleitner, 1966; Yildirim and Özbek, 1996b, 1999; Madl, 1997; Yildirim, 1998; Yildirim and Kojima, 1999; Yildirim and Gusenleitner, 2009, 2007, 2012; Gusenleitner, 2013a; Fateryga, 2017a, 2017b; Fateryga *et al.*, 2021).

Extralimital distribution: Armenia, Azerbaijan (Fateryga *et al.*, 2021), Europe, North Africa, Russia, Ukraine (Fateryga, 2017b, 2018; Fateryga *et al.*, 2021), Malta (Cassar *et al.*, 2022), Italy (Selis, 2023).

***Leptochilus* (*Lionotulus*) *ressli* Gusenleitner, 1985b**

Leptochilus (*Lionotulus*) *ressli* Gusenleitner, 1985b: 88, ♂; Iran.

Catalogues with Iranian records: Ebrahimi and Carpenter (2008), Rahmani *et al.* (2020).

Distribution in Iran: Lorestan (Gusenleitner, 1985b).

Distribution in the Middle East: Iran (Gusenleitner, 1985b; Gusenleitner, 2013a), Turkey (Gusenleitner, 1988; Yildirim and Kojima, 1999).

Extralimital distribution: None.

***Leptochilus* (*Neoleptochilus*) *salomon* Gusenleitner in Bytinski-Salz and Gusenleitner, 1971**

Leptochilus (*Neoleptochilus*) *salomon* Gusenleitner, 1971 [in Bytinski-Salz and Gusenleitner, 1971]: 239, 264, 292, ♀; Israel.

Distribution in the Middle East: Egypt (Zalat *et al.*, 1993), Israel-Palestine (Bytinski-Salz and Gusenleitner, 1971; Zalat *et al.*, 1993; Gusenleitner, 2013a).

Extralimital distribution: Morocco (Bytinski-Salz and Gusenleitner, 1971).

***Leptochilus* (*Neoleptochilus*) *schatzmayri* (Giordani Soika, 1938)**

Odynerus (*Neoleptochilus*) *schatzmayi* Giordani Soika, 1938: 13, ♀; "Sinai: Tor".

Distribution in the Middle East: Egypt (Giordani Soika, 1938 as *Odynerus* (*Leptochilus*) *schatzmayri*; Zalat *et al*., 1993; Gusenleitner, 2013a), Israel-Palestine (Giordani Soika, 1977 as *Leptochilus* (*Lionotulus*) *sulciventris*; Zalat *et al*., 1993).
Extralimital distribution: None.

***Leptochilus* (*Lionotulus*) *schindleri* (Guiglia, 1933)**
Odynerus schindleri Guiglia, 1933: 366, ♀; [replacement name for *Odynerus schulthessi* Guiglia, 1929].
Distribution in the Middle East: Egypt (Giordani Soika, 1938 as *Odynerus schindleri*; Zalat *et al*., 1993; Gusenleitner, 2013a).
Extralimital distribution: Libya (Gusenleitner, 2013a).

***Leptochilus* (*Lionotulus*) *signatus gueruensis* Gusenleitner, 1995b**
Leptochilus (*Lionotulus*) *signatus gueruensis* Gusenleitner, 1995b: 169, 172, ♀; "Turkei, Gürün".
Distribution in the Middle East: Known only from Turkey (Gusenleitner, 1995b, 2013a).
Extralimital distribution: None.

***Leptochilus* (*Lionotulus*) *signatus signatus* Gusenleitner, 1995b**
Leptochilus (*Lionotulus*) *signatus* Gusenleitner, 1995b: 169, 170, ♀; "N-Caucasus, Teberda".
Distribution in the Middle East: Turkey (Gusenleitner, 1995b, 2013a; Fateryga, 2017b).
Extralimital distribution: Russia (Gusenleitner, 2013a; Fateryga, 2017b).

***Leptochilus* (*Lionotulus*) *simurgh* Gusenleitner, 2016**
Leptochilus (*Lionotulus*) *simurgh* Gusenleitner, 2016: 91, ♀, ♂; Iran.
Catalogues with Iranian records: Rahmani *et al*. (2020).
Distribution in Iran: Alborz (Gusenleitner, 2016).
Distribution in the Middle East: Iran (Gusenleitner, 2016; Selis *et al*., 2025), Iraq (Selis *et al*., 2025).
Extralimital distribution: None.

***Leptochilus* (*Lionotulus*) *speciosus* Gusenleitner, 2002**
Leptochilus (*Lionotulus*) *speciosus* Gusenleitner, 2002: 335, 341, ♀; "Israel, Wadi Segur, 40 km NW Eilat".
Distribution in the Middle East: Known only from Israel-Palestine (Gusenleitner, 2002, 2013a).
Extralimital distribution: None.

***Leptochilus* (*Lionotulus*) *subtarsatellus* Gusenleitner, 2002**
Leptochilus (*Lionotulus*) *subtarsatellus* Gusenleitner, 2002: 335, 340, ♂; "Israel, Mt. Hermon".
Distribution in the Middle East: Known only from Israel-Palestine (Gusenleitner, 2002, 2013a).
Extralimital distribution: None.

***Leptochilus* (*Lionotulus*) *tarsatellus* Giordani Soika, 1970**
Leptochilus (*Lionotulus*) *tarsatellus* Giordani Soika, 1970: 73, ♂, ♀; "Libano: Kadish".
Distribution in the Middle East: Lebanon (Giordani Soika, 1970; Yildirim and Kojima, 1999; Gusenleitner, 2013a), Turkey (Gusenleitner, 1988; Yildirim and Kojima, 1999).
Extralimital distribution: None.

***Leptochilus* (*Lionotulus*) *tarsatiformis* (Giordani Soika, 1943)**
Odynerus (*Leptochilus*) *tarsatiformis* Giordani Soika, 1943: 14, ♂; "Palestina: Gerusalemme".
Distribution in the Middle East: Israel-Palestine (Giordani Soika, 1943 as *Odynerus* (*Leptochilus*) *tarsatiformis*; Bytinski-Salz and Gusenleitner, 1971; Yildirim and Kojima, 1999; Gusenleitner, 2013a), Turkey (Giordani Soika, 1970; Gusenleitner, 1988, 2013a; Yildirim and Gusenleitner, 2004).
Extralimital distribution: None.

***Leptochilus* (*Lionotulus*) *tarsatus* (de Saussure, 1855)**
Odynerus (*Odynerus*) *tarsatus* de Saussure, 1855: pl.XIII, fig. 1, ♂; Switzerland.

Distribution in the Middle East: Israel-Palestine (Giordani Soika, 1952; Bytinski-Salz and Gusenleitner, 1971; Yildirim and Kojima, 1999), Syria (Giordani Soika, 1970; Yildirim and Kojima, 1999), Turkey (Giordani Soika, 1970; Yildirim and Kojima, 1999).

Extralimital distribution: Austria, Croatia, Czech Republic, France, Germany, Italy, Portugal, Slovakia, Spain, Switzerland (Castro, 1997).

***Leptochilus* (*Lionotulus*) *torretassoi* (Giordani Soika, 1938)**

Odynerus (*Leptochilus*) *torre-tassoi* Giordani Soika, 1938: 12, ♂; «Egitto: Cairo».

Distribution in the Middle East: Egypt (Giordani Soika, 1938 as *Leptochilus* (*Leptochilus*) *torre-tassoi*; Zalat *et al.*, 1993 as *L. torretassoi torretassoi*; Gusenleitner, 2013a as *L.* (*L.*) *torretassoi torretassoi*).

Extralimital distribution: Algeria (Giordani Soika, 1938 as *Odynerus* (*Leptochilus*) *torre-tassoi*), Tunisia (Gusenleitner, 2013a).

***Leptochilus* (*Lionotulus*) *tricolor* Gusenleitner, 2016**

Leptochilus (*Lionotulus*) *tricolor* Gusenleitner, 2016: 89, ♂; Syria.

Distribution in the Middle East: Known only from Syria (Gusenleitner, 2016).

Extralimital distribution: None.

***Leptochilus* (*Lionotulus*) *turcicus* Gusenleitner in Bytinski-Salz and Gusenleitner, 1971**

Leptochilus (*Lionotulus*) *mimulus turcicus* Gusenleitner, 1971 [in Bytinski-Salz and Gusenleitner, 1971]: 239, 263, 291, ♀, ♂; "Turkey: Urfa".

Distribution in the Middle East: Iraq (Selis *et al.*, 2025), Israel-Palestine (Bytinski-Salz and Gusenleitner, 1971; Gusenleitner, 2013a both as *Leptochilus* (*Lionotulus*) *mimulus turcicus*; Selis *et al.*, 2025), Jordan (Castro and Dvořák, 2010 as *Leptochilus mimulus*; Yildirim and Kojima, 1999; Gusenleitner, 2013a; Selis *et al.*, 2025), Turkey (Bytinski-Salz and Gusenleitner, 1971; Yildirim and Kojima, 1999; Gusenleitner, 2013a; Selis *et al.*, 2025).

Extralimital distribution: None.

Comments. The subspecies, *L. mimulis turcicus* has been described by Gusenleitner in Bytinski-Salz and Gusenleitner (1971) from Turkey and Israel. He differentiated it from the nominotypical subspecies from the Balkan Peninsula based on few morphological characters (Selis *et al.*, 2025). However, examining the types as well as new material by Selis *et al.* (2025: 960), in addition to genetic data, revealed more characters that does not support the subspecific status of *turcicus* as a mere subspecies, leading to its elevation to the specific rank. This is also more supported by the geographic range (Selis *et al.*, 2025).

***Leptochilus* (*Neoleptochilus*) *tussaci* Giordani Soika, 1986**

Leptochilus (*Neoleptochilus*) *tussaci* Giordani Soika, 1986: 108, ♂; "Marocco: Errachidia".

Distribution in the Middle East: Israel-Palestine, Jordan (Gusenleitner, 2013a).

Extralimital distribution: Morocco, Tunisia (Gusenleitner, 2013a).

***Leptochilus* (*Lionotulus*) *villosus* Gusenleitner, 2001a**

Leptochilus (*Lionotulus*) *villosus* Gusenleitner, 2001a: 663, 667, ♂; "Jordanien: Rawayahid".

Distribution in the Middle East: Known only from Jordan (Gusenleitner, 2001a, 2013a).

Extralimital distribution: None.

Genus *Leptodynerus* Blüthgen, 1938

Leptodynerus Blüthgen, 1938: 448, 457. Type species: *Leptodynerus biskrensis* Blüthgen, 1938, by original designation and monotypy.

***Leptodynerus arabicus* Giordani Soika, 1970**

Leptodynerus arabicus Giordani Soika, 1970: 62, ♀; "Arabia Saudita: El Riyadh".

Distribution in the Middle East: Israel-Palestine (Gusenleitner, 2010b), Oman (Guichard, 1986; Gusenleitner, 2013a; Carpenter and Gadallah, 2020), Saudi Arabia (Giordani Soika, 1970;

Guichard, 1986; Gusenleitner, 2013a; Carpenter and Gadallah, 2020), United Arab Emirates (Gusenleitner, 2010b; Carpenter and Gadallah, 2020).
Extralimital distribution: None.

***Leptodynerus vanharteni* Gusenleitner, 2010b**

Leptodynerus vanharteni Gusenleitner, 2010b: 341, ♀, ♂; United Arab Emirates.
Distribution in the Middle East: Known only from United Arab Emirates (Gusenleitner, 2010b, 2013a; Carpenter and Gadallah, 2020).
Extralimital distribution: None.

Genus *Micreumenes* Ashmead, 1902

Micreumenes Ashmead, 1902: 208. Type species: *Micreumenes curriei* Ashmead, 1902, by monotypy.

***Micreumenes arabicus* Giordani Soika, 1979**

Micreumenes arabicus Giordani Soika, 1979: 276, ♀; "Oman: Nizwa".
Distribution in the Middle East: Oman (Giordani Soika, 1979; Guichard, 1986; Gusenleitner, 2013a; Carpenter and Gadallah, 2020), Yemen (Gusenleitner, 2007, 2013a; Carpenter and Gadallah, 2020).
Extralimital distribution: None.

***Micreumenes glaber glaber* Giordani Soika, 1983**

Micreumenes glaber Giordani Soika, 1983: 156, 172, sex not stated.
Distribution in the Middle East: Yemen (Gusenleitner, 2007, 2013a; Carpenter and Gadallah, 2020).
Extralimital distribution: Angola, Burkina Faso, Congo Republic, Ghana, Ivory Coast, Kenya, Namibia, Senegal, Sudan, Tanzania, Togo (Carpenter *et al.*, 2010; Carpenter and Gadallah, 2020).

Genus *Microdynerus* Thomson, 1874

Microdynerus Thomson, 1874: 58. Type species: *Odynerus exilis* Herrich-Schäffer, 1839; designated by Jones, 1937.

***Microdynerus* (*Alastorynerus*) *microdynerus* (Dalla Torre, 1889)**

Odynerus microdynerus Dalla Torre, 1889: 124, sex not stated; 124 [replacement name for *O. alastoroides* (Morawitz, 1885) *non* de Saussure, 1852].
Catalogues with Iranian records: Ebrahimi and Carpenter (2008), Rahmani *et al.* (2020).
Distribution in Iran: Iran (no specific locality cited) (Gusenleitner, 1997a; Fateryga, 2017b).
Distribution in the Middle East: Iran (see references above), Jordan, Turkey (Gusenleitner, 1967 as *Alastorynerus microdynerus*, 2001b; Yildirim and Özbek, 1996a, 1996b as *A. microdynerus*; Yildirim and Kojima, 1999; Yildirim and Gusenleitner, 2012; Yildirim and Tezcan, 2018; Fateryga, 2017b; Fateryga *et al.*, 2019).
Extralimital distribution: Armenia, Azerbaijan, Georgia, Kazakhstan, Russia, Tajikistan (Fateryga *et al.*, 2019), Caucasus, Europe (SE, EE) (Fateryga, 2017b).

***Microdynerus* (*Microdynerus*) *abdelkader abdelkader* (de Saussure, 1856)**

Odynerus (*Odynerus*) *Abd-El-Kader* de Saussure, 1856: 254, ♀, ♂; "La côte de l'Algérie, Oran".
Distribution in the Middle East: Syria (Giordani Soika, 1970; Yildirim and Kojima, 1999), Turkey (Giordani Soika, 1970; Gusenleitner, 1979; Yildirim and Kojima, 1999).
Extralimital distribution: North Africa, Southwest Europe (Gusenleitner, 2013a).

***Microdynerus* (*Microdynerus*) *abdelkader pharaonum* Giordani Soika, 1960**

Microdynerus abdelkader pharaonum Giordani Soika, 1960 (1958): 97, ♀; "Egitto, senza precisa localita".
Distribution in the Middle East: Egypt (Giordani Soika, 1960; Zalat *et al.*, 1993; Gusenleitner, 2013a).
Extralimital distribution: Tunisia (Gusenleitner, 1991).

***Microdynerus* (*Microdynerus*) *aegaeicus* Gusenleitner, 1998c**

Microdynerus aegaeicus Gusenleitner, 1998c: 155, 156, ♂, ♀; "Griechenland, Samos, Weg Pandroson/Karvounis, 800-1000 m, 37°44'N26°50'E".

Distribution in the Middle East: Turkey (Yildirim and Gusenleitner, 2001; Gusenleitner, 2013a).

Extralimital distribution: Greece (Gusenleitner, 2013a).

***Microdynerus* (*Microdynerus*) *appenninicus* Giordani Soika, 1960**

Microdynerus appenninicus Giordani Soika, 1960 (1958): 56, ♀; Italy.

Distribution in the Middle East: Iraq (Selis *et al.*, 2025), Jordan (Gusenleitner, 2013a; Selis *et al.*, 2025), Turkey (Gusenleitner, 1979, 2013a; Yildirim and Özbek, 1996a; Yildirim and Kojima, 1999; Selis *et al.*, 2025).

Extralimital distribution: Bulgaria, France, Greece, Italy, North Macedonia (Selis *et al.*, 2025), Italy (Gusenleitner, 2013a; Selis, 2023; Selis *et al.*, 2025).

***Microdynerus* (*Microdynerus*) *atriceps* Morawitz, 1895**

Microdynerus atriceps Morawitz, 1895: 485, ♀; Transcaucasia, Helenendorf [Azerbaijan].

Catalogues with Iranian records: Ebrahimi and Carpenter (2008), Rahmani *et al.* (2020).

Distribution in Iran: Qazvin (Gusenleitner, 1979).

Distribution in the Middle East: Iran (Gusenleitner, 1979, 2013a; Yildirim and Kojima, 1999), Israel-Palestine (Yildirim and Kojima, 1999; Gusenleitner, 2013a), Turkey (Gusenleitner, 1979, 2013a; Yildirim and Özbek, 1996a; Yildirim and Kojima, 1999).

Extralimital distribution: Azerbaijan, Turkmenistan (Morawitz, 1895), Greece (Gusenleitner, 2008).

***Microdynerus* (*Pseudomicrodynerus*) *anatolicus* (Blüthgen, 1938)**

Pseudomicrodynerus (*Pachymicrodynerus*) *anatolicus* Blüthgen, 1938: 456, ♀; Kleinasien [without details].

Catalogues with Iranian records: Ebrahimi and Carpenter (2008), Rahmani *et al.* (2020).

Distribution in Iran: Iran (no specific locality cited) (Ebrahimi and Carpenter, 2008; Gusenleitner, 2013a).

Distribution in the Middle East: Iran (see references above), Israel-Palestine (Yildirim and Kojima, 1999; Gusenleitner, 2013a), Syria (Giordani Soika, 1970 as *Pseudomicrodynerus anatolicus*; Yildirim and Kojima, 1999; Gusenleitner, 2013a), Turkey (Blüthgen, 1938 as *P.* (*Pachymicrodynerus*) *anatolicus*; Gusenleitner, 1979, 1988 both as *P.* (*Pachymicrodynerus*) *anatolicus*; Yildirim and Kojima, 1999; Gusenleitner, 2013a).

Extralimital distribution: None.

***Microdynerus* (*Microdynerus*) *confinis* Gusenleitner, 1979**

Microdynerus confinis Gusenleitner, 1979: 88, ♀, ♂; "Türkei, Mut, Sertavul, 1300 m".

Distribution in the Middle East: Iraq (Selis *et al.*, 2025), Turkey (Gusenleitner, 1979, 2013a; Yildirim and Özbek, 1996a; Yildirim and Kojima, 1999; Selis *et al.*, 2025).

Extralimital distribution: Bulgaria (Gusenleitner, 1979; Selis *et al.*, 2025).

***Microdynerus* (*Microdynerus*) *curdistanicus* Gusenleitner, 1988**

Microdynerus curdistanicus Gusenleitner, 1988: 713, 716, ♀, ♂; "Türkei, Hakkari, Sat Dag, Varegös SW Yüksekova, 1700 m".

Distribution in the Middle East: Known only from Turkey (Gusenleitner, 1988, 2007, 2013a; Yildirim and Kojima, 1999; Yildirim and Gusenleitner, 2007).

Extralimital distribution: None.

***Microdynerus* (*Microdynerus*) *erzincanensis* Yildirim and Özbek, 1995**

Microdynerus erzincanensis Yildirim and Özbek, 1995: 241, 243, ♀; "Turkey, Erzincan, Bahçe Kültürleri Araştırma Enstitüsü, 1250 m".

Distribution in the Middle East: Known only from Turkey (Yildirim and Özbek, 1995; Yildirim and Kojima, 1999; Gusenleitner, 2013a).

Extralimital distribution: None.

***Microdynerus* (*Pseudomicrodynerus*) *eurasius* (Blüthgen, 1938)**

Pseudomicrodynerus (*Pachymicrodynerus*) *eurasius* Blüthgen, 1938: 455, ♀, ♂; "Parnass" [Greece].

Distribution in the Middle East: Turkey (Gusenleitner, 1966, 1979 both as *Pseudomicrodynerus* (*Pachymicrodynerus*) *eurasius*; Giordani Soika, 1970 as *P.* (*P.*) *eurasius*; Yildirim and Özbek, 1996a, 1996b as *Pseudomicrodynerus eurasius*; Yildirim and Özbek, 1999 as *P.* (*P.*) *eurasius*; Yildirim and Kojima, 1999; Gusenleitner, 2013a; Fateryga *et al.*, 2021).

Extralimital distribution: Azerbaijan (Fateryga *et al.*, 2021), Bulgaria, Greece (Gusenleitner, 2013a).

***Microdynerus* (*Microdynerus*) *exilis* (Herrich-Schäffer, 1839)**

Odynerus exilis Herrich-Schäffer, 1839: 13, 30, ♀, ♂; locality not given.

Distribution in the Middle East: Turkey (Giordani Soika, 1970; Yildirim and Kojima, 1999; Fateryga, 2017b; Fateryga *et al.*, 2025).

Extralimital distribution: Azerbaijan (Fateryga *et al.*, 2025), Europe, North Africa, Russia (Fateryga, 2017b; Fateryga *et al.*, 2025), Italy (Selis, 2023), Slovenia (Gogala, 2022).

***Microdynerus* (*Pseudomicrodynerus*) *hoetzendorfi* (Dusmet, 1917)**

Odynerus (*Lionotus*) *höozendorfi* Dusmet, 1917: 367, ♂; "Marruecos, Mogador".

Distribution in the Middle East: Egypt (Giordani Soika, 1938 as *Pseudomicrodynerus felinus*; Zalat *et al.*, 1993 as *P. felinus*; Gusenleitner, 2013a).

Extralimital distribution: Morocco (Gusenleitner, 2013a).

***Microdynerus* (*Microdynerus*) *latro* Blüthgen, 1955b**

Microdynerus latro Blüthgen, 1955b: 27, ♀; Jericho.

Distribution in the Middle East: Israel-Palestine (Gusenleitner, 2013a), Syria (Castro and Dvořák, 2009), Turkey (Blüthgen, 1955b).

Extralimital distribution: None.

***Microdynerus* (*Microdynerus*) *longicollis longicollis* Morawitz, 1895**

Microdynerus longicollis Morawitz, 1895: 486, ♀; "Tauria: Aidanil".

Distribution in the Middle East: Turkey (Giordani Soika, 1970 as both *Microdynerus patagiatus* and *M. longicollis*; Gusenleitner, 2001a, 2013a; Yildirim and Kojima, 1999; Fateryga, 2017b, 2018).

Extralimital distribution: Europe, Russia, Turkmenistan (Fateryga, 2017b, 2018), Italy (Selis, 2023).

***Microdynerus* (*Microdynerus*) *mirandus* (Giordani Soika, 1947)**

Leptochilus (*Microdynerus*) *mirandus* Giordani Soika, 1947: 131, ♀; Sicilia.

Distribution in the Middle East: Turkey (Giordani Soika, 1970 as *Leptochilus* (*Microdynerus*) *mirandus*; Gusenleitner, 1979, 1988 as *L.* (*M.*) *mirandus*; Yildirim and Kojima, 1999; Fateryga, 2017b, 2018).

Extralimital distribution: France, Greece, Italy (Borsato and Ratti, 1999), Russia (Fateryga, 2017b, 2018).

***Microdynerus* (*Microdynerus*) *nugdunensis nugdunensis* (de Saussure, 1855)**

Odynerus (*Odynerus*) *nugdunensis* de Saussure, 1855: plate XIII, fig. 4, ♂; "environs de Genève et . . . de Paris".

Distribution in the Middle East: Turkey (Blüthgen, 1961; Giordani Soika, 1970; Giordani Soika, 1986 as *Leptochilus* (*Lionotulus*) *micans*; Gusenleitner, 1979, 2013a; Yildirim and Kojima, 1999; Yildirim and Gusenleitner, 2004; Fateryga, 2017b, 2018).

Extralimital distribution: Azerbaijan, Russia (Fateryga, 2017b, 2018), Middle to South Europe (Gusenleitner, 2013a), Italy (Selis, 2023), Slovenia (Gogala, 2022).

***Microdynerus* (*Pseudomicrodynerus*) *parvulus* (Herrich-Schäeffer, 1838)**

Odynerus parvulus Herrich-Schäeffer, 1838: 19, ♀, ♂; bei Regensburg [Germany].

Catalogues with Iranian records: Rahmani *et al.* (2020).

Distribution in Iran: Iran (no specific locality cited) (Gusenleitner, 1979 as *Pseudomicrodynerus* (*P.*) *parvuus*; Fateryga, 2017b, 2018).

Distribution in the Middle East: Iran (see references above), Turkey (Gusenleitner, 1966 as *P. parvulus*; Yildirima and Özbek, 1996a as *P. parvulus*; Yildirim, 1998 as *P. parvulus*; Yildirim and Kojima, 1999; Yildirim and Özbek, 1999 as *P. parvulus*; Yildirim and Gusenleitner, 2012; Yildirim and Tezcan, 2018; Fateryga, 2017b, 2018).

Extralimital distribution: Caucasus, Europe, Kazakhstan, Russia, Uzbekistan (Fateryga, 2017b, 2018), Italy (Selis, 2023), Slovenia (Gogala, 2022).

***Microdynerus* (*Microdynerus*) *rubronotatus rubiculus* Gusenleitner, 2000c**

Microdynerus rubiculus Gusenleitner, 2000c: 927, 930, ♀; "Türkei, Kahta (Adiyaman)".

Distribution in the Middle East: Iraq (Selis *et al.*, 2025), Turkey (Gusenleitner, 2000c, 2013a; Mokrousov, 2015; Selis *et al.*, 2025).

Extralimital distribution: None.

Comment: Originally described as *Microdynerus rubiculus* Gusenleitner, 2000c. In his regional revision, Mokrousov (2015) concluded that *M. rubronotatus* (Kostylev, 1940) represents a subspecies of *M. rubronotatus*. Thus, *rubiculus* was reduced to subspecific rank as *Microdynerus* (*Microdynerus*) *rubronotatus rubiculus*.

***Microdynerus* (*Microdynerus*) *saundersi saundersi* Blüthgen, 1955a**

Microdynerus saundersi Blüthgen, 1955a: 415, ♀; Algeria.

Distribution in the Middle East: Israel-Palestine (Blüthgen, 1955a).

Extralimital distribution: Algeria, Morocco (Gusenleitner, 2013a).

***Microdynerus* (*Microdynerus*) *syriacus* Gusenleitner, 2000c**

Microdynerus syriacus Gusenleitner, 2000c: 927, 931, ♀; "Syria mer., Ganawat".

Distribution in the Middle East: Known only from Syria (Gusenleitner, 2000a, 2013a).

Extralimital distribution: None.

***Microdynerus* (*Microdynerus*) *urdunus* Gusenleitner, 2004**

Microdynerus urdunus Gusenleitner, 2004: 1093, 1096, ♀; "Jordanien NW, Irbid reg., 200 m, Aaham vill.".

Distribution in the Middle East: Iraq (Selis *et al.*, 2025), Jordan, Syria (Gusenleitner, 2004, 2013a; Selis *et al.*, 2025), Turkey (Gusenleitner, 2006d, 2013a; Selis *et al.*, 2025).

Extralimital distribution: None.

Genus *Odynerus* Latreille, 1802

Odynerus Latreille, 1802: 362. Type species: *Vespa spinipes* Linnaeus, 1758; designated by Shuckard, 1837.

***Odynerus* (*Spinicoxa*) *albopictus* de Saussure, 1856**

Odynerus (*Epipona* div. *Hoplopus*) *albopictus* de Saussure, 1856: 312, ♂; "L'ile de Rhodes" [Greece].

Catalogues with Iranian records: Ebrahimi and Carpenter (2008) as *Odynerus* (*Spinicoxa*) *albopictus calcaratus* (Morawitz, 1885) - synonym), Rahmani *et al.* (2020) as *O. albopictus calcaratus*).

Distribution in Iran: East Azarbaijan (Ebrahimi and Carpenter, 2008; Modarres Awal, 2012 both as *O. albopictus calcaratus*), Khuzestan (Blüthgen and Gusenleitner, 1970 as *O. albopictus calcaratus*), Iran (no specific locality cited) (Blüthgen and Gusenleitner, 1970; Fateryga, 2018 both as *O. albopictus calcaratus*; Feteryga *et al.*, 2021; Selis *et al.*, 2025).

Distribution in the Middle East: Iran (see references above), Iraq (Gusenleitner, 1998a; 2013a as *O. albopictus calcaratus*; Fateryga, 2017b, 2018; Fateryga *et al.*, 2021; Selis *et al.*, 2025), Israel-Palestine, Jordan (Gusenleitner, 2013a as *O. albopictus calcaratus*; Fateryga, 2017b, 2018 as *O. albopictus calcaratus*; Fateryga *et al.*, 2021; Selis *et al.*, 2025), Syria (Giordani Soika, 1970; Gusenleitner, 2013a; Fateryga, 2018 all as *O. albopictus calcaratus*), Turkey (Gusenleitner, 1967, 2001a, 2013a; Giordani Soika, 1970 as *O. albopictus calcaratus*; Yildirim and Özbek, 1996a,

1996b, 1999; Yildirim and Kojima, 1999; Fateryga, 2017b, 2018; Fateryga *et al.*, 2021; Selis *et al.*, 2025).

Extralimital distribution: Armenia (Giordani Soika, 1970; Fateryga *et al.*, 2021; Selis *et al.*, 2025), Azerbaijan (Fateryga *et al.*, 2021; Selis *et al.*, 2025), Bulgaria, France, Greece, Kazakhstan, Morocco, North Macedonia, Spain, Portugal, Russia, Turkmenistan, Uzbekistan, (Fateryga *et al.*, 2021; Selis *et al.*, 2025), Italy (Selis, 2023; Selis *et al.*, 2025).

***Odynerus* (*Spinicoxa*) *alexandrinus* de Saussure, 1853**

Odynerus (*Oplopus*) *alexandrinus* de Saussure, 1853: 218, 225, ♀, ♂; «L'Egypte».

Distribution in the Middle East: Egypt (de Saussure, 1853; von Schulthess, 1928; Giordani Soika, 1935; Zalat *et al.*, 1993; Gusenleitner, 2001b, 2013a), Israel-Palestine (von Schulthess, 1928; Giordani Soika, 1952, 1977 as *O.* (*Odynerus*) *cauponalis*; Bytinski-Salz and Gusenleitner, 1972; Gusenleitner, 2001b, 2013a), Jordan (Gusenleitner, 2001b), Saudi Arabia (de Saussure, 1853 as *O.* (*Oplopus*) *savignyi*; Zalat *et al.*, 1993; Gusenleitner, 2013a; Carpenter and Gadallah, 2020), Syria (Giordani Soika, 1970; Gusenleitner, 2013a).

Extralimital distribution: None.

***Odynerus* (*Odynerus*) *armeniacus* (Morawitz, 1885)**

Hoplomerus armeniacus Morawitz, 1885: 157, ♀, ♂; Armenia.

Catalogues with Iranian records: Ebrahimi and Carpenter (2008), Rahmani *et al.* (2020).

Distribution in Iran: Alborz (Giordani Soika, 1970), East Azarbaijan, Golestan (Ebrahimi and Carpenter, 2008; Modarres Awal, 2012).

Distribution in the Middle East: Cyprus, Lebanon (Selis *et al.*, 2025), Iran (see references above, and du Buysson, 1912; Kostylev, 1929 as *Hoplomerus armeniacus*; Fateryga *et al.*, 2021; Selis *et al.*, 2025; Kaplan and Efil, 2025 as *Odynerus melanocephalus armeniacus* Morawitz, 1885), Iraq (Morice, 1921; Derwesh, 1965; Augul, 2017 as *Odynerus spinicus*; Selis *et al.*, 2025; Kaplan and Efil, 2025 as *O. m. armeniacus*), Israel-Palestine (Giordani Soika, 1952 as *Odynerus plurinotatus*; Yildirim and Kojima, 1999; Fateryga *et al.*, 2021; Selis *et al.*, 2025; Kaplan and Efil, 2025 as *O. m. armeniacus*), Jordan (Fateryga *et al.*, 2021; Selis *et al.*, 2025; Kaplan and Efil, 2025 as *O. m. armeniacus*), Syria (Giordani Soika, 1970 as *H. armeniacus*; Yildirim and Kojima, 1999; Fateryga *et al.*, 2021; Selis *et al.*, 2025; Kaplan and Efil, 2025 as *O. m. armeniacus*), Turkey (Blüthgen, 1951a; Gusenleitner, 1966, 1967 all as *H. armeniacus*; Giordani Soika, 1970; Yildirim and Özbek, 1996a, 1996b, 1999; Madl, 1997; Yildirim and Kojima, 1999; Yildirim and Gusenleitner, 2004, 2012; Gusenleitner, 2007; Yildirim and Tezcan, 2018; Fateryga *et al.*, 2021; Selis *et al.*, 2025).

Extralimital distribution: Afghanistan (Blüthgen, 1961 as *H. armeniacus*; Selis *et al.*, 2025; Kaplan and Efil, 2025 as *O. m. armeniacus*), Armenia (Blüthgen, 1941 as *Hoplomerus* (*Hoplomerus*) *armeniacus* var. *plurinotata*; Fateryga *et al.*, 2021; Selis *et al.*, 2025; Kaplan and Efil, 2025 as *O. m. armeniacus*), Azerbaijan (Fateryga *et al.*, 2021; Selis *et al.*, 2025; Kaplan and Efil, 2025 as *O. m. armeniacus*).

Comments. Based on morphological and ecological data, *Odynerus armeniacus*, was treated by Eck (1971) as a subspecies of *Odynerus melanocephalus* (Selis *et al.*, 2025). This classification was followed by several authors except Kurzenko (1977), who raised it as a valid species. Using a combined data of morphological and genetics by Selis *et al.* (2025), their results do not support Eck's treatment as a subspecies, and instead support *O. armeniacus* as a species.

***Odynerus* (*Monoplomerus*) *consobrinus consobrinus* Dufour, 1839**

Odynerus consobrinus Dufour, 1839: 91, ♀; "environs de Saint-Sever".

Distribution in the Middle East: Egypt (Giordani Soika, 1935).

Extralimital distribution: North Africa, Southwest Europe (Gusenleitner, 2013a), Italy (Selis, 2023).

***Odynerus* (*Odynerus*) *ezechiae* von Schulthess, 1923**

Odynerus (*Hoplopus*) *ezechiae* von Schulthess, 1923 [in Friese and von Schulthess, 1923]: 290, ♂; Constantinople [Turkey], Jerusalem [Israel].

Catalogues with Iranian records: Ebrahimi and Carpenter (2008), Rahmani *et al.* (2020).
Distribution in Iran: Iran (no specific locality cited) (Gusenleitner, 1998a, 2013a; Yildirim and Kojima, 1999).
Distribution in the Middle East: Iran (see references above), Israel-Palestine (Friese and von Schulthess, 1923 as *Odynerus* (*Hoplopus*) *ezechiae*; Blüthgen, 1941 as *Hoplomerus* (*Hoplomerus*) *ezechiae*; Yildirim and Kojima, 1999; Gusenleitner, 2013a), Jordan (Yildirim and Kojima, 1999; Gusenleitner, 2013a), Turkey (Giordani Soika, 1970; Yildirim and Özbek, 1996a; Yildirim and Kojima, 1999; Gusenleitner, 2013a).
Extralimital distribution: Greece, Uzbekistan (Gusenleitner, 2013a).

***Odynerus* (*Odynerus*) *femoratus* de Saussure, 1856**
Odynerus (*Epipona* div. *Hoplopus*) *femoratus* de Saussure, 1856: 310, ♀, ♂; «environs de Paris» [France].
Catalogues with Iranian records: Ebrahimi and Carpenter (2008), Rahmani *et al.* (2020).
Distribution in Iran: Alborz (Giordani Soika, 1970).
Distribution in the Middle East: Iran (see references above, and Gusenleitner, 2013a; Fateryga, 2017b, 2018; Fateryga *et al.*, 2021; Selis *et al.*, 2025), Iraq (Selis *et al.*, 2025), Israel-Palestine, Syria (Gusenleitner, 2013a; Fateryga, 2017b, 2018; Fateryga *et al.*, 2021; Selis *et al.*, 2025), Turkey (Blüthgen, 1961 as *Oplomerus* (*Oplomerus*) *femoratus*; Giordani Soika, 1970; Yildirim and Özbek, 1996a; Yildirim and Kojima, 1999; Yildirim and Özbek, 1999; Yildirim and Gusenleitner, 2004, 2012; Gusenleitner, 2013a; Fateryga, 2017b, 2018; Fateryga *et al.*, 2021; Selis *et al.*, 2025).
Extralimital distribution: Austria, Bulgaria, France, Hungary Portugal, Spain (Selis *et al.*, 2025), Azerbaijan (Fateryga *et al.*, 2021; Selis *et al.*, 2025), Russia (Fateryga, 2017b, 2018), Italy (Selis, 2023; Selis *et al.*, 2025).

***Odynerus* (*Spinicoxa*) *fulvitarsis* (Morawitz, 1895)**
Hoplomerus fulvitarsis Morawitz, 1895: 437, ♀; Azerbaijan.
Distribution in the Middle East: Turkey (Gusenleitner, 2001a; Fateryga, 2017a, 2017b).
Extralimital distribution: Azerbaijan, Kazakhstan, Russia, Turkmenistan, Uzbekistan (Fateryga, 2017a, 2017b).

***Odynerus* (*Odynerus*) *guichardi* Giordani Soika, 1977**
Odynerus (*Odynerus*) *guichardi* Giordani Soika, 1977 (1976): 154, ♂, ♀; "Israel: Gerusalemme".
Distribution in the Middle East: Known only from Israel-Palestine (Giordani Soika, 1977; Gusenleitner, 2013a).
Extralimital distribution: None.

***Odynerus* (*Monoplomerus*) *laticinctus laticinctus* (Białynicki-Birula, 1926)**
Hoplomerus laevipes laticinctus Białynicki-Birula, 1926: 904, 905, ♂; "Mittel-Asien, Umgebung der Stadt Semipalatinsk".
Catalogues with Iranian records: None.
Distribution in Iran: Kerman (Schmid-Egger *et al.*, 2021).
Distribution in the Middle East: Iran (Schmid-Egger *et al.*, 2021).
Extralimital distribution: Kazakhstan, Tajikistan, Turkmenistan, Uzbekistan (Castro and Dvořák, 2010; Schmid-Egger *et al.*, 2021).

***Odynerus* (*Odynerus*) *leucopus* (Blüthgen, 1941)**
Hoplomerus (*Hoplomerus*) *leucopus* Blüthgen, 1941: 310, 331, ♀, ♂; Iran.
Catalogues with Iranian records: Ebrahimi and Carpenter (2008), Rahmani *et al.* (2020).
Distribution in Iran: Bushehr (Blüthgen, 1941 as *Hoplomerus leucopus* Blüthgen, 1941), Ilam (Giordani Soika, 1970; Ebrahimi and Carpenter, 2008; Modarres Awal, 2012), Kermanshah, Khuzestan (Giordani Soika, 1970).

Distribution in the Middle East: Iran (see references above, and Gusenleitner, 2013a), Israel-Palestine (Gusenleitner, 2013a), Syria (Giordani Soika, 1970 as *Odynerus leucopus*; Gusenleitner, 2013a).
Extralimital distribution: None.

***Odynerus* (*Odynerus*) *melanocephalus melanocephalus* (Gmelin, 1790)**
Vespa melanocephalus Gmelin, 1790: 2760, sex not stated. Europa (destroyed).
Distribution in the Middle East: Cyprus (Eck, 1971 as *Odynerus melanocephalus cypriacus*; Fateryga, 2017b, 2018), Iraq (Fateryga, 2017b, 2018), Israel-Palestine (Giordani Soika, 1952; Yildirim and Kojima, 1999; Gusenleitner, 2013a; Fateryga, 2017b), Jordan (Yildirim and Kojima, 1999; Fateryga, 2017b, 2018), Syria, Turkey (Yildirim and Özbek, 1996a, 1999; Yildirim and Kojima, 1999; Fateryga, 2017b, 2018).
Extralimital distribution: Afghanistan, Caucasus, Central Asia, Europe, Kazakhstan, Russia (Fateryga, 2017b, 2018), Italy (Selis, 2023), Slovenia (Gogala, 2022).
Comments: Yildirim and Kojima, (1999) and Gusenleitner (2013a) recorded *Odynerus* (*Odynerus*) *melanocephalus melanocephalus* in Iran, without citing exact localities. Additionally, Fateryga (2017b, 2018) considered Iran (without exact locality) as a distributional region for *Odynerus* (*Odynerus*) *melanocephalus* (Gmelin) [unknown subspecies]. Therefore, this record should be treated with suspicion until otherwise confirmed.

***Odynerus* (*Spinicoxa*) *nigrospinosus* (Morawitz, 1895)**
Hoplomerus nigrospinosus Morawitz, 1895: 436, ♀, ♂; Transcaspia, Serachs [Turkmenistan].
Catalogues with Iranian records: Ebrahimi and Carpenter (2008), Rahmani *et al.* (2020).
Distribution in Iran: Khuzestan (Blüthgen and Gusenleitner, 1970).
Distribution in the Middle East: Egypt (Zalat *et al.*, 1993; Gusenleitner, 2013a; Fateryga *et al.*, 2023), Iran (Blüthgen and Gusenleitner, 1970; Fateryga *et al.*, 2023), Israel-Palestine (Gusenleitner, 2013a; Fateryga *et al.*, 2023), Jordan (Giordani Soika, 1952; Fateryga *et al.*, 2023), Turkey (Gusenleitner, 2013a; Fateryga *et al.*, 2023).
Extralimital distribution: Armenia, Turkmenistan (Gusenleitner, 2013a), Uzbekistan (Fateryga *et al.*, 2023).

***Odynerus* (*Monoplomerus*) *palaestinensis* Gusenleitner in Bytinski-Salz and Gusenleitner, 1971**
Odynerus (*Allogymnomerus*) *palaestinensis* Gusenleitner, 1971 [in Bytinski-Salz and Gusenleitner, 1971]: 239, 256, 288, ♂, ♀; "Israel, Deganyah".
Distribution in the Middle East: Known only from Israel-Palestine (Bytinski-Salz and Gusenleitner, 1971 as *Odynerus* (*Allogymnomerus*) *palaestinensis*; Gusenleitner, 2013a).
Extralimital distribution: None.

***Odynerus* (*Odynerus*) *poecilus* de Saussure, 1856**
Odynerus (*Epipona*) *poecilus* de Saussure, 1856: 314, ♀; "Le midi de la France".
Distribution in the Middle East: Turkey (Gusenleitner, 1966, 1967, 2013a; Giordani Soika, 1970; Yildirim and Özbek, 1996a, 1996b, 1999; Madl, 1997; Yildirim and Kojima, 1999; Fateryga, 2017b).
Extralimital distribution: Azerbaijan, Europe, Russia (Fateryga, 2017b), Italy (Selis, 2023), Slovenia (Gogala, 2022).

***Odynerus* (*Spinicoxa*) *reniformis* (Gmelin, 1790)**
Vespa reniformis Gmelin, 1790: 2760 [♂]; "in Europa".
Distribution in the Middle East: Israel-Palestine (Gusenleitner, 2013a; Fateryga, 2017b), Turkey (Blüthgen, 1959 as *Odynerus* (*Spinicoxa*) *reniformis balcanicus*; Giordani Soika, 1970; Yildirim and Özbek, 1996a as *O.* (*S.*) *reniformis balcanicus*; Gusenleitner, 2013a; Fateryga, 2017b; Fateryga *et al.*, 2020).
Extralimital distribution: Europe, Kazakhstan, Russia (Fateryga, 2017b; Fateryga *et al.*, 2020), North Africa (Gusenleitner, 2013a; Fateryga, 2017b), Italy (Selis, 2023), Slovenia (Gogala, 2022).

Comments: Based on Gusenleitner (2013a), various indications of the presence of this species should be checked.
Comments: *Odynerus reniformis* was excluded from the fauna of Iraq by Selis *et al.* (2025). It was recorded as the fauna of Iraq by Derwesh (1965).

Odynerus* (*Odynerus*) *rotundigaster citripes **Gusenleitner, 1970**
Odynerus (*Odynerus*) *rotundigaster citrinus* Gusenleitner, 1970: 55, ♀; "Jerusalem".
Distribution in the Middle East: Known only from Israel-Palestine (Gusenleitner, 1970, 2013a; van der Vecht and Fischer, 1972).
Extralimital distribution: None.

Odynerus* (*Odynerus*) *rotundigaster rotundigaster **de Saussure, 1853**
Odynerus (*Oplopus*) *rotundigaster* de Saussure, 1853: 221, ♀; "Alger".
Distribution in the Middle East: Turkey (Gusenleitner, 1966, 1967; Giordani Soika, 1970; Yildirim and Özbek, 1996a; Yildirim and Kojima, 1999).
Extralimital distribution: Algeria (de Saussure, 1853 as *O.* (*Oplopus*) *rotundigaster*), Greece (Morawitz, 1885 as *Hoplomerus congener*), Italy (Blüthgen, 1955b as *Hoplomeus rotundigaster extimus*; Selis, 2023).

Odynerus* (*Odynerus*) *serricrus **(Blüthgen, 1963)**
Oplomerus (*Hoplomerus*) *serricrus* Blüthgen, 1963: 205, sex not stated; Russia.
Distribution in the Middle East: Turkey (Yildirim and Özbek, 1996b, 1999; Yildirim and Kojima, 1999; Gusenleitner, 2013a; Fateryga, 2017b).
Extralimital distribution: Kazakhstan (Castro and Dvořák, 2009; Fateryga, 2017b), Kyrgyzstan (Fateryga, 2017b), Russia (Gusenleitner, 2013a; Fateryga, 2017b).

Odynerus* (*Spinicoxa*) *shestakovae **(Kostylev, 1935)**
Oplomerus shestakovae Kostylev, 1935 (1934): 124, ♀, ♂; "Bagir (Aschabad)".
Catalogues with Iranian records: None.
Distribution in Iran: Kerman (Schmid-Egger *et al.*, 2021).
Distribution in the Middle East: Iran (Schmid-Egger *et al.*, 2021).
Extralimital distribution: Turkmenistan (Kostylev, 1935 as *Oplomerus shestakovae*).

Odynerus* (*Spinicoxa*) *simillimus **Morawitz, 1867**
Odynerus (*Epipona*) *simillimus* Morawitz, 1867: 138, ♀, ♂; Russia.
Distribution in the Middle East: Israel-Palestine (Giordani Soika, 1952; Gusenleitner, 2013a; Fateryga, 2017b, 2018), Turkey (Yildirim and Özbek, 1996a, 1999; Yildirim and Kojima, 1999; Yildirim and Gusenleitner, 2004; Gusenleitner, 2013a; Fateryga, 2017b, 2018).
Extralimital distribution: Caucasus, Europe, Kazakhstan, Mongolia, Russia, Turkmenistan (Fateryga, 2017b, 2018).

Odynerus* (*Odynerus*) *spinipes spinipes **(Linnaeus, 1758)**
Vespa spinipes Linnaeus, 1758: 573 [♂]; "in Svecia".
Distribution in the Middle East: Turkey (Yildirim and Özbek, 1996a, 1996b; Madl, 1997; Yildirim and Kojima, 1999; Yildirim and Gusenleitner, 2004, 2012; Gusenleitner, 2013a; Fateryga, 2017b).
Extralimital distribution: Caucasus, Europe, Kazakhstan, Mongolia, Russia, Tajikistan (Fateryga, 2017b), Italy (Selis, 2023), Slovenia (Gogala, 2022).
Comments: *Odynerus spinipes spinipes* was excluded from the fauna of Iraq by Selis *et al.* (2025). It was listed as the fauna of Iraq by Khalaf (1958 as *Hoplomerus spinipes*), Derwesh (1965), Augul (2017), Al-Saffar and Augul (2021).

Odynerus* (*Odynerus*) *alpinus **von Schulthess, 1897**
Odynerus spinipes var. *alpinus* von Schulthess, 1897: 114, ♀, ♂; "Alpenregion" various localities.
Distribution in the Middle East: Turkey (Fateryga, 2017b).
Extralimital distribution: Europe (WE, SE, EE), Mongolia, Russia, Tajikistan (Fateryga, 2017b).

Genus *Omicrabulus* Giordani Soika, 1987

Omicrabulus Giordani Soika, 1987: 154. Type species: *Labus saganensis* Giordani Soika, 1944, by original designation.

***Omicrabulus arabicus* Gusenleitner, 2002**

Omicrabulus arabicus Gusenleitner, 2002: 307, 311, ♀; "N Yemen, Khamir, Bani Said".

Distribution in the Middle East: Yemen (Gusenleitner, 2002; Carpenter and Gadallah, 2020).

Extralimital distribution: Tanzania (Gusenleitner, 2003; Carpenter *et al.*, 2010; Carpenter and Gadallah, 2020).

Genus *Onychopterocheilus* Blüthgen, 1955b

Onychopterocheilus Blüthgen, 1955b: 407. Type species: *Odynerus (Hoplomerus) daw* Dusmet y Alonso, 1903, by original designation.

***Onychopterocheilus* (*Onychopterocheilus*) *albidus albidus* (Giordani Soika, 1941)**

Pterochilus albidus Giordani Soika, 1941: 266, ♂; «Egitto: Ikingi Mariut».

Distribution in the Middle East: Egypt (Giordani Soika, 1941 as *Pterochilus albidus*; Zalat *et al.*, 1993 as *P. albidus albidus*; Gusenleitner, 2013a).

Extralimital distribution: Libya (Blüthgen, 1955b as *Pterochilus albidus*).

***Onychopterocheilus* (*Onychopterocheilus*) *albidus aramaeus* (Blüthgen, 1955b)**

Pterocheilus albidus aramaeus Blüthgen, 1955b: 29, ♂, ♀; "Kfar Yeroham (Negev)".

Distribution in the Middle East: Known only from Israel-Palestine (Blüthgen, 1955b as *Pterocheilus albidus aramaeus*; van der Vecht and Fischer, 1972 as *P. a. aramaeus*; Gusenleitner, 2013a).

Extralimital distribution: None.

***Onychopterocheilus* (*Onychopterocheilus*) *albopictus* (Kriechbaumer, 1869)**

Pterochilus albopictus Kriechbaumer, 1869: 599, ♀; «Syria».

Distribution in the Middle East: Syria (Gusenleitner, 2013a), Turkey (Gusenleitner, 1967 as *Pterocheilus albopictus*, 2013a; Yildirim and Kojima, 1999).

Extralimital distribution: South and Southeast Europe (Gusenleitner, 2013a), Italy (Selis, 2023).

***Onychopterocheilus* (*Onychopterocheilus*) *anatolicus* (Blüthgen, 1955b)**

Pterocheilus anatolicus Blüthgen, 1955b: 95, ♀; "Ankara".

Distribution in the Middle East: Known only from Turkey (Blüthgen, 1955b as *Pterocheilus anatolicus*; Giordani Soika, 1970 as *P. anatolicus*; Gusenleitner, 1966 as *P. anatolicus*, 2013a; van der Vecht and Fischer, 1972 as *P. anatolicus*; Kurzenko, 1988; Yildirim and Kojima, 1999).

Extralimital distribution: None.

***Onychopterocheilus* (*Acutopterocheilus*) *atrohirtus* (Morawitz, 1885)**

Pterochilus atrohirtus Morawitz, 1885: 142, ♀, ♂; "Graecia, Syra".

Distribution in the Middle East: Turkey (Gusenleitner, 1988 as *Pterocheilus atrohirtus*, 2013a; Yildirim and Kojima, 1999).

Extralimital distribution: Greece (Morawitz, 1885 as *Pterochilus atrohirtus*), Turkmenistan (Gusenleitner, 2013a).

***Onychopterocheilus* (*Onychopterocheilus*) *bytinskii* (Gusenleitner, 1970)**

Pterocheilus bytinskii Gusenleitner, 1970: 56, ♀; Israel: Wadi Raman.

Distribution in the Middle East: Israel-Palestine (Gusenleitner, 1970 as *Pterocheilus bytinskii*, 2013a).

Extralimital distribution: Morocco (Gusenleitner, 1970 as *Pterochilus bytinskii*).

***Onychopterocheilus (Onychopterocheilus) coctus* (Giordani Soika, 1970)**

Pterocheilus coctus Giordania Soika, 1970: 42, ♂; «Sinai: Wadi Mitla».

Distribution in the Middle East: Known only from Egypt (Giordani Soika, 1970 as *Pterocheilus coctus*; van der Vecht and Fischer, 1972 as *P. coctus*; Kurzenko, 1988; Zalat *et al.*, 1993 as *P. coctus*; Gusenleitner, 2013a).
Extralimital distribution: None.

***Onychopterocheilus* (*Onychopterocheilus*) *cyanipennis* (André, 1895)**
Pterocheilus cyanipennis André, 1895: 355.
Distribution in the Middle East: Egypt, Israel-Palestine (Gusenleitner, 2013a).
Extralimital distribution: Algeria, Tunisia (Gusenleitner, 2013a).

***Onychopterocheilus* (*Asiapterocheilus*) *ecarinatus* (Morawitz, 1895)**
Pterochilus ecarinatus Morawitz, 1895: 426, ♀; "Turkestan: Jagnob: Schach-sara".
Distribution in the Middle East: Turkey (Yildirim and Özbek, 1996a as *Pterocheilus ecarinatus*; Yildirim and Kojima, 1999; Yildirim and Gusenleitner, 2007; Gusenleitner, 2013a).
Extralimital distribution: Armenia, Kazakhstan (Gusenleitner, 2013a), Turkmenistan (Morawitz, 1895 as *Pterochilus ecarinatus*).

***Onychopterocheilus* (*Neopterocheilus*) *fausti aurantiopictus* (Giordani Soika, 1970)**
Pterocheilus fausti aurantiopictus Giordani Soika, 1970: 49, ♀; "Turchia: Kayseri, Sultanhani, 1200 m".
Distribution in the Middle East: Known only from Turkey (Giordani Soika, 1970 as *Pterocheilus fausti aurantiopictus*; van der Vecht and Fischer, 1972 as *P. f. aurantiopictus*; Gusenleitner, 1988, 2013a as *P. f. aurantiopictus*; Kurzenko, 1988; Yildirim and Kojima, 1999).
Extralimital distribution: None.

***Onychopterocheilus* (*Neopterocheilus*) *fausti fausti* (Morawitz, 1873)**
Pterochilus fausti Morawitz, 1873: 296, ♀, ♂; "Krasnowodsk".
Distribution in the Middle East: Arabia (no specific locality cited; Giordani Soika, 1957 as *Pterocheilus fausti*), Egypt (Giordani Soika, 1942 as *P. lutorius*; Zalat *et al.*, 1993 as *P. lutorius*), Kuwait (Al-Houty, 1989, 1997 both as *P. fousti*; Carpenter and Gadallah, 2020), Saudi Arabia (Guichard, 1986 as *P. fausti*; Gusenleitner, 2013a; Carpenter and Gadallah, 2020), United Arab Emirates (Gusenleitner, 2010b; Carpenter and Gadallah, 2020).
Extralimital distribution: Turkmenistan (Gusenleitner, 2013a).

***Onychopterocheilus* (*Neopterocheilus*) *fausti paravespoides* (Blüthgen, 1955b)**
Pterocheilus paravespoides Blüthgen, 1955b: 29, ♀, ♂; "5 km east of Gvulot (Western Negev)".
Distribution in the Middle East: Known only from Israel-Palestine (Blüthgen, 1955b as *Pterocheilus paravespoides*; Giordani Soika, 1970 as *P. fausti paravespoides*; van der Vecht and Fischer, 1972 as *P. f. paravespoides*; Kurzenko, 1988; Gusenleitner, 2013a).
Extralimital distribution: None.

***Onychopterocheilus* (*Onychopterocheilus*) *fereniger* (Giordani Soika, 1952)**
Pterocheilus atrohirtus fereniger Giordani Soika, 1952: 60, ♀; Palestina.
Distribution in the Middle East: Israel-Palestine (Giordani Soika, 1952, 1970 both as *P. a. fereniger*; Blüthgen, 1955 as *Pterocheilus atrohirtus fereniger*; Gusenleitner, 2013a), Syria (Giordani Soika, 1970 as *P. a. fereniger*).
Extralimital distribution: None.

***Onychopterocheilus* (*Onychopterocheilus*) *flaviventris* Gusenleitner, 1991**
Onychopterocheilus flaviventris Gusenleitner, 1991: 297, ♀, ♂; "Saudi-Arabia, 40 km W of Hail".
Distribution in the Middle East: Known only from Saudi Arabia (Gusenleitner, 1991, 2013a; Carpenter and Gadallah, 2020).
Extralimital distribution: None.

***Onychopterocheilus* (*Onychopterocheilus*) *grandiceps* (Blüthgen, 1955b)**
Pterocheilus grandiceps Blüthgen, 1955b: 93, ♀; Palästina.
Distribution in the Middle East: Israel-Palestine (Gusenleitner, 1970 as *Pterocheilus grandiceps*, 2001b, 2013a), Jordan (Gusenleitner, 2001b), Palestine (Blüthgen, 1955b as *P. grandiceps*).
Extralimital distribution: None.

***Onychopterocheilus* (*Onychopterocheilus*) *hellenicus hellenicus* (Morawitz, 1885)**
Pterochilus hellenicus Morawitz, 1885: 137, ♀; Greece.
Distribution in the Middle East: Syria (Gusenleitner, 2013a), Turkey (Yildirim and Özbek, 1996b; Yildirim and Kojima, 1999; Gusenleitner, 2001a, 2013a; Yildirim, 2014).
Extralimital distribution: Italy (Gusenleitner, 2013a; Selis, 2023).

***Onychopterocheilus* (*Onychopterocheilus*) *hellenicus syriacus* (Blüthgen, 1952)**
Pterocheilus syriacus Blüthgen, 1952: 4, 9, ♀; "Beirut".
Catalogues with Iranian records: Rahmani *et al.* (2020).
Distribution in Iran: Iran (no specific locality cited) (Gusenleitner, 2013a).
Distribution in the Middle East: Iran, Syria (Gusenleitner, 2013a), Lebanon (Blüthgen, 1952 as *Pterocheilus syriacus*; Yildirim and Kojima, 1999; Gusenleitner, 2013a), Turkey (Blüthgen, 1955b as *P. hellenicus syriacus*; Yildirim and Özbek, 1996a; Yildirim and Kojima, 1999; Yildirim and Gusenleitner, 2012; Gusenleitner, 2013a).
Extralimital distribution: None.

***Onychopterocheilus* (*Onychopterocheilus*) *mavromoustakisi* (Giordani Soika, 1970)**
Pterocheilus mavromoustakisi Giordani Soika, 1970: 46, ♀; Iran.
Catalogues with Iranian records: Ebrahimi and Carpenter (2008), Rahmani *et al.* (2020).
Distribution in Iran: Alborz (Giordani Soika, 1970 as *Pterocheilus mavromoustakisi* Giordani Soika, 1970).
Distribution in the Middle East: Iran (see references above, and Yildirim and Kojima, 1999; Gusenleitner, 2013a), Lebanon (Gusenleitner, 2013a), Turkey (Giordani Soika, 1970 as *P. mavromoustakisi*; Gusenleitner, 1988 as *P. mavromoustakisi*, 2013a; Yildirim and Özbek, 1996a; Yildirim and Kojima, 1999).
Extralimital distribution: None.

***Onychopterocheilus* (*Onychopterocheilus*) *mena* (Giordani Soika, 1943)**
Pterochilus mena Giordani Soika, 1943: 38, ♀, ♂; «Egitto: W. Galala».
Distribution in the Middle East: Egypt (Giordani Soika, 1943 as *Pterochilus mena*; Zalat *et al.*, 1993 as *P. mena*), Jordan (Gusenleitner, 2001b), Palestine (Zalat *et al.*, 1993 as *P. mena*), Tukey (Gusenleitner, 2013a).
Extralimital distribution: None.

***Onychopterocheilus* (*Onychopterocheilus*) *mochii* (Giordani Soika, 1942)**
Pterochilus mochii Giordanai Soika, 1942: 57, ♂; «Sinai: Wadi Um Mitla».
Distribution in the Middle East: Known only from Egypt (Giordani Soika, 1942, 1973 both as *Pterochilus mochii*; van der Vecht and Fischer, 1972 as *Pterocheilus mochii*; Zalat *et al.*, 1993 as *P. mochii*; Borsato and Ratti, 1999 as *P. mochii*; Kurzenko, 1988; Gusenleitner, 2013a).
Extralimital distribution: None.

***Onychopterocheilus* (*Onychopterocheilus*) *pallasii* (Klug, 1805)**
Pterocheilus pallasii Klug, 1805: 150 [♀]; Russia.
Distribution in the Middle East: Turkey (Yildirim and Özbek, 1996a; Yildirim and Kojima, 1999; Fateryga, 2017a, 2017b, 2018).
Extralimital distribution: Kazakhstan, Mongolia, Russia (Fateryga, 2017a, 2017b, 2018).

***Onychopterocheilus* (*Asiapterocheilus*) *pamirensis pentheri* (Blüthgen, 1956)**
Pterocheilus pamirensis pentheri Blüthgen, 1956: 119, ♀; Pamir Mts.: "Kitschik-Mardjanai (westl. Täler), 3900 m".
Distribution in the Middle East: Known only from Turkey (Blüthgen, 1956 as *Pterocheilus pamirensis pentheri*; van der Vecht and Fischer, 1972 as *P. p. pentheri*; Yildirim and Kojima, 1999).
Extralimital distribution: None.

***Onychopterocheilus* (*Onychopterocheilus*) *rectus teta* (Giordani Soika, 1943)**
Pterochilus teta Giordani Soika, 1943: 40, ♀, ♂; «Palestina: Gericho».
Distribution in the Middle East: Israel-Palestine (Giordani Soika, 1943 as *P. teta*; Gusenleitner, 2013a).
Extralimital distribution: Morocco (Giordani Soika, 1986 as *Pterochilus daw teta*).

***Onychopterocheilus* (*Onychopterocheilus*) *rothi* (Dusmet y Alonso, 1928)**
Pterochilus rothi Dusmet, 1928: 111, ♀; "Ghardaia (Mzab, Sur de Argelia)".
Distribution in the Middle East: Egypt, Israel-Palestine (Zalat *et al.*, 1993 as *Pterocheilus rothi*), Israel-Palestine (Gusenleitner, 1970 as *Pterochilus rothi*, 2001b; Gusenleitner, 2013a), Jordan (Gusenleitner, 2001b).
Extralimital distribution: Algeria, Morocco (Gusenleitner, 2001b, 2013a).

***Onychopterocheilus* (*Onychopterocheilus*) *sarikamisensis* Yildirim, 2014**
Onychopterocheilus sarikamisensis Yildirim, 2014: 220, ♂; "Turkey, Kars, Sarıkamış, Karakurt, 1500 m".
Distribution in the Middle East: Known only from Turkey (Yildirim, 2014).
Extralimital distribution: None.

***Onychopterocheilus* (*Acutopterocheilus*) *spheciformis* (Gusenleitner, 1970)**
Pterocheilus spheciformis Gusenleitner, 1970: 9, ♀; Iran.
Catalogues with Iranian records: Ebrahimi and Carpenter (2008), Rahmani *et al.* (2020).
Distribution in Iran: Khuzestan (Blüthgen and Gusenleitner, 1970).
Distribution in the Middle East: Known only from Iran (Blüthgen and Gusenleitner, 1970; Gusenleitner, 2013a).
Extralimital distribution: None.

***Onychopterocheilus* (*Onychopterpocheilus*) *stiziformis* (Gusenleitner, 1970)**
Pterocheilus stiziformis Gusenleitner, 1970: 7, ♀; Iran.
Catalogues with Iranian records: Ebrahimi and Carpenter (2008), Rahmani *et al.* (2020).
Distribution in Iran: Sistan & Baluchestan (Blüthgen and Gusenleitner, 1970).
Distribution in the Middle East: Iran (Blüthgen and Gusenleitner, 1970; Gusenleitner, 2013a), United Arab Emirates (Gusenleitner, 2010a, 2013a, Carpenter and Gadallah, 2020).
Extralimital distribution: None.

***Onychopterocheilus* (*Onychopterocheilus*) *tertius* (Giordani Soika, 1970)**
Pterocheilus tertius Giordani Soika, 1970: 45, ♀; "Siria: Damasco, Ghuta".
Distribution in the Middle East: Known only from Syria (Giordani Soika, 1970 as *Pterocheilus tertius*; van der Vecht and Fischer, 1972 as *P. tertius*; Kurzenko, 1988; Gusenleitner, 2013a as *P. tertius*).
Extralimital distribution: None.

Genus *Paragymnomerus* Blüthgen, 1938
Paragymnomerus Blüthgen, 1938 (1937): 286. Type species: *Odynerus spiricornis* Spinola, 1808, by original designation.

***Paragymnomerus amitinorum* Blüthgen, 1952**
Paragymnomerus amitinorum Blüthgen, 1952: 4, ♀, ♂; Tiberias [Israel].
Catalogues with Iranian records: Rahmani *et al.* (2020).

Distribution in Iran: Fars (Schmid-Egger *et al.*, 2021).

Distribution in the Middle East: Iran (Gusenleitner, 2013a; Schmid-Egger *et al.*, 2021), Israel-Palestine (Blüthgen, 1952; Gusenleitner, 2013a), Turkey (Blüthgen, 1952; Gusenleitner, 1966, 2013a; Giordani Soika, 1970; Yildirim and Özbek, 1996b, 1999; Yildirim and Kojima, 1999).

Extralimital distribution: Croatia, Greece (Gusenleitner, 2013a).

***Paragymnomerus discoidalis* (de Saussure, 1856)**

Odynerus (Epipona) discoidalis de Saussure, 1856: 315, ♀, ♂; probably Spain.

Catalogues with Iranian records: None.

Distribution in Iran: Iran (no specific locality cited) (Selis, 2025b).

Distribution in the Middle East: Iran, Turkey (Selis, 2025b).

Extralimital distribution: Spain, France, Switzerland, Italy, Austria, Czech Republic, Slovenia, Hungary, Bulgaria, Greece, Armenia (Giraud, 1863; von Schulthess, 1887; Gribodo, 1919; Berland, 1928; Blüthgen, 1944, 1951, 1956; Castro, 1992; Gusenleitner, 1998a, 2008, 2013a; Woydak, 2001; Yildirim and Gusenleitner, 2009; Gogala, 2022; Selis, 2025b).

Comments: The real distribution of this species is uncertain, due to the numerous misinterpretations and misidentifications made by past authors; all the countries cited in the literature are reported, either in a certain or dubitative form, while in the distribution map only confirmed and reliable data are reported (Selis, 2025b).

For the Iranian fauna, no precise data are reported, but Gusenleitner (2013a) seems to be the first source that reports *P. discoidalis* (under the name *P. spiricornis*) for Iran. M. Selis (pers. comm.) has expressed doubts that this species occurs so far east, and the records from Turkey also is somewhat uncertain.

***Paragymnomerus mammillatus* (Morawitz, 1885)**

Hoplomerus mammillatus Morawitz, 1885: 149, ♀; Iran.

Catalogues with Iranian records: Ebrahimi and Carpenter (2008), Rahmani *et al.* (2020).

Distribution in Iran: Semnan (Morawitz, 1885 as *Hoplomerus mammillatus*).

Distribution in the Middle East: Known only from Iran (Morawitz, 1885; Gusenleitner, 2013a).

Extralimital distribution: None.

***Paragymnomerus signaticollis tauricus* (Kostylev, 1940)**

Odynerus (Hoplomerus) signaticollis tauricus Kostylev, 1940: 39, ♀; "Crimée, Karadag".

Distribution in the Middle East: Turkey (Gusenleitner, 2013a; Fateryga, 2018).

Extralimital distribution: Eastern Europe, Russia, Ukraine (Fateryga, 2017b, 2018).

***Paragymnomerus spiricornis spiricornis* (Spinola, 1808)**

Odynerus spiricornis Spinola, 1808: 257, ♂; Genuam [Italy].

Catalogues with Iranian records: Rahmani *et al.* (2020).

Distribution in Iran: Iran (no specific locality cited) (Gusenleitner, 2013a).

Distribution in the Middle East: Iran (Gusenleitner, 2013a; Selis, 2025b), Israel-Palestine (Selis, 2025b), Turkey (Yildirim and Gusenleitner, 2009; Gusenleitner, 2013a; Selis, 2025b).

Extralimital distribution: Armenia, Austria, Bulgaria, Croatia (Selis, 2025b), Czech Republic, France, Germany, Hungary, Italy, Slovenia (Gogala, 2022), Spain, Switzerland (Gusenleitner, 2013a).

Genus *Paraleptomenes* Giordani Soika, 1970

Paraleptomenes Giordani Soika, 1970: 79. Type species: *Paraleptomenes nurseanus* Giordani Soika, 1970, by original designation.

***Paraleptomenes miniatus miniatus* (de Saussure, 1855)**

Odynerus miniatus de Saussure, 1855: 249, ♀; Les Indes orientales.

Catalogues with Iranian records: Ebrahimi and Carpenter (2008), Rahmani *et al.* (2020).

Distribution in Iran: Kerman (Blüthgen and Gusenleitner, 1970; Pashaei Rad *et al.*, 2010).

Distribution in the Middle East: Iran (Blüthgen and Gusenleitner, 1970; Pashaei Rad *et al.*, 2010; Li *et al.*, 2019).
Extralimital distribution: China, Hong Kong, India, Mauritius, Pakistan, Taiwan (Li *et al.*, 2019).

Genus *Paravespa* Radoszkowski, 1886
Paravespa Radoszkowski, 1886: 46. Type species: *Hoplomerus komarowi* Radoszkowski, 1886 (= *Hoplomerus quadricolor* Morawitz, 1885), by monotypy.

***Paravespa* (*Paravespa*) *grandis caucasica* (Kokujev, 1912)**
Rhynchium ? *caucasicum* Kokujev, 1912: 3, ♀; "In steppa Boz, Distr. Ares, Prov. Elisavetpol".
Catalogues with Iranian records: Ebrahimi and Carpenter (2008), Rahmani *et al.* (2020).
Distribution in Iran: Iran (no specific locality cited) (Yildirim and Kojima, 1999).
Distribution in the Middle East: Iran (Yildirim and Kojima, 1999), Turkey (Giordani Soika, 1970; Yildirim and Özbek, 1996a; Yildirim and Kojima, 1999; Gusenleitner, 2013a).
Extralimital distribution: Georgia (Gusenleitner, 2013a).

***Paravespa* (*Paravespa*) *grandis grandis* (Morawitz, 1885)**
Hoplomerus grandis Morawitz, 1885: 159, ♂; Persia boreali, Scharud [Iran].
Catalogues with Iranian records: Ebrahimi and Carpenter (2008).
Distribution in Iran: Semnan (Morawitz, 1885 as both *Hoplomerus grandis* Morawitz, 1885 and *Hoplomerus persa* Morawitz, 1885), Sistan & Baluchestan (Hamzavi *et al.*, 2019).
Distribution in the Middle East: Iran (see reference above, and Esmaili and Rastegar, 1974 as *Odynerus grandis*), Lebanon (Blüthgen, 1953 as *Paravespa grandis albida*; Giordani Soika, 1970; Gusenleitner, 2013a), Turkey (Gusenleitner, 1967; Yildirim, 1998; Yildirim and Kojima, 1999; Gusenleitner, 2001b).
Extralimital distribution: Afghanistan (Gusenleitner, 2013a).

***Paravespa* (*Paravespa*) *mimetica* (von Schulthess in Friese and von Schulthess, 1923)**
Odynerus (*Hoplopus*) *mimeticus* von Schulthess, 1923 [in Friese and von Schulthess, 1923]: 284, ♀, ♂; Jericho [Israel].
Catalogues with Iranian records: Ebrahimi and Carpenter (2008), Rahmani *et al.* (2020).
Distribution in Iran: Khuzestan (Giordani Soika, 1970; Ebrahimi and Carpenter, 2008; Modarres Awal, 2012).
Distribution in the Middle East: Iran (see references above, and Yildirim and Kojima, 1999; Gusenleitner, 2013a), Israel-Palestine (Friese and von Schulthess, 1923 as *Odynerus* (*Hoplopus*) *mimeticus*; Giordani Soika, 1952; Yildirim and Kojima, 1999; Gusenleitner, 2013a; Fateryga *et al.*, 2019), Jordan, Syria (Fateryga *et al.*, 2019), Saudi Arabia (Giordani Soika, 1952; Guichard, 1986; Yildirim and Kojima, 1999; Gusenleitner, 2013a; Carpenter and Gadallah, 2020), Turkey (Gusenleitner, 1988; Yildirim and Kojima, 1999).
Extralimital distribution: Armenia, Azerbaijan, Turkmenistan (Fateryga *et al.*, 2019).

***Paravespa* (*Paravespa*) *quadricolor* (Morawitz, 1885)**
Hoplomera quadricolor Morawitz, 1885: 146, ♀; Territorio Transcaspico, Krasnowodsk [Turkmenistan].
Catalogues with Iranian records: Ebrahimi and Carpenter (2008), Rahmani *et al.* (2020).
Distribution in Iran: Hormozgan (Schmid-Egger *et al.*, 2021), Sistan & Baluchestan (Białynicki-Birula, 1926 as *Hoplomerus quadricolor* Morawitz, 1885; Ebrahimi and Carpenter, 2008; Modarres Awal, 2012).
Distribution in the Middle East: Iran (see references above, and Giordani Soika, 1970; Gusenleitner, 2013a; Selis *et al.*, 2025), Iraq (Selis *et al.*, 2025), Israel-Palestine (Giordani Soika, 1952), Turkey (Gusenleitner, 1988; Yildirim, 2012; Selis *et al.*, 2025).
Extralimital distribution: Afghanistan, Turkmenistan (Gusenleitner, 1988, 2013a; Selis *et al.*, 2025).

***Paravespa* (*Paravespa*) *rex* (von Schulthess, 1923)**

Odynerus (*Hoplopus*) *rex* von Schulthess, 1923 [in Friese and von Schulthess, 1923]: 284, 285, ♀, ♂; Turkestan, Kisilkum.

Catalogues with Iranian records: Ebrahimi and Carpenter (2008), Rahmani *et al.* (2020).

Distribution in Iran: Alborz (Giordani Soika, 1970).

Distribution in the Middle East: Iran (see references above, and Yildirim and Kojima, 1999; Gusenleitner, 2013a; Fateryga, 2017b, 2018), Turkey (Giordani Soika, 1970; Yildirim and Kojima, 1999; Fateryga, 2017b, 2018).

Extralimital distribution: Kazakhstan, Kyrgyzstan, Russia, Uzbekistan (Fateryga, 2018).

Genus *Pareumenes* de Saussure, 1855

Pareumenes de Saussure, 1855: 133. Type species: *Eumenes quadrispinosus* de Saussure, 1855, by subsequent designation of Bequaert, 1918.

***Pareumenes* (*Pareumenes*) *arabicus* Giordani Soika, 1981**

Pareumenes sansibaricus arabicus Giordani Soika, 1981 (1980): 116, ♀; Arabia.

Distribution in the Middle East: Oman (Giordani Soika, 1981; Guichard, 1986; Carpenter and Gadallah, 2020 (all as *P.* (*P.*) *sansibaricus arabicus*; Selis, 2025a), Yemen (Gusenleitner, 2013a; Carpenter and Gadallah, 2020 (both as *P.* (*P.*) *sansibaricus arabicus*).

Extralimital distribution: None.

***Pareumenes* (*Nortonia*) *enslini* (von Schulthess, 1931)**

Nortonia enslini von Schulthess, 1931: 49, ♂; "Palestina, Wadi el Kelt nahe Jericho".

Distribution in the Middle East: Israel-Palestine (von Schulthess, 1931 as *Nortonia enslini*; Giordani Soika, 1938; Giordani Soika, 1952; Gusenleitner, 2013a; Selis, 2025a), Oman (Giordani Soika, 1981 as *P. minus*; Guichard, 1986; Gusenleitner, 2013a; Carpenter and Gadallah, 2020; Selis, 2025a), Saudi Arabia (Giordani Soika, 1970; Guichard, 1986; Carpenter and Gadallah, 2020).

Extralimital distribution: None.

***Pareumenes* (*Nortonia*) *laminatus laminatus* (Kriechbaumer, 1879)**

Eumenis laminata Kriechbaumer, 1879: 59, ♂, ♀; "Dalmatien".

Distribution in the Middle East: Turkey (Giordani Soika, 1970; Yildirim and Özbek, 1996b; Yildirim and Kojima, 1999; Gusenleitner, 2013a).

Extralimital distribution: Southeast Europe (Gusenleitner, 2013a), Italy (Selis, 2023).

***Pareumenes* (*Nortonia*) *laminatus palaestinensis* Giordani Soika, 1952**

Pareumenes (*Nortonia*) *laminatus palaestinensia* Giordani Soika, 1952: 16, ♀; "Palestina: Tal el Nadi".

Distribution in the Middle East: Israel-Palestine (Giordani Soika, 1952; Yildirim and Kojima, 1999; Gusenleitner, 2013a), Turkey (Giordani Soika, 1952; Gusenleitner, 1966, 2013a; Yildirim and Kojima, 1999).

Extralimital distribution: None.

Genus *Parodontodynerus* Blüthgen, 1938

Parodontodynerus Blüthgen, 1938 (1937): 280. Type species: *Eumenes ephippium* Klug in Germar, 1817, by original designation.

***Parodontodynerus aramaeus* Blüthgen, 1955a**

Parodontodynerus ephippium aramaeus Blüthgen, 1955a: 27, ♀; Israel.

Catalogues with Iranian records: Rahmani *et al.* (2020).

Distribution in Iran: Fars (Schmid-Egger *et al.*, 2021; Giordani Soika, 1961 as *Odontodynerus ephippium eremicus* Giordani Soika, 1961), Khuzestan (Giordani Soika, 1961 as *O. ephippium eremicus*).

Distribution in the Middle East: Iran (see references above, and Yildirim and Kojima, 1999), Israel-Palestine (Blüthgen, 1955a as *Parodontodynerus ephippium aramaeus*; Gusenleitner, 2013a),

Syria, Turkey (Giordani Soika, 1961 1970 both as *P. eremicus*; Gusenleitner, 1966, 1967, 1970 all as *P. fulgidus*; Gusenleitner, 1972, 2013a; Yildirim and Kojima, 1999).
Extralimital distribution: None.

Parodontodynerus ephippium ephippium **(Klug in Germar, 1817)**
Eumenes ephippium Klug, 1817 [in Germar, 1817]: 264; Dalmatiae insulae [Croatia].
Catalogues with Iranian records: Ebrahimi and Carpenter (2008), Rahmani *et al.* (2020).
Distribution in Iran: Alborz (Giordani Soika, 1970; Pashaei Rad *et al.*, 2006), Mazandaran (Pashaei Rad *et al.*, 2006), Zanjan (Pashaei Rad *et al.*, 2006; Abbasi *et al.*, 2008a, 2008b).
Distribution in the Middle East: Iran (see references above, and Yildirim and Kojima, 1999; Gusenleitner, 2013a; Fateryga, 2017b, 2018; Fateryga and Proshchalykin, 2024; Selis *et al.*, 2025), Iraq (Selis *et al.*, 2025), Lebanon (Giordani Soika, 1970; Yildirim and Kojima, 1999; Fateryga, 2017b, 2018; Fateryga *et al.*, 2019; Fateryga and Proshchalykin, 2024; Selis *et al.*, 2025), Turkey (Giordani Soika, 1952 as *Odontodynerus ephippium anatoliae*, 1970; Gusenleitner, 1966, 1967, 1970; Yildirim and Özbek, 1996a, 1996b, 1999; Yildirim and Kojima, 1999; Yildirim and Gusenleitner, 2009, 2012; Fateryga, 2017b, 2018; Fateryga *et al.*, 2019; Fateryga and Proshchalykin, 2024; Selis *et al.*, 2025).
Extralimital distribution: Albania, Austria, Croatia, France, Greece, Kyrgyzstan, Morocco, Portugal, Spain, Switzerland, (Fateryga and Proshchalykin, 2024; Selis *et al.*, 2025), Armenia, Azerbaijan, Kazakhstan, Russia, Turkmenistan, Uzbekistan (Fateryga, 2018; Fateryga *et al.*, 2019; Selis *et al.*, 2025), Italy (Selis, 2023; Selis *et al.*, 2025), Malta (Cassar *et al.*, 2022).

Parodontodynerus rufinus **(Kostylev, 1940)**
Odynerus (*Parodontodynerus*) *ephippium rufinus* Kostylev, 1940: 30, ♀; Iran.
Catalogues with Iranian records: Ebrahimi and Carpenter (2008), Rahmani *et al.* (2020).
Distribution in Iran: Kerman (Kostylev, 1940 as *Odynerus* (*Parodontodynerus*) *ephippium rufinus*).
Distribution in the Middle East: Known only from Iran (Kostylev, 1940; Gusenleitner, 2013a).
Extralimital distribution: None.

Genus *Pseudepipona* de Saussure, 1856

Pseudepipona de Saussure 1856: 309. Type species: *Odynerus herrichii* de Saussure, 1856, by monotypy.

Pseudepipona **(*Deuterepipona*)** ***ankarensis*** **Giordani Soika, 1970**
Pseudepipona (*Deuterepipona*) *ankarensis* Giordani Soika, 1970: 148, ♀; "Turchia, Ankara, Idris Dagi".
Distribution in the Middle East: Known only from Turkey (Giordani Soika, 1970; van der Vecht and Fischer, 1972; Gusenleitner, 1988; Yildirim, 1998; Yildirim and Kojima, 1999; Yildirim and Gusenleitner, 2012; Gusenleitner, 2013a).
Extralimital distribution: None.

Pseudepipona **(*Pseudepipona*)** ***augusta*** **(Morawitz, 1867)**
Odynerus augustus Morawitz, 1867: 122 (= *Odynerus aurantiacus* Mocsary, 1877).
Distribution in the Middle East: Turkey (Fateryga, 2017b, 2018).
Extralimital distribution: Central Asia, China, Eastern Europe, Hungary, Kazakhstan, Mongolia, Russia, Ukraine (Fateryga, 2017b, 2018; Selis *et al.*, 2025).

Pseudepipona **(*Pseudepipona*)** ***herrichii herrichii*** **(de Saussure, 1856)**
Odynerus variegatus Herrich-Schäffer, 1839: 16, ♀♂; «bi Regensburg» [Germany].
Odynerus (*Epipona* div. *Pseudepipona*) *herrichi* de Saussure, 1856: 309, replacement name for *O. variegatus* herrich-Schäffer, 1839, not *O. variegatus* (Fabricius).
Catalogues with Iranian records: Rahmani *et al.* (2020).
Distribution in Iran: Iran (no specific locality cited) (Gusenleitner, 1998b).

Distribution in the Middle East: Iran (Gusenleitner, 1998b), Syria (Blüthgen, 1955a as *Pseudepipona variegata* var. *derufata*; Fateryga, 2017b, 2018; Fateryga *et al.*, 2020, 2021; Kochetkov, 2020), Turkey (Giordani Soika, 1970 as both *P.* (*P.*) *herrichii derufata* and *P.* (*P.*) *herrichii*; Yildirim and Özbek, 1996a, 1996b; Madl, 1997; Yildirim and Gusenleitner, 2009, 2012; Gusenleitner, 2013a; Fateryga, 2017b, 2018, 2022; Fateryga *et al.*, 2020, 2021; Kochetkov, 2020).

Extralimital distribution: China (Gusenleitner, 1998b; Fateryga, 2017b), Armenia, Azerbaijan, Caucasus, Europe, Kazakhstan, Kyrgyzstan, Mongolia, North Africa, North America, Russia, Turkmenistan, Uzbekistan (Fateryga, 2017b, 2018; Fateryga and Mokrousov, 2019; Fateryga *et al.*, 2020, 2021), Georgia (Fateryga *et al.*, 2020; Fateryga, 2022), Croatia, Ukraine (Fateryga, 2022), Italy (Selis, 2023).

Comments: The presence of this species in Iran needs confirmation (Fateryga, 2022).

***Pseudepipona* (*Deuterepipona*) *herzi enslini* (Blüthgen, 1955a)**

Deuterepipona herzi enslini Blüthgen, 1955: 28, ♂; "Jericho".

Distribution in the Middle East: Known only from Israel-Palestine (Blüthgen, 1955a as *Deuterepipona herzi enslini*; van der Vecht and Fischer, 1972; Gusenleitner, 2013a).

Extralimital distribution: None.

***Pseudepipona* (*Deuterepipona*) *herzi herzi* (Morawitz, 1895)**

Odynerus herzi Morawitz, 1895: 471, ♂, ♀, "Transcaspia: Sumbar" [Turkmenistan].

Distribution in the Middle East: Israel-Palestine (Blüthgen, 1955a as *Deuterepipona herzi enslini*).

Extralimital distribution: China, Kazakhstan, Kyrgyzstan, Mongolia, Turkmenistan, Uzbekistan (Fateryga *et al.*, 2023).

***Pseudepipona* (*Deuterepipona*) *ionia clausa* (Giordani Soika, 1943)**

Odynerus (*Rhynchium*) *ionius* var. *clausa* Giordani Soika, 1943: 11, ♀; "N. Siria: Ladikie".

Distribution in the Middle East: Syria (Giordani Soika, 1943 as *Odynerus* (*Rhynchium*) *jonius* var. *clausa*; Yildirim and Kojima, 1999; Gusenleitner, 2013a), Turkey (Gusenleitner, 1966, 2001a, 2013a; Yildirim and Kojima, 1999).

Extralimital distribution: None.

***Pseudepipona* (*Deuterepipona*) *ionia ionia* (de Saussure, 1855)**

Odynerus (*Odynerus*) *Ionius* de Saussure, 1855: 263, ♀; "L'île de Rhodes".

Distribution in the Middle East: Turkey (Giordani Soika, 1970 as *Pseudepipona jonia*; van der Vecht and Fischer, 1972; Yildirim and Özbek, 1999; Yildirim and Kojima, 1999; Gusenleitner, 2013a)

Extralimital distribution: Greece (Gusenleitner, 2013a).

***Pseudepipona* (*Pseudepipona*) *lativentris cypria* Blüthgen, 1942**

Pseudepipona (*Pseudepipona*) *lativentris cypria* Blüthgen, 1942: 309, ♀, ♂; «Limassol».

Distribution in the Middle East: Cyprus (Blüthgen, 1942; van der Vecht and Fischer, 1972; Gusenleitner, 2013a), Israel-Palestine (Gusenleitner, 2013a).

Extralimital distribution: None.

***Pseudepipona* (*Pseudepipona*) *lativentris lativentris* (de Saussure, 1855)**

Pseudepipona (*Odynerus*) *lativentris* de Saussure, 1855: 275, ♀; Le Midi de la France. Montpellier [France].

Catalogues with Iranian records: Ebrahimi and Carpenter (2008), Rahmani *et al.* (2020).

Distribution in Iran: Alborz (Ebrahimi and Carpenter, 2008; Modarres Awal, 2012).

Distribution in the Middle East: Iran (Ebrahimi and Carpenter, 2008; Modarres Awal, 2012), Israel-Palestine (Yildirim and Kojima, 1999; Gusenleitner, 2013a), Lebanon (Giordani Soika, 1970; Yildirim and Kojima, 1999; Gusenleitner, 2013a), Turkey (Blüthgen, 1951a as *Pseudepipona*

(*Trichepipona*) *lativentris*; Giordani Soika, 1970; Yildirim and Özbek, 1996a, 1999 as *P.* (*T.*) *lativentris*; Yildirim and Kojima, 1999; Yildirim and Gusenleitner, 2004; Gusenleitner, 2013a).
Extralimital distribution: Armenia, Europe, North Africa (Gusenleitner, 2013a), Italy (Selis, 2023).

***Pseudepipona* (*Deuterepipona*) *niveopicta* Giordani Soika, 1970**

Pseudepipona (*Pseudepipona*) *niveopicta* Giordani Soika, 1970: 144, ♀, ♂; "Turchia: Nigde, 25 km. Ulukisla-Aksaray," "3000 ft".
Distribution in the Middle East: Turkey (Giordani Soika, 1970 as *Pseudepipona* (*Pseudepipona*) *niveopicta*; van der Vecht and Fischer, 1972 as *P.* (*P.*) *niveopicta*; Yildirim and Kojima, 1999; Gusenleitner, 2013a; Fateryga, 2017b).
Extralimital distribution: Russia (Gusenleitner, 2013a; Fateryga, 2017b).
Comments: Fateryga *et al.* (2017) transferred this species to the subgenus *Deuterepipona* Blüthgen, 1951, based on the structure of its pronotum which has "obtuse lateral corners with lateral sides somewhat convergent basally".

***Pseudepipona* (*Pseudepipona*) *peculiaris* (Morawitz, 1895)**

Odynerus (*Lionotus*) *peculiaris* Morawitz, 1895: 469, ♀; Mons Bogdo [Russia].
Catalogues with Iranian records: Ebrahimi and Carpenter (2008), Rahmani *et al.* (2020).
Distribution in Iran: Semnan (Giordani Soika, 1970; Ebrahimi and Carpenter, 2008; Modarres Awal, 2012).
Distribution in the Middle East: Iran (see references above, and Gusenleitner, 2013a; Fateryga, 2017b).
Extralimital distribution: Kazakhstan (Fateryga, 2017b), Russia (Gusenleitner, 2013a; Fateryga, 2017b).

***Pseudepipona* (*Pseudepipona*) *priesneri* Gusenleitner, 1970**

Pseudepipona priesneri Gusenleitner, 1970: 10, ♀, ♂; Saudi Arabia.
Catalogues with Iranian records: Ebrahimi and Carpenter (2008), Rahmani *et al.* (2020).
Distribution in Iran: Khuzestan (Blüthgen and Gusenleitner, 1970).
Distribution in the Middle East: Iran (Blüthgen and Gusenleitner, 1970; Gusenleitner, 2013a), Saudi Arabia (Blüthgen and Gusenleitner, 1970; Carpenter and Gadallah, 2020).
Extralimital distribution: None.

***Pseudepipona* (*Pseudepipona*) *przewalskyi* (Morawitz, 1885)**

Lionotus przewalskyi Morawitz, 1885: 161, ♂; "In provincia chinensi Ordoss".
Distribution in the Middle East: Egypt (Giordani Soika, 1935 as both *Odynerus tripunctatus* and *O.* (*Rhynchium*) *tripunctatus convergens*; Giordani Soika, 1943 as *O.* (*Rhynchium*) *nekt*; Zalat *et al.*, 1993 as *Pseudepipona nekt*; Gusenleitner, 2013a; Carpenter and Gadallah, 2020), Israel-Palestine (Giordani Soika, 1952 as *Pseudepipona nekt*; Gusenleitner, 2013a; Carpenter and Gadallah, 2020), United Arab Emirates (Guichard, 1986 as *P. nekt*; Howarth and Gillett, 2008 as *P. nekt*; Gusenleitner, 2013a; Carpenter and Gadallah, 2020).
Extralimital distribution: Algeria, China, Mongolia (Gusenleitner, 2013a; Carpenter and Gadallah, 2020).

***Pseudepipona* (*Pseudepipona*) *pseudominuta* Gusenleitner in Bytinski-Salz and Gusenleitner, 1971**

Pseudepipona pseudominuta Gusenleitner, 1971 [in Bytinski-Salz and Gusenleitner, 1971]: 239, 273, 295, ♀; Israel, Jericho.
Distribution in the Middle East: Israel-Palestine (Bytinski-Salz and Gusenleitner, 1972; Yildirim and Kojima, 1999; Gusenleitner, 2001b, 2013a), Jordan (Gusenleitner, 2001b), Turkey (Gusenleitner, 1988, 2001b, 2013a; Yildirim and Kojima, 1999).
Extralimital distribution: None.

Pseudepipona (Pseudepipona) rubidus Gusenleitner, 1994

Pseudepipona rubidus Gusenleitner, 1994: 307, 313, ♀; "Israel, Arava, 4 km W of Hazeva".
Distribution in the Middle East: Known only from Israel-Palestine (Gusenleitner, 1994, 2013a).
Extralimital distribution: None.

Pseudepipona (Pseudepipona) sellata (Morawitz, 1885)

Lionotus sellatus Morawitz, 1885: 172, ♀♂; Deserto Kirgisorum. Mons Bogdo [Russia: Astrakhan Province].
Catalogues with Iranian records: Ebrahimi and Carpenter (2008), Rahmani *et al.* (2020).
Distribution in Iran: Southern Iran (no specific locality cited) (van der Vecht and Fischer, 1972).
Distribution in the Middle East: Iran (van der Vecht and Fischer, 1972; Fateryga, 2017a, 2017b; Fateryga *et al.*, 2023), Turkey (Fateryga, 2017a, 2017b; Fateryga *et al.*, 2023).
Extralimital distribution: Azerbaijan, Kazakhstan, Russia, Turkmenistan (Fateryga, 2017a, 2017b), Uzbekistan (Fateryga *et al.*, 2023).

Pseudepipona (Pseudepipona) tricarinata (Kokujev, 1912)

Odynerus (Lionotus) herrichii var. *tricarinatus* Kokujev, 1912: 4, ♀; "Caucasus (Geok-Tapa)".
Distribution in the Middle East: Turkey (Giordani Soika, 1970; Yildirim and Kojima, 1999; Gusenleitner, 2013a; Fateryga, 2017b, 2022).
Extralimital distribution: Azerbaijan, Kazakhstan, China, Russia (Fateryga, 2017b), Armenia, Georgia (Fateryga, 2022).

Pseudepipona (Pseudepipona) tripunctata (Fabricius, 1787)

Vespa tripunctata Fabricius, 1787: 920, ♀; "in Barbaria".
Distribution in the Middle East: Egypt (Zalat *et al.*, 1993; Gusenleitner, 1998b), Israel-Palestine (Giordani Soika, 1952; Gusenleitner, 1998b, 2013a).
Extralimital distribution: Algeria, Italy (Gusenleitner, 2013a), Malta (Schembri, 1980; Gusenleitner, 1998b).
Comments: The presence of this species in Egypt is doubtful, as it was cited to be present in North Africa except in Egypt (Gusenleitner, 2013a).

Genus _Pseudodontodynerus_ Blüthgen, 1939

Pseudodontodynerus Blüthgen, 1939: 249. Type species: *Odynerus pretiosus* Dusmet, 1928, by monotypy.

Pseudodontodynerus gambiensis gambiensis (Meade-Waldo, 1915)

Odynerus (Lionotus) gambiensis Meade-Waldo, 1915: 487, 492, 504, ♀; Gambia.
Odontodynerus (Lionotus) brittoni Giordani Soika, 1957: 479, ♀, ♂; "Yemen: Usaifira, 1 mile north of Ta'izz, 4500 ft".
Distribution in the Middle East: Saudi Arabia (Guichard, 1986 as *Pseudontodynerus brittoni*; Carpenter and Gadallah, 2020), Yemen (Giordani Soika, 1957 as *Odontodynerus brittoni*; Gusenleitner, 2013a).
Extralimital distribution: Congo, Eritrea, Gambia, Tanzania, Zimbabwe (Carpenter *et al.*, 2010).

Pseudodontodynerus peculiariventris (Giordani Soika, 1952)

Odontodynerus peculiariventris Giordani Soika, 1952: 50, ♀; Tiberias [Israel].
Catalogues with Iranian records: Ebrahimi and Carpenter (2008), Rahmani *et al.* (2020).
Distribution in Iran: Fars (Giordani Soika, 1970 as *Pseudodontodynerus peculiariventris salsus* (Giordani Soika, 1970)), Sistan & Baluchestan (Hamzavi *et al.*, 2019).
Distribution in the Middle East: Iran (see references above, and Gusenleitner, 2013a), Israel-Palestine (Giordani Soika, 1952 as *Odontodynerus peculiariventris*; Gusenleitner, 2013a).
Extralimital distribution: None.

***Pseudodontodynerus pretiosus* (Dusmet, 1928)**
Odynerus (*Lionotus*) *pretiosus* Dusmet, 1928: 101, ♀; Algeria.
Distribution in the Middle East: Egypt (Giordani Soika, 1939 as *Odynerus pretiosus*; Giordani Soika, 1970; Zalat *et al.*, 1993; Gusenleitner, 2013a), Israel-Palestine (Blüthgen, 1939; Giordani Soika, 1952 as *Odontodynerus pretiosus houskai*; Gusenleitner, 2013a).
Extralimital distribution: Algeria (Gusenleitner, 2013a).

Genus *Pseudonortonia* Giordani Soika, 1936
Pseudonortonia Giordani Soika, 1936: 268. Type species: *Odynerus defformis* de Saussure, 1853, by original designation.

***Pseudonortonia aberratica aberratica* (Morice, 1903)**
Odynerus (*Ancistrocerus*) *aberraticus* Morice, 1903: 613, ♂; Egypt.
Distribution in the Middle East: Egypt (Morice, 1903 as *Odynerus* (*Ancistrocerus abberaticus*; Zalat *et al.*, 1993); Giordani Soika, 1935 as *Nortonia abberatica*; Gusenleitner, 2013a).
Extralimital distribution: Eritrea (Giordani Soika, 1935 as *Nortonia abberatica*), Burkina Faso (Giordani Soika, 1987), Ethiopia, Sudan (Gusenleitner, 2013a).

***Pseudonortonia aberratica arabica* Giordani Soika, 1957**
Pseudonortonia aberratica arabica Giordani Soika, 1957: 476, ♀; "Western Aden Protectorate: Jebel Jihaf, 7100 ft.".
Distribution in the Middle East: Saudi Arabia (Guichard, 1986; Carpenter and Gadallah, 2020), Yemen (Giordani Soika, 1957; Gusenleitner, 2013a; Carpenter and Gadallah, 2020).
Extralimital distribution: None.

***Pseudonortonia aegyptiaca aegyptiaca* (de Saussure, 1863)**
Odynerus (*Ancistrocerus*) *aegyptiacus* de Saussure, 1863: 210, ♀; Egypt.
Distribution in the Middle East: Egypt (de Saussure, 1863 as *Odynerus* (*Ancistrocerus*) *aegyptiacus*; Giordani Soika, 1935 as *Nortonia aegyptiaca*; Zalat *et al.*, 1993; Gusenleitner, 2013a).
Extralimital distribution: Algeria, Eritrea, Ethiopia, Sudan (Gusenleitner, 2013a).

***Pseudonortonia bicarinatus* Guichard, 1986**
Pseudonortonia bicarinatus Guichard, 1986 (1985): 202, 209, 219, ♀; "Saudi Arabia: Wadi Majarish below Taif 800 m".
Distribution in the Middle East: Known only from Saudi Arabia (Guichard, 1986; Gusenleitner, 2013a; Carpenter and Gadallah, 2020).
Extralimital distribution: None.

***Pseudonortonia bisuturalis* (de Saussure, 1852)**
Odynerus bisuturalis de Saussure, 1852: 127, ♂; "le Sénégal ?".
Distribution in the Middle East: Yemen (Gusenleitner, 2007, 2013a; Carpenter and Gadallah, 2020).
Extralimital distribution: Cameroon, Congo, Ethiopia, Mali, Niger, ?Senegal, South Africa (Carpenter *et al.*, 2010; Carpenter and Gadallah, 2020).

***Pseudonortonia difformis* (de Saussure, 1853)**
Odynerus difformis de Saussure, 1853: 145, ♂; "le Sénégal".
Distribution in the Middle East: Oman (Gusenleitner, 2004, 2013a; Carpenter and Gadallah, 2020), Saudi Arabia (Guichard, 1986; Gusenleitner, 2013a; Carpenter and Gadallah, 2020).
Extralimital distribution: Burkina Faso, Ethiopia, Gambia, Niger, North Africa, Senegal (Carpenter and Gadallah, 2020).

***Pseudonortonia emirata* Gusenleitner, 2012**
Pseudonortonia emirata Gusenleitner, 2012: 307, 309, ♀; "Al Wasit, 25,36N 55,46E".
Distribution in the Middle East: Known only from United Arab Emirates (Gusenleitner, 2012, 2013a; Carpenter and Gadallah, 2020).
Extralimital distribution: None.

***Pseudonortonia omanensis* Giordani Soika, 1979**

Pseudonortonia omanensis Giordani Soika, 1979: 279, ♀, ♂; "Oman: Nizwa, 500 m".
Distribution in the Middle East: Known only from Oman (Giordani Soika, 1979; Guichard, 1986; Gusenleitner, 2013a; Carpenter and Gadallah, 2020).
Extralimital distribution: None.

***Pseudonortonia scotti* Giordani Soika, 1957**

Pseudonortonia scotti Giordani Soika, 1957: 475, ♀; "Yemen: Usaifira, 1 mile north of Ta'izz, 4500 ft".
Distribution in the Middle East: Known only from Yemen (Giordani Soika, 1957; Gusenleitner, 2013a; Carpenter and Gadallah, 2020).
Extralimital distribution: None.

***Pseudonortonia sudanensis* (von Schulthess, 1921)**

Labus sudanensis von Schulthess, 1921 (1920): 285, ♂, ♀; "Ägyptischer Sudan".
Distribution in the Middle East: Oman (Gusenleitner, 2004, 2005, 2013a; Carpenter and Gadallah, 2020), United Arab Emirates (Gusenleitner, 2010a, 2013a; Carpenter and Gadallah, 2020).
Extralimital distribution: Cameroon, Mali, Niger, Senegal, Sudan (Carpenter and Gadallah, 2020).

***Pseudonortonia tilkiani* Guichard, 1986**

Pseudonortonia tilkiani Guichard, 1986 (1985): 202, 209, 219, ♂, ♀; "Saudi Arabia: Wadi Majarish below Taif 800 m".
Distribution in the Middle East: Saudi Arabia (Guichard, 1986; Gusenleitner, 2013a; Carpenter and Gadallah, 2020), Yemen (Gusenleitner, 2007, 2013a; Carpenter and Gadallah, 2020).
Extralimital distribution: None.

Genus *Pseudosymmorphus* Blüthgen, 1938

Pseudosymmorphus Blüthgen, 1938 (1937): 293. Type species *Odynerus hindenburgi* Dusmet, 1917, by original designation.

***Pseudosymmorphus adnexus* (Gusenleitner, 1969)**

Acanthodynerus adnexus Gusenleitner, 1969 (1966): 17, ♂, ♀; Turkey.
Distribution in the Middle East: Known only from Turkey (Gusenleitner, 1969 as *Acanthodynerus adnexus*, 2013a; Yildirim and Kojima, 1999).
Extralimital distribution: None.

***Pseudosymmorphus angulatus* Gusenleitner and Ockermüller, 2018**

Pseudosymmorphus angulatus Gusenleitner and Ockermüller, 2018: 1092, ♀, ♂; Iran.
Catalogues with Iranian records: Rahmani *et al.* (2020).
Distribution in Iran: Lorestan (Gusenleitner and Ockermüller, 2018).
Distribution in the Middle East: Known only from Iran (Gusenleitner and Ockermüller, 2018).
Extralimital distribution: None.

***Pseudosymmorphus capillus* Gusenleitner and Ockermüller, 2018**

Pseudosymmorphus capillus Gusenleitner and Ockermüller, 2018: 1094, ♀; Iran.
Catalogues with Iranian records: Rahmani *et al.* (2020).
Distribution in Iran: Fars (Gusenleitner and Ockermüller, 2018).
Distribution in the Middle East: Known only from Iran (Gusenleitner and Ockermüller, 2018).
Extralimital distribution: None.

***Pseudosymmorphus moricei* (von Schulthess, 1925)**

Alastor (*Antalastor*) *moricei* von Schulthess, 1925: 60, 61, 198, ♂; "Palaestina, Jericho ad fontem Elisae".
Distribution in the Middle East: Egypt (Gusenleitner, 2013a), Israel-Palestine (von Schulthess, 1925 as *Alastor* (*Antalastor*) *moricei*; Blüthgen, 1954; van der Vecht and Fischer, 1972; Gusenleitner, 2013a).
Extralimital distribution: None.

***Pseudosymmorphus prophetus* (Giordani Soika, 1952)**

Leptomenes (*Eumenidiopsis*) *prohpheta* Giordani Soika, 1952: 55, ♂; "Palestina: El Hamme".

Distribution in the Middle East: Israel-Palestine (Yildirim and Kojima, 1999; Gusenleitner, 2013a), Jordan (Gusenleitner, 2013a), Syria (Giordani Soika, 1970 as *Pseudoleptochilus asiaticus*; Gusenleitner, 2013a), Turkey (Gusenleitner, 1988 as *Acanthodynerus prophetus*; Yildirim and Kojima, 1999).

Extralimital distribution: None.

Genus *Pterocheilus* Klug, 1805

Pterocheilus Klug, 1805: 143. Type species: *Vespa phalerata* Panzer, 1797; designated by Blanchard, 1840.

***Pterocheilus* (*Pterocheilus*) *arabicus* Giordani Soika, 1970**

Pterocheilus arabicus Giordani Soika, 1970: 55, ♀; "Arabia: Shakkot El. Irman".

Distribution in the Middle East: Saudi Arabia (Giordani Soika, 1970; Guichard, 1986; Gusenleitner, 2013a; Carpenter and Gadallah, 2020).

Extralimital distribution: Pakistan (Carpenter and Gadallah, 2020).

***Pterocheilus* (*Pterocheilus*) *chobauti calefactus* Giordani Soika, 1970**

Pterocheilus chobauti calefactus Giordani Soika, 1970: 54, ♀; "Giordania: Jericho".

Distribution in the Middle East: Egypt (Zalat *et al.*, 1993), Israel-Palestine (Giordani Soika, 1970; Bytinski-Salz and Gusenleitner, 1971; Zalat *et al.*, 1993; Gusenleitner, 2013a), Jordan (Gusenleitner, 2013a), Oman (Guichard, 1986; Carpenter and Gadallah, 2020), Saudi Arabia (Guichard, 1986; Gusenleitner, 2013a; Carpenter and Gadallah, 2020), Syria (Giordani Soika, 1970; Zalat *et al.*, 1993).

Extralimital distribution: None.

***Pterocheilus* (*Pterocheilus*) *chobauti chlorodyneroides* Gusenleitner, 1970**

Pterocheilus chlorodyneroides Gusenleitner, 1970 [in Blüthgen and Gusenleitner, 1970]: 4, 8, ♀; Iran.

Catalogues with Iranian records: Ebrahimi and Carpenter (2008), Rahmani *et al.* (2020).

Distribution in Iran: Sistan & Baluchestan [= Baluchistan] (Blüthgen and Gusenleitner, 1970; van der Vecht and Fischer, 1972).

Distribution in the Middle East: Known only from Iran (Blüthgen and Gusenleitner, 1970; van der Vecht and Fischer, 1972).

Extralimital distribution: None.

***Pterocheilus chobauti chobauti* Dusmet, 1928**

Pterocheilus chobauti Dusmet, 1928: 109, ♀; Algeria.

Distribution in the Middle East: Egypt (Blüthgen, 1959; Zalat *et al.*, 1993; Gusenleitner, 2013a), Israel-Palestine (Gusenleitner, 2013a), Oman (Guichard, 1986), Saudi Arabia (Gusenleitner, 2013a; Carpenter and Gadallah, 2020), Syria (Gusenleitner, 2013a).

Extralimital distribution: Algeria, Morocco (Giordani Soika, 1970), Tunisia (Gusenleitner, 2013a).

***Pterocheilus* (*Pterocheilus*) *coccineus* André, 1884**

Pterocheilus coccineus André, 1884: 802, ♀; "Algérie".

Distribution in the Middle East: Israel-Palestine, Jordan (Gusenleitner, 2001b, 2013a).

Extralimital distribution: Algeria, Spain, Tunisia (Gusenleitner, 2001b, 2013a).

***Pterocheilus* (*Pterocheilus*) *dives* Radoszkowski, 1876**

Pterocheilus dives Radoszkowski, 1876: 145, ♀; Egypt.

Distribution in the Middle East: Known only from Egypt (Radoszkowski, 1876; Dalla Torre, 1894; Schmiedecknecht, 1900 as *Pterochilus pharaonum* [sic]; Kohl, 1907 as *Pterocheilus iterabilis*; Bequaert, 1918 as both *P. dives* and *P. iterabilis*; Blüthgen, 1939; van der Vecht and Fischer, 1972; Zalat *et al.*, 1993; Gusenleitner *et al.*, 2013).

Extralimital distribution: None.

***Pterocheilus* (*Pterocheilus*) *kamanensis* Gusenleitner, 1967**

Pterocheilus kamanensis Gusenleitner, 1967: 665, ♀; Turkey.

Distribution in the Middle East: Known only from Turkey (Gusenleitner, 1967, 1976, 2013a; van der Vecht and Fischer, 1972; Yildirim and Kojima, 1999).

Extralimital distribution: None.

***Pterocheilus* (*Pterocheilus*) *merpeba* Giordani Soika, 1943**

Pterocheilus merpeba Giordani Soika, 1943: 42, ♀; "Sinai: W. um Mitla".

Distribution in the Middle East: Known only from Egypt (Giordani Soika, 1943, 1973; van der Vecht and Fischer, 1972; Zalat *et al.*, 1993; Borsato and Ratti, 1999; Gusenleitner, 2013a).

Extralimital distribution: None.

***Pterocheilus* (*Pterocheilus*) *multimaculatus* Gusenleitner, 2004**

Pterocheilus multimaculatus Gusenleitner, 2004: 1093, 1094, ♀; "Türkei: Merkez, Nevşehir, 1100 m".

Distribution in the Middle East: Known only from Turkey (Gusenleitner, 2004, 2013a).

Extralimital distribution: None.

***Pterocheilus* (*Pterocheilus*) *nigrobilineolatus* Giordani Soika, 1942**

Pterocheilus nigrobilineolatus Giordani Soika, 1942: 56, ♂; "Heggiaz: Gedda".

Distribution in the Middle East: Known only from Saudi Arabia (Giordani Soika, 1942; van der Vecht and Fischer, 1972; Borsato and Ratti, 1999; Gusenleitner, 2013a).

Extralimital distribution: None.

***Pterocheilus* (*Pterocheilus*) *perpunctatus* Giordani Soika, 1970**

Pterocheilus perpunctatus Giordani Soika, 1970: 51, ♀; "Heggiaz: Gedda".

Distribution in the Middle East: Known only from Turkey (Giordani Soika, 1970; van der Vecht and Fischer, 1972; Gusenleitner, 1976, 2004 both, 2004, 2013a; Yildirim and Kojima, 1999).

Extralimital distribution: None.

***Pterocheilus* (*Pterocheilus*) *persicus* Gusenleitner and Fallahzadeh in Gusenleitner *et al.*, 2013**

Pterocheilus persicus Gusenleitner and Fallahzadeh, 2013 [in Gusenleitner *et al.*, 2013]: 109, ♀; Iran.

Catalogues with Iranian records: Rahmani *et al.* (2020).

Distribution in Iran: Fars (Gusenleitner *et al.*, 2013).

Distribution in the Middle East: Known only from Iran (Gusenleitner *et al.*, 2013; Gusenleitner, 2013a).

Extralimital distribution: None.

***Pterocheilus* (*Pterocheilus*) *phaleratus phaleratus* (Panzer, 1797)**

Vespa phalerata Panzer, 1797: 21, ♂; "Semel capta ruri mens. Iun. 1796" (destroyed).

Distribution in the Middle East: Turkey (Yildirim and Özbek, 1996a; Yildirim and Kojima, 1999; Gusenleitner, 2013a; Fateryga, 2017b, 2018; Fateryga *et al.*, 2020).

Extralimital distribution; Azerbaijan, Europe, Georgia, Kazakhstan, Mongolia, Russia, Ukraine (Fateryga, 2017b, 2018; Fateryga *et al.*, 2020), Italy (Selis, 2023).

***Pterocheilus* (*Pterocheilus*) *schwarzi* Gusenleitner, 1994**

Pterocheilus schwarzi Gusenleitner, 1994: 307, ♀, ♂; "Türkei, Konya, Karaman".

Distribution in the Middle East: Known only from Turkey (Gusenleitner, 1994, 2013a; Yildirim and Kojima, 1999).

Extralimital distribution: None.

***Pterocheilus* (*Pterocheilus*) *tenebricosus* Gusenleitner, 1998c**

Pterocheilus tenebricosus Gusenleitner, 1998c: 503, ♀; Iran.

Catalogues with Iranian records: Ebrahimi and Carpenter (2008), Rahmani *et al.* (2020).

Distribution in Iran: Fars (Gusenleitner, 1998c).

Distribution in the Middle East: Known only from Iran (Gusenleitner, 1998c; Gusenleitner, 2013a).

Extralimital distribution: None.

***Pterocheilus* (*Pterocheilus*) *unipunctatus* (Lepeletier, 1841)**

Pterochilus [!] *unipunctatus* Lepeletier, 1841: 676, ♀; [Algeria] "Oran".

Distribution in the Middle East: Egypt (Zalat *et al.*, 1993).

Extralimital distribution: Algeria (Blüthgen, 1956).

***Pterocheilus* (*Pterocheilus*) *varius* Gusenleitner, 2002**

Pterocheilus varius Gusenleitner, 2002: 335, 337, ♀; "Israel: Arava Valley, Shizav Nature Reserve, Hazeva, 30°46.01'N35°15,37'E, -80 m".

Distribution in the Middle East: Known only from Israel-Palestine (Gusenleitner, 2002, 2013a).

Extralimital distribution: None.

Genus *Rhynchium* Spinola, 1806

Rhynchium Spinola, *1806*: 84. Type species: *Rygchium* (!) *europaeum* Spinola, *1806* (= *Vespa oculata* Fabricius, 1781), by monotypy.

***Rhynchium acromum* Giordani Soika, 1952**

Rhynchium acromum Giordani Soika, 1952: 49, ♀; Karachi [Pakistan].

Catalogues with Iranian records: Ebrahimi and Carpenter (2008), Rahmani *et al.* (2020).

Distribution in Iran: Kerman (Pashaei Rad *et al.*, 2010), Sistan & Baluchestan (Blüthgen and Gusenleitner, 1970).

Distribution in the Middle East: Iran (Blüthgen and Gusenleitner, 1970; Pashaei Rad *et al.*, 2010).

Extralimital distribution: Pakistan (Giordani Soika, 1952).

***Rhynchium cyanopterum cyanopterum* de Saussure, 1852**

Rygchium (!) *cyanopterum* de Saussure, 1852: 108, ♀, ♂; Egypt.

Rhynchium aenigmaticum von Schulthess, 1922: 32, 36, ♀; Senegal.

Catalogues with Iranian records: Rahmani *et al.* (2020).

Distribution in Iran: Iran (no specific locality cited) (du Buysson, 1912).

Distribution in the Middle East: Arabia (Giordani Soika, 1952 as *Rygchium cyanopterum*), Egypt (de Saussure, 1852 as *Rygchium cyanopterum*; Walker, 1871 as *Rhynchium cyanopterum*; Giordani Soika, 1935 as *Odynerus cyanopterus*; Zalat *et al.*, 1993 as both *Rhynchium aenigmaticum* and *R. cyanopterum*; Gusenleitner, 2013a; Carpenter and Gadallah, 2020), Iran (du Buysson, 1912), Israel-Palestine (Giordani Soika, 1952 as *Rygchium cyanopterum*; Gusenleitner, 2013a; Carpenter and Gadallah, 2020), Jordan (Gusenleitner, 2013a; Carpenter and Gadallah, 2020), Oman (Gusenleitner, 2005; Carpenter and Gadallah, 2020), Saudi Arabia (Guichard, 1986; Carpenter and Gadallah, 2020), United Arab Emirates (Guichard, 1986; Howarth and Gillett, 2008; Gusenleitner, 2010b; Carpenter and Gadallah, 2020), Yemen (Giordani Soika, 1957 as *Rygchium cyanopterum*; Gusenleitner, 2007, 2013a; Carpenter and Gadallah, 2020).

Extralimital distribution: Algeria, Botswana, Chad, Malawi, Mali, Namibia, Niger, Nigeria, Senegal, Somalia (Carpenter and Gadallah, 2020).

***Rhynchium oculatum adenense* Giordani Soika, 1952**

Rhynchium oculatum adenense Giordani Soika, 1952: 49 (key).

Distribution in the Middle East: Oman (Giordani Soika, 1957; Gusenleitner, 2004, 2005, 2013a; Carpenter and Gadallah, 2020), Saudi Arabia (Giordani Soika, 1957; Gadallah. 2003; Carpenter and Gadallah, 2020), United Arab Emirates (Gusenleitner, 2010b, 2012, 2013a; Carpenter and Gadallah, 2020), Yemen (Giordani Soika, 1957; Carpenter and Gadallah, 2020).

Extralimital distribution: None.

***Rhynchium oculatum distinguendum* du Buysson, 1913**

Rhynchium distinguendum du Buysson, 1913: 80, ♀; "Bender-Bouchir", Persia [= Bandar e Bushehr], "Mascate" [Muscat] and "Aden".

Catalogues with Iranian records: Rahmani *et al.* (2020).

Distribution in Iran: Bushehr (du Buysson, 1913 as *Rhynchium distinguendum*), Kerman (Pashaei Rad *et al.*, 2010).

Distribution in the Middle East: Iran (see references above, and Blüthgen and Gusenleitner, 1970; Giordani Soika, 1970; Gusenleitner, 2013a), Oman (du Buysson, 1913 as *Rhynchium distinguendum*; Carpenter and Gadallah, 2020), Saudi Arabia (Gusenleitner, 2013a; Carpenter and Gadallah, 2020), Yemen (du Buysson, 1913 as *R. distinguendum*; Gusenleitner, 2013a; Carpenter and Gadallah, 2020).

Extralimital distribution: None.

***Rhynchium oculatum hebraeum* Giordani Soika, 1952**

Rhynchium oculatum hebraeum Giordani Soika, 1952: 49, ♀, ♂; Tel Aviv, Palestina.

Distribution in the Middle East: Cyprus (Giordani Soika, 1952; Gusenleitner, 2013a), Israel-Palestine (Giordani Soika, 1952; Gusenleitner, 2013a), Jordan (Giordani Soika, 1970; Yildirim and Kojima, 1999; Tezcan *et al.*, 2005; Haddad *et al.*, 2007; Gusenleitner, 2013a); Yalniz and Tüzün, 2019, Turkey (Giordani Soika, 1970; Yildirim and Özbek, 1996a, 1996b, 1999; Yildirim and Kojima, 1999; Tezcan *et al.*, 2005; Gusenleitner, 2007, 2013a; Yildirim and Gusenleitner, 2007, 2012; Yildirim, 2012; Yildirim and Tezcan, 2018; Yalniz and Tüzün, 2019).

Extralimital distribution: Greece (Gusenleitner, 2013a).

***Rhynchium oculatum oculatum* (Fabricius, 1781)**

Vespa oculata Fabricius, 1781: 463, sex not stated. Italia.

Catalogues with Iranian records: Ebrahimi and Carpenter (2008), Rahmani *et al.* (2020).

Distribution in Iran: Hormozgan (Giordani Soika, 1970; Ebrahimi and Carpenter, 2008; Modarres Awal, 2012; Khoobdel *et al.*, 2014).

Distribution in the Middle East: Egypt (Walker, 1871; Giordani Soika, 1935 as *Odynerus oculatus*; Krombein, 1969; Zalat *et al.*, 1993; Yildirim and Kojima, 1999; Gusenleitner, 2013a), Iran (see references above), Oman (Yildirim and Kojima, 1999), Saudi Arabia (Guichard, 1986; Walker and Pittaway, 1987; Gusenleitner, 2013a; Carpenter and Gadallah, 2020), Turkey (Tüzün, 1986; Gusenleitner, 1988; Yildirim and Özbek, 1996a; Yildirim and Kojima, 1999), United Arab Emirates (Gess and Roosenschoon, 2016; Carpenter and Gadallah, 2020), Yemen (Guichard, 1986; Yildirim and Kojima, 1999; Carpenter and Gadallah, 2020).

Extralimital distribution: Afghanistan, Italy, Malta, Spain (Carpenter and Gadallah, 2020).

Genus *Stenancistrocerus* de Saussure, 1863

Stenancistrocerus de Saussure, 1863: 216. Type species: *Odynerus atropos* Lepeletier, 1841; designated by Bequaert, 1925.

***Stenancistrocerus* (*Paratropancistrocerus*) *biblicus* (Giordani Soika, 1952)**

Tachyancistrocerus biblicus Giordani Soika, 1952: 38, ♀, ♂; Israel.

Catalogues with Iranian records: Rahmani *et al.* (2020).

Distribution in Iran: Sistan & Baluchestan (Hamzavi *et al.*, 2019).

Distribution in the Middle East: Iran (Hamzavi *et al.*, 2019), Israel-Palestine (Giordani Soika, 1952 as *Tachyancistrocerus biblicus*; Gusenleitner, 2013a), Jordan (Giordani Soika, 1952 as *T. biblicus*).

Extralimital distribution: None.

***Stenancistrocerus* (*Stenancistrocerus*) *leonhardi* Gusenleitner, 2015**

Stenancistrocerus leonhardi Gusenleitner, 2015: 1327, ♀; "Oman-NE, 17 km W Sùr".

Catalogues with Iranian records: None.

Distribution in Iran: Kerman (Schmid-Egger *et al.*, 2021).

Distribution in the Middle East: Iran (Schmid-Egger *et al.*, 2021), Oman (Gusenleitner, 2015; Carpenter and Gadallah, 2020).

Extralimital distribution: None.

***Stenancistrocerus* (*Paratropancistrocerus*) *obstrictus* (Morawitz, 1895)**

Odynerus (*Ancistrocerus*) *obstrictus* (Morawitz, 1895): 447, ♀, ♂; «Transcaucasia: Derbent» [Russia].

Catalogues with Iranian records: None.

Distribution in Iran: Kerman (Schmid-Egger *et al.*, 2021).

Distribution in the Middle East: Iran (Schmid-Egger *et al.*, 2021; Fateryga *et al.*, 2021).

Extralimital distribution: Armenia, Azerbaijan (Fateryga *et al.*, 2021), Caucasus, Russia (Fateryga, 2017b; Fateryga *et al.*, 2021).

Genus *Stenodynerus* de Saussure, 1863

Stenodynerus de Saussure, 1863: 228. Type species: *Odynerus chinensis de* Saussure, 1863; designated by Bohart, 1939.

***Stenodynerus aequisculptus aequisculptus* (Kostylev, 1940)**

Odynerus (*Nannodynerus*) *aequisculptus* Kostylev, 1940: 29, ♀, ♂; "Ragousa (Dalmatie)".

Distribution in the Middle East: Syria (Giordani Soika, 1970; Yildirim and Kojima, 1999; Fateryga, 2017b), Turkey (Giordani Soika, 1952 as *Pseudepipona* (*Pseudepipona*) *constrictiventris*; Giordani Soika, 1970; Gusenleitner, 1981 as *P.* (*P.*) *constrictiventris*; Yildirim and Özbek, 1996b, 1999; Yildirim, 1998; Yildirim and Kojima, 1999; Yildirim and Gusenleitner, 2004; Gusenleitner, 2013a; Fateryga, 2017b).

Extralimital distribution: Bulgaria, Caucasus, Central Asia, Croatia, France, Greece, Pakistan, Tunisia (Gusenleitner, 2013a; Fateryga, 2017b), Italy (Selis, 2023).

***Stenodynerus aequisculptus inaequisculptus* Giordani Soika, 1979**

Stenodynerus aequisculptus inaequisculptus Giordani Soika, 1979: 252, ♂, ♀; "Cipro: Lania, 1800 ft".

Distribution in the Middle East: Known only from Cyprus (Giordani Soika, 1979; Gusenleitner, 1981, 2013a).

Extralimital distribution: None.

***Stenodynerus aequisculptus taurus* (Gusenleitner, 1966)**

Nannodynerus aequisculptus taurus Gusenleitner, 1966: 349, ♀, ♂; Turkey.

Distribution in the Middle East: Israel-Palestine (Bytinski-Salz and Gusenleitner, 1971; Yildirim and Kojima, 1999), Lebanon, Syria (Giordani Soika, 1970; Yildirim and Kojima, 1999; Gusenleitner, 2013a), Turkey (Gusenleitner, 1966 as *Nannodynerus aequisculptus taurus*, 2013a; Giordani Soika, 1970; Yildirim and Kojima, 1999).

Extralimital distribution: None.

***Stenodynerus bluethgeni* van der Vecht, 1971**

Stenodynerus bluethgeni van der Vecht, 1971: 131, ♀, ♂; France.

Catalogues with Iranian records: Ebrahimi and Carpenter (2008), Rahmani *et al.* (2020).

Distribution in Iran: Iran (no specific locality cited) (Yildirim and Kojima, 1999; Gusenleitner, 2000a; Fateryga, 2017b, 2018).

Distribution in the Middle East: Cyprus (Fateryga *et al.*, 2025), Iran (see references above, and Fateryga *et al.*, 2025), Turkey (Giordani Soika, 1970 as *Stenodynerus dentisquama*; Yildirim and Özbek, 1996b, 1999; Yildirim, 1998; Yildirim and Kojima, 1999; Yildirim and Gusenleitner, 2012; Gusenleitner, 2013a; Fateryga, 2017b, 2018; Fateryga *et al.*, 2025).

Extralimital distribution: Armenia, Azerbaijan, Georgia (Fateryga *et al.*, 2025), Caucasus, China, Europe, Kazakhstan, Russia, Serbia, Turkmenistan (Fateryga, 2017b, 2018; Fateryga *et al.*, 2025), Italy (Selis, 2023), Slovenia (Gogala, 2022).

***Stenodynerus chevrieranus* (de Saussure, 1855)**

Odynerus (*Odynerus* div. *Epsilon*) *chevrieranus* de Saussure, 1855: 268, ♀, ♂; Les environs de Genève [Switzerland].

Catalogues with Iranian records: Ebrahimi and Carpenter (2008), Rahmani *et al.* (2020).

Distribution in Iran: East Azarbaijan (Ebrahimi and Carpenter, 2008; Modarres Awal, 2012), Zanjan (Abbasi *et al.*, 2008a, 2008b; Pashaei Rad *et al.*, 2010).

Distribution in the Middle East: Iran (see references above, and Yildirim and Kojima, 1999; Gusenleitner, 2013a; Fateryga, 2017b, 2018), Turkey (Giordani Soika, 1952 as *Stenodyneroides chevrieranus*; Giordani Soika, 1970; Yildirim and Özbek, 1996a, 1996b, 1999; Yildirim, 1998; Yildirim and Kojima, 1999; Yildirim and Gusenleitner, 2004, 2007; Gusenleitner, 2007, 2013a; Fateryga, 2017b, 2018; Fateryga *et al.*, 2019).

Extralimital distribution: Albania, Austria, Belarus, Bulgaria, Czech Republic, France, Germany, Greece, Hungary, Italy, Kazakhstan, Poland, Romania, Russia, Serbia, Spain, Switzerland (Fateryga, 2017b, 2018), Ukraine (Gusenleitner, 2013a).

***Stenodynerus chitgarensis* Giordani Soika, 1970**

Stenodynerus chitgarensis Giordani Soika, 1970: 108, ♀, ♂; Iran.

Catalogues with Iranian records: Ebrahimi and Carpenter (2008), Rahmani *et al.* (2020).

Distribution in Iran: Alborz (Giordani Soika, 1970), Tehran (Giordani Soika, 1970; Ebrahimi and Carpenter, 2008), Zanjan (Abbasi *et al.*, 2008a, 2008b; Pashaei Rad *et al.*, 2010).

Distribution in the Middle East: Iran (see references above, and Yildirim and Kojima, 1999; Gusenleitner, 2013a; Fateryga *et al.*, 2020; Fateryga and Proshchalykin, 2024), Turkey (Gusenleitner, 1981, 2013a; Yildirim and Özbek, 1996b; Yildirim and Kojima, 1999; Yildirim and Gusenleitner, 2004, 2012; Fateryga *et al.*, 2019, 2020; Fateryga and Proshchalykin, 2024).

Extralimital distribution: Azerbaijan, Kyrgyzstan, Uzbekistan (Fateryga *et al.*, 2019), China, Kazakhstan, Mongolia, Russia (Fateryga *et al.*, 2020), Kazakhstan (Fateryga and Proshchalykin, 2024).

***Stenodynerus claviger* Gusenleitner, 1981**

Stenodynerus claviger Gusenleitner, 1981: 209, 214, 294, ♀; "Türkei, Horasan, Arastal".

Distribution in the Middle East: Known only from Turkey (Gusenleitner, 1981, 1988, 2013a; Yildirim and Kojima, 1999).

Extralimital distribution: None.

***Stenodynerus clyppeopictus* (Kostylev, 1940)**

Odynerus (*Nannodynerus*) *clypeopictus* Kostylev, 1940: 28, ♀; "Fleuve Ili, Semiritchié".

Distribution in the Middle East: Turkey (Fateryga, 2017b; Kochetkov, 2020).

Extralimital distribution: Caucasus, China, Europe (WE, SE, EE), Japan, Kazakhstan, Korea, Kyrgyzstan, Mongolia, Russia (Fateryga, 2017b; Kochetkov, 2020).

***Stenodynerus cyrus* Gusenleitner, 2012**

Stenodynerus cyrus Gusenleitner, 2012: 1131, ♀; Iran.

Catalogues with Iranian records: Rahmani *et al.* (2020).

Distribution in Iran: Kerman (Gusenleitner, 2012).

Distribution in the Middle East: Known only from Iran (Gusenleitner, 2012, 2013a).

Extralimital distribution: None.

***Stenodynerus difficilis* (Morawitz, 1867)**

Odynerus (*Leionotus*) *difficilis* Morawitz, 1867: 132, ♀, ♂; Russia.

Catalogues with Iranian records: Ebrahimi and Carpenter (2008), Rahmani *et al.* (2020).

Distribution in Iran: Kuhgiloyeh & Boyerahmad (Gusenleitner, 2012).

Distribution in the Middle East: Iran (Gusenleitner, 2000a, 2012, 2013a; Yildirim and Kojima, 1999; Fateryga, 2017a, 2018), Turkey (Giordani Soika, 1952 as *Pseudepipona* (*Pseudepipona*) *fastidiosissima*; Blüthgen, 1959 as *Nannodynerus fastidiosissimus difficilis*; Gusenleitner, 1966 as *N. difficilis*, 2000a, 2013a; Giordani Soika, 1970 as *Stenodynerus difficilis*; Yildirim and Özbek, 1996b as *Stenodynerus fastidiosissima*; Yildirim and Kojima, 1999; Gusenleitner, 2007; Yildirim and Gusenleitner, 2007; Fateryga, 2017a, 2018).

Extralimital distribution: Bulgaria, Caucasus, Central Asia, Europe, Kazakhstan, Russia, Serbia, Turkmenistan (Gusenleitner, 2013a; Fateryga, 2017a, 2018; Selis *et al.*, 2024).

***Stenodynerus fastidiosissimus fastidiosissimus* (de Saussure, 1855)**
Odynerus (*Odynerus* div. *Epsilon*) *fastidiosissimus* de Saussure, 1855: 265, ♀, ♂; L'Algérie ou l'Europe méridionale.
Distribution in the Middle East: Israel-Palestine (Giordani Soika, 1952; Fateryga, 2017b).
Extralimital distribution: Algeria, France, Italy, Morocco, Portugal, Spain, Tunisia (Gusenleitner, 2013a; Selis *et al.*, 2024), Malta (Cassar *et al.*, 2022).

***Stenodynerus fastidiosissimus judaicus* Gusenleitner, 1970**
Stenodynerus difficilis judaicus Gusenleitner, 1970: 59, ♀; "Israel, Maale Hachamisha".
Distribution in the Middle East: Known only from Israel-Palestine (Gusenleitner, 1970 as *Stenodynerus difficilis judaicusi*; van der Vecht and Fischer, 1972 as *S. d. judaicus*; Gusenleitner, 1981, 2013a).
Extralimital distribution: None.

***Stenodynerus haladaorum* Gusenleitner, 1999**
Stenodynerus haladaorum Gusenleitner, 1999: 421, ♂, ♀; "Syria, 60 km S Damaskus, Khabab".
Distribution in the Middle East: Jordan, Syria, Turkey (Gusenleitner, 1999, 2013a).
Extralimital distribution: None.

***Stenodynerus orenburgensis* (André, 1884)**
Odynerus orenburgensis André, 1884: 719, ♀; "Russie mérid. (Orenbourg)".
Distribution in the Middle East: Turkey (Yildirim and Özbek, 1996a, 1999; Yildirim and Kojima, 1999; Yildirim and Gusenleitner, 2004; Fateryga, 2017b, 2018; Fateryga *et al.*, 2020).
Extralimital distribution: Caucasus, China, Europe, Kazakhstan, Kyrgyzstan, Mongolia, Russia (Fateryga, 2017b, 2018; Fateryga *et al.*, 2020), Italy (Selis, 2023), Slovenia (Gogala, 2022).

***Stenodynerus pannosus* Gusenleitner, 1985a**
Stenodynerus pannosus Gusenleitner, 1985a: 108, ♂; "Türkei, Suvari-Halil Pass, 2900 m".
Distribution in the Middle East: Known only from Turkey (Gusenleitner, 1985a, 2013a; Yildirim and Özbek, 1999).
Extralimital distribution: None.

***Stenodynerus picticrus* (Thomson, 1874)**
Lionotus picticrus Thomson, 1874: 75, ♀, ♂; Sweden.
Distribution in the Middle East: Turkey (Fateryga, 2017b; Fateryga *et al.*, 2020).
Extralimital distribution: Europe, Kazakhstan, Mongolia, Russia (Fateryga, 2017b; Fateryga *et al.*, 2020), Italy (Selis, 2023).

***Stenodynerus pullus* Gusenleitner, 1981**
Stenodynerus pullus Gusenleitner, 1981: 209, 220, 246, ♀; Russia.
Distribution in the Middle East: Turkey (Gusenleitner, 2007; Yildirim and Gusenleitner, 2007; Fateryga, 2017b; Fateryga *et al.*, 2020; Kochetkov, 2020).
Extralimital distribution: China, Korea, Mongolia, Russia (Fateryga, 2017b; Fateryga *et al.*, 2020).

***Stenodynerus punctifrons* (Thomson, 1874)**
Odynerus (*Leionotus*) *punctifrons* Thomson, 1874: 57, ♀; Helvetia.
Catalogues with Iranian records: Rahmani *et al.* (2020).
Distribution in Iran: Iran (no specific locality cited) (Kim and Yamane, 2004; Fateryga, 2017b).
Distribution in the Middle East: Iran (see references above, and Kochetkov, 2020), Turkey (Fateryga, 2017b; Fateryga and Mokrousov, 2019; Kochetkov, 2020).
Extralimital distribution: Caucasus, Europe, Kazakhstan, Kyrgyzstan, Mongolia, Russia (Fateryga, 2017b; Fateryga and Mokrousov, 2019), Italy (Selis, 2023), Slovenia (Gogala, 2022).
Comments: The presence of *S. punctifrons* in Iran needs confirmation.

***Stenodynerus sapidus* Giordani Soika, 1970**

Stenodynerus sapidus Giordani Soika, 1970: 110, ♀; Iran.

Catalogues with Iranian records: Ebrahimi and Carpenter (2008), Rahmani *et al.* (2020).

Distribution in Iran: Alborz (Giordani Soika, 1970).

Distribution in the Middle East: Iran (Giordani Soika, 1970; Gusenleitner, 2013a).

Extralimital distribution: Russia (Fateryga and Popov, 2017).

***Stenodynerus simulatus* Gusenleitner, 1981**

Stenodynerus simulatus Gusenleitner, 1981: 209, 214, 257, ♂, ♀; "Türkei, Meram bei Konya".

Distribution in the Middle East: Turkey (Gusenleitner, 1981, 2013a; Yildirim and Özbek, 1996a, 1999; Yildirim and Kojima, 1999; Yildirim and Gusenleitner, 2004, 2012).

Extralimital distribution: Serbia (Gusenleitner, 1981).

***Stenodynerus steckianus* (von Schulthess, 1897)**

Odynerus dentisquama var. *stekianus* von Schulthess, 1897: 90, ♂; Switzerland.

Distribution in the Middle East: Turkey (Gusenleitner, 1967, 1981 as *Nannodynerus teutonicus*; Gusenleitner, 1981, 2013a; Yildirim and Kojima, 1999; Yildirim and Gusenleitner, 2012; Fateryga, 2017a, 2017b, 2018).

Extralimital distribution: Azerbaijan, Europe, Kazakhstan, Russia, Serbia, Ukraine (Fateryga, 2017a, 2017b, 2018), Italy (Selis, 2023), Slovenia (Gogala, 2022).

***Stenodynerus trotzinai* (Morawitz, 1895)**

Odynerus trotzinai Morawitz, 1895: 463, ♀; Transcaspia [Turkmenistan].

Catalogues with Iranian records: Rahmani *et al.* (2020).

Distribution in Iran: Sistan & Baluchestan (Hamzavi *et al.*, 2019).

Distribution in the Middle East: Iran (Hamzavi *et al.*, 2019), Turkey (Yildirim and Gusenleitner, 2012; Fateryga *et al.*, 2023).

Extralimital distribution: Pakistan (Gusenleitner, 2006c), Turkmenistan (Morawitz, 1895 as *Odynerus trotzinai*), Uzbekistan (Fateryga *et al.*, 2023).

***Stenodynerus xanthomelas* (Herrich-Schäffer, 1839)**

Odynerus xanthomelas Herrich-Schäffer, 1839: 13, ♀, ♂; no locality given.

Catalogues with Iranian records: Ebrahimi and Carpenter (2008), Rahmani *et al.* (2020).

Distribution in Iran: Mazanadaran (Schmid-Egger *et al.*, 2021).

Distribution in the Middle East: Iran (Gusenleitner, 2013a; Fateryga, 2017b, 2018; Schmid-Egger *et al.*, 2021), Turkey (Fateryga, 2017b, 2018).

Extralimital distribution: Caucasus, China, Europe, Russia, Serbia (Fateryga, 2017b, 2018), Italy (Selis, 2023), Slovenia (Gogala, 2022).

Genus *Symmorphus* Wesmael, 1836

Symmorphus Wesmael, 1836: 45. Type species: *Odynerus elegans* Wesmael, 1833; designated by Richards, 1935.

***Symmorphus* (*Symmorphus*) *allobrogus* (de Saussure, 1855)**

Odynerus (*Protodynerus*) *allobrogus* de Saussure, 1855: 187, 190, ♀; "Les montagnes de la Savoie".

Distribution in the Middle East: Turkey (Cumming, 1989; Yildirim and Kojima, 1999; Fateryga, 2017b).

Extralimital distribution: China, Europe, Kazakhstan, Kyrgyzstan, Russia (Fateryga, 2017b), Italy (Selis, 2023).

***Symmorphus* (*Symmorphus*) *angustatus* (Zetterstedt, 1838)**

Odynerus angustatus Zetterstedt, 1838: 457, ♀; "in Finmarkia occidentali ad Bossekop (Lapponia Norvegica)".

Distribution in the Middle East: Turkey (Castro and Dvořák, 2009).

Extralimital distribution: China, Europe, Japan, Kazakhstan, Korea, Mongolia, Russia (Fateryga, 2017b), Italy (Selis, 2023).

***Symmorphus* (*Symmorphus*) *bifasciatus* (Linnaeus, 1761)**

Vespa bisfasciata Linnaeus, 1761: 219, ♀; type locality unknown.

Distribution in the Middle East: Turkey (Cumming, 1989; Yildirim and Özbek, 1996a, 1999, 1996b; Madl, 1997; Yildirim and Kojima, 1999; Yildirim and Gusenleitner, 2012; Fateryga, 2017b, 2018).

Extralimital distribution: Caucasus, Central Asia, China, Europe, Japan, Kazakhstan, Korea, Mongolia, Russia (Fateryga, 2017b, 2018), Italy (Selis, 2023), Slovenia (Gogala, 2022).

***Symmorphus* (*Symmorphus*) *connexus* (Curtis, 1826)**

Odynerus connexus Curtis, 1826: 137, ♂; Great Britain.

Distribution in the Middle East: Turkey (Giordani Soika, 1970; Gusenleitner, 1988, 2013a; Yildirim and Kojima, 1999; Fateryga, 2017b).

Extralimital distribution: Europe, Kazakhstan, Russia (Fateryga, 2017b), Italy (Selis, 2023).

***Symmorphus* (*Symmorphus*) *crassicornis* (Panzer, 1798)**

Vespa crassicornis Panzer, 1798: 53, ♀; Austria.

Catalogues with Iranian records: Ebrahimi and Carpenter (2008), Rahmani *et al.* (2020).

Distribution in Iran: Iran (no specific locality cited) (Ebrahimi and Carpenter, 2008).

Distribution in the Middle East: Iran (Ebrahimi and Carpenter, 2008; Fateryga, 2017b, 2018), Turkey (Giordani Soika, 1970; Cumming, 1989; Yildirim and Özbek, 1996a, 1996b, 1999; Yildirim and Kojima, 1999; Yildirim and Gusenleitner, 2009; Gusenleitner, 2013a; Fateryga, 2017b, 2018; Yildirim and Tezcan, 2018).

Extralimital distribution: Caucasus, Central Asia, China, Europe, India, Kazakhstan, Mongolia, Russia (Fateryga, 2017b, 2018), Italy (Selis, 2023), Slovenia (Gogala, 2022).

***Symmorphus* (*Symmorphus*) *debilitatus* (de Saussure, 1855)**

Odynerus (*Protodynerus*) *debilitatus* de Saussure, 1855: 187, 194, ♀; "Les environs de Genève".

Distribution in the Middle East: Turkey (Cumming, 1989; Yildirim and Kojima, 1999; Fateryga, 2017b, 2018).

Extralimital distribution: Europe, Russia (Fateryga, 2017b, 2018), Italy (Selis, 2023), Slovenia (Gogala, 2022).

***Symmorphus* (*Symmorphus*) *declivis* Harttig, 1932**

Symmorphus declivis Harttig, 1932: 73, ♀, ♂; "Podcetrtek (Windisch-Landsberg), Jugoslawien".

Distribution in the Middle East: Cyprus (Giordani Soika, 1963 as *Symmorphus* (*Koptodynerus*) *declivis*; Yildirim and Kojima, 1999; Gusenleitner, 2013a), Turkey (Giordani Soika, 1970 as *S.* (*K.*) *declivis*; Gusenleitner, 1972 as *S.* (*K.*) *declivis*, 2013a; Cumming, 1989; Yildirim and Kojima, 1999).

Extralimital distribution: Central Europe, France (Gusenleitner, 2013a), Italy (Selis, 2023), Slovenia (Gogala, 2022).

***Symmorphus* (*Symmorphus*) *gracilis* (Brullé, 1832)**

Odynerus gracilis Brullé 1832: 362, ♂; "forêt de Koubeh" [Greece].

Catalogues with Iranian records: Ebrahimi and Carpenter (2008), Rahmani *et al.* (2020).

Distribution in Iran: Iran (no specific locality cited) (Ebrahimi and Carpenter, 2008).

Distribution in the Middle East: Iran (Ebrahimi and Carpenter, 2008; Fateryga, 2018; Yalniz and Tüzün, 2019; Fateryga *et al.*, 2021), Israel-Palestine, Lebanon (Gusenleitner, 2013a; Fateryga, 2017b, 2018; Yalniz and Tüzün, 2019; Fateryga *et al.*, 2021), Syria (Fateryga, 2018; Yalniz and Tüzün, 2019; Fateryga, 2017b, Fateryga *et al.*, 2021), Turkey (Giordani Soika, 1970; Cumming, 1989; Yildirim and Özbek, 1996a, 1996b; Madl, 1997; Yildirim and Kojima, 1999; Gusenleitner, 2001a, 2013a; Yildirim and Gusenleitner, 2004, 2012; Yildirim, 2012; Fateryga, 2017b, 2018; Yalniz and Tüzün, 2019).

Extralimital distribution: Armenia, Central Asia, Europe, Russia (Gusenleitner, 2013a; Fateryga, 2018; Fateryga *et al.*, 2021), Azerbaijan (Fateryga *et al.*, 2021), Italy (Selis, 2023), Caucasus, Kazakhstan, Tajikistan, Turkmenistan (Fateryga, 2017b), Slovenia (Gogala, 2022).

***Symmorphus* (*Symmorphus*) *murarius* (Linnaeus, 1758)**

Vespa muraria Linnaeus, 1758: 573, ♀; Insecten von Teutschland.

Catalogues with Iranian records: Ebrahimi and Carpenter (2008), Rahmani *et al.* (2020).

Distribution in Iran: Iran (no specific locality cited) (Ebrahimi and Carpenter, 2008).

Distribution in the Middle East: Iran (Ebrahimi and Carpenter, 2008; Fateryga, 2017b, 2018; Fateryga *et al.*, 2020; Kochetkov, 2020; Fateryga *et al.*, 2025), Turkey (Giordani Soika, 1970 as *Symmorphus murarius* var. *nitidulator*; Gusenleitner, 1988 as *S. m. nitidulatori*; Gusenleitner, 2013a; Cumming, 1989; Yildirim and Özbek, 1996a, 1999; Yildirim and Kojima, 1999; Fateryga, 2017b, 2018; Fateryga *et al.*, 2020; Kochetkov, 2020; Fateryga *et al.*, 2025; Kaplan and Efil, 2025).

Extralimital distribution: Armania, Azerbaijan (Fateryga *et al.*, 2020, 2025), Caucasus, Central Asia, Europe, Kazakhstan, Korea, North Africa, Russia (Fateryga, 2017b, 2018; Fateryga *et al.*, 2020), Italy (Selis, 2023), Korean Peninsula, Kyrgyzstan, Mongolia, Uzbekistan (Fateryga *et al.*, 2025), Slovenia (Gogala, 2022).

Genus *Synagris* Latreille, 1802

Synagris Latreille, 1802: 360. Type species: "*Vespa cornuta*", by monotypy.

***Synargis* (*Paragris*) *rubescens* (Giordani Soika, 1989)**

Paragris rubescens Giordani Soika, 1989: 19, 20, ♂, ♀; "Yemen: Ju Amlali, 17°07'N - 43°31 E, 26 km NW Sa.dah, m 1950".

Distribution in the Middle East: Saudi Arabia (Giordani Soika, 1989 as *Paragris rubescens*; Wahis, 2000; Carpenter and Gadallah, 2020), Yemen (Giordani Soika, 1989 as *P. rubescens*; Wahis, 2000; Dvořák and Carpenter, 2010; Carpenter and Gadallah, 2020).

Extralimital distribution: None.

***Synargis* (*Paragris*) *spiniventris* (Illiger, 1802)**

Vespa spiniventris Illiger, 1802: 190, ♂ [described as a ♀]; "Sierra Leone".

Distribution in the Middle East: Saudi Arabia (Guichard, 1986; Gusenleitner, 2013a; Carpenter and Gadallah, 2020), Yemen (Gusenleitner, 2007, 2013a; Carpenter and Gadallah, 2020).

Extralimital distribution: Tropical Africa (Carpenter and Gadallah, 2020).

***Synargis (Paragris) spinosuscula* de Saussure, 1852**

Synagris spinosuscula de Saussure, 1852: 85, ♀, ♂; "l'Abyssinie".

Distribution in the Middle East: Yemen (Gusenleitner, 2007, 2013a; Carpenter and Gadallah, 2020).

Extralimital distribution: Eritrea, Ethiopia, Kenya (Carpenter and Gadallah, 2020).

Genus *Syneuodynerus* Blüthgen, 1951

Syneuodynerus Blüthgen, 1951b: 67. Type species: *Odynerus egregius* Herrich-Schäffer, 1839, by original designation.

***Syneuodynerus egregius egregior* (Gusenleitner, 1970)**

Euodynerus (*Syneuodynerus*) *egregius egregior* Gusenleitner, 1970: 59, ♀, ♂; "Wadi Qilt, Israel".

Catalogues with Iranian records: Ebrahimi and Carpenter (2008), Gusenleitner (1997b), Rahmani *et al.* (2020).

Distribution in Iran: Hormozgan (Gusenleitner, 1970, 1997b).

Distribution in the Middle East: Iran (see references above, and Yildirim and Kojima, 1999), Israel-Palestine (Gusenleitner, 1970 as *Euodynerus* (*Syneuodynerus*) *egregius egregior*, 2013a; Yildirim and Kojima, 1999), Jordan (Gusenleitner, 2001b), Syria (Yildirim and Kojima, 1999; Gusenleitner, 2013a), Turkey (Gusenleitner, 1970 as *E.* (*S.*) *egregius egregior*, 2001b, 2013a; Yildirim and Kojima, 1999).

Extralimital distribution: None.

***Syneuodynerus egregius egregius* (Herrich-Schäffer, 1839)**

Odynerus egregius Herrich-Schäffer, 1839: 10, ♀; Spain.

Distribution in the Middle East: Cyprus, Israel-Palestine (Gusenleitner, 1970 as *Euodynerus (Syneuodynerus) egregius egregius*; Fateryga, 2017b, 2018; Fateryga *et al.*, 2021), Jordan, Syria (Fateryga, 2017b, 2018; Fateryga *et al.*, 2021), Turkey (Gusenleitner, 1966 as *E.* (*Synodynerus*) *egregiusi*, 2013a; Giordani Soika, 1970 as *E.* (*S.*) *egregius*; Yildirim and Özbek, 1996a, b, 1999 as *E.* (*S.*) *egregius*; Yildirim and Kojima, 1999; Gusenleitner, 2001a; Yildirim and Gusenleitner, 2012; Fateryga, 2017b, 2018; Fateryga *et al.*, 2021).

Extralimital distribution: Armenia, Azerbaijan (Fateryga *et al.*, 2021), Caucasus, Europe, North Africa, Russia, Turkmenistan (Fateryga, 2017b, 2018; Fateryga *et al.*, 2021), Italy (Selis, 2023).

***Syneuodynerus erichi* (Gusenleitner, 1967)**

Euodynerus (*Syneuodynerus*) *erichi* Gusenleitner, 1967: 672, ♂; Turkey.

Distribution in the Middle East: Known only from Turkey (Gusenleitner, 1967 as *Euodynerus* (*Syneuodynerus*) *erichi*; Gusenleitner, 1988, 1997b, 2013a; Giordani Soika, 1970 as *E. erichi*; van der Vecht and Fischer, 1972 as *E.* (*S.*) *erichi*; Yildirim and Özbek, 1996a, 1999 as *E.* (*S.*) *erichi*; Yildirim and Kojima, 1999).

Extralimital distribution: None.

***Syneuodynerus fouadi dhofariensis* Giordani Soika, 1979**

Syneuodynerus fouadi dhofariensis Giordani Soika, 1979: 284, ♀; "Dhofar: Ayun Pools, 700 m".

Distribution in the Middle East: Oman (Giordani Soika, 1979; Gusenleitner, 2013a; Carpenter and Gadallah, 2020), United Arab Emirates (Gusenleitner, 2010b, 2012, 2013a; Carpenter and Gadallah, 2020).

Extralimital distribution: None.

***Syneuodynerus fouadi fouadi* (Giordani Soika, 1939)**

Odynerus (*Rhynchium*) *fouadi* Giordani Soika, 1939: 1, ♀; "Egitto; Helwan and North Galala".

Distribution in the Middle East: Known only from Egypt (Giordani Soika, 1939, 1973 both as *Odynerus* (*Rhynchium*) *fouadi*, 1970; van der Vecht and Fischer, 1972 as *E.* (*Syneuodynerus*) *fouadi*; Borsato and Ratti, 1999 as *E.* (*S.*) *fouadi fouadi*; Gusenleitner, 2013a).

Extralimital distribution: None.

***Syneuodynerus fritzi* (Gusenleitner, 1997b)**

Intereuodynerus fritzi Gusenleitner, 1997b: 763, 765, ♂, ♀; "Syrien (Süd), 50 km Se Suwayda, Anata".

Distribution in the Middle East: Iraq (Selis *et al.*, 2025), Syria, Turkey (Gusenleitner, 1997b, 2013a both as *Intereuodynerus fritzi*; Yildirim and Kojima, 1999 as *I. fritzi*).

Extralimital distribution: None.

Comments: By comparing type species of the genus *Intereuodynerus*, including *I. fritzi* of Gusenleitner (1997b), with various species of *Syneuodynerus* by Selis *et al.* (2025), no sufficient distinct differences were found. Therefore, the genus *Intereuodynerus* Gusenleitner, 1997b is synonymised with *Syneuodynerus* (Selis *et al.*, 2025).

***Syneuodynerus gebi* (Gusenleitner, 1970)**

Euodynerus (*Syneuodynerus*) *gebi* Gusenleitner, 1970: 59, ♀, ♂; "Israel, Sussita, near En Geb".

Distribution in the Middle East: Israel-Palestine, Syria (Gusenleitner, 1970 as *Euodynerus* (*Syneuodynerus*) *gebi*; Gusenleitner, 2013a; van der Vecht and Fischer, 1972 as *E.* (*S.*) *gebi*).

Extralimital distribution: None.

***Syneuodynerus insolitus* Gusenleitner, 1994**

Syneuodynerus insolitus Gusenleitner, 1994: 307, 315, 323, ♀; "Jordanien, 80 km NE Aqaba (Straße nach Amman)".

Distribution in the Middle East: Egypt, Jordan (Gusenleitner, 2013a), Israel-Palestine (Gusenleitner, 1994 as *Syneuodynerus insolatus*).
Extralimital distribution: None.

***Syneuodynerus siegberti* (Gusenleitner, 1967)**
Euodynerus (*Syneuodynerus*) *siegberti* Gusenleitner, 1967: 674, ♂; Turkey.
Distribution in the Middle East: Syria (Gusenleitner, 1967 as *Euodynerus* (*Syneuodynerus*) *siebergi*; Gusenleitner, 1988 as *Syneuodynerus siegberti*; Gusenleitner, 1997b, 2013a as *Intereuodynerus siegberti*), Turkey (Gusenleitner, 1967 as *E.* (*S.*) *siebergi*; Gusenleitner, 1997b, 2013a; Giordani Soika, 1970 as *E.* (*S.*) *siebergi*; Gusenleitner, 1997b, 2013a as *I. siegberti*; Yildirim and Kojima, 1999 as *I. siegberti*).
Extralimital distribution: None.

Genus *Tachyancistrocerus* Giordani Soika, 1952

Tachyancistrocerus Giordani Soika, 1952: 37. Type species: *Odynerus rhodensis* de Saussure, 1855, by original designation.

***Tachyancistrocerus aereus* Giordani Soika, 1952**
Tachyancistrocerus aereus Giordani Soika, 1952 (1951): 378, 380, ♂, ♀; "Anatolia: Toros, Bürücek".
Distribution in the Middle East: Known only from Turkey (Giordani Soika, 1952, 1973; van der Vecht and Fischer, 1972; Borsato and Ratti, 1999; Yildirim and Kojima, 1999; Gusenleitner, 2013a).
Extralimital distribution: None.

***Tachyancistrocerus cyprius* (Pittioni, 1950)**
Subancistrocerus cyprius Pittioni, 1950: 12, ♂, ♀; Cyprus.
Distribution in the Middle East: Known only from Cyprus (Pittioni, 1950 as *Subancistrocerus cyprius*; Giordani Soika, 1952 (1951) as *Tachyancistrocerus cypricus* [sic]; van der Vecht and Fischer, 1972; Gusenleitner, 2012, 2013a).
Extralimital distribution: None.

***Tachyancistrocerus komarowi derufata* Gusenleitner, 1971**
Tachyancistrocerus komarowi derufata Gusenleitner, 1971 [in Bytinski-Salz and Gusenleitner, 1971]: 239, 271, 294, ♀, ♂; "Israel ... Ein Gedi".
Distribution in the Middle East: Known only from Israel-Palestine (Bytinski-Salz and Gusenleitner, 1971; Gusenleitner, 2013a).
Extralimital distribution: None.

***Tachyancistrocerus komarowi komarowi* (Morawitz, 1885)**
Ancistrocerus komarowi Morawitz, 1885: 175, ♀; Territorio transcaspico. Asschabad [Ashkhabad, Turkmenistan].
Catalogues with Iranian records: Ebrahimi and Carpenter (2008), Rahmani *et al.* (2020).
Distribution in Iran: Bushehr (du Buysson, 1912 as *Odynerus komarowi* Morawitz, 1885; Giordani Soika, 1943 as *Pseudonortonia bushirensis* Giordani Soika, 1943), Fars, Khuzestan (Giordani Soika, 1970), Kerman (Pashaei Rad *et al.*, 2010; Schmid-Egger *et al.*, 2021).
Distribution in the Middle East: Iran (see references above, and Gusenleitner, 2013a; Fateryga *et al.*, 2023), Israel-Palestine (Fateryga *et al.*, 2023), Oman (Giordani Soika, 1970 as *Tachyancistrocerus komarowi bushirensis*; Gusenleitner, 1988; Yildirim and Özbek, 1999; Yildirim and Kojima, 1999; Carpenter and Gadallah, 2020; Fateryga *et al.*, 2023), Turkey (Gusenleitner, 1988, 2006a; Fateryga *et al.*, 2023).
Extralimital distribution: Afghanistan, Kazahkstan, Pakistan, Turkmenistan (Gusenleitner, 2006a), Uzbekistan (Fateryga *et al.*, 2023).

***Tachyancistrocerus liliput* (Gusenleitner, 1993)**
Stenancistrocerus (*Paratropancistrocerus*) *liliput* Gusenleitner, 1993: 43, ♂; "Türkei, Urfa".
Distribution in the Middle East: Known only from Turkey (Gusenleitner, 1993 as *Stenancistrocerus* (*Paratropancistrocerus*) *lilliput*, 2015; Yildirim and Kojima, 1999 as *S.* (*P.*) *lilliput*; Gusenleitner, 2013a as *S.* (*S.*) *liliput*).
Extralimital distribution: None.

***Tachyancistrocerus pallidenotatus* Giordani Soika, 1970**
Tachyancistrocerus pallidenotatus Giordani Soika, 1970: 90, ♀; Iran.
Catalogues with Iranian records: Ebrahimi and Carpenter (2008), Rahmani *et al.* (2020).
Distribution in Iran: Fars (Giordani Soika, 1970).
Distribution in the Middle East: Iran (Giordani Soika, 1970; Gusenleitner, 2013a), Turkey (Fateryga *et al.*, 2020).
Extralimital distribution: Caucasus, China, Europe, Kazakhstan, Mongolia, North Africa, Russia (Fateryga *et al.*, 2020).

***Tachyancistrocerus quabosi* Giordani Soika, 1979**
Tachyancistrocerus quabosi Giordani Soika, 1979: 278, ♀, ♂; "Oman: Behla".
Catalogues with Iranian records: Rahmani *et al.* (2020).
Distribution in Iran: Sistan & Baluchestan (Hamzavi *et al.*, 2019).
Distribution in the Middle East: Iran (Hamzavi *et al.*, 2019), Oman (Giordani Soika, 1957; Gusenleitner, 2013a; Carpenter and Gadallah, 2020), United Arab Emirates (Guichard, 1986; Howarth and Gillett, 2008; Gusenleitner, 2013a; Carpenter and Gadallah, 2020).
Extralimital distribution: None.

***Tachyancistrocerus rhodensis* (de Saussure, 1855)**
Odynerus (*Ancistrocerus*) *rhodensis* de Saussure, 1855: 208, ♀; "L'île de Rhodes" [Greece].
Catalogues with Iranian records: None.
Distribution in Iran: Mazandaran (Schmid-Egger *et al.*, 2021).
Distribution in the Middle East: Cyprus (Giordani Soika, 1942 as *Ancistrocerus rhodensis*; Yildirim and Kojima, 1999; Gusenleitner, 2013a; Fateryga *et al.*, 2021), Iran (Fateryga *et al.*, 2021; Schmid-Egger *et al.*, 2021), Israel-Palestine (Giordani Soika, 1952; Yildirim and Kojima, 1999; Gusenleitner, 2013a; Fateryga *et al.*, 2021), Syria (Giordani Soika, 1970; Gusenleitner, 2013a), Turkey (Giordani Soika, 1952, 1970; Yildirim and Özbek, 1996a, 1996b, 1999; Yildirim and Kojima, 1999; Gusenleitner, 2007, 2013a; Yildirim and Gusenleitner, 2007, 2009, 2012; Yildirim and Tezcan, 2018; Fateryga *et al.*, 2021).
Extralimital distribution: Azerbaijan (Fateryga *et al.*, 2021), Greece, Italy (Gusenleitner, 2013a), Malta (Erlandsson, 1972; Schembri, 1980).

***Tachyancistrocerus schmidti* (Kokujev, 1913)**
Odynerus (*Ancistrocerus*) *schmidti* Kokujev, 1913: 4, ♀; Caucasus (Geok-tapa) [Azerbaijan].
Catalogues with Iranian records: Ebrahimi and Carpenter (2008), Rahmani *et al.* (2020).
Distribution in Iran: Alborz, Razavi Khorasan, Tehran (Giordani Soika, 1970).
Distribution in the Middle East: Iran (Giordani Soika, 1970; Gusenleitner, 2013a), Turkey (Blüthgen, 1951a as *Subancistrocerus schmidti*; Yildirim and Özbek, 1996a, b, 1999; Yildirim and Kojima, 1999; Yildirim and Gusenleitner, 2004; Fateryga *et al.*, 2019).
Extralimital distribution: Afghanistan, Armenia, Azerbaijan, Kyrgyzstan, Uzbckistan (Fateryga *et al.*, 2019).

***Tachyancistrocerus serenus* (Giordani Soika, 1935)**
Ancistricerus (*Subancistrocerus*) *serenus* Giordani Soika, 1935: 169, 177, ♂, ♀; Egypt.
Distribution in the Middle East: Egypt (Giordani Soika, 1935 as *Ancistrocerus* (*Subancistrocerus*) *serenus*; Zalat *et al.*, 1993; Gusenleitner, 2013a), Israel-Palestine (Giordani Soika, 1952;

Gusenleitner, 2013a), Oman (Gusenleitner, 2004, 2005, 2013a; Carpenter and Gadallah, 2020), Saudi Arabia (Guichard, 1986; Gadallah, 2003; Gusenleitner, 2013a; Carpenter and Gadallah, 2020), United Arab Emirates (Gusenleitner, 2012, 2013a; Carpenter and Gadallah, 2020).
Extralimital distribution: Algeria, Chad, Libya, Western Sahara (Carpenter and Gadallah, 2020).

***Tachyancistrocerus syriacus* Giordani Soika, 1970**
Tachyancistrocerus syriacus Giordani Soika, 1970: 87, ♂; "Siria: W. Raggad".
Distribution in the Middle East: Syria (Giordani Soika, 1970; van der Vecht and Fischer, 1972; Yildirim and Kojima, 1999; Gusenleitner, 2013a), Turkey (Madl, 1997; Yildirim and Kojima, 1999).
Extralimital distribution: None.

Genus *Tricarinodynerus* Giordani Soika 1952
Tricarinodynerus Giordani Soika, 1952: 73, 79. Type species: *Odynerus geyrinii* de Saussure, 1852, by original designation.

***Tricarinodynerus arabicus* Guichard, 1986**
Tricarinodynerus arabicus Guichard, 1986: 202, 213, 221, ♀, ♂; "Saudi Arabia: Hada Asahm near Jeddah".
Distribution in the Middle East: Saudi Arabia (Guichard, 1986; Gusenleitner, 2013a; Carpenter and Gadallah, 2020), Yemen (Gusenleitner, 2007, 2013a; Carpenter and Gadallah, 2020).
Extralimital distribution: None.

***Tricarinodynerus magrettii* (Gribodo, 1884)**
Odynerus magrettii Gribodo, 1884: 290 in footnote, ♀, ♂; "Abissinia".
Distribution in the Middle East: Egypt (Gribodo, 1884 as *Odynerus magrettii*), Saudi Arabia (Giordani Soika, 1957 as *Pseudepipona magrettii*; Guichard, 1986 as *P. magrettii*; Gusenleitner, 2013a; Carpenter and Gadallah, 2020), Yemen (Kohl, 1906 as *Rhynchium forticulum*; Giordani Soika, 1957 as *P. magrettii*; Gusenleitner, 2013a; Carpenter and Gadallah, 2020).
Extralimital distribution: Eritrea, Ethiopia, Kenya, Somalia, Sudan (Carpenter and Gadallah, 2020).

Genus *Tropidodynerus* Blüthgen, 1939
Tropidodynerus Blüthgen, 1939: 259, 260. Type species: *Hoplonerus interruptus* (Brullé, 1832) = *H. mandibularis* Morawitz, 1885 [= *Polistes interrupta* Brullé, 1832], by original designation.

***Tropidodynerus* (*Tropidodynerus*) *flavus semiticus* (Giordani Soika, 1952)**
Odynerus (*Ptropidodynerus*) *variegatus semiticus* Giordani Soika, 1952: 57, ♀, ♂; "Palestina: Gerico".
Distribution in the Middle East: Known only from Israel-Palestine (Giordani Soika, 1952 as *Odynerus* (*Tropidodynerus*) *variegatus semiticus*; van der Vecht and Fischer, 1972; Borsato and Ratti, 1999; Gusenleitner, 2013a; Kumar and Carpenter, 2013).
Extralimital distribution: None.

***Tropidodynerus* (*Tropidodynerus*) *interruptus interruptus* (Brullé, 1832)**
Polistes interrupta Brullé, 1832: 360, ♀; "bois de Lygoudista".
Distribution in the Middle East: Turkey (Kostylev, 1929 as *Hoplomerus mandibularis*; Giordani Soika, 1970; Yildirim, 1998; Yildirim and Kojima, 1999; Gusenleitner, 2013a; Fateryga, 2017b, 2018).
Extralimital distribution: Armenia, Azerbaijan, Europe, Kazakhstan, Russia (Fateryga, 2017b, 2018).

Genus *Xanthodynerus* Blüthgen, 1954
Xanthodynerus Blüthgen, 1954: 255, 265, subgenus of *Euodynerus*. Type species: *Euodynerus octavus* (Giordani Soika, 1943), by original designation.

***Xanthodynerus castrorum* (Blüthgen, 1954)**

Euodynerus (*Xanthodynerus*) *castrorum* Blüthgen, 1954: 268, 269, ♀; Egypt.

Distribution in the Middle East: Known only from Egypt (Blüthgen, 1954 as *Euodynerus* (*Xanthodynerus*) *castrorum*; Giordani Soika, 1958 as *Pseudepipona castrorum*; van der Vecht and Fischer, 1972 as *E. castrorum*; Zalat *et al*., 1993; Gusenleitner, 2013a).

Extralimital distribution: None.

***Xanthodynerus jordanicus* Gusenleitner, 1990**

Xanthrodynerus jordanicus Gusenleitner, 1990: 193, 209, ♀, ♂; "Jordanien, 70 km NE Aqaba (Straße nach Amman)".

Distribution in the Middle East: Known only from Jordan (Gusenleitner, 1990, 2013a).

Extralimital distribution: None.

***Xanthodynerus octavus* (Giordani Soika, 1943)**

Odynerus (*Rhynchium*) *octavus* Giordani Soika, 1943: 37, ♂; (lectotype ♂ Zürich).

Distribution in the Middle East: Egypt (von Schulthess, 1928 as *Odynerus* (*Lionotus*) *testaceus*; Giordani Soika, 1934 as *O.* (*Rhynchium*) *testaceus*; Zalat *et al*., 1993; Gusenleitner, 2013a), Israel-Palestine (Gusenleitner, 2013a), Oman, Saudi Arabia (Guichard, 1986; Gusenleitner, 2004, 2013a; Carpenter and Gadallah, 2020), United Arab Emirates (Gusenleitner, 2010b, 2012, 2013a; Carpenter and Gadallah, 2020).

Extralimital distribution: Morocco (Giordani Soika, 1953 both as *Pseudepipona octava* and *P. nona*; Carpenter and Gadallah, 2020), Niger (Carpenter and Gadallah, 2020).

***Xanthodynerus pilosus* Gusenleitner, 1990**

Xanthodynerus pilosus Gusenleitner, 1990: 192, 203, ♀, ♂; "Israel, Hazeva, 140 m".

Distribution in the Middle East: Known only from Israel-Palestine (Gusenleitner, 1990, 2013a).

Extralimital distribution: None.

***Xanthodynerus trichopygus* (Blüthgen, 1954)**

Euodynerus (*Xanthodynerus*) *trichopygus* Blüthgen, 1954: 267, 269, ♂; Egypt.

Distribution in the Middle East: Known only from Egypt (Blüthgen, 1954 as *Euodynerus* (*Xanthodynerus*) *trichopygus*; van der Vecht and Fischer, 1972 as *E.* (*X.*) *trichopygus*; Zalat *et al*., 1993; Gusenleitner, 2013a).

Extralimital distribution: None.

Subfamily Raphiglossinae Ashmead, 1902

Raphiglossinae are distinguished from the Eumeninae and Zethinae by the position of the recurrent veins in the fore wing: the first recurrent vein is received by the second submarginal cell, and the second recurrent vein by the third submarginal cell (Gusenleitner, 2013a).

In the Middle East, the subfamily Raphiglossinae is represented by two genera: *Psiliglossa* S. Saunders and *Raphiglossa* S. Saunders. A third genus, the monotypic *Elisella*, occurs in South Africa. The elevation of the raphiglossines to subfamily status is based on Bank *et al.* (2017). Information on nesting and plant-feeding behaviors for this group is provided by Mauss *et al.*, (2019).

Genus *Psiliglossa* S.S. Saunders, 1872

Psiliglossa S.S Saunders, 1872: 42. New name for *Stenoglossa* de Saussure, 1852. Type species: *Raphiglossa odyneroides* S.S. Saunders, 1850, by monotypy as type species of *Stenoglossa* de Saussure, 1852.

***Psiliglossa algeriensis* E. Saunders, 1905**

Psiloglossa algeriensis E. Saunders, 1905: 400, ♀, ♂; Biskar [Algeria].

Catalogues with Iranian records: Ebrahimi and Carpenter (2008), Rahmani *et al.* (2020).

Distribution in Iran: Hormozgan (Gusenleitner, 1973, 2011a).

Distribution in the Middle East: Iran (see references above, and Gusenleitner, 2013a as *Psiliglossa anatolica* Giordani Soika, 1979), Israel-Palestine, Jordan (Gusenleitner, 2013a as *P. anatolica*), Turkey (Giordani Soika, 1979 as *Psiloglossa anatolica*; Gusenleitner, 1988, 2013a as *P. anatolica*, 2011a; Yildirim and Özbek, 1996a, 1996b as *P. anatolica*; Yildirim and Kojima, 1999 as *P. anatolica*), United Arab Emirates (Gusenleitner, 2010b).

Extralimital distribution: Algeria, Libya, Tunisia (Gusenleitner, 2011a), Morocco (Gusenleitner, 2013a).

Comments: *P. anatolica* Giordani Soika, 1979 is a junior synonym of *P. algeriensis* (Gusenleitner, 2011a).

***Psiliglossa odyneroides odyneroides* (S.S. Saunders, 1850)**

Raphiglossa odyneroides S.S. Saunders, 1850: 17, 72, ♀, ♂; Greece.

Distribution in the Middle East: Turkey (Giordani Soika, 1970; Yildirim and Özbek, 1996a, 1999; Yildirim, 1998; Yildirim and Kojima, 1999; Gusenleitner, 2013a; Fateryga, 2017b; Fateryga and Proshchalykin, 2024).

Extralimital distribution: Bulgaria, Greece, Italy (Gusenleitner, 2013a), Caucasus, Europe (EE, SE), Kazakhstan, Uzbekistan (Fateryga and Proshchalykin, 2024), Russia, Turkmenistan (Fateryga, 2017b).

***Psiliglossa pulchra zhelochovtsevi* Panfilov, 1968**

Psiliglossa zhelochovtsevi Panfilov, 1968: 42, ♀; Tyan Shan Mts., central Asia.

Catalogues with Iranian records: Ebrahimi and Carpenter (2008), Rahmani *et al.* (2020).

Distribution in Iran: Kuhgiloyeh & Boyerahmad (Gusenleitner, 2004 as *Psiliglossa zhelochovtsevi* Panfilov, 1968).

Distribution in the Middle East: Iran (Gusenleitner, 2004).

Extralimital distribution: Afghanistan, Central Asia (Gusenleitner, 2013a as *P. zhelochovtsevi*), Kyrgyzstan (Panfilov, 1968).

Genus *Raphiglossa* S.S. Saunders, 1850

Raphiglossa S.S. Saunders, 1850: 71. Type species: *Raphiglossa eumenoides* Saunders, 1850; designated by Ashmead, 1902.

***Raphiglossa bytinskii* Giordani Soika, 1974**

Raphiglossa bytinskii Giordani Soika, 1974: 111, 124, ♀; "Israele: 5 km E di Arad".

Distribution in the Middle East: Known only from Israel-Palestine (Giordani Soika, 1974; Gusenleitner, 2013a).

Extralimital distribution: None.

***Raphiglossa eumenoides caucasica* Giordani Soika, 1970**

Raphiglossa eumenoides caucasica Giordani Soika, 1970: 28, ♂; "Armenia, Monastero di Gherard".

Catalogues with Iranian records: Ebrahimi and Carpenter (2008), Rahmani *et al.* (2020).

Distribution in Iran: Alborz (Gusenleitner, 2000b).

Distribution in the Middle East: Iran (Gusenleitner, 2000b, 2013a), Turkey (Gusenleitner, 1988; Yildirim and Kojima, 1999).

Extralimital distribution: None.

***Raphiglossa eumenoides eumenoides* S.S. Saunders, 1850**

Raphiglossa eumenoides S.S. Saunders, 1850: 17, ♀, ♂; Greece.

Distribution in the Middle East: Lebanon, Turkey (Giordani Soika, 1970; Yildirim and Özbek, 1996b, 1999; Yildirim and Kojima, 1999; Gusenleitner, 2001a, 2013a; Fateryga *et al.*, 2019).

Extralimital distribution: Albania, Greece (Gusenleitner, 2013a), Armenia, Azerbaijan, Europe (Fateryga *et al.*, 2019), Italy (Gusenleitner, 2013a; Selis, 2023).

***Raphiglossa formosa* Kostylev, 1940**
Raphiglossa formosa Kostylev, 1940: 139, ♂; «Kabadian» [Tajikistan].
Catalogues with Iranian records: Ebrahimi and Carpenter (2008), Rahmani *et al.* (2020).
Distribution in Iran: Fars (Giordani Soika, 1970).
Distribution in the Middle East: Iran, Syria (Giordani Soika, 1970; Gusenleitner, 2013a; Fateryga *et al.*, 2021).
Extralimital distribution: Azerbaijan (Fateryga *et al.*, 2021), Tajikistan (Gusenleitner, 2013a; Fateryga *et al.*, 2021).

***Raphiglossa irana* Giordani Soika, 1970**
Raphiglossa irana Giordani Soika, 1970: 28, ♂; Iran.
Catalogues with Iranian records: Ebrahimi and Carpenter (2008), Rahmani *et al.* (2020).
Distribution in Iran: Alborz (Giordani Soika, 1970), Hormozgan, Kerman (Schmid-Egger *et al.*, 2021).
Distribution in the Middle East: Known only from Iran (see references above, and Gusenleitner, 2013a).
Extralimital distribution: None.

***Raphiglossa spinosa rufescens* (E. Saunders, 1905)**
Raphidoglossa (!) *rufescens* E. Saunders, 1905: 402, ♀, ♂; Algeria.
Distribution in the Middle East: Israel-Palestine (von Schulthess, 1931).
Extralimital distribution: Algeria, Libya, Morocco, Tunisia (Gusenleitner, 2013a).

Subfamily Zethinae de Saussure, 1855

The Zethinae are a subfamily of wasps within the family Vespidae, formerly included in the subfamily Eumeninae (Piekarski *et al.*, 2018). The "Zethini"» *sensu* Hermes *et al.* (2014) were recovered as a paraphyletic grouping comprising two major clades, Raphiglossinae and Zethinae (Bank *et al.*, 2017). Recognizing Zethinae as a separate subfamily renders the traditional Eumeninae *sensu* Carpenter (1981) paraphyletic (Bank *et al.*, 2017; Piekarski *et al.*, 2018). The subfamily status of Zethinae, as a sister to Polistinae + Vespinae, is supported by both nuclear and mitochondrial DNA phylogenies (Hines *et al.*, 2007; Piekarski *et al.*, 2018; Luo *et al.*, 2022). As a distinct subfamily, current evidence suggests that Zethinae is the sister group to eusocial subfamilies Vespinae and Polistinae (Bank *et al.*, 2017; Piekarski *et al.*, 2018).

Species of Zethinae resemble Eumeninae *s.s.* in having both recurrent veins of the forewing received in the second submarginal cell (Gusenleitner, 2013a). However, they differ from eumenines in possessing two spurs at the apex of the middle tibia.

Many Zethinae construct their nests from masticated and salivated plant material (usually leaves), which is pasted together with resinus substances (Bohart and Stange, 1965). Genra such as *Discoelius*, *Elimus*, *Ischnocoelia*, *Protodiscoelius*, and *Zethus* use vegetable matter to form cell partitions (Bohart and Stange, 1965; Hines *et al.*, 2007). Other members of Zethinae utilize old insect burrows in twigs, branches, or in the ground, a behaviour that may represent the ancestral nesting condition (Benda *et al.*, 2024). A form of social behavior has evolved in some zethine species, in which multiple females share a nest, but each female tends her own brood, yet they exhibit complex interactions such as nest usurpation, nest defense, or adoption of orphaned nests (Lopez and Noll, 2014; Kelstrup *et al.*, 2023).

In the Middle East, the subfamily Zethinae is represented by three genera: *Discoelius* Latreille, *Paramischocyttarus* Magretti, and *Zethus* Fabricius. These wasps provide important ecosystem services, including pollination, biological control, and trophic regulation (Brock *et al.*, 2021).

Genus *Discoelius* Latreille, 1809

Discoelius Latreille, 1809: 140. Type species: *Vespa zonalis* Panzer, 1801, by monotypy.

***Discoelius dufourii dufourii* Lepeletier, 1841**
Discoelius dufourii Lepeletier, 1841: 605, ♀; Saint Séver département des Landes [France].

Catalogues with Iranian records: Ebrahimi and Carpenter (2008), Rahmani *et al.* (2020).
Distribution in Iran: North of Iran (van der Vecht and Fischer, 1972).
Distribution in the Middle East: Iran (van der Vecht and Fischer, 1972; Yildirim and Kojima, 1999; Gusenleitner, 2013a; Fateryga, 2017b, 2018; Kochetkov, 2020), Turkey (Gusenleitner, 1988; Yildirim and Kojima, 1999; Fateryga, 2017b, 2018; Kochetkov, 2020).
Extralimital distribution: Central to South Europe (Gusenleitner, 2013a), China, Kazakhstan, Korea, Mongolia, Russia (Fateryga, 2017b, 2018), Italy (Selis, 2023), Slovenia (Gogala, 2022).

***Discoelius pictus* Kostylev, 1940**
Discoelius pictus Kostylev, 1940: 140, ♀, ♂; Ak-Tach, chaîne Ala-Tau de Talas, près de Tachkent [Uzbekistan].
Catalogues with Iranian records: Rahmani *et al.* (2020).
Distribution in Iran: Iran (no specific locality cited) (Kurzenko, 1978).
Distribution in the Middle East: Iran (Kurzenko, 1978).
Extralimital distribution: Afghanistan (Kurzenko, 1978), Kazakhstan, Kyrgyzstan, Russia, Tajikistan, Turkmenistan, Uzbekistan (Fateryga and Mokrousov, 2019).

***Discoelius zonalis* (Panzer, 1801)**
Vespa zonalis Panzer 1801: pl. 18, ♀; Nürnberg [Germany].
Catalogues with Iranian records: Rahmani *et al.* (2020).
Distribution in Iran: North of Iran (Kurzenko, 1977, 1978).
Distribution in the Middle East: Iran (Kurzenko, 1977, 1978; Fateryga, 2017b, 2018; Kochetkov, 2020; Rahmani *et al.*, 2020), Turkey (Giordani Soika, 1970; Yildirim and Kojima, 1999; Gusenleitner, 2013a; Fateryga, 2017b, 2018; Kochetkov, 2020).
Extralimital distribution: Caucasus, Europe, China, Japan, Korea, Mongolia, Russia (Fateryga, 2017b, 2018), Italy (Selis, 2023), Slovenia (Gogala, 2022).

Genus *Paramischocyttarus* Magretti, 1884

Paramischocyttarus Magretti, 1884: 250. Type species: *Paramischocyttarus subtilis* Magretti, 1884, by monotypy.

***Paramischocyttarus subtilis* Magretti, 1884**
Paramischocyttarus subtilis Magretti, 1884 (1883): 251, ♂; "lungo le rive del Gasch, nei dintorni di Kassala (Sudan orientale)".
Distribution in the Middle East: Israel-Palestine (Giordani Soika, 1952; van der Vecht and Fischer, 1972; Gusenleitner, 2013a; Carpenter and Gadallah, 2020), Oman (Guichard, 1986; Gusenleitner, 2012, 2013a; Madl, 2018; Carpenter and Gadallah, 2020), Saudi Arabia (Guichard, 1986; Gadallah, 2003; Gusenleitner, 2013a; Madl, 2018; Carpenter and Gadallah, 2020), United Arab Emirates (Gusenleitner, 2010b; Carpenter and Gadallah, 2020), Yemen (Gusenleitner, 2007, 2013a; Madl, 2018; Carpenter and Gadallah, 2020).
Extralimital Distribution: Djibouti, Eritrea, Ethiopia, Niger, Sudan, Tanzania (Carpenter and Gadallah, 2020).

Genus *Zethus* Fabricius, 1804

Zethus Fabricius, 1804: xii, 282. Type species: *Zethus coeruleo-pennis* Fabricius, 1804, by subsequent designation.

***Zethus (Zethus) favillaceus* Walker, 1871**
Zethus favillaceus Walker, 1871: 28, ♀, ♂; "Tajoura".
Distribution in the Middle East: Saudi Arabia (Walker, 1871; Guichard, 1986; Gusenleitner, 2013a; Madl, 2018; Carpenter and Gadallah, 2020).
Extralimital distribution: Burundi, Djibouti, Kenya, South Africa, Somalia, Sudan, Tanzania, Zimbabwe (Carpenter and Gadallah, 2020).

Species excluded from the fauna of Iran, and the Middle East

***Euodynerus aurantiacus* (Cam.)**

Records from Iran: Hormozgan (Modarres Awal, 2012).

Comments: Another species was collected, from Hormozgan: Bandar-Abbas: 29.iii.1965, leg. Safavi, and determined as "*Syneuodynerus aurantiacus* (Cm.)" by Giordani Soika, but this is a mistake, as *Syneuodynerus aurantiacus* Giordani Soika, is an Australian species (Ebrahimi and Carpenter, 2008).

***Pseudepipona (Deuterepipona) tricolor* Gusenleitner, 1976**

Pseudepipona (Pseudepipona) tricolor Gusenleitner, 1976: 115, ♀, ♂; Daghestan [Russia].

Records from Iran: Iran (no specific locality cited) (Gusenleitner, 2013a; Fateryga 2017b).

Comments: *Pseudepipona tricolor* was originally described from the Caucasus region (Russia: Dagestan). This species is listed in Gusenleitner (2013a) as an element of the Iranian fauna. Since this area is now separated from Iran, the above-mentioned species is also excluded from the fauna of Iran (Rahmani *et al.*, 2020). *Pseudepipona tricolor* is distributed in Kazakhstan and Russia (Fateryga *et al.*, 2017).

***Stenancistrocerus (Paratropancistrocerus) transcaspicus* (Kostylev, 1935)**

Odynerus transcaspicus Kostylev, 1935 (1934): 134, ♀, ♂; Kazakhstan.

Records from Iran: Iran (no specific locality cited) (Gusenleitner, 2013a).

Comments: *Stenancistrocerus transcaspicus* was originally described from the Transcaspien region (Farab). This species is listed in Gusenleitner (2013a) as an element of the Iranian fauna. Since this area is now separated from Iran, the above-mentioned species is also excluded from the fauna of Iran (Rahmani *et al.*, 2020). *Stenancistrocerus transcaspicus* is distributed in China, Kazakhstan, Kyrgyzstan (Kostylev, 1935 as *Odynerus transcaspicus*)

CONCLUSION

The fauna of the Middle Eastern Eumeninae *s.l.* comprises 605 species and subspecies, representing 15.5% of the world species, across 57 genera. Among the genera recorded from the region, *Leptochilus* with 72 species is the most diverse, followed by *Euodynerus* (39 species) species), *Antepipona* (32 species), *Onychopterocheilus* (28 species), *Ancistrocerus* and *Stenodynerus* (23 species each), and *Alastor*, *Eumenes*, *Microdynerus*, and *Odynerus* (22 species each).

A total of 211 species and subspecies in 45 genera have been recorded from Iran to date, with 38 species and subspecies endemic or subendemic. Iranian Eumeninae have been documented from 29 provinces, excluding Chaharmahal & Bakhtiari, and West Azarbaijan. Among these Kerman hosts the highest number of recorded species (44), followed by Alborz (39 species), Fars, Sistan & Baluchestan (each with 37 species), Khuzestan (30 species), Tehran (25 species), East Azarbaijan (19 species), Hormozgan (18 species), Zanjan (16 species), Golestan, Markazi, and Mazandaran (each with 15 species), Razavi Khorasan and Semnan (each with 10 species), Isfahan, Lorestan, and Southern Khorasan (each with seven species), Kordestan (six species), Bushehr, Guilan (each with five species), Qazvin (four species), Hamadan, Kermanshah, and Kuhgiloyeh & Boyerahmad (each with three species), Ardabil and Northern Khorasan (each with two species), Ilam, Qom, and Yazd (each with one).

Additionally, three parasitoid species of Ichneumonoidea have been collected from the nests of *Delta dimidiatipenne*: *Cotesia vanessae* (Reinhard) (Braconidae), *Netelia fuscicornis* (Holmgren) (Ichneumonidae), and *Ophion similis* (Szépligeti) (Ichneumonidae). Future research on host-parasitoid interactions may provide deeper insight into parasitoid life histories, nutritional requirements during development, host-parasitoid network dynamics, and mechanisms used to successfully penetrate host nests (Henri and van Veen, 2011; Miller *et al.*, 2021; Johnson *et al.*, 2023).

A comparison of the Eumeninae fauna from 16 countries in the Middle East suggests that the fauna of Turkey, with 278 recorded species, has the highest diversity. This is followed by Iran (211 species), Israel-Palestine (168 species), Egypt (109 species), Saudi Arabia (84 species), Syria (80 species), Jordan (77 species), Oman and Yemen (each with 61 species), Iraq (53 species), United Arab Emirates (47 species), Lebanon (33 species), Cyprus (31 species), Qatar (four species), Kuwait (three species), and Bahrain (two species). Additionally, among the Middle Eastern and adjacent countries, Turkey shares the most number of species with Iran (113 species), followed by Russia (67 species), Kazakhstan (61 species), Israel-Palestine (59 species), Azerbaijan (56 species), Turkmenistan (53 species), Armenia (47 species), Syria (41 species), Pakistan (32 species), Iraq (31 species), Jordan (30 species), Afghanistan (26 species), Lebanon (25 species), United Arab Emirates (20 species), Saudi Arabia (19 species), Oman (18 species), Egypt (17 species), Cyprus (14 species), Yemen (eight species), Qatar (four species), Bahrain and Kuwait (two species each).

BIBLIOGRAPHY

Abbasi, R., Mashhadikhan, M., Abbasi, M. and Kiabi, B. 2008a. Biodiversity of vespid wasps in spatial and temporal dimension in northern Zanjan province of Iran. *Chinese Journal of Ecology* 27(5): 797–802.

Abbasi, R., Pashaei Rad, S. and Ebrahimi, E. 2006a. New record for the fauna of vespid wasps (Insecta: Hymenoptera) of Iran. *Proceedings of the 14th National and 2nd International Conference of Biology*, p. 208.

Abbasi, R., Pashaei Rad, S. and Ebrahimi, E. 2006b. New record for the fauna of the Vespidae wasps in Iran. *Proceedings of the 17th Iranian Plant Protection Congress*, 2–5 September 2006, Campus of Agriculture and Natural Resources, University of Tehran, Karaj, p. 62.

Abbasi, R., Pashaei Rad, S., Ebrahimi, E. and Sheidaei, M. 2008b. Faunistic study of vespid wasps in Zanjan province (Northwest of Iran) with some ecological measures. *Environmental Sciences* 6(1): 65–74.

Abdul-Rassoul, M.S. 1976. Checklist of Iraq Natural History Museum insect collections. *Iraq Natural History Research Center* 30: 1–41.

Abo Kaf, N.H. and Allewa, A.N. 2016. Identification of some species from wasps family (Vespidae; Hymenoptera) in Lattakia, Syria. *Tishreen University Journal for Research and Scientific Studies* 38(4): 224–235.

Al-Houty, W. 1989. *Insect fauna of Kuwait.* Fahad Al-Marzouk Printing and Publishing, University of KuwaitEstablishment, Kuwait City, Kuwait, 186 pp.

Al-Houty, W. 1997. Checklist of the insect fauna of Kuwait. *Kuwait Journal of Science and Engineering* 24(1): 145–162.

Al-Saffar, H.H., and Augul, R.S. 2021. Survey of insects in some Southern Iraqi marshes. *Bulletin of the Iraq Natural History Museum* 16(4): 571 621.

André, E. 1884. *Spécies des Hyménoptères d'Europe et Algérie.* Vol. 2. André and André, Beaune, 304 pp.

André, E. 1895. Sur quelques Vespides africains nouveaux ou peu connus. *Revue d'Entomologie* 14: 352–356.

Ashmead, W.H. 1902. Classification of the fossorial, predaceous and parasitic wasps, or the superfamily Vespoidea. Paper no. 6. Family XXVIII. Vespidae. *Canadian Entomologist* 34: 163–166.

Augul, R.S. 2017. Revision of the vespid wasps (Hymenoptera; Vespidae) in Iraq. *Journal of Biodiversity and Environmental Sciences* 10(3): 99–111.

Bank, S., Sann, M., Mayer, C., Meusemann, K., Donath, A., Podsiadlowski, L., Kozlov, A., Petersen, M., Krogmann, L., Meier, R., Rosa, P., Schmitt, T., Wurdack, M., Liu, S., Zhou, X., Misof, B., Peters, R.S. and Niehuis, O. 2017. Transcriptome and target DNA enrichment sequence data provide new insights into the phylogeny of vespid wasps (Hymenoptera: Aculeata: Vespidae). *Molecular Phylogenetics and Evolution* 116: 213–226.

Barker-Webb, P. & Berthelot, S. 1839. *Miscellanées Canariennes. Histoire naturelle des Iles Canaries*, 2 livr. 44: 89.

Benda, D., Pohl, H., Beutel, R. and Straka, J. 2024. Solitary folded-winged wasps of the genus *Zethus* Fabricius (Vespidae: Zethinae) parasitized by two new species of Stresiptera on different continents. *Journal of Hymenoptera Research* 97: 721–739.

Bequaert, J. 1918. A revision of the Vespidae of the Belgian Congo Expedition, with a list of the Ethiopian diplopterous wasps. *Bulletin of the American Museum of Natural History* 39(1): 1–384.

Bequaert, J. 1925. The genus *Ancistrocerus* (*Hymenoptera*, Vespidae) in North America, with a partial key to the species. *Transactions of the American Entomological Society* 51: 57–117.

Bequaert, J. 1926. The genus *Eumenes* Latreille in South Africa, with a revision of the Ethiopian species (Hymenoptera). *Annals of the South African Museum* 23: 483–577.

Berland, L. 1928. *Faune de France*, Vol. 19. Hyménoptères vespiformes. II (Eumenidae, Vespidae, Masaridae, Bethylidae, Dryinidae, Embolemidae). Paul Lechevalier, Paris. viii + 208 pp.

Białynicki-Birula, A. 1926. Ein Beitrag zur Kenntnis der paläarktischen *Hoplomerus*-Arten (*Hymenoptera*: Diploptera). *Bulletin de l'Académie des Sciences de l'URSS* 20(10–11): 885–906.

Bingham, C.T. 1897. *Fauna of British India, including Ceylon and Burma-Hymenoptera- Vol.I. Wasps and Bees*. London: Taylor and Francis, 579 pp.

Bingham, C.T. 1898. The aculeate Hymenoptera procured at Aden by Col. Yerbury, R.A. and Capt. Nurse, I.S.C. *Journal of the Bombay Natural History Society* 12: 101–114.

Blanchard, E. 1840. *Histoire naturelle des insectes. Animaux Articules par M. Brullé*. La cour et Cie. Paris, 672 pp.

Blüthgen, P. 1938 [1937]. Systematisches Verzeichnis der Faltenwespen Mitteleuropas, *Skandanaviens und Englands. Konowia* 16: 270–295.

Blüthgen, P. 1939. Beiträge zur Kenntnis der paläarktischen Eumeniden. *Deutsche Entomologische Zeitschrift* 1938: 43–96.

Blüthgen, P. 1940a. Neue oder bemerkenswerte paläarktische Faltenwespen aus der Zoologischen Staatssammlung in München. *Mitteilungen aus den Königl. Naturwissenschaftlichen Instituten in Sofia* 13: 166–210.

Blüthgen, P. 1940b. Über *Odynerus meyeri* CAM., *Od. meyeri* CAM. var. *albolimbatus* SCHULTHESS und *Od. pseudolateralis* M.W. (*Hym. Vespidae*). *Entomologisk Tidskrift* 61(1–2): 41–45.

Blüthgen, P. 1941. Die Untergattungen Hoplomerus und Monoplomerus der Gattung Hoplomerus (Hym. Vespidae Eumenidinae). *Archiv für Naturgeschichte (N.F.)* 10(3): 305–344.

Blüthgen, P. 1942. Taxonomische und biologische Notizen über paläarktische Faltenwespen. *Stettiner Entomologische Zeitung* 104: 149–158.

Blüthgen, P. 1943. Neue paläarktische *Eumenes*-Arten oder -Rassen (*Hym.*, Vespidae, Eumenidinae). *Mitteilungen der Münchener Entomologischen Gesellschaft* 38: 297–306.

Blüthgen, P. 1944. Neue oder für Deutschland neue Bienen und Wespen und neue deutsche Fundorte einiger Arten (*Hym.*, Apid., Sphecid., Vespid.). *Mitteilungen der Deutschen Entomologischen Gesellschaft* 12: 24–31.

Blüthgen, P. 1951a. Neues oder Wissenswersstes über mitteleuropäische Aculeaten und Goldwespen II (*Hym.*). *Bonner Zoologische Beiträge* 2: 229–234.

Blüthgen, P. 1951b. Weitere Beiträge zur Kenntnis der paläarktischen Faltenwespen (Hym. Vespidae). *Mitteilungen der Münchner Entomologischen Gesellschaft*, 41: 158–165.

Blüthgen, P. 1952. Bemarkenswerte Aculeaten funde aus Schwaben, insbesondere aus dem Allgäu. *Bereicht der Naturforschenden Gesellschaft zur Augsburg* 5: 125–130.

Blüthgen, P. 1953. Portuguese and Spanish wasps (*Hymenoptera*, Vespoidea). *Memórias e Estudos do Museu Zoológico da Universidade de Coimbra*, 218: 1–23.

Blüthgen, P. 1954. Untersuchungen über paläarktische Eumenidae (Hymenoptera: Diploptera) *Deutsche Entomologische Zeitschrift*. III/IV/V: 224–268.

Blüthgen, P. 1955a. New Diploptera from Israel and the Near East. *Bulletin of the Research Council of Israel* 58: 24–31.

Blüthgen, P. 1955b. Untersuchung über paläarktische Eumenidae. III Ztr Kenntnis der paläarktischen Pterocheilus-Arten. *Mitteilungen aus dem Zoologische Museum in Berlin* 31: 59–103.

Blüthgen, P. 1956. Untersuchungen über paläarktische Eumenidae (Hym. Diploptera) II. Die Untergattung *Alastor* Lep. und *Paralastor* Blüth. Die Gattung *Alastor* Lep. *Mitteilungen aus dem Zoologische Museum zur Berlin* 32: 120–149.

Blüthgen, P. 1957. Vespidae and Eumenidae. In: Bytinski-Salz, H. Coleoptera and Hymenoptera from a journey through Asia Minor. Descriptions of new species and forms. *Review of the Faculty of Science University of Istanbul (B)* 22: 163–169.

Blüthgen, P. 1959. Wissenschaftliche Ergebnisse der zoologischen Expedition des National-Museums in Prag nach der Türkei 27. Ueber einige Faltenwespen aus Anatolien. *Acta Entomologica Musei Nationalis Pragae* 33: 269–272.

Blüthgen, P. 1961. Faltenwespen Mitteleuropas. *Abhandlungen der Deutschen Akademie der Wesenschaften zu Berlin, Klasse für Chemie, Geologie und Biologie* 2: 1–252.

Blüthgen, P. 1963. 8. Beitrag zur Synonymie der paläarktischen Faltenwespen (*Hym. Diploptera*). *Deutsche Entomologische Zeitschrift, NF* 10: 201–205.

Blüthgen, P. and Gusenleitner, J. 1970. Über Faltenwespen aus dem Iran (Hym., Diploptera), *Stuttgarter Beiträge zur Naturkunde* 223: 1–13.
Bohart, R.M. 1939. Notes on *Odynerus* with a key to the North American subgenera and description of a new subgenus. *The Pan-Pacific Entomologist* 15(3): 97–104.
Bohart, R.M. and Stange, L.A. 1965. A revision of the genus *Zethus* in the western hemisphere (Hymenoptera: Eumenidae). *University of California Press*, 40: 1–208.
Borsato, W. and Ratti, E. 1999. Antonio Giordani Soika (1913–1997): l'opera scientifica. *Memorie della Societá Entomologica Italiana* 77: 43–103.
Bosly, H.A. 2021. A preliminary faunistic survey of hymenopteran insects at Jazan Lake Dam region, Southewest of Saudi Arabia. *Saudi Journal of Biological Sciences* 28: 2342–2351.
Boyeri Shemi, F. and Pashaei Rad, S. 2017. Faunistic study of Vespidae (Hymenoptera) in Izeh, Khuzestan province. *Proceedings of the 3rd National Congress on Development and Promotion of Agricultural Engineering and Soil Sciences of Iran*, 3 September 2017, Tehran, 11 pp. [in Persian, English summary]
Brock, R.E., Cini, A. and Sumner, S. 2021. Ecosystem services provided by aculeate wasps. *Biological Review* 96: 1645–1675.
Brullé, M.A. 1832. *Expédition scientifique de Morée*. Section des Sciences physiques, Tom III. 1re partie. Zoologie. F.G. Levrault, Paris, 400 pp.
Bytinski-Salz, H. and Gusenleitner, J. 1971. The Vespoidea of Israel. *Israel Journal of Entomology* 6: 239–250.
Cameron, P. 1905. Hymenoptera and Vespidae of the Museum of the Royal Zoological Society "Natura Artis Magistra" at Amsterdam. *Tijdschrift voor entomologie* 48: 48–208.
Cameron, P. 1907. Description of a new genus and some new species of Hymenoptera captured by Lieut. Col. C.G. Nurse at Deesa, Matheran and Ferozepore. *Journal of the Bombay Natural History Society* 17: 1001–1012.
Carpenter, J.M. 1982. The phylogenetic relationships and natural classification of the Vespoidea (Hymenoptera). *Systematic Entomology* 7: 11–38.
Carpenter, J.M. 2023. A catalogue of the pacific Eumeninae (Hymenoptera: Vespidae). *Zootaxa* 5315(5): 401–455.
Carpenter, J.M. and Brown, G.R. 2021. A key to the Australian genera of Eumeninae (Hymenoptera: Vespidae). *Records of the Australian Museum* 73(3): 87–101.
Carpenter, J.M. and Cumming, J.M. 1985. A character analysis of the North American potter wasps (Hymenoptera: Vespidae: Eumeninae). *Journal of Natural History* 19(5): 877–916.
Carpenter, J.M. and Gadallah, N.S. 2020. Biodiversity of the aculeate wasps (Hymenoptera: Aculeata) of the Arabian Peninsula: Vespoidea, *Vespidae. Zootaxa* 4754(1): 191–216.
Carpenter, J.M. and Garcete-Barrett, B.R. 2002. A key to the Neotropical genera of Eumeninae (Hymenoptera: Vespidae). *Boletin Del Museo Nacional de Historia Natural del Paraguay* 14(1–2): 52–73.
Carpenter, J.M., Gusenleitner, J. and Madl, M. 2009. A catalogue of the Eumeninae (Hymenoptera: Vespidae) of the Ethiopian region excluding Malagasy subregion. Part I: Introduction, key to genera, genera *Aethiopicodynerus* Gusenleitner 1997 to *Cyrtolabulus* van der Vecht 1969. *Linzer Biologische Beiträge* 41(1): 513–638.
Carpenter, J.M., Gusenleitner, J. and Madl, M. 2010. A catalogue of the Eumeninae (Hymenoptera: Vespidae) of the Ethiopian region excluding Malagasy subregion. Part II: genera *Delta* de Saussure 1885 to *Zethus* Fabricius 1804 and species incertae sedis. *Linzer Biologische Beiträge* 42(1): 95–315.
Cassar, T., Mifsud, D. and Selis, M. 2022. The potter wasps of the Maltese Islands (Hymenoptera: Vespidae: Eumeninae). *Journal of Hymenoptera Research* 90: 201–212.
Castro, L. 1986. Contribución al conocimiento de los Eumenidae del sur de la Península Ibérica. *Boletin de la Asociación Española de Entomología* 10: 293–301.
Castro, L. 1992. Sobre los euménidos (*Hym.*, Vespoidea) del valle medio del Ebro. *Zapateri, Revista Aragonesa de Entomología* 2: 21–34.
Castro, L. 1997. Familia Vespidae: Subfamilia Eumeninae. *Catalogus de la Entomofauna Aragonesa* 16: 3–8.
Castro, L. and Dvořák, L. 2009. New and noteworthy records of vespid wasps (Hymenoptera: Vespidae) from the Palaearctic region (II). *Boletín de la Sociedad Entomológica Aragonesa* 44: 295–304.
Castro, L. and Dvořák, L. 2010. New and noteworthy records of vespid wasps (Hymenoptera: Vespidae) from the Palaearctic region (III). *Acta Musei Moraviae, Scientiae Biologicae (Brno)* 95(2): 37–53.
Christ, J.L. 1791. *Naturgeschichte, Classification und Nomenclatur der Insekten vom Bienen, Wespen, und Ameisengeschlecht*, etc., Frankfurt-am-Main, 535 pp., 60 pis.
Costa, A. 1888. Miscellanea entomologica. Memoria prima. *Rendiconto dell'Accademia della Scienze Fisiche e Matematiche, Napoli*, (2)I: 242–244.

Cowan, D.P. 1991. The solitary and presocial Vespidae, pp. 33–73. In: Ross, K.G. and Matthews, R.W. (eds.), *The social biology of wasps*. Comstock Publishing Associates, Cornell University Press, Ithaca, 696 pp.

Cumming, J.M. 1989. Classification and evolution of the eumenine wasp genus *Symmorphus* Wesmael (Hymenoptera: Vespidae). *The Memoirs of the Entomological Society of Canada* 121: 5–168.

Curtis, J. 1826. *British Entomology; being illustrations and descriptions of the genera of insects in Greta Britain and Ireland* 3: 120–141.

Dai, M., He, S.-L., Chen, B. and Li, T.-J. 2023. Phylogeny of *Rhynchium* and its related genera (Hymenoptera: Eumeninae) based on universal single-copy orthologs and ultraconserved elements. *Insects*, 14(9): 775.

de Saussure, H. 1852–1853. *Études sur la famille des Vespides I: Monographie des Guêpes solitaires, ou de la tribu des Eumèniens.* Genève: Cherbuliez & Paris, Masson, 268 pp.

de Saussure, H. 1852–1858. *Études sur la Famille des Vespides*, 1, 2 and 3. V. Masson, Paris, J. Kessmann, Genève, and J. Cherbuliez, Genève, 350 + 470 + 364 pp., 22 + 39 + 16 pls. [see Griffin (1939) for publication dates of specific pages]

de Saussure, H. 1854–1856. *Études sur la famille des Vespides. 3. Monographie des fausses guêpes, ou de la tribu des Masariens. Troisième partie comprenant la monographie des Masariens et un supplément à la monographie des Eumeniens.* Masson, Kessmann & Cherbuliez, Paris and Genève, 352 pp.

de Saussure, H. 1863. Melange Hymémoptèrologiques II. *Mémoires de la Société de Physique d'Histoire Naturelle de Genève* 17: 171–244.

DeGeer, C. 1773. *Memoires pour servir a l'histoire des insectes.* Tome Troisième, Stockholm, 696 pp.

Derwesh A.I. 1965. A preliminary list of identified insects and arachnids of Iraq. *Director General Agriculture Research Projections Baghdad, Bulletin* 121: 123 pp.

Dover, C. 1925 [1924]. Further notes on the Indian Diplopterous wasps. *Journal of the Asiatic Society of Bengal (new series)* 20: 289–305.

Dover, C. 1929. Wasps and bees in the Raffles Museum, Singapore. *Bullettin of Raffles Museum* 2: 43–70.

Dover, C. and Rao, H.S. 1922. A note on the Diplopterous Wasps in the Collection of the Indian Museum. *Journal of Asiatic Society of Bengal*, 18: 235–249.

du Buysson, R. 1908. Voyage de M. René Chudeau du Sud-Algérien au Tchad (1905-1906) [Hym.]. *Bulletin de la Société entomologique de France* 1908: 131–135.

du Buysson, R. 1912. Les Vespides, Euménides et Scoliides. *Annales d'Histoire Naturelle Délégation en Perse II. Entomologie* 1: 77–83.

du Buysson, R. 1913. Les Vespides, Euménides, et Scoliides. *Annales d'Histoire Naturelle* (Délégation Perse) 2 (Entomologie), pp. 77–83.

Dufour, L. 1839. Mémoire pour server à l'histoire de l'insustrie et des metamorphoses des Odynères, et description de quelques Nouvelles espèces de ce genre d'insectes. *Annales des Sciences Naturelles, série* 2, 11, 85–103.

Dusmet y Alonso, J.M. 1903. Vespidos, Eumenidos y Masäridos de España. *Memorias de la Real Sociedad Española de Historia Natural* 2: 119–225.

Dusmet y Alonso, J.M. 1917. Véspidos, Euménoidos y Masáridos de Marreucos. *Memorias de la Real Sociedad Española de Historia Naturale* 8: 363.

Dusmet y Alonso, J.M. 1928. Algunos Euménodos y Maśaridos del Notre de Africa (Hym. Vesp.). *Eos* 4(1): 97–112.

Dusmet, J.M. 1917. Véspidos, Euménidos y Masáridos de Marruecos. *Memorias de la Real Sociedad Española de Historia Natural* 8: 343–383.

Dvořák, L. & Carpenter, J.M. 2010. New records of vespid wasps from Yemen with synonymy in *Belonogaster* (*Hymenoptera*: Vespidae: Polistinae and Eumeninae). *Linzer Biologische Beiträge* 42(1): 561–563.

Ebadi, R. 1995. Collection, identification and preliminary study of pollinator insects fauna in Isfahan province. *Proceedings of the 12th Iranian Plant Protection Congress*, 2–7 September 1995, Karaj Junior College of Agriculture, Karaj, p. 319.

Ebrahimi, E. 2014. Aculeate wasps and their role in natural equilibrium and pest control (with a review of the Iranian species). *Biocontrol in Plant Protection* 2(2): 59–68.

Ebrahimi, E. and Carpenter, J.M. 2008. Catalog of the vespid wasps of Iran (Hymenoptera, Vespidae). *Zootaxa* 1785: 1–42.

Eck, R. 1971. Untersuchung des Art status von *Oplemerus melanocephalus* (Gmelin) und *O. armeniacus* (Morawitz) mit Besonderer Berücksichtigung der morphologisch-geographischen Feingliederung im europäischen Raume (Hymenoptera, Vespidae). *Entomologische Abhandlungen des Stäalichen Museums für Tierkunde in Dresden* 37(5): 363–391.

El-Hawagry, M.S., Khalil, M.W., Sharaf, M.R., Fadl, H.H. and AlDawood, A.S. 2013. A preliminary study on the insect fauna of Al-Baha Province, Saudi Arabia, with descriptions of two new species. *ZooKeys* 274: 1–88.

Erlandsson, S. 1972. Hymenoptera Aculeata from the European parts of the Mediterranean countries. *Eos Madrid* 48: 11–93.

Esmaili, M. and Rastegar, R. 1974. Identified species of aculeate Hymenoptera of Iran. *Journal of the Entomological Society of Iran* 2(2): 41–52.

Eversmann, E. 1854. Beiträge zur Lepidopterologie Russlands, und Beshreibund einiger anderer Insecten aus den südlichen Kirgisensteppen, der nördlichen ufern des Aralsees und des Sir-Darja's. *Bulletin de la Société impériale des naturalists de Moscou* 27(3–4): 174–200.

Fabricius, J.C. 1775. Syswtema Entomologiae, sistens Insectorum classes, ordines, genera, species. *Flensburgi et Lipsae*, 832 pp.

Fabricius, J.C. 1781. *Species insectorum*, Tom I. Hamburgii et Kilonii, 552 pp.

Fabricius, J.C. 1787. *Mantissa insectorum Tom.* I. Hafniae, 348 pp.

Fabricius, J.C. 1793. *Entomologia systematica emendate et aucta*, Tom II. *Hafniae* 519 pp.

Fabricius, J.C. 1798. *Supplementum entomologiae systematicae*. Hafniae, 572 pp.

Fabricius, J.C. 1804. *Systema Piezatorum. Secundum ordines, genera, species, adjectis synonymis, locis, observationibus, descriptionibus*. Carolum Reichard, Brunsvigae, 439 + 32 pp.

Fateryga, A.V. & Popov, I.B. 2017. New records of *Vespidae* and *Megachilidae* (*Hymenoptera*) in Russia. *Ekosistemy* 9: 86–89.

Fateryga, A.V. 2010. Landscape distribution of potter wasps of the subfamily Eumeninae (*Hymenoptera*: Vespidae) of Crimea. *Proceedings of the Russian Entomological Society* 81(2): 74–82.

Fateryga, A.V. 2017a. New records of solitary vespid wasps (Hymenoptera: Vespidae: Eumeninae, Masarinae) from Russia and adjacent countries. *Far Eastern Entomologist* 334: 1–16.

Fateryga, A.V. 2017b. Subfamily Eumeninae – Potter wasps, pp. 178–192. In: Belokobylskij, S.A. and Lelej, A.S. (eds), Annotated catalogue of the Hymenoptera of Russia. Volume I. Symphyta and Apocrita: Aculeata. *Proceedings of the Zoological Institute of the Russian Academy of Sciences*, Supplement No. 6, 475 pp.

Fateryga, A.V. 2018. Wasps of the family Vespidae (Hymenoptera) of the Crimean Peninsula. *Entomofauna* 39 (1): 193–233.

Fateryga, A.V. 2022. Revision of the *Pseudepipona herrichii*-group of the eumenine wasps (Hymenoptera: Vespidae: Eumeninae) with the description of two new species from China and Russia. *Zootaxa* 5154(2): 101–126.

Fateryga, A.V. and Mokrousov, M.V. 2019. New records of eumenine wasps (Hymenoptera: Vespidae: Eumeninae) from Russia with description of a new species of *Leptochilus* de Saussure, 1853. *Zootaxa* 4612(3): 412–422.

Fateryga, A.V. and Proshchalykin, M.Y. 2024. New records of the solitary Vespidae (Hymenoptera: Vespidae: Masarinae and Eumeninae *s.l.*) from Central Asia, with new synonymys and description of a new species of *Eumenes* Latreille. *Zootaxa* 5529(5): 551–569.

Fateryga, A.V., Mokrousov, M.V. and Danilov, Y.N. 2017. New and noteworthy records of solitary vespid wasps (Hymenoptera: Vespidae: Masarinae, Eumeninae) in Russia. *Proceedings of the Russian Entomological Society* 88(2): 31–37.

Fateryga, A.V., Proshchalykin, M.Y. and Maharramov, M.M. 2025. New records of the eumenine wasps (Hymenoptera: Vespidae: Eumeninae) from Azerbaijan, with description of hitherto unknown male of *Ancistrocerus terekensis* Kostylev, 1940 and female of *Eustenancistrocerus iranicus* Gusenleitner, 2013. *Zootaxa* 5633(1): 110–122.

Fateryga, A.V., Proshchalykin, M.Y., Aliyev, K.A. and Maharramov, M.M. 2019. To the knowledge of eumenine wasps (Hymenoptera: Vespidae: Eumeninae) of the Nakhchivan Autonomous Republic of Azerbaijan. *Far Eastern Entomologist* 379: 25–32.

Fateryga, A.V., Proshchalykin, M.Y., Kochetkov, D.N. and Buyanjargal, B. 2020. New records of eumenine wasps (Hymenoptera, Vespidae, Eumeninae) from Russia, with description of a new species of Stenodynerus de Saussure, 1863. *Journal of Hymenoptera Research* 79: 89–109.

Fateryga, A.V., Proshchalykin, M.Y., Maharramov, M.M. and Astafurova, Y.V. 2021. New records of solitary vespid wasps (Hymenoptera: Vespidae: Masarinae and Eumeninae *s.l.*) from the Nakhchivan Autonomus Republic of Azerbaijan. *Zootaxa* 5027(1): 36–60.

Fateryga, A.V., Proshchalykin, M.Y., Mokrousov, M.V. and Akhmedov, A.G. 2023. To the knowledge of the solitary vespid wasps (Hymenoptera: Vespidae: Masarinae and Eumeninae *s.l.*) of Uzbekistan. *Zootaxa* 5278(3): 439–460.

Ferton, C. 1909 [1908]. Notes détachées sur l'instinct des Hyménoptères mellitères et ravisseurs. *Annales de la Société entomologique de France* 77: 535–586.

Friese, H. & Schulthess, A.V. 1924 ("1923"). Neue Hymenopteren aus paläarktisch Afrika und Asien. *Konowia: Zeitschrift für systematische Insektenkunde* 2: 279–294.

Friese, H. and von Schulthess, A. 1923. Neue Hymenopteren aus paläarktische Afrika und Asien. *Konowia (Vienna)* 2: 279–294.

Gadallah, N.S. 2003. Eumenidae from the Jeddah region of Saudi Arabia (Hymenoptera: Aculeata). *Zoology in the Middle East* 30: 77–90.

Germar, E.F. 1817. *Reise nach Dalmatien und in das Gebiet von Ragusa*. F.A. Brockhaus, Leipzig und Altenburg. xii + 323 pp.

Gess, S.K. and Roosenschoon, P.A. 2016. A preliminary survey of flower visiting by aculeate wasps and bees in the Dubai Desert Conservation Reserve, United Arab Emirates. *Journal of Hymenoptera Research* 52: 81–141.

Ghahari, H., Lehr, P.A., Lavigne, R.J., Hayat, R. and Ostovan, H. 2007. New records of robber flies (Diptera, Asilidae) for the Iranian fauna with their prey records. *Far Eastern Entomologists* 179: 1–9.

Gillett, M. and Gillett, C. 2005. Insects and other arthropods, pp. 168–195 + 352–356 + 377–379. In: Hellyer, P. and Aspinall, S. (eds.), *The Emirates: a Natural History*. Trident Press Ltd., London, 428 pp.

Gillett, M.P.T. and Howarth, B. 2004. The Insects of Jebel Hafit (including a short note on representatives of other arthropod groups), pp. 94–143. In: Aspinall, S. and Hellyer, P. (eds.), *Jebel Hafit, a Natural History, Emirates Natural History Group*. Abu Dhabi, 220 pp.

Giordani Soika, A. 1934. Labus ed *Eumenes* nuovi o poco noti. *Memorie della Società Entomologica Italiana* 12: 215–228.

Giordani Soika, A. 1935. Su alcuni Eumenini del Deutsches entomologishes Institut di Berlin-Dahlem. *Arbeiten über Morphologische und Taxonomische Entomologie, Berlin-Dahlem* 2(4): 242–252.

Giordani Soika, A. 1936. Caratteri del genere Nortonia Sauss and descrizione di due nuove especie (Hymen. Vepidae). *Annali del Museo Civico di Storia Naturale di Genova* 60(15): 267–271.

Giordani Soika, A. 1938. Le specie del sottogenere Nortonia Sauss. (Hym. Vespidae). *Annali del Museo di storia naturale di Genova* 60(17): 111–116.

Giordani Soika, A. 1939 (1938). Sulgenere Pachymenese descrizione di un nuovogenere e di nouve specie. *Memorie della Societa Entomologica Italiana* 17: 85–96.

Giordani Soika, A. 1940. Monografia degli *Odynerus* etiopici. Supplemento primo (*Hymen. Vespidae*). *Annali del Museo Civico di Storia Naturale di Genova* 60: 469–484.

Giordani Soika, A. 1941. Studi sui Vespidi solitari. *Bollettino della Societá Veneziana di Scienze Naturali e del Museo Civico di Storia Naturale* 2(3): 130–279.

Giordani Soika, A. 1942. Monografia degli Alastor etiopici (Hym. Vesp.). *Mitteilungen der Münchner Entomologischen Gesellschaft* 32: 399–445.

Giordani Soika, A. 1943. Eumenini paleartici nuovi o poco noti (Hym.). *Bollettino del Museo Civico di Storia Naturale di Venezia* 3(1): 1–14.

Giordani Soika, A. 1944. Risultati di raccolte imenotterologiche in Sicilia. *Atti della Società Italiana di Scienze naturali* 83: 5–21.

Giordani Soika, A. 1947. Descrizione di tre nuovi *Leptochilus* (*Hymenoptera* - Vespidae) della fauna italiana. *Bollettino Istituto Entomologico Università di Bologna* 16: 129–133.

Giordani Soika, A. 1949. Revisione degli *Eumenes* appartenenti al sottogenere *Katamenes* (Hymenoptera: Vespidae). *Bollettino della Società Entomologica Italiana* 79 (3–4): 40–47.

Giordani Soika, A. 1952. Sulle caratteristiche biogeografiche della Palestina, Arabia ed Egitto, con un contributo alla conoscenza degli Zethini ed Eumenini della Palestina. *Bollettino del Museo Civico di Storia Naturale di Venezia* 6(1): 5–62.

Giordani Soika, A. 1953 (1952). Hyménoptères récoltés par une mission Suisse au Maroc (1947). Eumeninae. *Bulletin de la Société Scientifique Naturelle de Maroc* 32: 235–267.

Giordani Soika, A. 1954. Contributo allo studio del popolamento del Sahara Centrale: Vespidi dell'Hoggar, Tassili n'Agger e Tibesti. *Bollettino del Museo Civico di Storia Naturale di Venezia* 7: 7–37.

Giordani Soika, A. 1957. Expedition to South-West Arabia 1937–8. 31- Hymenoptera: Vespidae. *British Museum (Natural History) Expedition to South-West Arabia* 1(31): 471–484.

Giordani Soika, A. 1958 [1957]. Biogeografia, Evoluzione e Sistematica dei Vespidi Solitari della Polinesia Meridionale. *Bollettino del Museo Civico di Storia Naturale di Venezia* 10: 183–221.

Giordani Soika, A. 1960 ["1958"]. Notulae vespidologicae, V, Biogeografia e sistematica del sottogenere Katamenes (M. W.). *Bollettino del Museo Civico di Storia Naturale di Venezia* 11: 57–68.

Giordani Soika, A. 1961. Notulae Vespidologicae XVII-XVIII. *Atti della Societa Italiana di Scienze Naturali e del Museo Civico di Storia Naturale in Milano* 100: 373–388.

Giordani Soika, A. 1963. Notulae vespidologicae XIX, *Symmorphus* palearctici nuovi o poco noti. *Bollettino della Società Entomologica Italiana* 93(7–8): 122–126.

Giordani Soika, A. 1966. Notulae vespidologicae XXIII. Descrizione dei tre nuovi Eumenidi dell'Africa occidentale. *Bollettino della Società Entomologica Italiana* 95(3–4): 46–51.

Giordani Soika, A. 1968. Le Specie Paleartiche del Genere *Cyrtolabus* V.d.Vecht (Hym. Eumenidae), *Bollettino del Museo Civico di Storia Naturale di Venezia* 18: 109–124.

Giordani Soika, A. 1970. Contributo alla conoscenza degli Eumenidi del Medio Oriente. *Bollettino del Museo Civico di Storia Naturale di Venezia* 41: 27–183.

Giordani Soika, A. 1971. Notulae vespidologicae – XXXI: Nuovi *Leptochilus* e *Microdynerus* della Spagna. *Bollettino della Società Entomologica Italiana* 103(5–6): 112–115.

Giordani Soika, A. 1973. Descrizione di nuovi Eumenidi. *Bollettino del Museo Civico di Storia Naturale di Venezia* 24: 97–131.

Giordani Soika, A. 1974 [1972]. Revisione della Sottofamiglia Raphiglossinae (Hym. Vesp.). *Bollettino del Museo Civico di Storia Naturale di Venezia* 25: 107–146.

Giordani Soika, A. 1979. Notulae vespidologicae, XL: Eumenidi paleartici nuovi o poco noti. *Bollettino del Museo Civico di Storia Naturale di Venezia* 30: 247–263.

Giordani Soika, A. 1981 [1980]. Eumenidi raccolti nell'Arabia meridionale da K. Guichard (Hym.). Nota II. *Bollettino del Museo Civico di Storia Naturale di Venezia* 31: 111–116.

Giordani Soika, A. 1982a ["1981"]. Contributo alla conoscenza dei Delta afrotropicali (*Hymenoptera*: Eumenidae). *Memorie della Società Entomologica Italiana* 60(2): 209–213.

Giordani Soika, A. 1982b. Revisione del genere *Microdynerus* Ashm. (*Hym. Vesp.*). *Bollettino Museo Civico di Storia Venezia* 33: 153–175.

Giordani Soika, A. 1986 (1984). Eumenidi paleartici nuovi o poco noti. *Bollettino del Museo Civico di Storia Naturale di Venezia* 35: 91–162.

Giordani Soika, A. 1987. Eumenidi raccolti in Gambia e Kenya dal dr. L.A. Janzon (*Hymenoptera*, Eumenidae). *Bollettino Museo Civico di Storia Venezia* 37: 151–156.

Giordani Soika, A. 1989. Terzo contributo alla conoscenza degli Eumenidi afrotropicali (Hymenoptera). *Lavori Società Veneziana di Scienze Naturali* 14(1): 19–68.

Giordani Soika, A. 1996 (1994). Notulae vespidologicae XLVIII. Descrizione di due nuovi Eumenidi afrotropicali. *Bollettino del Museo Civico di Storia Naturale di Venezia* 45: 33–34.

Giordani Soika, A. 1977. Nuovi Eumenidi paleartici. *Notulae vespidologicae-XXXIX. Bollettino del Museo di Storia Naturale di Venezia* 28(1976): 153–178.

Girard, M. 1879. Traité élémentaire d'Entomologie. Hyménoptères porte-aiguillon. *Paris, Baillière* 2: 577–1028.

Giraud, J. 1863. Hyménoptères recueillis aux environs de Suse, en Piémont, et dans le département des Hautes-Alpes, en France. *Verhandlungen der Zoologisch-Botanischen Gesellschaft in Wien* 13: 11–46.

Giraud, J. 1866. Mémoire sur les insectes qui habitent les tiges sèches de la Ronce. *Annales de la Société Entomologique de France* 4(6): 443–500.

Gmelin, J.F. 1790. Caroli a Linné Systema Naturae (Ed. XIII) Tom I. G.E. Beer. *Lipsiae*, 2225–3020.

Gogala, A. 2022. Vespid wasps of Slovenia (Hymenoptera: Vespoidae: Vespidae). *Scopolia* 102: 1–79.

Goulet, H. and Huber, J.T. 1993. *Hymenoptera of the world: An identification guide to families*. Agriculture Canada Publication, Ottawa, 668 pp.

Gribodo, G. 1884. Diagnosi di nuovo specie di Imenotteri scavatori ed osservazione sopra alcune specie poco conosciute. *Bullettino della Società Entomologica Italiana* 16: 275–284.

Gribodo, G. 1886. Hymenopterorum novorum diagnoses proecursoriae [sic]. III. Miscellanea. *Nuntius Entomologicus Internationalis* 4: 12–14.

Gribodo, G. 1919. Gli "Imenotteri" delle oasi xerotropiche di Val di Susa. *Atti della Reale Accademia delle Scienze di Torino* 54: 846–863.

Guérin-Méneville, F.E. 1844. *Iconographie du règne animal de G. Cuvier, ou representation d'aprés nature de l'une des espèces les plus remarquable et souvent encore figure, de chaque genre d'animaux Insectes.* Paris, J.B. Baillière, 576 pp.

Guichard, K.M. 1986 [1985]. Wasps of the family Eumenidae (Hymenoptera: Vespoidea) of the Arabian Peninsula. *Fauna of Saudi Arabia* 7: 202–229.

Guiglia, D. 1929. Risultati zoologici della missione inviata dalla R. Società Geografica Italiana per l'esplorazione dell'oasi de Giarabub (1926–1927), Hymenoptera Aculeata. *Annali del Museo Civico di Storia Naturale di Genova* 53: 393–421.

Guiglia, D. 1933. Su tre specie d'Imenotteri della Cyrenaica e su due casi di omonimia. *Annali del Museo Civico di Storia Naturale di Genova* 56: 358–366.

Gusenleitner, F. and Ockermüller, E. 2024. Dr. Josef Gusenleitner, ein bedeutender oberösterreichischer Entomologe, hat uns verlassen (*17.9.1929 †6.7.2023). *Linzer biologische Beiträge* 56(1): 379–482.

Gusenleitner, J. 1966. Vespidae, Eumenidae und Masaridae aus der Türkie. Teil I. *Polskie Pismo Entomologiczne* 36: 343–363.

Gusenleitner, J. 1967. Vespidae, Eumenidae und Masaridae aus der Türkie. Teil I. *Polskie Pismo Entomology* 37: 663–676.

Gusenleitner, J. 1969. Eine neue Eumenidengattung. *Bollettino del Museo Civico Stororia Naturale di Venezia* 19: 13–19.

Gusenleitner, J. 1970. Bemerkenswertes über Faltenwespen II (Diploptera, Hymenopt.). *Nachrichtenblatt der Bayerischen Entomologen* 18(4–6): 112–116.

Gusenleitner, J. 1972. Übersicht über die derzeit bekannten westpaläarktischen Arten der Gattung *Eumenes* Latreille 1802 (Hym. Vespoidea). *Bolletino del Museo Civico di Storia Naturale di Venezia* 22 (23): 67–117.

Gusenleitner, J. 1973. Bemerkenswertes über Faltenwespen. V. (Diploptera, Hymenoptera). *Nachrichtenblatt der Bayerischen Entomologen* 22: 118–120.

Gusenleitner, J. 1976. Bemerkenswertes uber Faltenwespen. VI. (Diploptera, Hymenoptera). *Nachrichtenblatt der Bayerischen Entomologen* 25(6): 112–119.

Gusenleitner, J. 1977. Neue *Leptochilus*-Arten aus Anatolien und Nordafrika. *Linzer biologische Beiträge* 9(1): 163–178.

Gusenleitner, J. 1979. Die *Microdynerus*- und *Pseudomicrodynerus*-Arten des Balkans und der Türkei. *Linzer biologische Beiträge* 11(1): 75–94.

Gusenleitner, J. 1981. Revision der paläarktischen *Stenodynerus*-Arten (*Hymenoptera*: Eumenidae). *Polskie Pismo Entomologiczne* 51(2): 209–305.

Gusenleitner, J. 1984. Zwei Neue Subspezies der Art *Euodynerus* (*Pareuodynerus*) *quadrifasciatus* (Fabricius, 1793) (Hymenoptera, Eumenidae). *Entomofauna* 5(14): 165–169.

Gusenleitner, J. 1985a. Bemerkenswertes über Faltenwespen VIII (Hymenoptera, Vespoidea). *Nachrichtenblatt der Bayerischen Entomologen* 34: 105–110.

Gusenleitner, J. 1985b. Neue *Leptochilus*-Arten aus dem Mittelmeergebiet und dem Iran (Hymenoptera, Eumenidae). *Entomofauna* 6(9): 81–101.

Gusenleitner, J. 1986a. Beitrag zur Kenntnis der paläarktischen *Antepipona*-Arten (Hym., Eumenidae). *Annalen des Naturhistorischen Museums in Wien Serie B*, 88/89: 565–577.

Gusenleitner, J. 1986b. Drei neue *Alastor*-Arten aus dem Iran (Hymenoptera, Eumenidae). *Entomofauna* 7(2): 29–39.

Gusenleitner, J. 1986c. Zwei Neue *Antepipona*-Arten aus der Turkei (Hym.,Eumenidae). *Linzer biologische Beiträge* 18(2): 361–371.

Gusenleitner, J. 1988. Neue und bemerkenswerte Faltenwespen aus der Türkei (Hymenoptera, Vespoidea). *Linzer biologische Beiträge* 20(2): 713–737.

Gusenleitner, J. 1990. Neue Eumeniden aus Nordafrika und dem Nahen Osten (Hymenoptera, Vespoidea, Eumenidae). *Linzer biologische Beiträge* 22(1): 193–211.

Gusenleitner, J. 1991. Über Vespoidea (Hymenoptera) aus der Mongolei und der Sowjetunion. *Linzer biologische Beiträge* 23(2): 631–641.

Gusenleitner, J. 1992. Funddaten von Eumeniden und Masariden aus Elfenbeinküste, Ghana, Mali, Senegal und Togo (*Hymenoptera*, Vespoidea, Eumenidae und Masaridae). *Linzer Biologische Beiträge* 24(2): 671–682.

Gusenleitner, J. 1993. Drei neue paläarktische Eumeniden-Arten (Hymenoptera: Vespoidea). *Linzer biologische Beiträge* 25(1): 43–55.

Gusenleitner, J. 1994. Über Eumeniden aus dem Nahen Osten und Arabien sowie Beschreibung einer neuen Subspecies aus Zanzibar. *Linzer biologische Beiträge* 26(1): 307–324.

Gusenleitner, J. 1995a. Bemerkungen über die Gattung *Leptochilus* Saussure 1852 und Beschreibung neuer Arten dieser Gattung (Hymenoptera, Vespoidea, Eumenidae). *Linzer biologische Beiträge* 27(1): 169–181.

Gusenleitner, J. 1995b. Bestimmungstabellen mittel- und südeuropäischer Eumeniden (Vespoidea, Hymenoptera). Teil 4: Die Gattung *Ancistrocerus* Wesmael 1836 mit einem Nachtrag zum Teil 1: Die Gattung *Leptochilus* Saussure. *Linzer biologische Beiträge* 27(2): 753–775.

Gusenleitner, J. 1997a. Bestimmungstabellen mittel- und südeuropäischer Eumeniden (Vespoidea, Hymenoptera) Teil 7. Die Gattungen *Microdynerus* Thomson 1874 und *Eumicrodynerus* Gusenleitner, 1972. *Linzer biologische Beiträge* 29(2): 779–797.

Gusenleitner, J. 1997b. Die Gattungen *Syneuodynerus* Blüthgen 1951 und *Intereuodynerus* gen. nov. Im Nahen Osten (Hymenoptera, Eumenidae). *Linzer biologische Beiträge* 29(2): 763–769.

Gusenleitner, J. 1997c. Bestimmungstabellen mittel- und südeuropäischer Eumeniden (Vespoidea, Hymenoptera) Teil 6. Die Gattungen *Euodynerus* Dalla Torre 1904, *Syneuodynerus* Bluethgen 1951 und *Chlorodynerus* Bluethgen 1951. *Linzer Biologische Beiträge* 29(1): 117–135.

Gusenleitner, J. 1998a. Bestimmungstabellen mittel- und südeuropäischer Eumeniden (Vespoidea, Hymenoptera), Teil 8. Die Gattungen *Odynerus* Latreille 1802, *Gymnomerus* Blüthgen 1938, *Paragymnomerus* Blüthgen 1938 und *Tropidodynerus* Blüthgen 1939. *Linzer biologische Beiträge* 30(1): 163–181.

Gusenleitner, J. 1998b. Bestimmungstabellen mittel- und südeuropäischer Eumeniden (Vespoidea, Hymenoptera), Teil 9. Die Gattung *Pseudepipona* Saussure. *Linzer biologische Beiträge* 30(2): 487–495.

Gusenleitner, J. 1998c. Über Faltenwespen aus dem asiatischen Raum (Hymenoptera, Eumenidae, Masaridae). *Linzer biologische Beiträge* 30(2): 503–513.

Gusenleitner, J. 1998d. Vespoidea and Sapygidae gesammelt auf der Insel Samos (*Hymenoptera*: Eumenidae, Masaridae, Sapygidae). *Linzer Biologische Beiträge* 30(1): 155–161.

Gusenleitner, J. 1999. Bestimmungstabellen mittel- und südeuropäischer Eumeniden (Vespoidea, Hymenoptera) Teil 11. Die Gattungen *Discoelius* Latreille 1809, *Eumenes* Latreille 1802, *Katamenes* Meade-Waldo 1910, *Delta* Saussure 1855, *Ischnogasteroides* Magretti 1884 und *Pareumenes* Saussure 1855. *Linzer biologische Beiträge* 31(2): 561–584.

Gusenleitner, J. 2000a. Bestimmungstabellen mittel- und südeuropäischer Eumeniden (Vespoidea, Hymenoptera), Teil 13. Die Gattung *Stenodynerus* Saussure 1863. *Linzer biologische Beiträge* 32(1): 29–41.

Gusenleitner, J. 2000b. Bestimmungstabellen mittel- und südeuropäischer Eumeniden (Vespoidea, Hymenoptera) Teil 14. Der Gattungsschlüssel und die bisher in dieser Reihe nicht behandelten Gattungen und Arten. *Linzer biologische Beiträge* 32(1): 43–65.

Gusenleitner, J. 2000c. Neue Eumenidae aus Syrien, der Türkei und dem Iran (Hymenoptera, Vespoidea). *Linzer biologische Beiträge* 32(2): 927–938.

Gusenleitner, J. 2001a. Über die Gattung *Labochilus* Blüthgen 1939 (Hymenoptera, Vespoidea, Eumenidae). *Linzer biologische Beiträge* 33(2): 677–687.

Gusenleitner, J. 2001b. Über Eumeniden-Arten aus Jordanien (Hymenoptera, Vespoidea, Eumenidae). *Linzer biologische Beiträge* 33(1): 233–238.

Gusenleitner, J. 2002. Über bemerkenswerte Faltenwespen aus der äthiopischen Region (Hymenoptera, Vespoidea). *Linzer biologische Beiträge* 34(1): 307–320.

Gusenleitner, J. 2003. Über Eumenidae aus Kenya und Tanzania (Hymenoptera: Eumenidae). *Linzer biologische Beiträge* 35(1): 167–172.

Gusenleitner, J. 2004. Ein Beitrag zur Kenntnis und Verbreitung paläarktischer Eumenidae (Hymenoptera: Vespoidea). *Linzer biologische Beiträge* 36(2): 1093–1105.

Gusenleitner, J. 2005. Über Faltenwespen aus dem Oman Teil 2 (Hymenoptera, Vespidae, Eumenidae, Masaridae). *Linzer biologische Beiträge* 37(2): 1199–1201.

Gusenleitner, J. 2006a. Über Eumeninae, aufgesammelt in Pakistan (Hymenoptera: Vespidae). *Linzer biologische Beiträge* 38(2): 1295–1305.

Gusenleitner, J. 2006b. Eine neue Eumenes-Art aus dem Iran (Hymenoptera: Vespidae, Eumeninae). *Linzer biologische Beiträge* 38(2): 1361–1363.

Gusenleitner, J. 2006c. Über Aufsammlungen von Faltenwespen in Indien (Hymenoptera: Vespidae). *Linzer biologische Beiträge* 38(1): 677–695.

Gusenleitner, J. 2006d. Über Eumeniden aus dem Mittelmeergebiet mit einem Anhang über eine äthiopische Art (Hymenoptera: Vespidae: Eumeninae). *Linzer biologische Beiträge* 38(1): 663–669.

Gusenleitner, J. 2007. Vespidae aus Jemen in Sammlung des Oberösterreichischen Landesmuseums (Hymenoptera, Vespidae). *Linzer biologische Beiträge* 39(1): 119–127.

Gusenleitner, J. 2008. Bestimmungstabellen mittel- und südeuropäischer Eumeniden (Vespoidea, Hymenoptera) 4. *Nachtrag. Linzer biologische Beiträge* 40(2): 1505–1507.

Gusenleitner, J. 2010a. Bemerkenswertes über Faltenwespen X (Hymenoptera: Vespidae: Eumeninae, Masarinae). *Linzer biologische Beiträge* 42(2): 1347–1352.

Gusenleitner, J. 2010b. Order Hymenoptera, family Vespidae, 422–467. In: van Harten, A. (ed.), *Arthropod fauna of the UAE*, Vol. 3, 700 pp.

Gusenleitner, J. 2011a. Bemerkenswertes über Faltenwespen XI (Hymenoptera: Vespidae: Eumeninae: Masarinae). *Linzer biologische Beiträge* 43(1): 323–329.

Gusenleitner, J. 2011b. Bemerkenswertes über Faltenwespen XII (Hymenoptera: Vespidae: Eumeninae). *Linzer biologische Beiträge* 43(2): 1127–1134.

Gusenleitner, J. 2011c. Eine Aufsammlung von Faltenwespen aus Laos im Biologiezentrum Linz (Hymenoptera: Vespidae: Vespinae, Stenogastrinae, Polistinae, Eumeninae). *Linzer biologische Beiträge* 43(2): 1351–1368.

Gusenleitner, J. 2012. Bemerkenswertes über Faltenwespen XII (Hymenoptera: Vespidae: Eumeninae). *Linzer biologische Beiträge* 44(2): 1129–1136.

Gusenleitner, J. 2013a. Die Gattungen der Eumeninae im Nahen Osten, in Nordafrika und in Arabien (Hymenoptera: Vespidae: Eumeninae). *Linzer biologische Beiträge* 45(1): 5–107.

Gusenleitner, J. 2013b. Die Gattungen der Vespidae im Nahen Osten, in Nordafrika und in Arabien (Hymenoptera: Vespidae). *Linzer biologische Beiträge* 45(2): 1025–1046.

Gusenleitner, J. 2014. Eine neue Gattung und Art der Eumeninae aus dem Iran (Hymenoptera: Vespidae, Eumeninae). *Linzer biologische Beiträge* 46(1): 537–540.

Gusenleitner, J. 2015. Eine neue Eumeninae aus dem Oman (Hymenoptera: Vespidae: Eumeninae). *Linzer biologische Beiträge* 47(2): 1327–1330.

Gusenleitner, J. 2016. Über Eumeninae aus dem Nahen Osten (Hymenoptera: Vespidae: Eumeninae). *Linzer biologische Beiträge* 48(1): 89–97.

Gusenleitner, J. 2021. *Antepipona levis* nov. sp. – eine neue Art aus dem Iran (Hymenoptera, Vespidae, Eumeninae). *Linzer biologische Beiträge* 53(1): 61–63.

Gusenleitner, J. 2023. Über paläarktische Eumeninae (Hymenoptera, Vespidae). *Linzer biologische Beiträge* 54(2): 417–421.

Gusenleitner, J. and Gusenleitner, F. 2010. *Antepipona assmanni* nov. sp. und *Antepipona aubrechti* nov. sp., zwei neue Arten aus Kenia (Hymenoptera: Vespidae: Eumeninae). *Linzer biologische Beiträge* 42(1): 711–723.

Gusenleitner, J. and Ockermüller, E. 2018. Neue Eumeninae aus der Paläarktischen Region (Hymenoptera: Vespidae: Eumeninae). *Linzer biologische Beiträge* 50(2): 1091–1098.

Gusenleitner, J., Fallahzadeh, M., Haghighi, A. & Dousti, A.F. 2013. Two new species of Eumeninae from Iran (*Hymenoptera*: Vespidae). *Linzer Biologische Beiträge* 45(1): 109–116.

Haddad, N.J., Dvořák, L., Adwan, O., Mdanat, H. and Bataynah, A. 2007. New data on vespid wasp fauna of Jordan (Hymenoptera: Vespidae). *Linzer biologische Beiträge* 39(1): 137–142.

Hamer, I. 1985. Bees and wasps of the Ruus al Jibal. *Bulletin of the Abu Dhabi Natural History Group* 25, 24.

Hamzavi, F., Carpenter, J.M. and Ting-Jing, L. 2019. A contribution to the study of Eumeninae (Hymenoptera, Vespidae) in South-Eastern Iran. *Journal of Insect Biodiversity and Systematics* 5(4): 393–398.

Harttig, G. 1932. Eine neue paläarktische *Symmorphus*-Art (*Hym. Vesp.*). *Mitteilungen der Münchner Entomologischen Gesellschaft* 3: 73–75.

Hayat, R., Ghahari, H., Lavigne, R.J. and Ostovan, H. 2008. Iranian Asilidae (Insecta: Diptera). *Turkish Journal of Zoology* 32: 175–195.

Henri, D.C. and van Veen, F.J.F. 2011. Body Size, Life History and the Structure of Host–Parasitoid Networks, pp. 135–180. In: Belgrano, A. and Reiss, J. (eds.), *Advances in ecological research*. Vol. 45, Amsterdam, The Netherlands, 318 pp.

Hermes, M.G., Melo, G.A. and Carpenter, J.M. 2014. The higher-level phylogenetic relationships of the Eumeninae (Insecta, Hymenoptera, Vespidae), with emphasis on *Eumenes sensu lato*. *Cladistics* 30: 453–484.

Herrich-Schäeffer, G.A.W. 1838. *Fauna Insectorum Germaniae initia, oder, Deutschlands Insects* 154: 19.

Herrich-Schäffer, G.A.W. 1839. *Fauna insectorum Germaniae initia oder Deutschlands Insecten. Nürnberg* 173, 32 pp.

Herrich-Schäffer, G.A.W. 1841. Fauna insectorum Germaniae 173: 13.

Hines, H.M., Hunt, J.H., O'Connor, T.K., Gillespie, J.J. and Cameron, S.A. 2007. Multigene phylogeny reveals eusociality evolved twice in vespid wasps. *Proceedings of the National Academy of Sciences, USA* 104: 3295–3299.

Howarth, B. and Gillett, M.P.T. 2008. The terrestrial and freshwater arthropods of Abu Dhabi Emirate, pp. 380–463. In: Perry, R.J. (ed.), *Terrestrial Environment of Abu Dhabi Emirate*. Environment Agency, Abu Dhabi, 463 pp.

ICZN, 1970. Opinion 893. Eumenidae names of Saussure (Hymenoptera): Grant of availability of certain names proposed for divisions of Genera. *Bulletin of Zoological Nomenclature* 26: 187–191.

Illiger, K. 1802. Neue Insekten. *Magazin für Insektenkunde* 1: 163–208.

Jennings, D.T. and Houseweart, M.W. 1984. Predation by eumenid wasps (Hymenoptera: Eumenidae) on source budworm (Lepidoptera: Tortricidae) and other lepidopterous larvae in Spurce Fir forests of Maine. *Annals of the Entomological Society of America* 49: 39–45.

Johnson, A.D., Rozenberg, T. and Segoli, M. 2023. Notes on the parasitoids found within the nests of *Delta dimidiatipenne* (Hymenoptera, Vespidae). *Journal of Hymenoptera Research* 96: 925–936.

Jones, H.P. 1937. A new British "Mason wasp". *Entomologist Monthly Magazine* 73: 13–15.

Jurine, L. 1807. *Nouvelle méthode de classer les Hyménoptères et les Diptères*. 1. Hyménoptères Paschoud, Geneva J.J. ed., 319 pp.

Kaplan, E. and Efil, L. 2025. New contributions to families of Bembicidae, Megachilidae, Scoliidae, Tiphiidae, Mutillidae, Vespidae (Hymenoptera: Aculeata) from western Türkiye. *Munis Entomology and Zoology* 20(2): 2610–2618.

Kelstrup, H., West-Eberhard, M.J., Nascimento, F.S., Riddiford, L.M. and Hartfelder, K. 2023. Behavior, ovarian status, and juvenile hormone titer in the emblematic social wasp *Zethus miniatus* (Hymenoptera: Vespidae: Eumeninae). *Behavioral Ecology and Sociobiology* 77: 58.

Kelstrup, H.C., Hartfelder, K., Lopes, T.F. and Wossler, T.C. 2018. The behavior and reproductive physiology of a solitary progressive provisioning vespid wasp: Evidence for a solitary-cycle origin of reproductive castes. *The American Naturalist* 191(2): doi:10.1086/695336

Kemal, M. and Koçak, A.Ö. 2018. Observations and notes on Hymenoptera and Coleoptera of Bahçesaray district (Van Province, East Turkey). *Cesa News* 154: 1–23.

Khalaf, A.N. and Al-Omar, M.A. 1974. *A second list of insects from Iraq*. Baghdad: Biological Research Centre, Scientific Research Foundation, 41 pp.

Khalaf, K.T. 1958. Some Hymenoptera and Coleoptera from Iraq. *Iraq Natural History Museum Publication* 14: 1–3.

Khoobdel, M., Tavassoli, M., Salari, M. and Firozi, F. 2014. The stinging Apidae and Vespidae (Hymenoptera: Apocrita) in Iranian islands, Qeshm, Abu–Musa, Great Tunb and Lesser Tunb on the Persian Gulf. *Asian Pacific Journal of Tropical Biomedicine* 4: 258–262.

Kim, J.K. and Yamane, S. 2004. Revision of the genus *Stenodynerus* Saussure (Hymenoptera, Vespidae, Eumeninae) in the Far East, with descriptions of a new species and a new subspecies from Taiwan. *Japanese Journal of Systematic Entomology* 10(2): 235–264.

Kirby, W.F. 1900. Descriptions of new species of Hymenoptera. *Bulletin of the Liverpool Museums under the City Council* 3: 13–24.

Kirby, W.F. 1903. Insecta: Hymenoptera, pp. 233–260. In: Forbes, H.O. (ed.), *The Natural History of Sokotra and Abd-el Kuri being the Report upon the Results of the Conjoint Expedition to these Islands in 1898–9 by Mr. W.R. Ogilvie-Grant, of the British Museum, and Dr. H.O. Forbes, of the Liverpool Museum, together with Information from Other Available Sources forming a Monograph of the Islands*. The Free Public Museum, Liverpool, 598 pp.

Klug, F. 1805. *Pterocheilus*, eine neue Insectengattung aus der Klasse der Piezaten, pp. 143–156. In: Weber, F. and Mohr, D.M.H. (eds.), *Beiträge zur Naturkunde*. Kiel, 1, 356 pp.

Kochetkov, D.N. 2020. Contribution to the potter wasps fauna (Hymenoptera: Vespidae, Eumeninae) of the Kinganskiy Nature Reserve, Amur Province. *Readings in Memory of Alexey Ivanovich Kurentsov* 31: 53–66.

Kohl, F. and Handlirsch, A. 1889. Transcaspische Hymenopteren. *Verhandlungen der kaiserlichköniglichen Zoologisch-Botanischen Gesellschaft in Wien* 39: 267–286.

Kohl, F.F. 1905. Hymenopteren. In: Penther, A. and Zederbauer, E. (eds.), *Ergebnisse einer naturwissenschftlichen Reise zur Erdschias-Dagh (Kleinasien). Annalen des Naturhistorischen Museum in Wien* 20: 220–246.

Kohl, F.F. 1906. Zoologische Ergebnisse der Expedition der kaiserlichen Akademie der Wissenschaften nach Südarabien und Sokótra in Jahre 1898–1899. Hymenopteren. *Denkschriften der Kaiserlichen Akademie der Wissenschaften in Wien, Mathematisch-Naturwissenschaftlische Klasse* 71(1): 169–301.

Kohl, F.F. 1907. Zoologische Ergebnisse der Expedition der kaiselichen Akademie der Wissenschaften nach südarabien und Sokotra in Jahre 1898–1899. *Denkschriften der Akademie der Wissenschaften math. Naturw. Wien* 71: 169–300.

Kokujev, N.R. 1912. Hymenoptera nova e Caucaso, collecta a dom. A.B. Schelkonikov. *Mitteilungen des Kaukasischen Museums* 7(1): 1–6.

Kokujev, N.R. 1913. Contribution à la faune des Hyménoptères de la Russie III. *Revue Russe d'Entomologie* 13: 161–170.

Kostylev, G. 1929. Materialien zur Kenntnis der Vespiden-Fauna Armeniens. *Revue Russe d'Entomologie* 23 (1–2): 76–82.

Kostylev, G. 1935 [1934]. Materialien zur Kenntnis der Masariden—Fauna der Paläarktis. *Archives du Musée Zoologique de l'Université de Moscou* 2: 85–116.

Kostylev, G. 1937 [1936]. Quelques especes nouvelles ou peu connues dela famille des Vespides. *Archives du Musee Zoologique de l'Universite de Moscou* 3: 221–226.

Kostylev, G. 1939. Nouveilles especes des genres *Eumenes* Latr. et *Labus* Sauss. du sudouest de I' Asie. *Archives du Musee Zoologique de I'Universite de Moscou* 5: 161–168.

Kostylev, G. 1940. Espèces nouvelles et peu connues de Vespides, d'Euménides et de Masarides paléarctiques (Hymenoptera). II. *Bulletin de la Société des Naturalistes de Moscou, Section Biologique (NS)* 49: 137–154.

Kriechbaumer, J. 1869. Hymenopterologische Beiträge. *Verhandlungen der Zoologisch-Botanischen Gesellschaft in Wien*, XIX: 587–600.

Kriechbaumer, J. 1879. Eumeniden-Studien. IV. Eine neue Art. *Entomologische Nachrichten* 5(7): 85–89.

Krombein, K.V. 1969. Life history notes on some Egyptian solitary wasps and bees and their associates (Hymenoptera: Aculeata). *Smithsonian Contributions to Zoology* 19: 1–18.

Kumar, G.P. and Carpenter, J.M. 2013. A taxonomic review of the genus *Antodynerus* de Saussure, 1855 (Hymenoptera: Vespidae: Eumeninae) from the Indian subcontinent. *Zootaxa* 3731(2): 267–278.

Kumar, G.P., Carpenter, J.M., Castro, L. and Sureshan, P.M. 2017. A taxonomic review of the Indian species of the genus *Eumenes* Latreille (Hymenoptera: Vespidae: Eumeninae). *Zootaxa* 4317(3): 469–498.

Kumar, P.G., Castro, L. and Sheikh, A.H. 2019a. Additions to the knowledge of the genus *Katamenes* Meade-Waldo, 1910 (Hymenoptera, Vespidae, Eumeninae) in India with a key and a checklist to the region's species. *Boletín de la Asociación Española de Entomología* 43 (1–2): 15–26.

Kumar, P.G., Pannure, A. and Carpenter, J.M. 2019b. Potter wasps (Hymenoptera: Vespidae: Eumeninae) of India, pp. 187–200. In: Ramani, S., Mohanraj, P. and Yeshlieanth, H.M. (eds.), *Indian Insects. Diversity and science.* Afestschrift for Professor C.A. Viraktamath's 75th birthday. CRC Press, Boca Raton, London, New York, 472 pp.

Kurzenko, N.V. 1977. Eumenid wasps (Hymenoptera, Eumenidae) of the Mongolian People's Republic and adjacent regions of China and Southern Siberia. *Insects of Mongolia* 5: 537–582.

Kurzenko, N.V. 1978. *Solitary Vespid Wasps of the Family Eumenidae (Hymenoptera) of the fauna of the USSR.* Thesis manuscript to acquire a scientific degree of Candidate of Biological Science. Institute of Biology and Soil Science of the Far East Center of the Academy of Sciences of USSR, Vladivostok, 326 pp.

Kurzenko, N.V. 1988. On the question of the taxonomic status and classification of *Onychopterocheilus* Blüthgen (Hymenoptera, Vespidae, Eumeninae). *Revue d'Entomologie de l'URSS* 70: 87–100.

Latreille, P.A. 1802. *Histoire naturelle, Générale et particulière des Crustacés et des Insectes.* Tome troisième, Paris, 468 pp.

Latreille, P.A. 1809. *Genera Crustaceorum et Insectorum secundum ordinem naturalem in familias disposita inconibus exemplisque plurimis explicata.* Tomus Quartus. Parisiis & Argentorati [Paris and Strasbourg], 339 pp.

Latreille, P.A. 1810. *Considérations générales sur l'ordre naturel des animaux composant les classes des Crustacés, des Arachnides et des Insectes.* F. Schoell, Paris, 444 pp.

Leach, W.E. 1815. Entomology. In: Brewster, D. 'Edinburgh Encyclopaedia, volume 9'.

Lehr, P.A., Ghahari, H. and Ostovan, H. 2007. A contribution to the robber flies of subfamilies Stenopogoninae and Asilinae (Diptera: Asilidae) from Iran. *Far Eastern Entomologist* 173: 1–14.

Lepeletier, A. 1841. *Histoire Naturelle des Insectes Hyménoptères* 2: 1–680.

Li, T.-J. and Chen, B. 2023. A color atlas of the Chinese vespid wasps. Zhengzhou (China), Henan Science and Technology Press, 622 pp. [in Chinese]

Li, T.-J., Barthélémy, C. and Carpenter, J.M. 2019. The Eumeninae (Hymenoptera, Vespidae) of Hong Kong (China), with description of two new species, two new synonymies and a key to the known taxa. *Journal of Hymenoptera Research* 72: 127–176.

Lichtenstein, M.J. in Desmarest, M.E. 1884. Bulletin des séances. *Annales de la Société Entomologique de France, Série* 6(4): XLIX–L.

Longair, R.W. 2004. Tusked males, male dimorphism and nesting behavior in a subsocial afrotropical wasp, *Synagris cornuta*, and weapons and dimorphism in the genus (Hymenoptera: Vespidae: Eumeninae). *Journal of the Kansas Entomological Society* 77: 528–557.

Lopez, R.B. and Noll, F.B. 2014. Notes on the Neotropical *Zethus* Fabricius, 1804 (Hymenoptera, Vespidae, Eumeninae) with the description of two new species from Brazil. *Zootaxa* 3784: 179–186.

Luo, L., Carpenter, J.M., Chen, B. and Li, T. 2022. First comprehensive analysis of both mitochondrial characteristics and mitogenomic-based phylogenetics in the subfamily Eumeninae (Hymenoptera: Vespidae). *Insects* 13: 1–16.

Madl, M. 1997. Uber Vespidae, Pompilidae und Tiphiidae der Turkei (Hymenoptera). *Linzer biologische Beiträge* 29(2): 823–827.

Madl, M. 2018. A preliminary catalogue of the Hymenoptera (Insecta) of the Republic of Djibouti. *Linzer biologische Beiträge* 50(2): 907–967.

Magretti, P. 1884. Resultati di Raccolte imenotterologische nell'Africa Orientale. *Annali del Museo Civico Storia Nat. Giocomo Doria* 2(1): 523–636.

Matthews, R.W. and Matthews, J.R. 2010. *Insect Behavior*. 2nd edition. Springer, Dordrecht, 527 pp.

Mauss, V., Kuba, K. & Krenn, H.W. 2019. Evolution of the multifunctional mouthparts of adult Vespidae. In: Krenn, H.W. (Ed.), *Insect Mouthparts. Zoological Monographs* 5. Springer, Cham, pp. 443–478.

Meade-Waldo, G. 1910. V. New species of Diploptera in the collection of the British Museum. *Journal of Natural History* 5(25): 30–51.

Meade-Waldo, G. 1915. A revision of the species of the genus *Odynerus* (Hymenoptera) occurring in the Ethiopian Region. *Transactions of the Entomological Society of London* 1914 (3–4): 485–520.

Miller, K.E., Polaszek, A. and Evans, D.M. 2021. A dearth of data: fitting parasitoids into ecological networks. *Trends in Parasitology* 37(10): 863–874.

Mocsáry, A. 1877–1878. Adatok Zolyóm és Liptó megyék faunajához – Data ad Faunam Hungariae Septentrionalis comitatum: Zolyóm et Liptó. *Mathematikai és Természetudományi Közlemények* 15: 223–263.

Mocsáry, A. 1883. Hymenoptera nova europaea et exotica. *Magyar Tudomanyos Akademia Ertekezesek Termeszet. Korebol* 13(11): 1–72.

Modarres Awal, M. 2012. Family Vespidae (Hymenoptera), pp. 525–527. In: Modarres Awal, M. (ed.), *List of agricultural pests and their natural enemies in Iran*. 3rd edition. Ferdowsi University of Mashhad Press, Iran, 759 pp.

Mokrousov, M.V. 2015. Description of the male and redescription of the female of *Microdynerus rubronotatus* (Kostylev, 1940) (*Hymenoptera*: Vespidae). *Eversmannia, Supplement* 5: 51–53.

Mokrousov, M.B. and Zryanin, V.A. 2015. Materials on the early spring wasps and ants fauna of Uzbekistan (Hymenoptera: Vespomorpha: Chrysidoidea, Scolioidea, Pompiloidea, Vespoidea, Apoidea [Spheciformes], Formicoidea). *Eversmannia* 5: 36–48.

Monks, J., Ross, S., Geiser, M., De Prins, J., Sharaf, M., Wyatt, N., Al Rijeibi, S. and Polaszek, A. 2019. A preliminary survey of the insect fauna of the Hajar Mountain Range, *Oman. Journal of Natural History* 53 (15–16): 939–963.

Morawitz, F. 1867. Übersicht der im Gouvernement von Saratow und um St. Petersburg vorkommenden Odynerus-Arten. *Horae Societatis Entomologicae Rossicae* 4: 109–144.

Morawitz, F. 1873. Nachtrag zur Bienenfauna des Gouvernements von St. Petersburg. *Horae Societatis Entomologicae Rossicae* 9: 151–159.

Morawitz, F. 1885. Eumenidarum species novae. *Horae Societatis Entomologicae Rossicae* 19: 135–181.

Morawitz, F. 1888. Hymenoptera aculeata nova. *Horae Societatis Entomologicae Rossicae* 22: 224–302.

Morawitz, F. 1895. Materialien zu einer Vespidenfauna des Russischen Reiches. *Horae Societatis Entomologicae Rossicae* 29: 407–493.

Morice, F. 1903. New Hymenoptera Aculeata taken by the Swedish zoological expedition to the White Nile in the spring of 1901. *The Annals and Magazine of Natural History (Series 7)* 12: 610–615.

Morice, F. 1921. Annotated lists of aculeate Hymenoptera (except Heterogyna) and chrysids recently collected in Mesopotamia and north-west Persia. *The Journal of the Bombay Natural History Society* 28: 192–203.

Müller, O.F. 1776. Zoologiae Danicae prodromus seu animalum Daniae et Norvegiae indigenarum characteres, nomina et synomyma imprimis popularium. *Hafniae*, 282 pp.

Nguyen, L.T., Dang, H.T., Kojima, J.I. and Carpenter, J.M. 2014. An annotated distributional checklist of solitary wasps of the subfamily Eumeninae (Hymenoptera: Vespidae) of Vietnam. *Entomologica Americana* 120(1): 7–18.

Nurse, C.G. 1903. New species of Indian aculeate Hymenoptera. *Annals and Magazine of Natural History* (7)11: 528–549.

Nurse, C.G. 1904. New species of Indian aculeate Hymenoptera. *Journal of the Bombay Natural History Society* 16: 19–26.

Panfilov, D.V. 1968. New species of Middle Asian and Transcaucasian hymenopterans (Hymenoptera: Chrysididae, Masaridae, Vespidae). *Archives du Muzée Zoologique de l'Universite de Moscow* 11: 36–43.

Pannure, A., Belavadi, V.V. and Carpenter, J.M. 2016. Taxonomic studies of potter wasps (Hymenoptera: Vespidae: Eumeninae) of South India. *Zootaxa* 4171(1): 1–51.

Panzer, G.W.F. 1797. Faunae Insectorum Germanicae. *Heft* 45: 14–15, 47: 19.

Panzer, G.W.F. 1798. *Faunae Insectorum Germanicae initia oder Deutschlands Insecten*. Felseckersche Buchhandlung, Heft 60: 1 May 1798; Heft 61–70.

Panzer, G.W.F. 1799. Faunae Insectorum Germaniae initia oder Deutschland' Insecten, Nürnberg. *Heft* 61–72.

Panzer, G.W.F. 1801. Faunae Insectorum Germanicae. *Heft* 80: 7–15, 81: 13, 83: 13, 84: 14–15.

Pashaei Rad, S., Abassi, R. and Ebrahimi, E. 2006. Two new records for the fauna of solitary wasps of the subfamily Eumeninae (Hymenoptera: Vespidae) in Iran. *Zoology in the Middle East* 39: 117–119.

Pashaei Rad, S., Abbasi, R., Soleimani, G. and Dvořák, L. 2010. New and supplementary information on the vespid fauna of Iran (Hymenoptera: Vespidae). *Zoology in the Middle East* 50(1): 95–100.

Pérez, J. 1907. Mission J. Bonnier et ch. Pérez (Golfe Persique, 1901). II. Hyménoptères. *Bulletin Scientifique de la France et de la Belgique Série* 6(41): 485–505.

Pickett, K.M. and Carpenter, J.M. 2010. Simultaneous analysis and the origin of eusociality in the Vespidae (Insecta: Hymenoptera). *Arthropod Systematics and Phylogeny* 68: 3–33.

Piekarski, P.K. 2017. Advanced castes at the outset of eusociality in wasps (Vespidae). M. Sc. Thesis, University of Central Florida, Orlando. Electronic Theses and Dissertations, 2004–2019. 5356.

Piekarski, P.K., Carpenter, J.M., Lemmon, A.R. and Sharanowski, B.J. 2018. Phylogenomic evidence overturn current conceptions of social evolution in wasps (Vespidae). *Molecular Biology and Evolution* 35: 2097–2109.

Pittioni, B. 1950. On the insect fauna of Cyprus. Results of the expedition of 1938 by Harald, Håkan and P.H. Lindberg. V. Hymenoptera aculeata. I. Diploptera, Fossores und Apoidea der Insel Cypern. *Commentationes Biologicae Societatis Scientiarum Fennicae* 10(2): 1–94.

Qin, J., Chen, B. & Li, T. 2024. A taxonomic study of the genus *Antepipona* de Saussure, 1855 (*Hymenoptera*: Vespidae: Eumeninae) from China, with two new species and three new records. *Zootaxa* 5501(2): 358–370.

Radoszkovski, O. 1893. Faune hyménopterologique Transcaspienne. *Horae Societatis Entomologicae Rossicae* 27: 38–81.

Radoszkowski, O. 1865. Description d'un genre nouveau *Pseudomelecta*, et de quelques espèces du genre *Eumenes*. *Horae Societatis Entomologicae Rossicae* 3: 53–60.

Radoszkowski, O. 1876. Comte-rendu des Hyménoptères recueillis en Egypt et Abyssinie in 1873. *Horae Societatis Entomologicae Rossicae* 12(1): 111–116; (2): 117–150.

Radoszkowski, O. 1886. Faune Hyménoptèrologique. Transcaspienne. *Horae Societatis Entomologicae Rossicae* 20: 3–56.

Rafi, M.A., Carpenter, J.M., Qasim, M., Shehzad, A., Zia, A., Khan, M.R., Mastoi, M.I., Naz, F., Yas, M., Shah, M. and Bhati, A.R. 2017. The vespid fauna of Pakistan. *Zootaxa* 4362(1): 1–28.

Rahmani, Z., Rakhshani, E. and Carpenter, J.M. 2020. Updated checklist of Vespidae (Hymenoptera: Vespoidea) in Iran. *Journal of Insect Biodiversity and Systematics* 6(1): 27–86.

Ratzeburg, J.T.C. 1844. Die Forstinsekten oder Abbildung und Beschreibung in den Wäldern Preussens und der Nachbarstäaten als schädlich oder nützlich bekannt gewordenen Insecten; in systematischer Folge und mit besonderer Rücksicht auf die Vertilgung der Schädlichsten Berlin, 314 pp.

Richards, O.W. 1935. Notes on the nomenclature of the Aculeate Hymenoptera, with special reference to British genera and species. *Transactions of the Entomological Society of London* 83: 143–176.

Richards, O.W. 1962. A revisional study of the masarid wasps (Hymenoptera, Vespoidea). British Museum (Natural History), London, UK, 294 pp.

Roche, C.G. 1981. Notes on the bees and wasps of the UAE. *Bulletin of the Abu Dhabi Natural History Group* 15: 12–20.

Rossi, P. 1790. *Fauna Etrusca: sistens insecta quae in provinciis Florentina et Pisana praesertim collegit. Tomus secundus*. Thomas Masi & Sociorum, Liburni, 348 pp.

Saghaei, N., Hayat, R., Ostovan, H. and Fallahzadeh, M. 2010. Prey of some robber flies (Diptera: Asilidae) in Fars province, *Iran. Journal of Entomological Research Society* 12(1): 17–26.

Sakenin, H., Raheb, J., Imani, S., Havaskary, M., Shirdel, F. and Mohseni, H. 2008. A preliminary survey on dipteran predators and parasitoids and Odonata in Iranian rice fields. *Proceedings of National Conference of Agronomical Rice Breeding, Young Researchers Club Islamic Azad University of Ghaemshahr*, Iran, 26–27 November, 2008, 14 pp. [in Persian, English Summary]

Salehi Fard, M., Modarres Awal, M. and Gusenleitner, J. 2013. First Record of *Allodynerus dignotus* Morawitz, 1895 (Hym: Vespidae: Eumeninae) from Iran. *Journal of Plant Protection* 26(4): 486–488. [in Persian, English summary]

Saunders, E. 1905. Hymenoptera aculeata collected in Algeria by the Rev. A.E. Eaton, M.A., F.E.S., and the Rev. Francis David Morice, M.A., F.E.S. part II. Diploptera. *Transactions of the Entomological Society of London* 1905: 399–417.

Saunders, S.S. 1850. Descriptions of some new aculeate Hymenoptera from Epirus. *Transactions of the Entomological Society of London* 2(1): 69.

Saunders, S.S. 1872. I. Stylopidarum, ordinem Strepsipterorum Kirbii constituentium, mihi tamen potius Coleopterorum Familiae, Rhipiphoridis Meloïdisque propinquae Monographia. *Transactions of the Entomological Society of London* 1872: 1–48.

Schembri, S.P. 1980. Vespoidea from the maltese Islands (Hymenoptera Aculeata). *Bollettino della Societa Entomologica Italiana* 112 (4–6): 90–93.

Schmid-Egger, C., Gusenleitner, J., Liebig, W.-H. and Jacobs, H.-J. 2021. New records of Vespidae (Hymenoptera) from Iran with descriptions of three species. *Linzer biologische Beiträge* 52(2): 1105–1121.

Schmiedeknecht, O. 1900. *Neue Hymenopteren aus Nord-Afrika. Természetrajzi Füzetek* 23: 220–247.

Selis, M. 2023. Illustrated key to the genera and a checklist of Italian Vespidae (Hymenoptera). *Fragmenta Entomologica* 55(1): 63–88.

Selis, M. 2024. Revision of the Afrotropical species of *Euodynerus* Dalla Torre (Hymenoptera: Vespidae: Eumeninae). *Zootaxa* 5399(4): 365–380.

Selis, M. 2025a. Revision of the Afrotropical species of *Pareumenes* de Saussure (Hymenoptera, Vespidae, Eumeninae). *Zootaxa* 5584(2): 205–227.

Selis, M. 2025b. Solving the identity of *Odynerus spiricornis* Spinola, with notes on the South European species of *Paragymnomerus* Blüthgen (Hymenoptera, Vespidae, Eumeninae). *Zootaxa* 5601(3): 558–570.

Selis, M. 2025c. The solitary vespid wasps of Madagascar (Hymenoptera: Vespidae: Eumeninae, Raphiglossinae and Zethinae). *Zootaxa* 5705: 1–171.

Selis, M., Fateryga, A.V. and Cilia, G. 2024. The genus *Euodynerus* Dalla Torre in Europe and the Maghreb (Hymenoptera: Vespidae: Eumeninae). *Zootaxa* 5537(2): 151–194.

Selis, M., Ismael, H.R., Hamad, M.I., Baiocchi, D. and Cilia, G. 2025. A first update to the vespid fauna of Iraq (Hymenoptera: Vespidae), with description of a new species. *Journal of Natural History* 59(13–16): 941–981.

Shuckard, W.E. 1837. Description of a new British wasp; with an account of its development. *Magazine of the Natural History and Journal of Zoology, Botany, Mineralogy, Geology, and Meteorology, New series* 1: 490–496.

Smith, A.P. and Alcock, J. 1980. A comparative study of the mating systems of Australian eumenid wasps (Hymenoptera). *Zeitschrift für Tierpsychologie* 53: 41–60.

Smith, F. 1857. *Catalogue of Hymenopterous insects in the collection of the British Museum*. London, Printed by order of the Trustees, 5: 1–147.

Spinola, M. 1808. *Insectorum Liguriae species novae aut rariores, quas in agro Ligustico nuper detexit, descripsit, et iconibus illustravit* (*Hymenoptera*), 2. Genua, 262 pp.

Spinola, W.A.M. 1806. Insectorum Liguriae species novae aut rariores, quas in agro Ligustico nuper detexit, descripsit, et inconibus illustravit (Hymenoptera). 2. *Genua* 1806: 262 pp., 1808: pp. 207–262.

Spinola, W.A.M. 1838. Des Hyménoptères recueillis par M. Fischer Pendant son voyage en Égypte, et communiqués. *Annalés de la Société Entomologique de France* 7: 437–575.
Spradbery, J.P. 1973. *Wasps: and account of the biology and natural history of social and solitary wasps*. University of Washington Press, Seattle, 235 pp.
Srinivasan, G. and Kumar, P.G. 2010. New records of the potter wasps (Hymenoptera: Vespidae: Eumeninae) from Arunachal Pradesh, India: five genera and ten species. *Journal of Threatened Taxa* 2(12): 1313–1322.
Strand, E. 1914. Neue Namen verschiedener Tiere. *Archiv für Naturgeschichte* 80(1): 163–164.
Tan J.L., van Achterberg C. and Chen X.-X. 2015. *Potentially lethal social wasps: the Fauna of the Chinese Vespinae (Hymenoptera: Vespidae)*. Science Press, Beijing, China, 198 pp.
Tan, J.L., Carpenter, J.M. and van Achterberg, C. 2018. An illustrated key to the genera of Eumeninae from China, with a checklist of species (Hymenoptera, Vespidae). *ZooKeys* 740: 109–149.
Tezcan, S., Karsavuran, Y., Pehlivan, E., Anlaş, S. and Yildirim, E. 2005. Contribution to the knowledge of the Vespidae (Hymenoptera: Aculeata) fauna of Turkey. *Turkish Journal of Entomology* 29(2): 101–110.
Thomson, C.G. 1874. *Skandinaviens Hymenoptera, 3: e delen. 1: a Häfet innehålande slägtet Vespa Lin.; 2: a häftet innehålande slägtet Mutilla och Sphex Lin.* Fr. Berlings Bocktryckeri, Lund, 295 pp.
Tüzün, A. 1986. *Ege Bölgesi Vespidae ve Eumenidae (Hymenoptera: Insecta) Türlerinin Tespiti*. Doktora Tezi, Ankara Universitesi Fen Bilimleri Enstutüsü, 46 pp.
Tüzün, A. and Tanyolag, T. 1988. Ege Bölgesinde yasayan *Odynerus* (Hyrnenoptera: Eumenidae) türleri üzerinde araştirmar. *Cumhuriyet Üniversitesi Fen-Edebiyat Fakültesi Fen Bilimleri Dergisi* 7: 105–128
van Achterberg, C., Smit, J.T. and Ljubomirov, T. 2023. Review of the European *Eumenes* Latreille (Hymenoptera, Vespidae)using morphology and DNA barcodes, with an illustrated key to species. *ZooKeys* 1143: 93–163.
van der Vecht, J. 1959. On *eumenes arcuatus* (Fabricius) and some allied Indo-Australian wasps (Hymenoptera, Vespidae). *Zoologische Verhandelingen. Leiden* 41: 1–71.
van der Vecht, J. 1963. Studies on the Indo-Australia and East Asiatic Eumenidae (Hymenoptera: Vespoidea). *Zoologische Verhandelingen. Leiden* 60: 1–116.
van der Vecht, J. 1969. A new name for the genus *Cyrtolabus* van der Vecht (Hymenoptera, Eumenidae). *Entomologische Berichten, Amsterdam* 29: 1–2.
van der Vecht, J. 1971. Notes on the Palaearctic Eumenidae (Hymenoptera) (Part 2). *Entomologische Berichten* 31: 127–134.
van der Vecht, J. and Fischer, F.C.J. 1972. Palaearctic Eumenidae. *Hymenopterum Catalogus (Novo Editio)* 8: 1–197.
Vernier, R. 1997. Essai d'analyse cladistique des genres d'Eumeninae (Vespidae, Hymenoptera) représentés en Europe septentrionale, occidentale et centrale. *Bulletin de la Société Neuchâteloise des Sciences Naturelles* 120: 87–98.
Villers, C. 1789 *Caroli Linnaei entomologia, Faunae Suecicae descriptionibus*. Tomus tertius, Ligduni, 657 pp.
von Dalla Torre, K.W. 1889. Hymenopterologische Notizen. *Wiener Entomologische Zeitung* 8: 124–125.
von Dalla Torre, K.W. 1894. Catalogus Hymenopterorum hucusque descriptorum systematicus et synonymicus. *Lipsiae* 9, 181 pp.
von Dalla Torre, K.W. 1904. Vespidae. *Genera Insectorum* 19: 1–108.
von Linnaeus, C. 1758. Systema naturae per regna tria naturae, secundum classes ordines, genera, species cum characteribus, differentis, synonymis locis. Tomus I. Editio decima, reformata. *Laurentii Salvii, Holmiae*, 824 pp.
von Linnaeus, C. 1761. Fauna Svecica. Editio altera. Laurentii Salvii, Stockholmiae, 578 pp.
von Schulthess, A. 1887. *Fauna Insectorum Helvetiae. Hymenoptera. Fam. Diploptera Lat.* (Vespidae aut.). Bd. 7.
von Schulthess, A. 1922. Vespiden aus Südafrika (*Hym.*), gesammelt von Professor Dr. Leonhard Schulze. *Deutsche Entomologische Zeitschrift*, 1922(4): 399–405.
von Schulthess, A. 1923. Hymenoptera VI: Vespidae. In: Michaelsen, W. (ed.), *Beiträge zur Kenntnis der Land- und Süßwasserfauna Deutsch-Südwestafrikas (zur Zeit Mandat Südwest-Afrika)*. Ergebnisse der Hamburger deutsch-südwestafrikanischen Studienreise 1911. *Band* 2(2): 135–140.
von Schulthess, A. 1929. Voyage au Congo de S.A.R. le Prince Léopold de Belgique (1925). Hymenoptera. I. Apides, Vespides, Sphégides, Chrysidides, Scoliides, Mutillides, Formicides. *Revue de Zoologie et Botanique Africaines* 17(2): 184–190.

von Schulthess, J.A. 1925. Nochmals südafrikanische *Alastor* (Hym. Vesp.). *Konowia (Vienna)* 4: 310–312.

von Schulthess, J.A. 1928. Beiträge zur Kenntnis nordafrikanischer Hymenoptera. Unter Mitarbeit von Specialisten. *Eos, Revista Española de Entomologia* 4: 65–92 + 405–410.

von Schulthess, J.A. 1931. Neue Vespiden (Hym.). *Mitteilungen der Schweizerischen Entomologischen Gesellschaft* 15(2): 49–52.

von Schulthess-Rechberg, A. 1897. *Fauna Insectorum Helvetiae. Hymenoptera. Fam. Diploptera* (Vespidae). *Mitteilungen der Schweizerischen Entomologischen Gesellschaft* 10: 1–91.

Wahis, R. 2000. Hyménoptères aculéates du Yemen (Pompilidae, Vespidae). *Notes Fauniques de Gembloux* 41: 73–100.

Walker, D.H. and Pittaway, A.R. 1987. *Insects of Eastern Arabia.* Macmillan, London, 191 pp.

Walker, F. 1871. *A list of Hymenoptera collected by J.K. Lord, Esq. in Egypt, in the Neighbourhood of the Red Sea, and in Arabia, with descriptions of the new species.* E.W. Janson, London, 63 pp.

Wesmael, C. 1833. Monographie des Odyneères de la Belgique. *Bruxelles* 51 pp.

Wesmael, C. 1836. Supplementa la monographie des Odyneres. *Bulletins de l'Académie royale des Sciences et belles-lettres de Bruxelles* 3: 44–54.

West-Eberhard, M.J. 1987. Flexible strategy and social evolution, pp. 35–51. In: Ito, Y., Brown, J.L. and Kikkawa, J. (eds.), *Animal societies: Theories and Facts.* Japan Scientific Societies Press, 291 pp.

West-Eberhard, M.J. 2005. Behavior of the primitively social wasp *Montezumia cortesioides* Willink (Vespidae, Eumeninae) and the origins of vespid sociality. *Ethology Ecology and Evolution* 17: 201–215.

Woydak, H. 2001. Die solitären Faltenwespen: Eumenidae (Lehmwespen) und Masaridae (Honigwespen) (*Hymenoptera*, Vespoidea) im Westfälischen Museum für Naturkunde Münster. *Natur und Heimat* 61 (3): 85–95.

Yalniz, S.E. and Tüzün, A. 2019. The faunistic studies of the Vespidae species (Hymenoptera: Vespoidea) of Adana province, Turkey. *Biological Diversity and Conservation* 1212: 15–22.

Yamane, S.K. 1990. A revision of the Japanese Eumenidae (Hymenoptera, Vespoidea). *Insecta Matsumurana, New series* 43: 1–189.

Yildirim, E. & Gusenleitner, J. 2004. The present situation of the Sapygidae and Vespidae-fauna (*Hymenoptera*, Aculeata) of Turkey. *Linzer Biologische Beiträge* 36(1): 131–135.

Yildirim, E. 1998. Türkiye Vespidae (Hymenoptera, Aculeata) faunasinin son durumu (Bazi türler için yeni lokaliteler ve yeni kayit). *Atatürk Universitesi Ziraat Fakultesi Dergisi* 29(2): 243–249.

Yildirim, E. 2012. Contribution to the knowledge of the Vespidae (Hymenoptera, Aculeata) of Turkey, with a checklist of the Turkish species. *Turkish Journal of Zoology* 36(3): 361–374.

Yildirim, E. 2014. A new species of *Onychopterochilus* Blüthgen (Hymenoptera: Vespidae: Eumeninae) from Turkey. *Zoology in the Middle East* 60(3): 222–225.

Yıldırım, E. and Gusenleitner, J. 2001. The present situation of the Vespidae-fauna (Hymenoptera, Aculeata) of Turkey. *Linzer biologische Beiträge* 33(1): 217–221.

Yıldırım, E. and Gusenleitner, J. 2007. The present situation of the Vespidae-fauna of Turkey (Hymenoptera: Vespidae). *Linzer bologische Beiträge* 39(1): 105–109.

Yildirim, E. and Gusenleitner, J. 2009. The present situation of the Vespidae-fauna (Hymenoptera, Aculeata) of Turkey. *Linzer biologische Beiträge* 41(1): 937–941.

Yildirim, E. and Gusenleitner, J. 2012. Contribution to the knowledge of the Vespidae (Hymenoptera, Aculeata) of Turkey, with a checklist of the Turkish species. *Turkish Journal of Zoology* 36(3): 361–374.

Yildirim, E. and Kojima, J. 1999. Distributional checklist of the species of the family Vespidae (Insecta: Hymenoptera; Aculeata) of Turkey. *Natural History Bulletin of Ibaraki University* 3: 19–50.

Yildirim, E. and Özbek, H. 1994. Türkiye'nin *Eumenes* Latreille (Hymenoptera: Eumenidae) türlerinin tespiti Checklist of Turkish Vespidae üzerinde araştırmalar, Türkiye 3. *Biyolojik Mücadele Kongresi*, 25–28 Ocak 1994: 109–121.

Yildirim, E. and Özbek, H. 1995. New species of Eumenidae (Hymenoptera: Vespoidea) from Turkey. *Türkie Entomoloji Dergisi* 19(4): 241–246.

Yildirim, E. and Özbek, H. 1996a. Erzurum ve çevre iller ile diğer bazı yörelerdeki Eumenidae (Hymenoptera, Vespoidea) türleri üzerinde faunistik ve sistematik çalışmalar. *Turkish Journal of Zoology* 20(1): 113–147.

Yildirim, E. and Özbek, H. 1996b. Eumenidae (Hymenoptera: Vespoidea) of Turkey. *Turkish Journal of Zoology* 20(2): 189–209.

Yildirim, E. and Özbek, H. 1999. Türkiye'nin Vespoidea (Hymenoptera, Aculeata) faunası üzerinde bir değerlendirme (yeni kayıtlar ve bazt türler igin yeni lokaliteler). *Turkish Journal of Zoology* 23(6): 591–604.

Yildirim, E. and Tezcan, S. 2018. New locality records of Crabronidae, Evaniidae, Ichneumonidae, Leucospidae, Mutillidae, Pompilidae, Sapygidae, Sphecidae, Tiphiidae and Vespidae (Hymenoptera: Apocrita) from Turkey. *Linzer biologische Beiträge* 50(2): 1707–1719.

Zalat, S., Abo-Ghalia, A., Gilbert, F. and Shaumar, N. 1993 ["1991–1992"]. II. Taxonomic studies of the Egyptian Eumenidae (Order: Hymenoptera). *Bulletin of the Entomological Society of Egypt* 70: 99–127.

Zavattari, E. 1909. Imenotteri del Kashmir. *Bollettino dei Musei di Zoologia ed Anatomia Comparata della R. Università di Torino* 605: 1–5.

Zetterstedt, J.W. 1838. Insects Lapponica. Sectio secunda. Hymenoptera. *Hymenoptera. Lipsiae*, pp. 358–408.

Celonites kozlovi (Kostylev, 1935) (Masarinae). (Photo courtesy of Alexander Fateryga)

Celonites ivanovi (Mauss & Fateryga, 2022) (Masarinae). (Photo courtesy of Alexander Fateryga)

Celonites kozlovi (Kostylev, 1935) (Masarinae). (Photo courtesy of Alexander Fateryga)

Celonites tauricus (Kostylev, 1935) (Masarinae). (Photo courtesy of Alexander Fateryga)

Celonites ivanovi (Mauss & Fateryga, 2022) (Masarinae). (Photo courtesy of Alexander Fateryga)

Celonites tauricus (Kostylev, 1935) (Masarinae). (Photo courtesy of Alexander Fateryga)

3 Subfamily Masarinae Latreille, 1802

Neveen S. Gadallah, Erol Yildirim, Hassan Ghahari, and Patrick K. Piekarski

INTRODUCTION

The pollen wasps (sometimes referred to as the honey wasps) of the subfamily Masarinae Latreille, 1802, comprising 14 genera and 374 species worldwide (Pickett and Carpenter, 2010), represent a relatively small subfamily. They are most speciose in the Mediterranean and in temperate to hot semi-arid and arid areas outside the tropics (Gess, 1992; Goulet and Huber, 1993). This distribution indicates that masarines generally prefer warm to hot regions with relatively low rainfall, typically in areas of open, scrubby vegetation (Gess, 1992). Gayelline wasps, traditionally considered a tribe within the subfamily (Gayellini *sensu* Carpenter, 1982), are restricted to the Neotropical region, whereas members of the tribe Masarini are more widespread, occurring in the Nearctic, Neotropical, Palaearctic, and Australian regions (Gess, 1992).

These unusual wasps are typically treated as a subfamily within the family Vespidae. In the past, they were recognized as a separate family, "Masaridae" (Richards, 1962). Since Richards' monograph (1962), numerous changes have been made to the classification of the masarine wasps (Carpenter, 2001). In 1996, Gess reviewed the Masarinae, listing the described species and providing distributional summaries. However, Gess's list lacked citations of the original references, as well as descriptions of new species and new synonymies, factors that limit its taxonomic value (Carpenter, 2001). An updated list of all known species of Masarinae was published by Carpenter (2001), which revealed a few errors in both Richards (1962) and Gess (1996).

The family Masaridae, as defined by Richards (1962), comprised three subfamilies: Euparagiinae, Gayellinae, and Masarinae (Gess, 1996). It was considered as a sister group to the families Eumenidae and Vespidae within the superfamily Vespoidea. In a cladistic study conducted by Carpenter (1982), the phylogenetic relationships of the vespoid taxa (*sensu* Richards) were reassessed. He recognized a single family, Vespidae, comprising six subfamilies: Euparagiinae, Masarinae, Eumeninae, Stenogastrinae, Polistinae, and Vespinae. This classification distinguished those that provision their nest cells with grubs (beetle larvae) (Euparagiinae) from those that provision their nests with pollen and nectar (Richard's Gayellinae and Masarinae) (Gess, 1992). Meanwhile, he treated the closely related subfamilies Gayellinae (*sensu* Giordani Soika, 1974) and Masarinae (*sensu* Richards, 1962), as the tribes Gayellini and Masarini within the subfamily Masarinae (Carpenter, 1988; Gess, 1992).

More recently, phylogenomic molecular evidence failed to recover Gayellini and Masarini as sister tribes (Piekarski *et al.*, 2018). Instead, the study inferred that Euparagiinae is sister to Gayellini, which would restrict the Masarinae to include only the former Masarini and elevate Gayellini to subfamily rank (= Gayellinae). This result has important implications for understanding the evolution of pollen feeding within the group, as both Gayellinae and Masarinae are pollen feeders.

Adults are small to moderate in size, with a sessile metasoma (Goulet and Huber, 1993). Most species are black or brown, marked with strikingly contrasting patterns of yellow, white, red, or a

 DOI: 10.1201/9781003740438-3

combination of all three. Unlike other solitary aculeate wasps that provision their nests with paralyzed insects as a protein source, masarine wasps feed their larvae exclusively with pollen and nectar, similar to many solitary bees (Eickwort and Ginsberg, 1980; Gess and Gess, 1989). Hence, they are commonly referred to as pollen or honey wasps. Pollen wasps and bees have independently evolved pollen-feeding behavior from predatory ancestors through convergent evolution. They have developed specialized morphological structures for collecting pollen from flowers (Richards, 1962; Thorp, 1979; Westerkamp, 1987; Westrich, 1989). A pollen-collecting apparatus on the face of *Celonites abbreviatus* was described by Schremmer (1959). It is composed of apically knobbed setae that enable the wasp to collect pollen by rubbing its head against the anthers of Lamiaceae (Müller, 1996). In addition, they possess elongated mouthparts (proboscis) formed only by the glossae, for uptaking the floral nectar (Krenn *et al.*, 2002).

All species of pollen wasps are solitary mass provisioners (Goulet and Huber, 1993) with the possible exception of *Trimeria howardi*, which has been documented to nest communally and progressively provision larvae (Zucchi *et al.*, 1976). All species provision their nests with a mixture of pollen and nectar that they transport in their crop and regurgitate to brood (Cowan, 1991; Goulet and Huber, 1993). Their nests are often constructed of mud or consist of burrows in the ground and may contain one or multiple individual cells in which larvae and provisions are placed. Other than *Trimeria howardi*, nest construction, egg laying, and provisioning are carried out by a single female per nest. The nests are commonly located in concealed places, either excavated in the ground or constructed aerially on stones, plant stems, or within pre-existing cavities (Carpenter, 2001; Gess and Gess, 2010).

Checklists of the Iranian Masarinae: Ebrahimi and Carpenter (2008) listed 18 species in five genera: *Celonites* Latreille, 1802, *Ceramius* Latreille, 1810, *Jugurtia* de Saussure, 1854, *Masaris* Fabricius, 1793, and *Quartinia* André, 1884. Modarres Awal (2012) reported five species in three genera: *Ceramius*, *Jugurtia* and *Celonites*. Samin *et al.* (2020) catalogued 27 species within the same five genera. More recently, Rahmani *et al.* (2020) listed 24 species within five genera. The present catalogue includes 66 species and subspecies in five genera for the fauna of Middle East, and 25 species and subspecies in five genera for Iran.

Identification key of the genera of the subfamily Masarinae in the Middle East (modified from Gusenleitner, 2013)

1. Clypeus with straight apical margin; posterior margin of scutellum without lamella .... *Ceramius* Latreille, 1810

• Clypeus with medially emarginate apical margin; posterior margin of scutellum more or less lamellate 2

2. Small species, with transparent veins in longitudinally folded forewing; large hind tibial spur simple, pointed.. *Quartinia* André, 1884

• Larger species with simple forewings; large hind tibial spur bifid or trifid..3

3. Sides of gaster conspicuously flattened; antennal flagellum short and knobbed, with five unfused segments; occipital carina well developed dorsally and ventrally..*Celonites* Latreille, 1802

• Sides of gaster rounded, not flattened; antennal flagellum conical or cylindrical and longer than pedicel; occipital carina is lacking dorsally, partly present ventrally in *Masaris*..4

4. Hind coxa with inner dorsal keel; scutellum separated from mesoscutum by a deep furrow; tegula about twice as long as its width; gaster of male ventrally unmarked.. *Jugurtia* de Saussure, 1854

• Hind coxa without inner dorsal keel; scutellum separated from mesoscutum by a narrow and shallow furrow; tegula about four times as long as its width; gaster of male ventrally specified or modified..*Masaris* Fabricius, 1793

List of Species of the Subfamily Masarinae in the Middle East

Subfamily Masarinae Latreille, 1802
Tribe Masarini Latreille, 1802
Genus *Celonites* Latreille, 1802

Celonites Latreille, 1802: 368. Type species: *Masaris apiformis* Fabricius, 1793 (= *Vespa abbreviata* de Villers, 1789), by monotypy.

***Celonites* (*Celonites*) *abbreviatus abbreviatus* (de Villers, 1789)**

Celonites abbreviatus de Villers, 1789: 281, sex not stated; "Europa. In Gallia Australiori" (destroyed).

Catalogues with Iranian records: Samin *et al.* (2020).

Distribution in Iran: West Azarbaijan (Samin *et al.*, 2020, in vineyard).

Distribution in the Middle East: Cyprus (Yildirim and Kojima, 1999), Iran (Samin *et al.*, 2020), Israel (Bytinski-Salz and Gusenleitner, 1971), Turkey (Gusenleitner, 1967, 1997 both as *C.* (*C.*) *abbreviatus*, 2013; Yildirim and Özbek, 1997; Yildirim and Kojima, 1999; Yildirim and Gusenleitner, 2012).

Extralimital distribution: Albania, Austria, Bulgaria, Croatia, Germany, Greece, Hungary, France, Italy, Morocco, Portugal, Spain, Switzerland (Carpenter, 2001; Gusenleitner, 2013), Slovenia (Gogala, 2022).

***Celonites* (*Celonites*) *andreasmuelleri* Mauss, 2013**

Celonites andreasmuelleri Mauss, 2013: 80, ♀, ♂; "Jordan, Wadi al Hasa S Al-Karak, 30°54'N, 35° 41'E".

Distribution in the Middle East: Israel-Palestine, Jordan (Mauss, 2013).

Extralimital distribution: None.

***Celonites* (*Eucelonites*) *cagrii* Mauss and Yildirim in Mauss *et al.*, 2022**

Celonites cagrii Mauss and Yildirim, 2022 [in: Mauss *et al.*, 2022]: 115–130, ♀, ♂; Turkey.

Distribution in the Middle East: Turkey (Mauss *et al.*, 2022).

Extralimital distribution: Armenia (Mauss *et al.*, 2022).

***Celonites* (*Eucelonites*) *clarus* Gusenleitner, 1973**

Celonites clarus Gusenleitner, 1973: 60, 66, ♀; Iran.

Catalogues with Iranian records: Ebrahimi and Carpenter (2008), Samin *et al.* (2020), Rahmani *et al.* (2020).

Distribution in Iran: Tehran (Gusenleitner, 1973).

Distribution in the Middle East: Known only from Iran (Gusenleitner, 1973, 2013; Carpenter, 2001).

Extralimital distribution: None.

***Celonites* (*Eucelonites*) *cyprius* de Saussure, 1854**

Celonites cyprius de Saussure, 1854: 4, ♂; Cyprus.

Catalogues with Iranian records: Ebrahimi and Carpenter (2008 as *Celonites cyprius smyrnensis* Richards, 1962), Samin *et al.* (2020), Rahmani *et al.* (2020).

Distribution in Iran: Alborz, Fars, Razavi Khorasan (Gusenleitner, 1973 as *Celonites cyprius smyrnensis* Richards, 1962), Tehran (Ebrahimi and Carpenter, 2008; Modarres Awal, 2012 both as *C. cyprius smyrnensis*).

Distribution in the Middle East: Cyprus (de Saussure, 1854; Richards, 1962 as *C. cyprius smyrnensis*; Carpenter, 2001; Gusenleitner, 2013; Fateryga, 2025), Iran (see references above, and Gusenleitner, 2013 as *C. c. smyrnensis*), Israel-Palestine (Richards, 1962; Bytinski-Salz and Gusenleitner, 1971; Gusenleitner, 2013; Carpenter, 2001, all as *C. c. smyrnensis*), Lebanon, Syria (Carpenter, 2001; Gusenleitner, 2013, both as *C. c. smyrnensis*), Turkey (Richards, 1962; Gusenleitner, 1966, 1997, 2013; Yildirim and Özbek, 1997; Yildirim and Kojima, 1999; Carpenter, 2001; Yildirim and Gusenleitner, 2009, 2012, all as *C. c. smyrnensis*; Fateryga, 2025).

Extralimital distribution: Armenia, Greece (Carpenter, 2001; Gusenleitner, 2013, both as *C. c. smyrnensis*).

Comments: *Celonites cyprius smyrnensis* Richards, 1962 was treated as a synonym of *Celonites cyprius* by Fateryga (2025).

***Celonites* (*Celonites*) *discretus* Gusenleitner, 1973**

Celonites discretus Gusenleitner, 1973: 62, ♀; Iran.

Catalogues with Iranian records: Ebrahimi and Carpenter (2008), Samin *et al.* (2020), Rahmani *et al.* (2020).

Distribution in Iran: Fars (Gusenleitner, 1973).

Distribution in the Middle East: Known only from Iran (Gusenleitner, 1973, 2013; Carpenter, 2001).

Extralimital distribution: None.

***Celonites* (*Celonites*) *fischeri* Spinola, 1838**

Celonites fischeri Spinola, 1838: 505, ♀; Egypt.

Distribution in the Middle East: Cyprus (Carpenter, 2001; Gusenleitner, 2013; Mauss and Müller, 2014; Fateryga, 2025), Egypt (Spinola, 1838; de Saussure, 1855 as *Celonites savignyi*; Carpenter, 2001; Gusenleitner, 2013), Israel-Palestine (Spinola, 1838; de Saussure, 1855 as *Celonites savignyi*; Carpenter, 2001; Gusenleitner, 2013), Jordan (Gusenleitner, 2013), Saudi Arabia (Carpenter, 2001; Carpenter and Gadallah, 2020).

Extralimital distribution: Algeria, Libya, Tunisia (Carpenter, 2001).

***Celonites* (*Eucelonites*) *foveolatus nigrior* Richards, 1962**

Celonites foveolatus nigrior Richards, 1962: 218, 234, ♀; Israel: Jericho.

Distribution in the Middle East: Israel-Palestine (Richards, 1962; Bytinski-Salz and Gusenleitner, 1971,; Carpenter, 2001), Turkey (Gusenleitner, 1992; Yildirim and Özbek, 1997; Yildirim and Kojima, 1999; Carpenter, 2001).

Extralimital distribution: None.

Comments: Based on Fateryga (2025), the taxonomic status of the subspecies *Celonites foveolatus nigrior* Richards, 1962 described from Israel-Palestine requires a special study. Because of its small size (Richards, 1962), it may represent a distinct species (Fateryga, 2025).

***Celonites* (*Eucelonites*) *hamanni* Gusenleitner, 1973**

Celonites hamanni Gusenleitner, 1973: 60, 65, 66, ♀, ♂; "Türkei Mut".

Distribution in the Middle East: Iraq (Selis *et al.*, 2025), Israel-Palestine (Gusenleitner, 2013), Jordan (Gusenleitner, 1992, 2013; Carpenter, 2001; Selis *et al.*, 2025), Syria (Gusenleitner, 1973; Carpenter, 2001; Selis *et al.*, 2025), Turkey (Gusenleitner, 1973, 2013; Yildirim and Özbek, 1997 as *C. hammani* [sic]; Yildirim and Kojima, 1999; Carpenter, 2001; Selis *et al.*, 2025).

Extralimital distribution: None.

***Celonites* (*Celonites*) *hermon* Gusenleitner, 2002**

Celonites (*Celonites*) *hermon* Gusenleitner, 2002: 335, ♀; "Israel: Mt. Hermon, 1900 m".

Distribution in the Middle East: Israel-Palestine (Gusenleitner, 2002, 2013), Lebanon, Syria (Mauss and Prosi, 2018).

Extralimital distribution: None.

***Celonites* (*Eucelonites*) *iranus* Gusenleitner, 2018**

Celonites iranus Gusenleitner, 2018: 303, 306, ♀; Iran.

Catalogues with Iranian records: Samin *et al.* (2020), Rahmani *et al.* (2020).

Distribution in Iran: Kerman (Gusenleitner, 2018).

Distribution in the Middle East: Known only from Iran (Gusenleitner, 2018).

Extralimital distribution: None.

***Celonites* (*Eucelonites*) *jousseaumei jousseaumei* du Buysson, 1906**

Celonites jousseaumei du Buysson, 1906: 104, ♀, ♂; Obock [Djibouti].

Catalogues with Iranian records: Ebrahimi and Carpenter (2008), Samin *et al.* (2020), Rahmani *et al.* (2020).

Distribution in Iran: Hormozgan (Ebrahimi and Carpenter, 2008; Modarres Awal, 2012).

Distribution in the Middle East: Israel (Carpenter, 2001), Oman (Richards, 1985; Carpenter, 2001; Gusenleitner, 2005, 2013; Carpenter and Gadallah, 2020), Qatar (Richards, 1985; Carpenter, 2001; Carpenter and Gadallah, 2020), Saudi Arabia (Richards, 1985; Carpenter, 2001; Gusenleitner, 2013; Carpenter and Gadallah, 2020), United Arab Emirates (Richards, 1985; Carpenter, 2001; Gillett and Gillett, 2005; Howarth and Gillett, 2008; Gusenleitner, 2010, 2013; Gess and Roosenschoon, 2016; Schmid-Egger, 2017; Carpenter and Gadallah, 2020), Yemen (Giordani Soika, 1957 as *C. jousseaumei asrensis*; Richards, 1985; Carpenter, 2001; Gusenleitner, 2013; Carpenter and Gadallah, 2020).

Extralimital distribution: Algeria, Djibouti, Mali, Senegal, Sudan (Carpenter, 2001; Gusenleitner, 2013).

***Celonites* (*Celonites*) *laetus* Panfilov, 1968**

Celonites laetus Panfilov, 1968: 40, ♀; "Нахичевань, окрестности Джульфы, пустынные предгорья" [Azerbaijan, Nakhchivan, desert foothills].

Distribution in the Middle East: Turkey (Gusenleitner, 1973, 2013; Yildirim and Özbek, 1997; Yildirim and Kojima, 1999; Carpenter, 2001 all as *Celonites* (*Celonites*) *phlomis* Gusenleitner, 1973; Fateryga, 2025 – synonymy).

Extralimital distribution: Azerbaijan.

Comments: *Celonites phlomis* Gusenleitner, 1973 = *Celonites laetus* (syn. nov. by Fateryga, 2025).

***Celonites* (*Celonites*) *longipilis* Gusenleitner, 1973**

Celonites longipilis Gusenleitner, 1973: 59, ♀; Iran.

Catalogues with Iranian records: Ebrahimi and Carpenter (2008), Samin et al. (2020), Rahmani et al. (2020).

Distribution in Iran: Fars (Gusenleitner, 1973).

Distribution in the Middle East: Known only from Iran (Gusenleitner, 1973, 2013; Carpenter, 2001).

Extralimital distribution: None.

***Celonites* (*Eucelonites*) *osseus* Morawitz, 1888**

Celonites osseus Morawitz, 1888: 268–269, ♀; "in territorio transcaspico (Tschikischljar)" [Turkmenistan].

Catalogues with Iranian records: None.

Distribution in Iran: Alborz [Elburz], Fars, Golestan (Mauss et al., 2022).

Distribution in the Middle East: Iran (Mauss et al., 2022; Fateryga, 2025).

Extralimital distribution: Armenia, Turkmenistan (Carpenter, 2001; Mauss et al., 2022; Fateryga, 2025).

***Celonites* (*Celonites*) *persicus* Richards, 1962**

Celonites persicus Richards, 1962: 221, 241, ♀; Iran.

Catalogues with Iranian records: Ebrahimi and Carpenter (2008), Samin et al. (2020), Rahmani et al. (2020).

Distribution in Iran: Khuzestan (Richards, 1962).

Distribution in the Middle East: Known only from Iran (Richards, 1962; Carpenter, 2001; Gusenleitner, 2013; Fateryga, 2025).

Extralimital distribution: None.

***Celonites* (*Celonites*) *pictus* Richards, 1962**

Celonites pictus Richards, 1962: 226, ♂; Senegal.

Catalogues with Iranian records: Ebrahimi and Carpenter (2008 as Celonites pictus rufiventris Gusenleitner, 1992), Samin et al. (2020), Rahmani et al. (2020).

Distribution in Iran: Hormozgan (Gusenleitner, 1992 as Celonites pictus rufiventris).
Distribution in the Middle East: Iran (Gusenleitner, 1992), Oman, Saudi Arabia (Carpenter, 2001; Carpenter and Gadallah, 2020), United Arab Emirates (Gusenleitner, 2010, 2013).
Extralimital distribution: Morocco, Senegal (Carpenter, 2001; Schmid-Egger, 2017).

***Celonites* (*Eucelonites*) *rugiceps* Bischoff, 1928**
Celonites rugipes Bischoff, 1928: 86, ♀; Kreta.
Distribution in the Middle East: Cyprus (Richards, 1962; Carpenter, 2001; Gusenleitner, 2013), Jordan (Carpenter, 2001), Turkey (Bischoff, 1928; Richards, 1962; Gusenleitner, 1997, 2013; Yildirim and Özbek, 1997; Yildirim and Kojima, 1999; Carpenter, 2001).
Extralimital distribution: Bulgaria, Croatia, Greece, Serbia (Carpenter, 2001; Mauss and Prosi, 2013).

***Celonites* (*Eucelonites*) *semenovi* Kostylev, 1935**
Celonites (Eucelonites) semenovi Kostylev, 1935: 112 [♀]; "Персия" [Persia = Iran].
Catalogues with Iranian records: Ebrahimi and Carpenter (2008), Samin et al. (2020), Rahmani et al. (2020).
Distribution in Iran: Hormozgan (Gusenleitner, 1992 as *Celonites (Celonites) pictus rufiventris* Gusenleitner, 1992 - synonymy), Kerman (Kostylev, 1935; Fateryga, 2025).
Distribution in the Middle East: Iran (see references above, and Carpenter, 2001), United Arab Emirates (Carpenter, 2001; Gusenleitner, 2010, 2013; Schmid-Egger, 2017 all as *Celonites (Celonites) pictus rufiventris* Gusenleitner, 1992).
Extralimital distribution: Morocco (Schmid-Egger, 2017 as *C. pictus rufiventris).*
Comments: *Celonites semenovi* is senior synonym of *Celonites pictus rufiventris* Gusenleitner, 1992 (Fateryga, 2025).

***Celonites* (*Celonites*) *spinosus* Gusenleitner, 1966**
Celonites spinosus Gusenleitner, 1966: 359, ♀; Turkey.
Distribution in the Middle East: Known only from Turkey (Gusenleitner, 1966, 2013).
Extralimital distribution: None.

***Celonites* (*Eucelonites*) *syriacus* Gusenleitner, 2012**
Celonites (Eucelonites) syriacus Gusenleitner, 2012: 319, 322, ♀; "Syrien: 40 km W Homs".
Distribution in the Middle East: Jordan, Syria (Gusenleitner, 2012, 2013)
Extralimital distribution: None.

***Celonites* (*Celonites*) *tauricus* Kostylev, 1935**
Celonites (Celonites) abbreviatus tauricus Kostylev, 1935: 108, ♀; [Crimea].
Catalogues with Iranian records: Samin et al. (2020), Rahmani et al. (2020).
Distribution in Iran: Alborz (Mauss et al., 2016).
Distribution in the Middle East: Cyprus (Carpenter, 2001; Fateryga, 2017a, b, 2018; Fateryga et al., 2021, 2022), Iran (Mauss et al., 2016; Fateryga, 2017a, b, 2025; Fateryga et al., 2021), Syria (Carpenter, 2001; Fateryga, 2017a, b, 2018; Fateryga et al., 2021, 2022), Turkey (Gusenleitner, 1966, 1967 both as *C. (C.) spinosus,* 1973 as *C. abbreviatus invitus* and as *C. (C.) spinosus*; Yildirim and Özbek, 1997 as *abbreviatus invitus* and as *C. spinosus*; Yildirim and Kojima, 1999 as *C. spinosus*; Yildirim and Gusenleitner, 2004 as *C. spinosus*; Yildirim and Gusenleitner, 2012 as *C. spinosus*; Fateryga, 2017a, b, 2018; Fateryga et al., 2021, 2022).
Extralimital distribution: Armenia, Azerbaijan, Greece, Russia (Fateryga, 2017a, b, 2018, 2025; Fateryga et al., 2021, 2022), Georgia (Fateryga et al., 2022; Fateryga, 2025), Kazakhstan, Uzbekistan (Fateryga, 2025), Slovenia (Gogala, 2022).

***Celonites* (*Eucelonites*) *tenebricosus* Gusenleitner, 2012**
Celonites (Eucelonites) tenebricosus Gusenleitner, 2012: 319, 320, ♀, ♂; "Türkei: Tanin-Tanin-Paß, 2500 m".

Distribution in the Middle East: Known only from Turkey (Gusenleitner, 2012, 2013).
Extralimital distribution: None.

***Celonites* (*Eucelonites*) *yemenensis yemenensis* Giordani Soika, 1957**
Celonites yemenensis Giordani Soika, 1957: 472, ♂, ♀; "Yemen: Ta'izz, on road to Mocha, 4100 ft.".
Distribution in the Middle East: Oman (Carpenter, 2001; Gusenleitner, 2005, 2013; Carpenter and Gadallah, 2020), Saudi Arabia (Richards, 1985; Carpenter, 2001; Carpenter and Gadallah, 2020), United Arab Emirates (Carpenter, 2001; Gusenleitner, 2010; Gess and Roosenschoon, 2016; Schmid-Egger, 2017; Carpenter and Gadallah, 2020), Yemen (Giordani Soika, 1957; Richards, 1962, 1985; Gusenleitner, 2013; Carpenter and Gadallah, 2020).
Extralimital distribution: None.

Genus *Ceramius* Latreille, 1810

Ceramius Latreille, 1810: 329, 438. Type species: *Ceramius fonscolombei* Latreille, 1810, by original designation.

***Ceramius bureschi* Atanassov, 1938**
Ceramius bureschi Atanassov, 1938: 114, ♂; "Bulgarien, im Park des Palais Wrana bei Sofia".
Distribution in the Middle East: Turkey (Richards, 1962; Gusenleitner, 1966 as *C. bureschi* and as *C. bureschi lycaonis*; Yildirim, 1998; Yildirim and Kojima, 1999; Carpenter, 2001).
Extralimital distribution: Bulgaria, Greece (Carpenter, 2001).

***Ceramius caucasicus* André, 1884**
Ceramius caucasicus André, 1884: 820, ♂; Caucase [Armenia].
Catalogues with Iranian records: Ebrahimi and Carpenter (2008), Samin *et al.* (2020), Rahmani *et al.* (2020).
Distribution in Iran: Alborz (Gusenleitner, 1973), East Azarbaijan (Ebrahimi and Carpenter, 2008; Modarres Awal, 2012).
Distribution in the Middle East: Iran (see references above; Carpenter, 2001; Fateryga *et al.*, 2021, 2022), Turkey (Richards, 1962; Gusenleitner, 1966, 1967, 1973; Yildirim and Özbek, 1997; Yildirim and Kojima, 1999; Carpenter, 2001; Mauss *et al.*, 2005; Yildirim and Gusenleitner, 2012; Gusenleitner, 2013; Fateryga *et al.*, 2021, 2022).
Extralimital distribution: Armenia (Carpenter, 2001; Fateryga *et al.*, 2021, 2022), Azerbaijan (Fateryga *et al.*, 2021, 2022).

***Ceramius fonscolombei* Latreille, 1810**
Ceramius fonscolombei Latreille, 1810: 438, sex not stated [♂]; "aux environs d'Aix".
Catalogues with Iranian records: Samin *et al.* (2020).
Distribution in Iran: Isfahan, Kerman (Abd-Rabou *et al.*, 2005 as *Ceramius capicola* Brauns, 1902; *Ceramius hispanicus* Dusmet, 1909 – misidentifications), Markazi (Abd-Rabou *et al.*, 2005), Semnan (Samin *et al.*, 2020).
Distribution in the Middle East: Iran (see references above).
Extralimital distribution: France, Spain, Portugal (Carpenter, 2001).

***Ceramius palaestinensis* (Giordani Soika, 1957)**
Paraceramius palaestinensis Giordani Soika, 1957: 172, ♀; "Palestine: Ein Geb".
Distribution in the Middle East: Israel-Palestine (Giordani Soika, 1957 as *Paraceramius paleastinensis*; Bytinski-Salz and Gusenleitner, 1971; Carpenter, 2001), Syria (Carpenter, 2001), Turkey (Gusenleitner, 1992; Yildirim and Özbek, 1997; Yildirim and Kojima, 1999; Carpenter, 2001; Yildirim and Gusenleitner, 2004, 2007; Mauss *et al.*, 2022; Kaplan and Yildirim, 2020).
Extralimital distribution: None.

***Ceramius tuberculifer* de Saussure, 1854**

Ceramius tuberculifer de Saussure, 1854: pl. iii, fig. 6, ♂; "L'Espagne, le midi de la France".
Catalogues with Iranian records: Samin *et al.* (2020).
Distribution in Iran: East Azarbaijan, Southern Khorasan (Samin *et al.*, 2020), Sistan & Baluchestan (Abd-Rabou *et al.*, 2005).
Distribution in the Middle East: Iran (Abd-Rabou *et al.*, 2005; Samin *et al.*, 2020).
Extralimital distribution: France (de Saussure, 1854; Carpenter, 2001), Spain (Carpenter, 2001).

Genus *Jugurtia* de Saussure, 1854

Jugurtia de Saussure, 1854: 19, 37, 42–44. Type species: *Celonites oraniensis* Lepeletier, 1841, by subsequent designation of Ashmead, 1902.

***Jugurtia escalerae* Meade-Waldo, 1910**

Jugurtia escalerae Meade-Waldo, 1910: 33, ♀; Iran.
Catalogues with Iranian records: Ebrahimi and Carpenter (2008), Samin *et al.* (2020), Rahmani *et al.* (2020).
Distribution in Iran: Khuzestan (Meade-Waldo, 1910), Kordestan (Ebrahimi and Carpenter, 2008; Modarres Awal, 2012), Sistan & Baluchestan (Blüthgen and Gusenleitner, 1970), Southwest of Iran (Gusenleitner, 1992).
Distribution in the Middle East: Iran (see references above, and Richards, 1962; Carpenter, 2001; Fateryga, 2025), Turkey (Gusenleitner, 1988, 1992, 2013; Yildirim and Kojima, 1999; Carpenter, 2001).
Extralimital distribution: None.

***Jugurtia eurycara* Kostylev, 1935**

Jugurtia eurycara Kostylev, 1935: 97, ♀, ♂; "Caucasus, Armen. Gab." [Armenia].
Catalogues with Iranian records: Ebrahimi and Carpenter (2008), Samin *et al.* (2020), Rahmani *et al.* (2020).
Distribution in Iran: Alborz (Gusenleitner, 1973), Hamadan (Abd-Rabou *et al.*, 2005 as *Jugurtia braunsi* (von Schulthess, 1922) – misidentification), Markazi (Ebrahimi and Carpenter, 2008; Modarres Awal, 2012).
Distribution in the Middle East: Iran (see references above; Gusenleitner, 1992; Carpenter, 2001; Fateryga *et al.*, 2021, 2022), Turkey (Gusenleitner, 1988, 1992, 2013; Yildirim, 1998; Yildirim and Kojima, 1999; Carpenter, 2001; Yildirim and Gusenleitner, 2004; Fateryga *et al.*, 2021, 2022).
Extralimital distribution: Armenia (Carpenter, 2001; Fateryga *et al.*, 2021, 2022; Fateryga, 2025), Azerbaijan (Fateryga *et al.*, 2021, 2022).

***Jugurtia irana* Kostylev, 1935**

Jugurtia irana Kostylev, 1935: 99, ♀; "Via Karvandar-Bampur, S.-O. Persien", "Tshanshen-Sadk, Bampur, S.-O. Persien", "Kuusha-Larumba, Bampur, S.-O. Persien", "Kurljut-Langer, Chorasan, O. Persien" [Iran].
Catalogues with Iranian records: Ebrahimi and Carpenter (2008), Samin *et al.* (2020), Rahmani *et al.* (2020).
Distribution in Iran: Alborz (Gusenleitner, 1973), Razavi Khorasan, Sistan & Baluchestan (Kostylev, 1935; Fateryga, 2025).
Distribution in the Middle East: Iran (see references above, and Carpenter, 2001; Fateryga *et al.*, 2021, 2022).
Extralimital distribution: Azerbaijan (Fateryga *et al.*, 2021, 2022).

***Jugurtia jemenensis* Kostylev, 1935**

Jugurtia jemenensis Kostylev, 1935: 99, ♀; "Jemen, Sanaa" [Yemen].
Distribution in the Middle East: Israel (Carpenter, 2001), Oman (Richards, 1985; Carpenter, 2001; Gusenleitner, 2010; Schmid-Egger, 2017; Monks *et al.*, 2019; Carpenter and Gadallah, 2020),

Saudi Arabia (Richards, 1985; Carpenter, 2001; Carpenter and Gadallah, 2020), United Arab Emirates (Richards, 1985; Carpenter, 2001; Gillett and Gillett, 2005; Howarth and Gillett, 2008; Gusenleitner, 2010; Schmid-Egger, 2017; Carpenter and Gadallah, 2020), Yemen (Kostylev, 1935; Richards, 1985; Carpenter, 2001; Carpenter and Gadallah, 2020; Fateryga, 2025).
Extralimital distribution: Algeria (Carpenter, 2001).

***Jugurtia punctuosa* Gusenleitner, 2011**
Jugurtia punctuosa Gusenleitner, 2011: 223, 226, ♀; "Syria S, Kafr, Suwayda".
Distribution in the Middle East: Known only from Syria (Gusenleitner, 2011).
Extralimital distribution: None.

***Jugurtia zarudnyi* Kostylev, 1935**
Jugurtia zarudnyi Kostylev, 1935: 94, ♀, ♂; "Chorasan", "Kermän, Prov. Mokran" [Iran].
Catalogues with Iranian records: Ebrahimi and Carpenter (2008), Samin *et al.* (2020), Rahmani *et al.* (2020).
Distribution in Iran: Hormozgan (Gusenleitner, 1973, 1992), Kerman, Razavi Khorasan (Kostylev, 1935; Gusenleitner, 1992; Fateryga, 2025).
Distribution in the Middle East: Known only from Iran (see references above, and Carpenter, 2001).
Extralimital distribution: None.

Genus *Masaris* Fabricius, 1793

Masaris Fabricius, 1793: 6, 283, 284. Type species: *Masaris vespiformis* Fabricius, 1793, by subsequent monotypy.

***Masaris aegyptiacus aegyptiacus* Meade-Waldo, 1911**
Masaris vespiformis aegyptiacus Meade-Waldo, 1911: 447, ♂; Cairo: Egypt.
Distribution in the Middle East: Egypt (Meade-Waldo, 1911 as *M. vespiformis aegyptiacus*; Carpenter, 2001), Israel-Palestine, Jordan (Carpenter, 2001).
Extralimital distribution: None.

***Masaris aegyptiacus arabicus* Giordani Soika, 1957**
Masaris vespiformis arabicus Giordani Soika, 1957: 472, ♀; "Saudi Arabia: Jebel Shammar Province, Hâil district".
Distribution in the Middle East: Known only from Saudi Arabia (Giordani Soika, 1957 as *M. vespiformis arabicus*; Carpenter, 2001; Carpenter and Gadallah, 2020).
Extralimital distribution: None.

***Masaris agnolii* Selis, 2019**
Masaris agnolii Selis, 2019: 395, ♂; Iran.
Catalogues with Iranian records: Samin *et al.* (2020), Rahmani *et al.* (2020).
Distribution in Iran: Chaharmahal & Bakhtiari (Selis, 2019).
Distribution in the Middle East: Known only from Iran (Selis, 2019).
Extralimital distribution: None.

***Masaris vespiformis* Fabricius, 1793**
Masaris vespiformis Fabricius, 1793: 283, sex not stated; "in Barbariae floribus".
Distribution in the Middle East: Egypt (de Saussure, 1853 as *Erynnis romandi*; Carpenter, 2001).
Extralimital distribution: Algeria, Morocco (Carpenter, 2001).

Genus *Quartinia* André, 1884

Quartinia André, 1884: 811, 822. Type species: *Quartinia dilecta* André, by monotypy.

***Quartinia amiae* Carpenter, 2003**
Quartinia amiae Carpenter, 2003: 143, ♀; "Egypt: Wadi Degla, 29°56'21"N 31°25'18"E . . . 209 m".

Distribution in the Middle East: Known only from Egypt (Carpenter, 2003).
Extralimital distribution: None.

***Quartinia arabica* Gusenleitner, 2010**

Quartinia arabica Gusenleitner, 2010: 460, ♀, ♂; [United Arab Emirates:] "N of Ajman, 25°26'N 55°29'E".

Distribution in the Middle East: Known only from United Arab Emirates (Gusenleitner, 2010; Schmid-Egger, 2017).

Extralimital distribution: None.

***Quartinia araxana* Giordani Soika, 1960**

Quartinia araxana Giordani Soika, 1960: 133, ♀, ♂; Caucaso.

Catalogues with Iranian records: Ebrahimi and Carpenter (2008), Samin *et al.* (2020), Rahmani *et al.* (2020).

Distribution in Iran: Alborz (Gusenleitner, 1985).

Distribution in the Middle East: Iran (Gusenleitner, 1985; Carpenter, 2001), Turkey (Richards, 1962; Gusenleitner, 1973, 2013; Yildirim and Özbek, 1997; Yildirim and Kojima, 1999; Carpenter, 2001).

Extralimital distribution: Armenia (Carpenter, 2001).

***Quartinia atrata* Gusenleitner, 1998**

Quartinia atrata Gusenleitner, 1998: 503, ♀; Iran.

Catalogues with Iranian records: Ebrahimi and Carpenter (2008), Samin *et al.* (2020), Rahmani *et al.* (2020).

Distribution in Iran: Fars (Gusenleitner, 1998).

Distribution in the Middle East: Known only from Iran (Gusenleitner, 1998; Carpenter, 2001).

Extralimital distribution: None.

***Quartinia candida* Gusenleitner, 1997**

Quartinia candida Gusenleitner, 1997: 137, 147, ♀; "Israel: 45 km N Elat, Sand tunes [dunes] E Qetura, 29°58′/35°06′".

Distribution in the Middle East: Known only from Israel-Palestine (Gusenleitner, 1997; Carpenter, 2001).

Extralimital distribution: None.

***Quartinia gusenleitneri* Schmid-Egger, 2017**

Quartinia gusenleitneri Schmid-Egger, 2017: 341, 343, 344, ♂; [United Arab Emirates] "Wadi Shawkah, 25.13°N 56.02°E".

Distribution in the Middle East: United Arab Emirates (Schmid-Egger, 2017).

Extralimital distribution: None.

***Quartinia haemorrhoa* Gusenleitner, 1997**

Quartinia haemorrhoa Gusenleitner, 1997: 137, 144, ♀; "Israel: Dead Sea, En Zeelim, 5 km N Massada, 31°23′/35°20′".

Distribution in the Middle East: Known only from Israel-Palestine (Gusenleitner, 1997; Carpenter, 2001).

Extralimital distribution: None.

***Quartinia halicticeps* Giordani Soika, 1939**

Quartinia halicticeps Giordani Soika, 1939: 9, 12, ♀; "Egitto: Gabal Elba (Wadi Abrag)".

Distribution in the Middle East: Known only from Egypt (Giordani Soika, 1939; Carpenter, 2001).

Extralimital distribution: None.

***Quartinia irana* Gusenleitner, 2012**

Quartinia irana, Gusenleitner, 2012: 319, ♀; Iran.

Catalogues with Iranian records: Samin *et al.* (2020), Rahmani *et al.* (2020).
Distribution in Iran: Markazi (Gusenleitner, 2012).
Distribution in the Middle East: Known only from Iran (Gusenleitner, 2012).
Extralimital distribution: None.

***Quartinia libanica* Richards, 1964 (1962)**
Quartinia libanica Richards, 1964 (1962): 83, ♂, ♀; "N. Lebanon: Becharré, 5000 ft.".
Distribution in the Middle East: Cyprus, Lebanon (Richards, 1962 (1964); Carpenter, 2001).
Extralimital distribution: None.

***Quartinia longiceps* Gusenleitner, 2010**
Quartinia longiceps Gusenleitner, 2010: 462, ♀; "United Arab Emirates, Wadi Maidaq, 25°18'N 56°07'E".
Distribution in the Middle East: Known only from United Arab Emirates (Gusenleitner, 2010; Schmid-Egger, 2017).
Extralimital distribution: None.

***Quartinia mochii* Giordani Soika, 1939**
Quartinia mochii Giordani Soika, 1939: 9, 10, ♂; "Egitto: Abu Rawash".
Distribution in the Middle East: Known only from Egypt (Giordani Soika, 1939; Carpenter, 2001).
Extralimital distribution: None.

***Quartinia monnerati* Gusenleitner, 2016**
Quartinia monnerati Gusenleitner, 2016: 405, ♀; "Oman, Qitbit (St 0), 19.15560N 54.50807E".
Distribution in the Middle East: Known only from Oman (Gusenleitner, 2016; Carpenter and Gadallah, 2020).
Extralimital distribution: None.

***Quartinia nilotica* Fischer, 1964**
Quartinia nilotica Fischer, 1964: 611, ♀; "Khor Musa Pascha S. v. Wadi-Halfa".
Distribution in the Middle East: Known only from Egypt (Fischer, 1964; Carpenter, 2001).
Extralimital distribution: None.

***Quartinia nitidiuscula* Gusenleitner, 2010**
Quartinia nitens Gusenleitner, 2010: 464, ♀; "United Arab Emirates, Wadi Shawkah [Al Kahri, Shoka lake], 310 m, 25°06,41′N 56°02,6′E".
Quartinia nitidiuscula Gusenleitner, 2010: 1350, replacement name for *nitens*.
Quartinia nitida Gusenleitner, 2013: 1025, 1035, replacement name for *nitens*.
Distribution in the Middle East: Known only from United Arab Emirates (Gusenleitner, 2010 as *Q. nitens*).
Extralimital distribution: None.
Comments: Modern checklists treat the valid name as *Q. nitidiuscula*, with *Q. nitida* as a later, unnecessary replacement.

***Quartinia nubiana* Richards, 1962**
Quartinia nubiana Richards, 1962: 133, 134, 138, ♀; "Egypt: Nubian Desert, Korosko to Abu Hamed".
Distribution in the Middle East: Egypt (Richards, 1962; Carpenter, 2001), Saudi Arabia (Richards, 1985; Carpenter, 2001; Carpenter and Gadallah, 2020), United Arab Emirates (Richards, 1985; Carpenter, 2001; Gillett and Gillett, 2005; Howarth and Gillett, 2008; Gusenleitner, 2010, 2012; Gess and Roosenschoon, 2016; Schmid-Egger, 2017; Carpenter and Gadallah, 2020).
Extralimital distribution: Libya, Sudan, Tunisia (Carpenter, 2001).

Quartinia orientalis **Gusenleitner, 1973**

Quartinia orientalis Gusenleitner, 1973: 56, ♂, ♀; "Türkei Ürgüp".

Distribution in the Middle East: Turkey (Gusenleitner, 1973; Yildirim and Özbek, 1997; Yildirim and Kojima, 1999; Carpenter, 2001).

Extralimital distribution: Afghanistan (Carpenter, 2001).

Quartinia popovi **Richards, 1962**

Quartinia popovi Richards, 1962: 143, ♀; "District of Kondara" [Tajikistan].

Distribution in the Middle East: Turkey (Carpenter, 2001; Yildirim and Gusenleitner, 2001, 2012).

Extralimital distribution: Kyrgyzstan (Fateryga and Proshchalykin, 2024), Tajikistan (Carpenter, 2001; Fateryga and Proshchalykin, 2024; Fateryga, 2025).

Comments: Based on Fateryga and Proshchalykin (2024), the presence of *Q. popovi* in Turkey was doubtful based on a misidentification. Re-examination of specimens recorded by Yildirim and Gusenleitner (2001) confirmed the presence of this species in Turkey.

Quartinia separata **Gusenleitner, 1997**

Quartinia separata Gusenleitner, 1997: 137, 146, ♀, ♂; "Syria west, 50 km S Homs".

Distribution in the Middle East: Iraq (Selis *et al.*, 2025), Syria (Gusenleitner, 1997, 2013; Carpenter, 2001; Selis *et al.*, 2025), Turkey (Gusenleitner, 1997, 2013; Yildirim and Kojima, 1999; Carpenter, 2001; Yildirim and Gusenleitner, 2012; Selis *et al.*, 2025).

Extralimital distribution: None.

Quartinia soikai **Richards, 1964 (1962)**

Quartinia soikai Richards, 1964 (1962): 84, ♀; Turkey.

Catalogues with Iranian records: Samin *et al.* (2020), Rahmani *et al.* (2020).

Distribution in Iran: Alborz (Gusenleitner, 1973).

Distribution in the Middle East: Iran (Gusenleitner, 1973; Carpenter, 2001), Turkey (Richards, 1962; Gusenleitner, 1973, 2013; Yildirim and Kojima, 1999; Carpenter, 2001).

Extralimital distribution: Greece (Carpenter, 2001).

Quartinia syriaca nitens **Gusenleitner, 1973**

Quartinia syriaca nitens Gusenleitner, 1973: 57, ♂; "Iran, Fars, Monti steppici 5 km N di Persepolis".

Catalogues with Iranian records: Ebrahimi and Carpenter (2008), Samin *et al.* (2020), Rahmani *et al.* (2020).

Distribution in Iran: Fars (Gusenleitner, 1973).

Distribution in the Middle East: Known only from Iran (Gusenleitner, 1973; Carpenter, 2001).

Extralimital distribution: None.

Quartinia syriaca syriaca **Richards, 1964 (1962)**

Quartinia syriaca Richards, 1964 (1962): ♀, ♂; Syria.

Distribution in the Middle East: Lebanon, Syria (Richards, 1964; Carpenter, 2001).

Extralimital distribution: None.

Quartinia thebaica **du Buysson, 1902**

Quartinia thebaica du Buysson, 1902: 141, ♀; "Égypte: Louqsor".

Distribution in the Middle East: Egypt (du Buysson, 1902; Giordani Soika, 1939 as *Q. ochracea*; Carpenter, 2001), Oman (Gusenleitner, 2004).

Extralimital distribution: Algeria, Libya, Morocco (Carpenter, 2001).

Quartinia tricolorata **Giordani Soika, 1954**

Quartinia tricolorata Giordani Soika, 1954: 22, ♀; "Egitto: Gebel Asfar".

Distribution in the Middle East: Known only from Egypt (Giordani Soika, 1954; Carpenter, 2001).

Extralimital distribution: None.

***Quartinia tripolitana sinaitica* Richards, 1964 (1962)**

Quartinia tripolitana sinaitica Richards, 1964 (1962): 83, ♂, ♀; "United Arab Republic, Sinai: Kun Mitla".

Distribution in the Middle East: Known only from Egypt (Richards, 1964 (1962); Carpenter, 2001).

Extralimital distribution: None.

***Quartinia tripolitana tripolitana* Richards, 1962**

Quartinia tripolitana tripolitana Richards, 1962: 134, 140, ♀, ♂; [Libya] "Tripolitania: Tripoli".

Distribution in the Middle East: Egypt (Carpenter, 2001).

Extralimital distribution: Libya (Carpenter, 2001).

***Quartinia tuareg* Giordani Soika, 1954**

Quartinia tuareg Giordani Soika, 1954: 21, ♂, ♀; [Algeria] "Guelta Idjef Melène".

Distribution in the Middle East: Egypt (Carpenter, 2001), United Arab Emirates (Carpenter, 2001; Gusenleitner, 2010; Schmid-Egger, 2017), Yemen (Carpenter, 2001; Gusenleitner, 2007).

Extralimital distribution: Algeria (Carpenter, 2001).

Species Excluded from the Fauna of Iran and the Middle East

In total, eight species of Masarinae are excluded from the fauna of Iran because of prior errors and/or misidentifications.

***Celonites (Eucelonites) crenulatus* Morawitz, 1888**

Celonites crenulatus Morawitz, 1888: 267, ♀; "Territorio Transcaspico (Kiltitschinar)" [Turkmenistan].

Records from Iran: Iran (no specific locality cited) (Carpenter, 2001; Ebrahimi and Carpenter, 2008; Samin *et al.*, 2020; Rahmani *et al.*, 2020; Fateryga and Proshchalykin, 2024).

Comments: Morawitz (1888) recorded *Celonites crenulatus* from Turkmenistan but not Iran while the above-mentioned authors mistakenly reported it from Iran. So, we exclude this species from the fauna of Iran. *Celonites crenulatus* has been recorded from Tajikistan (Fateryga and Proshchalykin, 2024), and Turkmenistan (Morawitz, 1888; Carpenter, 2001).

***Celonites rothschildi* du Buysson, 1906**

Celonites rothschildi du Buysson, 1906: 105, ♂; Afrique orientale anglaise: Lesammise, Rendile [Kenya].

Records from Iran: Isfahan (Gusenleitner, 1992), Iran (Carpenter, 2001; Samin *et al.*, 2020; Rahmani *et al.*, 2020).

Comments: The report of *Celonites rothschildi* in Iran by Carpenter (2001), Samin *et al.* (2020) and Rahmani *et al.* (2020) were based on Gusenleitner's (1992) record. This species was described from Kenya and after that has not been collected from any other country. *C. rothschildi* was considered doubtful for the Iranian fauna by Ebrahimi and Carpenter (2008). Since there are no confirmed records outside Kenya, we exclude it from the fauna of Iran.

***Ceramius bicolor* (Thunberg, 1815)**

Philanthus bicolor Thunberg, 1815: 289, sex not stated (male Uppsala).

Records from Iran: Mazandaran (Abd-Rabou *et al.*, 2005).

Comments: Re-examination of the specimen by Samin *et al.* (2020) as well as in this study, proved identification of *Ceramius bicolor* by Abd-Rabou *et al.* (2005) was misidentification of *Ceramius caucasicus* and we therefore exclude *C. bicolor* from the fauna of Iran and record the data under *C. caucasicus*.

***Ceramius capicola* Brauns, 1902**

Ceramius capicola Brauns, 1902: 278, ♀, ♂ "Willowmore" ["Hier"].

Records from Iran: Isfahan (Abd-Rabou *et al.*, 2005).

Comments: Re-examination of the specimen by Samin *et al.* (2020) as well as in this study, proved identification of *Ceramius capicola* by Abd-Rabou *et al.* (2005) was misidentification of *Ceramius fonscolombei* and we therefore exclude *C. capicola* from the fauna of Iran and record the data under *C. fonscolombei*.

Ceramius hispanicus **Dusmet, 1909**

Ceramius hispanicus Dusmet, 1909: 180, ♀, ♂, "Madrid! y Montarco! (en Madrid). Jaramiel! (en Valladolid)" (Madrid).

Records from Iran: Kerman (Abd-Rabou *et al.*, 2005).

Comments: Re-examination of the specimen by Samin *et al.* (2020), and subsequently in this study, proved that identification of *Ceramius hispanicus* by Abd-Rabou *et al.* (2005) was a misidentification of *Ceramius fonscolombei*. We therefore exclude *C. hispanicus* from the fauna of Iran and record the data under *C. fonscolombei*.

Jugurtia braunsi **(von Schulthess, 1922)**

Ceramiellus braunsi von Schulthess, 1922: 402, ♀, ♂; "Südwestafrika: Groß- und Klein-Namaland, Grootfontein ... Steinkopf" (Berlin).

Records from Iran: Hamadan (Abd-Rabou *et al.*, 2005).

Comments: Re-examination of the specimen by Samin *et al.* (2020), as well as in this study, proved that identification of *Jugurtia braunsi* by Abd-Rabou *et al.* (2005) was a misidentification of *Jugurtia eurycara*. We therefore exclude *J. braunsi* from the fauna of Iran and record the data under *J. eurycara*.

Masarina familiaris **Richards, 1962**

Masarina familiaris Richards, 1962: 270, ♂, ♀; "Cape Province: Camps Bay".

Records from Iran: Golestan (Abd-Rabou *et al.*, 2005).

Comments: Only one female specimen of *Masarina familiaris* was collected by Z. Karimian (Iran Cotton Research Institute) from Gonbad (around cotton fields) in September 2003. Identification of *M. familiaris* by Abd-Rabou *et al.* (2005) was a misidentification. The single specimen was destroyed and can't be a reliable voucher specimen for re-examination. Thus, we exclude this species from the fauna of Iran (see Samin *et al.*, 2020).

Masaris carli **von Schulthess, 1922**

Masaris saussurei Carl, 1921: 41, ♂, ♀; "Zarawchan; Sutkend. Turkestan".

Masaris carli von Schulthess, 1922: 404; replacement name for *saussurei* Carl, 1921.

Records from Iran: Fars (Gusenleitner, 1992, 2002).

Comments: The reports of *Masaris carli* from Iran by Carpenter (2001), Ebrahimi and Carpenter (2008), Samin *et al.* (2020) and Rahmani *et al.* (2020) were based on Gusenleitner's (1992, 2002) record. This species was excluded from the fauna of Iran by Fateryga and Proshchalykin (2024), and so we do not consider it as a part of the Iranian fauna until finding new specimens. That study also established that *Masaris carli* von Schulthess, 1922 = *M. elegans* Gusenleitner, 2002 (syn. nov. by Fateryga and Proshchalykin, 2024). *M. carli* is distributed in Kazakhstan, Uzbekistan (Carpenter, 2001), and Turkmenistan (Fateryga and Proshchalykin, 2024).

CONCLUSION

The Middle Eastern fauna of Masarinae comprises 66 species and subspecies (17.6% of the world species) within five genera. Among them, *Quartinia* and *Celonites*, with 27 and 24 recorded species, respectively, are more diverse than the others, followed by *Jugurtia* (six species), *Ceramius* (five species), and *Masaris* (four species). Additionally, 25 species and subspecies in five genera have been recorded from Iran to date (21.7% of the Palaearctic species, which includes 115 recorded species in five genera (Fateryga, 2017b)). Of these, 10 species are known only from Iran (endemic): *Celonites clarus* Gusenleitner, 1973; *Celonites discretus* Gusenleitner, 1973; *Celonites iranus* Gusenleitner, 2018; *Celonites longipilis* Gusenleitner, 1973; *Celonites persicus* Richards, 1962; *Jugurtia zarudnyi* Kostylev, 1935; *Masaris agnolii* Selis, 2019; *Quartinia atrata* Gusenleitner, 1998; *Quartinia irana* Gusenleitner, 2012; and *Quartinia syriaca nitens* Gusenleitner, 1973.

In total, eight species are excluded from the fauna of Iran as well as the Middle East in this treatment: 1) five species, *Ceramius bicolor*, *Ceramius capicola*, *Ceramius hispanicus*, *Jugurtia braunsi*, and *Masarina familiaris* were recorded by Abd-Rabou *et al.* (2005). These species were erroneously identified due to the low number of specimens examined, use of unreliable identification keys, and incorrect confirmation by specialists based on low-quality photos (see Samin *et al.*, 2020); 2) two species, *Celonites rothschildi* and *Masaris carli*, recorded by Gusenleitner (1992) (see previous section); 3) *Celonites crenulatus* (see previous section).

Among the 31 provinces of Iran, Alborz has the highest diversity with eight recorded species, followed by Fars with six species. Hormozgan and Kerman each have four species; Markazi, Razavi Khorasan, and Sistan & Baluchestan each have three species; East Azerbaijan, Khuzestan, and Tehran, each have two species; and Chaharmahal & Bakhtiari, Golestan, Hamadan, Isfahan, Kordestan, Semnan, Southern Khorasan, and West Azerbaijan each have one species. Twelve remaining provinces have not been carefully sampled and therefore lack any recorded species. We expect many more species to be discovered after faunistic surveys across all regions of Iran, especially in the northern provinces (Golestan, Guilan, and Mazandaran provinces), where the moderate and humid climate supports a rich and diverse flora.

Among the 16 countries of the Middle East, Iran has the most diverse Masarinae fauna, with 25 recorded species and subspecies, followed by Turkey (20 species), Egypt and Israel-Palestine (each with 13 species), United Arab Emirates (11 species), Syria (nine species), Cyprus and Saudi Arabia (each with seven species), Jordan and Oman (each with six species), Lebanon and Yemen (each with four species), Iraq (two species), and Qatar (one species). No species have been recorded from Bahrain and Kuwait to date. Additionally, among the 23 countries of the Middle East and those adjacent to Iran, Turkey shares the greatest number of species with Iran (eight species), followed by Armenia (six species), Azerbaijan and Cyprus (each with four species), Israel-Palestine and United Arab Emirates (each with three species), Oman, Saudi Arabia, Syria, and Turkmenistan (each with two species), and Kazakhstan, Lebanon, Qatar, Russia, and Yemen (each with one species). Currently, no species are known to be shared between Iran and the remaining neighboring countries: Afghanistan, Bahrain, Egypt, Iraq, Jordan, Kuwait, and Pakistan.

BIBLIOGRAPHY

Abd-Rabou, S., Ghahari, H., Mauss, V. and Plant, J. 2005. New records of Apidae, Andrenidae, Sphecidae and Vespidae (Hymenoptera) from Iran. *Egyptian Journal of Agricultural Research* 83(4): 1613–1618.

André, E. 1884. *Species des Hyménoptères d'Europe et Algérie*. Vol. 2. André and André, Beaune.

Ashmead, W.H. 1902. Classification of the fossorial, predaceous and parasitic wasps of the superfamily Vespoidea. *Canadian Entomologist* 34: 219–240.

Atanassov, H. 1938. *Ceramius bureschi*, eine neue Masariden art (Insecta, Hymenoptera) aus der bulgarischen Fauna. *Mitt Kgl Naturw Inst Sofia* 2: 114–118.

Bischoff, H. 1928. Hymenoptera (Vespidae, Scoliidae, Mutillidae, Chrysididae and Evaniidae). In: Roewer, C.F. (ed.), *Zoologische Streifzüge in Attika, Morea und besonders auf der Insel Kreta. II. Abhandlungen des Naturwissenschaftlichen Verein zu Bremen* 27: 85–90.

Blüthgen, P. and Gusenleitner, J. 1970. Über Faltenwespen aus dem Iran (Hym., Diploptera), *Stuttgarter Beiträge zur Naturkunde* 223: 1–13.

Brauns, H. 1902. Die Apidengattung Fidelia Friese (Hym.). *Zeitschrift für Systematische Hymenopterologie und Dipterologie* 2: 374–376.

Bytinski-Salz, H. and Gusenleitner, J. 1971. The Vespoidea of Israel (Hymenoptera). *Israel Journal of Entomology* 6: 239–299.

Carpenter, J.M. 1982. The phylogenetic relationships and natural classification of the Vespoidea (Hymenoptera). *Systematic Entomology* 7: 11–38.

Carpenter, J.M. 1988. The phylogenetic system of the Gayellini (Hymenoptera: Vespidae; Masarinae). *Psyche* 95: 211–241.

Carpenter, J.M. 2001. Checklist of species of the subfamily Masarinae (Hymenoptera: Vespidae). *American Museum Novitates* 3325: 1–39.

Carpenter, J.M. 2003. A new species of pollen wasp from Egypt (Hymenoptera: Vespidae: Masarinae). *Journal of the Kansas Entomological Society* 76(2): 143–146.

Carpenter, J.M. and Gadallah, N.S. 2020. Biodiversity of the aculeate wasps (Hymenoptera: Aculeata) of the Arabian Peninsula: Vespoidea, Vespidae. *Zootaxa* 4754(1): 191–216.

Cowan, D.P. 1991. The solitary and presocial Vespidae, pp. 33–73. In: Ross, K.G. and Matthews, R.W. (eds.), *The social biology of wasps*. Comstock Publishing Associates, Cornell University Press, Ithaca, 696 pp.

de Saussure, H. 1852–1858. Étude sur la Famille des Vespides, 1, 2 & 3. V. Masson, Paris; J. Kessmann, Genève & J. Cherbuliez, Genève, 250 + 470 + 364 pp., 22 + 39 + 16 pls.

de Saussure, H.F. (1853) [Sans titre]. *In*: Bulletin entomologique. *Annales de la Société Entomologique de France*, Série 3: 1, XVII–XXI.

de Saussure, H.F. (1854) Mélanges hyménoptérologiques. *Mémoires de la Société de Physique et d'Histoire Naturelle de Genève*, 14(1): 1–67.

de Villers, C. 1789. Caroli Linnaei entomologia, Faunae Suecicae descriptionibus. Tomus tertius. Piestre et Delamollière, Lugduni [Lyon], 657 pp.

De Villers, C. 1789. *Caroli Linnaei Entomologia, Faunae Suecicae descriptionibus aucta, DD. Scopoli, Geoffroy, de Geer, Fabricii, Schrank, &c. speciebus vel in Systemate non enumeratis, vel nuperrime detectis, vel speciebus Galliae Australis locupleata, generum specierumque rariorum iconibus ornata.* Piestre et Delamollière, Lugduni [= Lyon]. 657 pp.

du Buysson, R. 1902. Descriptions des trois guêpes nouvelles. *Bulletin de la Société Entomologique de France* 1902: 140–141.

du Buysson, R. 1906. Hyménoptêres nouveaux. *Revue d'Entomologie* 25(5): 103–112.

Dusmet, J.M. 1909. Véspidos, Euménidos y Masáridos de España, suplemento segundo. *Actas y Memorias del Primer Congreso de Naturalistas Españoles: Zaragoza*, pp. 163–184.

Ebrahimi, E. and Carpenter, J.M. 2008. Catalog of the vespid wasps of Iran (Hymenoptera, Vespidae). *Zootaxa* 1785: 1–42.

Eickwort, G.C. and Ginsberg, H.S. 1980. Foraging and mating behavior in Apoidea. *Annual Review of Entomology* 25: 421–446.

Fabricius, J.C. 1793. *Entomologia systematica emendata et aucta*. Tom III. Hafniae, 519 pp.

Fateryga, A.V. 2017a. New records of solitary vespid wasps (Hymenoptera: Vespidae: Eumeninae, Masarinae) from Russia and adjacent countries. *Far Eastern Entomologist* 334: 1–16.

Fateryga, A.V. 2017b. Subfamily Masarinae – Pollen wasps, pp. 177–178. In: Belokobylskij, S.A. and Lelej, A.S. (eds), Annotated catalogue of the Hymenoptera of Russia. Volume I. Symphyta and Apocrita: Aculeata. *Proceedings of the Zoological Institute of the Russian Academy of Sciences*, Supplement No. 6, 475 pp.

Fateryga, A.V. 2018. Wasps of the family Vespidae (Hymenoptera) of the Crimean Peninsula. *Entomofauna* 39(1): 193–233.

Fateryga, A.V. 2025. Type specimens of the pollen wasps (Hymenoptera, Vespidae: Masarinae) deposited in the Zoological Institute of the Russian Academy of Sciences and the Zoological Museum of the M.V. Lomonosov Moscow State University. *Zootaxa* 5632(2): 201–242.

Fateryga, A.V., Mauss, V. and Fateryga, V.V. 2022. New distributional records of *Celonites tauricus* (Hymenoptera, Vespidae, Masarinae) and new data on its behaviour at flowers. *Journal of Hymenoptera Research* 92: 241–256.

Fateryga, A.V. and Proshchalykin, M.Y. 2024. New records of the solitary Vespidae (Hymenoptera: Vespidae: Masarinae and Eumeninae s. l.) from Central Asia, with new synonyms and description of a new species of *Eumenes* Latreille. *Zootaxa* 5529(3): 551–569.

Fateryga, A.V., Proshchalykin, M.Y., Maharramov, M.M. and Astafurova, Y.V. 2021. New records of solitary vespid wasps (Hymenoptera: Vespidae: Masarinae and Eumeninae s. l.) from the Nakhchivan Autonomous Republic of Azerbaijan. *Zootaxa* 5027(1): 36–60.

Fischer, M. (1964) Ergebnisse der Zoologischen Nubien-Expedition 1962. Teil XXV Hymenoptera: Masaridae. *Annalen des Naturhistorischen Museums in Wien*, 67: 611–613.

Fischer, M. 1994. Ergebnisse der Zoologischen Nubien-Expedition 1962. Teil XXV. Hymenoptera: Masaridae. *Annalen des Naturhistorischen Museums in Wien* 67: 611–613.

Gess, S.K. 1992. Biogeography of the masarine wasps in southern Africa (Hymenoptera: Vespidae: Masarinae), with particular emphasis on the southern African taxa and on correlations between masarine and forage plant distributions. *Journal of Biogeography* 19: 491–503.

Gess, S.K. 1996. The pollen wasps: ecology and natural history of the Masarinae. Harvard University Press, Cambridge, MA, 340 pp.

Gess, S.K. and Gess, F.W. 1989. Flower visiting by masarid wasps in southern Africa (Hymenoptera: Vespoidea: Masaridae). *Annals of the Cape Provincial Museums (Natural History)* 18: 95–134.

Gess, S.K. and Gess, F.W. 2010. Pollen wasps and flowers in southern Africa, SANBI *Biodiversity Series* 18: 1–147.

Gess, S.K. and Roosenschoon, P.A. 2016. A preliminary survey of flower visiting by aculeate wasps and bees in the Dubai Desert Conservation Reserve, United Arab Emirates. *Journal of Hymenoptera Research* 52: 81–141.

Gillett, M. and Gillett, C. 2005. Insects and other arthropods. In: Hellyer, P. and Aspinall, S. (eds.), *The emirates: A natural history*. Trident Press Ltd., London, pp. 168–195.

Giordani Soika, A. 1939. Le specie egiziane del genere *Quartinia* Grib. (Hymenoptera: Vespidae). *Bulletin de la Société Fouad Ier d'Entomologie* 1939: 9–14.

Giordani Soika, A. 1954. Contributo allo studio dell'Hoggar, Tassili d'Ajjer e Tibesti. Viaggi di A. Giordani Soika nel Sahara VI. *Bollettino del Museo Civico di Storia Naturale di Venezia* 7: 7–37.

Giordani Soika, A. 1957. Hymenoptera: Vespidae, pp. 471–484. In: *Expedition to South-West Arabia 1937–1938*. Volume I. British Museum (Natural History), London.

Giordani Soika, A. 1960. Notule vespidologiche. *Bollettino del Museo Civico di Storia Naturale di Venezia* 11: 35–102.

Giordani Soika, A. 1974. Revisione della sottofamiglia Raphiglossinae (Hym., Vesp.). *Bollettino del Museo Civico di Storia Naturale di Venezia* 25: 107–146.

Gogala, A. 2022. Vespid wasps of Slovenia (Hymenoptera: Vespoidea: Vespidae). *Scopolia* 102: 1–79.

Goulet, H. and Huber, J.T. 1993. *Hymenoptera of the world: An identification guide to families*. Agriculture Canada Publication, Ottawa, 668 pp.

Gusenleitner, J. 1966. Vespidae, Eumenidae und Masaridae aus der Türkei. Teil I. *Polskie Pismo Entomologiczne* 36: 343–363.

Gusenleitner, J. 1967. Vespidae, Eumenidae und Masaridae aus der Türkei. Teil II. *Polskie Pismo Entomologiczne* 37: 663–676.

Gusenleitner, J. 1973. Über Masaridae aus dem Nahen Osten (Vespoidea, Hymenoptera). *Bolletino del Museo Civico di Storia Naturale di Venezia* 24: 55–69.

Gusenleitner, J. 1985. Bemerkenswertes über Faltenwespen VIII (Hymenoptera, Vespoidea). *Nachrichtenblatt der Bayerischen Entomologen* 34: 105–110.

Gusenleitner, J. 1988. Neue und Bemerkenswert Faltenwespen aus der Türkei (Hymenoptera, Vespoidea). *Linzer biologische Beiträge* 20(2): 713–737.

Gusenleitner, J. 1992. Bemerkenswerte Funde von Masaridae in Griechenland und im Nahen Osten (Vespoidea, Hymenoptera). *Linzer biologische Beiträge* 24(1): 59–63.

Gusenleitner, J. 1997. Die europäischen Arten der Gattung *Celonites* Latreille 1802 (Hymenoptera, Masaridae). *Linzer biologische Beiträge* 29(1): 109–115.

Gusenleitner, J. 1998. Über Faltenwespen aus dem asiatischen Raum (Hymenoptera, Eumenidae, Masaridae). *Linzer biologische Beiträge* 30(2): 503–513.

Gusenleitner, J. 2002. Über die Arten der Gattung *Masaris* Fabricius 1793 aus Zentralasien (Hymenoptera, Vespoidea, Masaridae). *Linzer biologische Beiträge* 34(1): 321–333.

Gusenleitner, J. 2004. Über Faltenwespen aus dem Oman (Hymenoptera: Vespidae, Eumenidae, Masaridae). *Linzer biologische Beiträge*, 36 (1): 151–158.

Gusenleitner, J. 2005. Über Faltenwespen aus dem Oman. Teil 2 (Hymenoptera, Vespidae, Eumenidae, Masaridae). *Linzer biologische Beiträge* 37(2): 1199–1201.

Gusenleitner, J. 2007. Vespidae aus Jemen in Sammlung des Oberösterreichischen Landesmuseums (Hymenoptera, Vespidae). *Linzer biologische Beiträge*, 39(1): 119–127.

Gusenleitner, J. 2010. Order Hymenoptera, family Vespidae. *Arthropod Fauna of the UAE*, 3: 422–467.

Gusenleitner, J. 2011. Bemerkenswertes über Faltenwespen XI (Hymenoptera: Vespidae: Eumeninae: Masarinae). *Linzer biologische Beiträge* 43(1): 323–329.

Gusenleitner, J. 2012. Neue Masarinae aus der paläarktischen Region (Hymenoptera: Vespidae: Masarinae). *Linzer biologische Beiträge* 44(1): 319–326.

Gusenleitner, J. 2013. Die Gattungen der Vespidae im Nahen Osten, in Nordafrika und in Arabien (Hymenoptera: Vespidae). *Linzer biologische Beiträge* 45(2): 1025–1046.

Gusenleitner, J. 2016. *Quartinia monnerati*, eine neue Art aus dem Oman (Hymenoptera: Vespidae: Masarinae). *Linzer biologische Beiträge* 48(1): 403–407.

Gusenleitner, J. 2018. Neue asiatische Faltenwespen (Hymenoptera: Vespidae: Eumeninae, Masarinae). *Linzer biologische Beiträge* 50(1): 303–308.

Howarth, B. and Gillett, M.P.T. 2008. The terrestrial and freshwater arthropods of Abu Dhabi Emirate, pp. 380–463. In: Perry, R.J. (ed.), *Terrestrial environment of Abu Dhabi Emirate*. Environment Agency, Abu Dhabi, 463 pp.

Kaplan, E. and Yildirim, E. 2020. Contribution to the knowledge of the Leucospidae, Sapygidae, and Vespidae (Apocrita: Hymenoptera) fauna from Bingöl, Diyarbakır and Muş provinces of Turkey. *Natura Somogyiensis* 35: 103–118.

Kostylev, G. 1935. Materialien zur Kenntnis der Masariden-Fauna der Paläarktis. *Archives du Musée Zoologique de l'Université de Moscow* 2: 85–116.

Krenn, H.W., Mauss, V. and Plant, J. 2002. Evolution of the suctorial proboscis in pollen wasps (Masarinae, Vespidae). *Arthropod Structure and Development* 31: 103–120.

Latreille, P.A. 1802. *Histoire naturelle, générale et particulière, des Crustacés et des Insectes*. Tome troisième. Paris, 468 pp.

Latreille, P.A. 1810. *Considérations générales sur l'ordre naturel des animaux composant les classes des Crustacés, des Arachnides et des Insectes, avec un tableau méthodique de leurs genres, disposés en familles*. F. Schoell, Paris, 444 pp.

Lepeletier, A. 1841. Histoire naturelle des Insectes. *Hyménoptères. Tome* 2: 585 pp.

Mauss, V. 2007. Evolution verschiedener Lebensformtypen innerhalb basaler Teilgruppen der Faltenwespen (Hymenoptera, Vespidae). *Denisia 20, zugleich Kataloge der Oberösterreichischen Landesmuseen, Neue Serie* 66: 701–722.

Mauss, V. 2013. Description of *Celonites andreasmuelleri* sp. n. (Hymenoptera, Vespidae, Masaridae) from the Middle East with a key to the Palaearctic species of *C. abbreviatus*-complex of the subgenus *Celonites s. str. Journal of the Hymenoptera Research* 31: 79–95.

Mauss, V., Fateryga, A. and Prosi, R. 2016. Taxonomy, distribution and bionomics of *Celonites tauricus* Kostylev, 1935, stat. n. (Hymenoptera, Vespidae, Masarinae). *Journal of Hymenoptera Research* 48: 33–66.

Mauss, V., Fateryga, A.V., Yildirim, E. and Carpenter, J.M. 2022. Contribution to the taxonomy, bionomics and distribution of the Palaearctic *Celonites cyprius*-group (Hymenoptera, Vespidae, Masarinae), with the description of two new species from the North Caucasus and East Anatolia. *Journal of Hymenoptera Research* 89: 109–155.

Mauss, V. and Müller, A. 2014. First contribution to the bionomics of the pollen wasp *Celonites fischeri* Spinola, 1838 (Hymenoptera, Vespidae, Masarinae) in Cyprus. In memory of Friedrich W. Gess. *Journal of Hymenoptera Research* 39: 119–153.

Mauss, V. and Prosi, R. 2013. First record of the pollen wasp *Celonites rugiceps* Bischoff, 1928 (Hymenoptera, Vespidae, Masarinae) from Central Europe. *Linzer biologische Beiträge* 45(1): 697–701.

Mauss, V. and Prosi, R. 2018. Identity and distribution of *Celonites hermon* Gusenleitner, 2002 (Hymenoptera, Vespidae, Masarinae) from the Middle East with a description of the hitherto unknown male. *Journal of Hymenoptera Research* 66: 55–70.

Meade-Waldo, G. 1910. V. New species of Diploptera in the collection of the British Museum. *Annals and Magazine of Natural History* 5(25): 30–51.

Meade-Waldo, G. 1911. X. New species of Diploptera in the collection of the British Museum. *Annals and Magazine of Natural History* 7(37): 98–113.

Modarres Awal, M. 2012. Family Vespidae (Hymenoptera), pp. 525–527. In: Modarres Awal, M. (ed.), *List of agricultural pests and their natural enemies in Iran*. 3rd edition. Ferdowsi University of Mashhad Press, Mashad, Iran, 759 pp.

Monks, J., Ross, S., Geiser, M., De Prins, J., Sharaf, M., Wyatt, N., Al Rijeibi, S. and Polaszek, A. 2019. A preliminary survey of the insect fauna of the Hajar Mountain Range, Oman. *Journal of Natural History* 53(15–16): 939–963.

Morawitz, F. 1888. Hymenoptera aculeata nova. *Horae Societatis Entomologicae Rossicae* 22: 224–302.

Müller, A. 1996. Convergent evolution of morphological specializations in Central European bee and honey wasps species as an adaptation to the updake of pollen from nototribic flowers (Hymenoptera, Apoidea and Masaridae). *Biological Journal of the Linnean Society* 57: 235–252.

Panfilov, D.V. 1968. New species of Middle Asian and Transcaucasian hymenopterans (Hymenoptera: Chrysididae, Masaridae, Vespidae). *Archives du Musée Zoologique de l'Université de Moscou* 11: 36–43. [in Russian]

Pickett, K.M. and Carpenter, J.M. 2010. Simultaneous analysis and the origin of eusociality in the Vespidae (Insecta: Hymenoptera). *Arthropod Systematics and Phylogeny* 68(1): 3–33.

Piekarski, P.K., Carpenter, J.M., Lemmon, A.R. and Sharanowski, B.J. 2018. Phylogenomic evidence overturns current conceptions of social evolution in wasps (Vespidae). *Molecular Biology and Evolution* 35: 2097–2109.

Rahmani, Z., Rakhshani, E. and Carpenter, J.M. 2020. Updated checklist of Vespidae (Hymenoptera: Vespoidea) in Iran. *Journal of Insect Biodiversity and Systematics* 6(1): 27–86.

Richards, O.W. 1962 (1964). A revisional study of the masarid wasps (Hymenoptera, Vespoidea). *British Museum (Natural History)*, 294 pp.

Richards, O.W. 1985 (1984). Insects of Saudi Arabia. Hymenoptera: Fam. Masaridae (The Arabian Species). *Fauna of Saudi Arabia* 6: 413–422.

Samin, N., Gusenleitner, J., Bagriacik, N., Gess, S.K. and Abd-Rabou, S. 2020. Annotated checklist of Masarinae (Hymenoptera: Vespidae) of Iran. *Calodema* 750: 1–8.

Schmid-Egger, C. 2017. Order Hymenoptera, family Vespidae, subfamily Masarinae. *Arthropod fauna of the UAE* 6: 341–349.

Schremmer, F. 1959. Der bisher unbekannte Pollensammelapparat der Honigwespe *Celonites abbreviatus* (Vespidae, Masaridae). *Zeitschrift für Morphologie und Ökologie der Tiere* 48: 424–438.

Selis, M. 2019. A new species of *Masaris* Fabricius from Iran (Hymenoptera: Vespidae: Masarinae). *Zootaxa* 4567(2): 395–400.

Selis, M., Ismael, H.R., Hamad, M.I., Baiocchi, D. and Cilia, G. 2025. A first update to the vespid fauna of Iraq (Hymenoptera: Vespidae), with description of a new species. *Journal of Natural History* 59(13–16): 941–981.

Spinola, W.A.M. 1838. Des Hyménoptères recueillis par M. Fischer pendant son voyage en Égypte, et communiqués. *Annalés de la Société Entomologique de France* 7: 437–575.

Thorp, R.W. 1979. Structural, behavioral and physiological adaptations of bees (Apoidea) for collecting pollen. *Annals of the Missouri Botanical Garden* 66: 788–812.

Thunberg, C.P. 1815. Philanthi, generis insecti Hymenopteri, monographia. *Nova Acta K. Vetenskaps Societeten i Uppsala (Nova Acta Regiae Societatis Scientiarum Upsaliensis)* 7: 126–139.

von Schulthess, A. 1922. Vespiden aus Südafrika (Hym.). *Deutsche Entomologische Zeitschrift* 401: 399–405.

Westerkamp, C. 1987. Das Pollensammeln der sozialen Bienen in Bezug auf die Anpassungen der Blüten. *Doctoral thesis, Universität Mainz.*

Westrich, P. 1989. *Die Wildbienen Baden-Württembergs*. Vol. 1–2. Eugen Ulmer Verlag, Stuttgart.

Yildirim, E. 1998. Türkiye Vespidae (Hymenoptera, Aculeata) faunasının son durumu (Bazı türler için yeni lokaliteler ve yeni kayıt). *Atatürk Üniversitesi Ziraat Fakültesi Dergisi* 29(2): 243–249.

Yildirim, E. and Gusenleitner, J. 2001. The present situation of the Vespidae fauna (Hymenoptera, Aculeata) of Turkey. *Linzer biologische Beiträge* 33(1): 217–221.

Yildirim, E. and Gusenleitner, J. 2004. The present situation of the Sapygidae and Vespidae fauna (Hymenoptera, Aculeata) of Turkey. *Linzer biologische Beiträge* 36(1): 131–135.

Yildirim, E. and Gusenleitner, J. 2007. The present situation of the Vespidae-fauna of Turkey (Hymenoptera: Vespidae). *Linzer biologische Beiträge* 39(1): 105–109.

Yildirim, E. and Gusenleitner, J. 2009. The present situation of the Vespidae-fauna (Hymenoptera, Aculeata) of Turkey. *Linzer biologische Beiträge* 41(1): 937–941.

Yildirim, E. and Gusenleitner, J. 2012. Contribution to the knowledge of the Vespidae (Hymenoptera, Aculeata) of Turkey, with a checklist of the Turkish species. *Turkish Journal of Zoology* 36(3): 361–374.

Yildirim, E. and Kojima, J. 1999. Distributional checklist of the species of the family Vespidae (Insecta: Hymenoptera: Aculeata) of Turkey. *Natural History Bulletin of Ibaraki University* 3: 19–50.

Yildirim, E. and Özbek, H. 1997. Masaridae (Hymenoptera, Vespoidea, Vespidae) of Turkey. *Atatürk Üniversitesi Ziraat Fakültesi Dergisi* 28(2): 222–226.

Zucchi, R., Yamane, S. and Sakagami, S.F. 1976. Preliminary notes on the habits of *Trimeria howardi*, a neotropical communal masarid wasp, with description of the mature larva. *Insecta Matsumurana, Series entomology, New Series* 8: 47–57.

Mischocyttarus rufidens (de Saussure, 1854) (Polistinae). (Photo courtesy of Angella Moorehouse)

Polistes rubiginosus (Lepeletier, 1836) (Polistinae). (Photo courtesy of Angella Moorehouse)

Polistes wattii (Cameron, 1900) (Polistinae). (Photo courtesy of Alexander Fateryga)

Polistes dominula (Christ, 1791) (Polistinae). (Photo courtesy of Alexander Fateryga)

Synoeca septentrionalis (Richards, 1978) (Polistinae). (Photo courtesy of Angella Moorehouse)

Metapolybia sp. (Polistinae). (Photo courtesy of Angella Moorehouse)

4 Subfamily Polistinae Latreille, 1802

Hassan Ghahari, Neveen S. Gadallah, Christopher K. Starr, Patrick K. Piekarski, and Alexander Somavilla

INTRODUCTION

Polistinae (paper wasps) is a cosmopolitan subfamily of eusocial wasps within the family Vespidae. It is the second most diverse of the six vespid subfamilies, comprising 25 genera and nearly 1,050 species (Arevalo *et al.*, 2004; Brito *et al.*, 2024). These genera are classified into four tribes: Polistini, Epiponini, Mischocyttarini, and Ropalidiini. Although Polistinae species are found worldwide, from cool temperate to tropical regions, they are most diverse in tropical or subtropical regions, particularly in the Neotropics (Goulet and Huber, 1993; Somavilla and Carpenter, 2021). In comparison to Africa, Europe, and Asia, the Americas exhibit a high level of endemism, nearly 70%, with most species restricted to the Neotropical region (Carvalho *et al.*, 2015; Jeanne, 1991, Silveira, 2008).

Adults range in body length from less than 1 cm (approximately 6 mm) to about 5 cm, and are characterized by the following combination of features: clypeus is not projecting, either acute, rounded ventrally, or not; pronotal lobe is usually separated from the tegula by a distance equal to or less than the length of the lobe, though rarely greater; the mesoscutum lacks a parategula, with a lateral longitudinal carina opposite to the tegula, or with a projecting lobe at the postero-lateral apex; abdomen is spindle-shaped; male antennae at least somewhat curled; the metacoxae are usually smooth dorsally; the mesofemur with or without a basal ring; tarsal claws are simple and smooth ventrally; forewing longitudinally folded at rest, usually reaching at least to the posterior margin of fourth gastral segment, apex of marginal cell narrowly or not separated from anterior margin of wing; the forewing is longitudinally folded at rest, typically reaching at least to the posterior margin of the fourth gastral segment, with the apex of the marginal cell narrowly separated or not separated from anterior wing margin; the radial vein is less than half the width of the stigma, or absent; vein 1cu-a is straight and much shorter than vein 2/1A; the hindwing may or may not have a jugal lobe; the gaster is petiolate or more or less sessile; first gastral tergum is gradually declivous anteriorly (Goulet and Huber, 1993).

Based on their mode of colony founding and mechanisms of reproductive dominance, polistines can be divided into two major behavioral groups: independent-founding and swarm-founding species (Jeanne, 1980, 1991; Arevalo *et al.*, 2004). In the first group, which includes the genera *Belonogaster*, *Mischocyttarus*, *Parapolybia*, *Polistes*, and some species of *Ropalidia*, colonies are founded by a single or a small group of gynes without worker assistance. These colonies are typically small (rarely exceeding 50 adult wasps) and are associated with simple nests that lack a protective paper envelope around the comb. In at least some species, soon after nest initiation one fertilized female becomes the sole egg-layer (Strassmann *et al.*, 1987; Gadagkar, 1991; Jeanne, 1991, 2003; Reeve, 1991; Zakharov, 1991; Keeping, 2002; Noll and Wenzel, 2008; Noll, 2013).

DOI: 10.1201/9781003740438-4

In contrast, the second group, including 20 genera of the tribe Epiponini, *Polybioides*, and some species of *Ropalidia*, establishes colonies through swarm-founding, where multiple queens are assisted by many workers. These colonies are usually much larger, often consisting of hundreds or thousands of individuals, and in most species, the nests are surrounded by a protective envelope (Jeanne, 1991, 2021; Strassmann *et al.*, 2003). Reproductive dominance in these species is maintained through worker aggression toward potential queens, and there is a tendency for the development of morphological castes in some taxa (Jeanne, 1991; Strassmann *et al.*, 2003; Arevalo *et al.*, 2004; Noll and Wenzel, 2008).

The independent-founding pattern is considered the ancestral condition (ground plan) of eusocial Vespidae, while swarm-founding appears to have evolved independently in both the Palaeotropics and Neotropics (Wenzel and Carpenter, 1994; Piekarski et al., 2018). In either case, the larvae are typically fed masticated insects, especially caterpillars, and sometimes also alate termites or ants (Goulet and Huber, 1993).

The specific features of polistine wasp population organization are closely related to the individual development and fecundity of their foundresses. Due to the high oviposition rates of queens, up to 800 eggs/day, and relatively short preimaginal development times (as short as 23 days), colonies of swarm-founding species can rapidly recover and grow to large sizes following establishment (Richards, 1978). This capability underlies the use of swarming as the primary strategy for responding to adverse abiotic and biotic conditions. Swarming may be viewed as the reproduction of stereotyped elements (in this case, colonies), providing a basis for considering polistine wasps populations as a corpuscular system (Rusina, 2015).

Phylogenetic inferences have provided crucial insights into the evolutionary history of the Polistinae (Carvalho *et al.*, 2015; Silveira *et al.*, 2021). Carpenter (1991) conducted the first cladistic analysis of relationships among the genera, based on morphological and nest characteristics. This analysis identified the cosmopolitan Polistini as the sister group of all other tribes (Carvalho *et al.*, 2015). Initial molecular studies using microsatellite data were consistent with, or at least did not contradict, this hypothesis (Ezenwa *et al.*, 1998; Arevalo *et al.*, 2004). However, Pickett and Carpenter (2010), after reviewing both molecular and morphological data for the family Vespidae, proposed a different arrangement: Ropalidiini as the most basal group, followed by Mischocyttarini, with Polistini + Epiponini as sister tribes.

More recent phylogenetic studies, employing diverse sources of evidence, including phylogenomic, molecular, morphological, and behavioral data, have corroborated this pattern of relationships. These studies support the placement of Ropalididiini as the most basal lineage, followed by Mischocyttarini, with Polistini + Epiponini forming a sister-group relationship (Piekarski *et al.*, 2018; Noll *et al.*, 2021; Menezes *et al.*, 2020).

There are limited data on the paleodistribution of paper wasps based on fossil records (Carvalho *et al.*, 2015). Carpenter (1993) estimated that this group diverged from other Vespidae during the breakup of Gondwana, approximately 140 million years ago (Mya). The fossil nest of *Brownichnus favosites* Brown, 1941 is thought to be an ancient social vespid (polistine-like) nest likely dating back ~66 Mya, implying that the split between Polistinae and Vespinae must be at least that old (Wenzel, 1990; Genise, 2000; Perrard *et al.*, 2017). Menezes *et al.* (2020) estimated that the Epiponini (New world swarm-founders) originated during the Eocene ~45 Mya (37.6–51.3 Mya), while the split between Polistini and Epiponini occurred ~51 Mya (45.2–56.8 Mya). This phylogenomic study also dated the most recent common ancestor of all extant Polistinae to be ~66 Mya (59–74 Mya), and of Polistinae + Vespinae to be ~87 Mya (76–94 Mya). An independent molecular dating study focusing on the Epiponini recovered similar results, with Epiponini originating ~48 Mya (37.9–55.1 Mya), Polistinae originating ~63 Mya (57.5–68.8 Mya), and the split between Polistinae and Vespinae occurring in the late Cretaceous at least ~70 Mya (Noll *et al.*, 2021). Regarding Polistini (i.e., *Polistes*), the oldest known fossil, which was unfortunately lost, is *Polistes vergnei* Piton, 1940 from the Paleocene (56–66 Mya). This would put the origin of *Polistes* to at least ~55 Mya (Perrard *et al.*, 2017).

Checklists of the Iranian Polistinae: Farahbakhsh (1961) reported a single species, *Polistes caspica* Fabricius [nomen nudum]. Modarres Awal (1997) listed three species, *P. caspica*, *P. gallicus* (Linnaeus, 1767), and *P. hebraeus* Fabricius 1787. In later work, Modarres Awal (2012) reported six species belonging to two genera: *Parapolybia* de Saussure, 1854 (one species) and *Polistes* Latreille, 1802 (five species). Ebrahimi and Carpenter (2008) catalogued 13 species in two genera. More recently, Rahmani *et al.* (2020) listed 17 species in two genera.

The present catalogue includes 32 species in four genera (*Belonogaster* de Saussure, 1854; *Parapolybia* de Saussure, 1854; *Polistes* Latreille, 1802; and *Ropalidia* Guérin-Méneville, 1831) for the Middle East and 17 species in two genera (*Parapolybia* and *Polistes*) for Iran.

Identification key to genera of Polistinae in the Middle East (modified from Carpenter and Nguyen, 2003 and Somavilla and Carpenter, 2021)

1. Metasomal segment I subsessile: evenly conical in dorsal view, smoothly rounded in lateral view, less than twice as long as wide and more than half the width of segment II; propodeum with orifice dorsally narrowed, acute; pronotum with a carina posterior to lateral fovea when present (Tribe Polistini) .... *Polistes* Latreille

• Metasomal segment I basally petiolate to long petiolate in dorsal view; propodeum with orifice often broadly rounded dorsally; pronotum with a carina anterior to lateral fovea when present (Tribe Ropalidiini) 2

2. Antenna 11-segmented in female and 12 in male; pronotum without dorsal carina; metasomal segment I nearly linearly petiolate in dorsal view .... *Belonogaster* de Saussure

• Antenna 12-segmented in female, 13 in male; pronotum with at least dorsal carina; metasomal segment I not linearly (parallel-sided) petiolate in dorsal view 3

3. Pronotum without pretegular carina; mesepisternum without scrobal sulcus .... *Ropalidia* Guérin-Méneville

• Pronotum with pretegular carina; mesepisternum with dorsal sulcus .... *Parapolybia* de Saussure

List of Species of the Subfamily Polistinae in the Middle East

Subfamily Polistinae Latreille, 1802
Tribe Polistini Latreille, 1802
Genus *Polistes* Latreille, 1802

Polistes Latreille, 1802, 363. Type species: *Vespa gallica* Linnaeus, 1767, by subsequent designation of Latreille, 1810.

Comments: The genus *Polistes* is one of the most species-rich, familiar, and widespread taxa of social wasps, comprising 238 recognized species in eight subgenera worldwide. In the Middle East, three subgenera are present: *P.* (*Gyrostoma*) Kirby, 1828, *P.* (*Polistella*) Ashmead, and *P.* (*Polistes*) *s.s.* Latreille, 1802. Species of this genus are found on all habitable continents (Silveira *et al.*, 2021; Somavilla *et al.*, 2021; Brito *et al.*, 2024).

Subgenus *Gyrostoma*: Mesopleuron more or less shiny with the structure of very fine punctures, among them several coarse punctures are present; pronotal keel less strong and invisible below the fovea; epicnemial carina strong and well visible; large species.

Subgenus *Polistella*: Small to moderately large wasps; clypeus produced dorsally beyond the level of the anterior tentorial pits; pronotal fovea absent; first abdominal tergum widens gradually from base to apex, its width longer than half length of second tergum; mesepisternum without epicnemial carina and incomplete dorsal episternal groove; prestigma short, distinctly shorter than half of pterostigma.

Subgenus *Polistes s.s.*: Mesopleuron matte with coarse punctures; pronotal keel strong and visible below the fovea; epicnemial carina present, but shallow; small to medium-sized species.

Identification key to *Polistes* species of the Middle East (modified from Schmid-Egger *et al.*, 2017)

1. Parastigma long compared with the first radial sector (1-SR) of the forewing; second and third gastral segments with dark curved posterior margins (Subgenus *Gyrostoma*)..2

– Parastigma short comapred with vein 1-SR of the forewing; second and third gastral segments with straight posterior margin (Subgenus *Polistes*)..3

2. Frontoclypeal suture of female dark brown to black; male clypeus evenly convex ....*Polistes olivaceus* (DeGeer, 1773)

– Frontoclypeal suture of female yellowish; male clypeus depressed..............*Polistes wattii* (Cameron, 1900)

3. Female: antenna 12-segmented; gaster with six visible tergites and sternites; face and/or clypeus with yellow and black pattern..4

– Male: antenna 13-segmented; gaster with seven visible tergites and sternites; face and/or clypeus yellow......12

4. Mandible conspicuously stout; clypeal apex distinctly convex and acute medially..............................5

– Mandible slender; clypeal apex straight and truncate medially..6

5. Basal half of mandible distinctly angulate and flattened; clypeus gradually depressed toward apex...*Polistes semenowi* (Morawitz, 1889)

– Basal half of mandible gradually curved and convex; clypeus suddenly depressed at apex...*Polistes atrimandibularis* (Zimmermann, 1930)

6. Malar space mainly yellow and continuous with the yellow ventral part of temple; mandible usually black; clypeus wider than high medially; hindcoxa black dorsally..7

– Malar space and ventral part of temple black; mandible with yellow; clypeus about as wide as high medially; hindcoxa variable dorsally 10

7. Apical half of sixth sternite yellow or largely yellow; mesopleuron coarsely sculptured ventrally............8

– Apical half of sixth sternite black (sometimes paler apically in *P. associus*); mesopleuron finely sculptured ventrally...9

8. Temple with wide yellow band that is wider than half width of temple medially; propodeum and tergum 2 largely yellow; eye sinus mainly yellow...*Polistes bucharensis* (Erichson, 1849)

– Temple with separated yellow spots, if continuous band, then narrower than half width of temple medially; propodeum and tergum 2 largely black; eye sinus black.............................*Polistes dominula* (Christ, 1791)

9. Posterolateral yellow bands of pronotum usually not connected to anterior transverse band; mesoscutum with two yellow spots...*Polistes associus* Kohl, 1898

– Posterolateral yellow bands of pronotum usually connected to anterior transverse band; mesoscutum sometimes lacking yellow spots..*Polistes nimpha* (Christ, 1791)

10. Pronotum with narrow yellow transverse band; yellow spots of propodeum narrow or absent...*Polistes biglumis* (Linnaeus, 1758)

– Pronotum with broad yellow transverse band; yellow spots of propodeum wide and larger.....................11

11. Fourth sternite with medially interrupted apical yellow band; clypeus often with a wide black transverse band or with more or less transverse trapezoidal black patch....................*Polistes bischoffi* (Weyrauch, 1937)

– Fourth sternite with complete and broad apical yellow band; clypeus with less developed central dark patch...*Polistes gallicus* (Linnaeus, 1767)

12. Mandible conspicuously stout, with distinct depression on outer surface; clypeus depressed medio-apically, sharply tapered apically; malar space comparatively long and wide..13

– Mandible slender, with flattened to slightly convex outer surface; clypeus flat medio-apically, with broadly triangular apical margin; malar space narrow to median-sized..14

13. Basal half of mandible situate dorsally; mandibular depression distinctly concave...*P. semenowi* (Morawitz, 1889)

– Basal half of mandible straight dorsally, with shallow mandibular depression...*Polistes atrimandibularis* (Zimmermanni, 1930)

14. Temple bulged behind eye in dorsal view, slightly convex; head trapezoidal in frontal view; apical margin of clypeus triangular...15

– Temple narrowed behind eye in dorsal view, more or less straight; apical margin of clypeus rounded or subtruncate...18

15. Clypeus distinctly depressed medially, conspicuously ridged laterally; face with U-shaped depression medially; a short sharp, yellow carina could be seen between antennal bases...*Polistes nimpha* (Christ, 1791)

– Clypeus without or with slight depression medially, not ridged laterally; face flattened, without a median depression; an obtuse short carina could be seen between antennal bases...16

16. Mesosternum entirely black; lateral ridges of clypeus and ridge between antennal sockets and clypeus more pronounced..*Polistes bucharensis* (Erichson, 1849)

– Mesosternum with at least some yellow spots posteriorly; clypeus laterally nearly flat, and ridge between antennal sockets and clypeus indistinct...17

17. Clypeus densely punctate and mostly with long bristles; apical half of antenna orange-yellow; clypeus slightly wider than long, distinctly flattened medio-ventrally.......................*Polistes dominula* (Christ, 1791)

– Basal 0.7 of clypeus impunctate, glabrous; apical half of antenna usually darker dorsally than ventrally; clypeus as wide as its length, less flattened medio-ventrally....................*Polistes biglumis* (Linnaeus, 1758)

18. Clypeus distinctly depressed on apical half, with distinct lateral ridges, more or less truncate apically; sternites III-VII usually with transverse yellow bands basally; frons with distinct longitudinal depression medially...*Polistes associus* (Kohl, 1898)

– Clypeus slightly depressed on apical half, without lateral ridges, rounded apically; sternites III-VII black at base, at least last sternite entirely black of nearly so..19

19. Yellow transverse pronotal band narrow laterally, if widened dorso-laterally then narrowed medio-laterally; apical half of antenna slightly infuscate; mesoscutum with or without minute paired yellow spots; scutellum usually with comparatively small paired yellow spots...*Polistes bischoffi* (Weyrauch, 1937)

– Yellow transverse pronotal band moderately to strongly widened laterally; apical half of antenna uniformly orange to yellow; mesoscutum nearly always with minute to large paired yellow spots; scutellum with large yellow spots ....*Polistes gallicus* (Linnaeus, 1767)

***Polistes* (*Polistes*) *associus* Kohl, 1898**

Polistes associus Kohl, 1898: 89, ♂; "Griechenland (Poros)".

Catalogues with Iranian records: Rahmani *et al.* (2020).

Distribution in Iran: Guilan (Castro and Dvořák, 2010).

Distribution in the Middle East: Iran (Castro and Dvořák, 2010; Maryam *et al.*, 2024; Selis *et al.*, 2025), Iraq (Selis *et al.*, 2025), Israel-Palestine (Yildirim and Kojima, 1999; Carpenter, 1996; Gusenleitner, 2013; Schmid-Egger *et al.*, 2017; Selis *et al.*, 2025), Turkey (Gusenleitner, 1967 as *P.* (*Leptopolistes*) *associus*; Yildirim and Özbek, 1993 as *P.* (*Leptopolistes*) *associus*, 1999; Carpenter, 1996; Madl, 1997; Yildirim and Kojima, 1999; Yildirim and Gusenleitner, 2012; Gusenleitner, 2013; Schmid-Egger *et al.*, 2017; Selis *et al.*, 2025).

Extralimital distribution: Austria, Slovakia, Slovenia, Russia, India (Selis *et al.*, 2025), Azerbaijan, Bulgaria, China, Croatia, France, Greece, Italy, Malta, Montenegro, North Macedonia, Pakistan, Spain, Switzerland, Tajikistan (Carpenter, 1996; Castro and Dvořák, 2010; Gusenleitner, 2013; Schmid-Egger *et al.*, 2017; Selis *et al.*, 2025).

Comments: One record published from Iran by du Buysson (1912) has been ignored by some authors dealing with the Iranian fauna. Its occurrence in Iran was confirmed by Castro and Dvořák (2010).

***Polistes* (*Polistes*) *atrimandibularis* Zimmermann, 1930**

Polistes atrimandibularis Zimmermann, 1930: 611, ♂, ♀; "Tirol Toblach".

Catalogues with Iranian records: Rahmani *et al.* (2020).

Distribution in Iran: Golestan (Lehr *et al.*, 2007).

Distribution in the Middle East: Egypt (Carpenter, 1996; Yildirim and Kojima, 1999; Gusenleitner, 2013), Iran (Lehr *et al.*, 2007; Carpenter, 1996; Gusenleitner, 2013; Selis *et al.*, 2025), Iraq (Selis *et al.*, 2025), Turkey (Blüthgen, 1961 as *Sulcopolistes atrimandibularis*; Yildirim and Özbek, 1993 as *P.* (*Sulcopolistes*) *atrimandibularis*; Carpenter, 1996; Yildirim and Özbek, 1999; Yildirim and Gusenleitner, 2012; Gusenleitner, 2013; Selis *et al.*, 2025).

Extralimital distribution: Armenia (Carpenter, 1996; Gusenleitner, 2013; Antropov, 2017; Schmid-Egger *et al.*, 2017; Selis *et al.*, 2025), Austria, Czech Republic, Germany, Hungary (Selis *et al.*, 2025), Bulgaria, France, Greece, Italy, Russia, Spain, Switzerland (Gusenleitner, 2013; Antropov, 2017; Schmid-Egger *et al.*, 2017; Selis *et al.*, 2025).

Comments: *Pycnopogon fasciculatus* (Loew) (Diptera: Asilidae) has been recorded by Lehr *et al.* (2007) as being a predator of *P. atrimandibularis*. It was also recorded as being a social parasite of *P. biglumis* (Schmid-Egger *et al.*, 2017).

***Polistes* (*Polistes*) *badius* Gerstaecker, 1871**

Polistes badia Gerstaecker, 1871: 351, sex not stated.

Distribution in the Middle East: Yemen (Giordani Soika, 1957; Carpenter, 1996; Gusenleitner, 2013; Carpenter and Gadallah, 2020).

Extralimital distribution: Congo, Ethiopia, Gabon, Gambia, Guinea-Bissau, Kenya, Senegal, Somalia, Sudan, Togo (Gusenleitner, 2013; Carpenter and Gadallah, 2020).

***Polistes* (*Polistes*) *biglumis* (Linnaeus, 1758)**

Vespa biglumis Linnaeus, 1758: 573, ♀; «Europa».

Catalogues with Iranian records: Rahmani *et al.* (2020).

Distribution in Iran: Semnan (Blüthgen and Gusenleitner, 1970).

Distribution in the Middle East: Iran (Blüthgen and Gusenleitner, 1970; Guiglia, 1972; Carpenter, 1996; Yildirim and Kojima, 1999; Gusenleitner, 2013), Israel-Palestine, Syria (Castro and Dvořák, 2010), Turkey (Blüthgen, 1956 as *P.* (*P.*) *biglumis alpium*; Yildirim and Özbek, 1993 as *P.* (*P.*) *biglumis alpium*; Carpenter, 1996; Yildirim and Kojima, 1999; Yildirim and Gusenleitner, 2012; Gusenleitner, 2013; Schmid-Egger *et al.*, 2017).

Extralimital distribution: Austria, Azerbaijan, Belgium, France, Georgia, Germany, Greece, Italy, Kazakhstan, Korea, Kyrgyzstan, Netherlands, Pakistan, Sweden, Switzerland, Tajikistan, Uzbekistan (Castro and Dvořák, 2010; Schmid-Egger *et al.*, 2017), Mongolia (Carpenter, 1996), Russia (Antropov and Proshchalykin, 2014).

Comments: *Polistes biglumis*, which was recorded from Iraq by Blüthgen and Königsmann (1969), has been excluded from the fauna of Iraq by Selis *et al.* (2025).

***Polistes* (*Polistes*) *bischoffi* Weyrauch, 1937**

Polistes bischoffi Weyrauch, 1937: 274, neotype ♀.

Catalogues with Iranian records: Rahmani *et al.* (2020).

Distribution in Iran: East Azarbaijan (Sakenin *et al.*, 2010), Lorestan (Abd-Rabou *et al.*, 2005), Tehran (Samin *et al.*, 2010).

Distribution in the Middle East: Iran (see references above, and Antropov, 2017), Turkey (Blüthgen, 1961 as *P.* (*Leptopolistes*) *bischoff*; Gusenleitner, 1967 as *P.* (*Leptopolistes*) *bischoff*; Carpenter, 1996; Madl, 1997; Yildirim and Kojima, 1999; Gusenleitner, 2013; Schmid-Egger *et al.*, 2017).

Extralimital distribution: Austria, Croatia, Gibraltar, Greece, France, Italy, Kazakhstan, Malta, Mongolia, North Africa, Russia, Spain, Switzerland (Antropov, 2017; Gusenleitner, 2013; Schmid-Egger *et al.*, 2017).

Comments: According to Neumeyer *et al.* (2015), *P. albellus* Giordani Soika, 1976 is a junior synonym of *P. bischoffi* Weyrauch, 1937. It has been recorded by Samin *et al.* (2010) as being prey of *Stenopogon avus* (Loew) (Diptera: Asilidae).

***Polistes* (*Polistes*) *bucharensis* Erichson, 1849**

Polistes bucharensis Erichson, 1849: 307, ♀; [Uzbekistan] "Bukhara".

Catalogues with Iranian records: Ebrahimi and Carpenter (2008 as *Polistes iranus* Guiglia, 1976b), Rahmani *et al.* (2020).

Distribution in Iran: Alborz, Razavi Khorasan (Guiglia, 1976b as *Polistes iranus*), Fars (Guiglia, 1976b as *P. iranus*), Golestan (Weyrauch, 1939; Schmid-Egger *et al.*, 2021), Guilan, Kerman, Semnan (Schmid-Egger *et al.*, 2021), Khuzestan (Boyeri Shemi and Pashaei Rad, 2017), Kordestan (Ebrahimi and Carpenter, 2008; Modarres Awal, 2012 both as *P. iranus*), Markazi (Ebrahimi and Carpenter, 2008; Modarres Awal, 2012 both as *P. iranus*; Schmid-Egger *et al.*, 2017), Mazandaran (Guiglia, 1976b; Schmid-Egger *et al.*, 2021), Southern Khorasan (Blüthgen and Gusenleitner, 1970 as *P. gallicus bucharensis* Erichson, 1849), Tehran (Blüthgen and Gusenleitner, 1970 as *P. gallicus bucharensis*; Guiglia, 1976b), Zanjan (Abbasi, 2009 as *P. iranus*; Abbasi *et al.*, 2008a, 2008b as both *P. dominulus bucharensis* and *P. iranus*; Pashaei Rad *et al.*, 2010 as both *P. iranus* and *P.* (*Polistes*) *dominulus bucharensis*).

Distribution in the Middle East: Cyprus, Egypt, Israel-Palestine (Schmid-Egger *et al.*, 2017; Selis *et al.*, 2025), Iran (see references above, and Ebrahimi, 2014; Selis *et al.*, 2025), Iraq (Derwesh, 1965; Selis *et al.*, 2025), Turkey (Yildirim and Özbek, 1993; Carpenter, 1996 both as *P. dominulus bucharensis*; Schmid-Egger *et al.*, 2017; Selis *et al.*, 2025).

Extralimital distribution: Azerbaijan, China, Greece, Mongolia, Uzbekistan (Schmid-Egger *et al.*, 2017; Selis *et al.*, 2025).

***Polistes* (*Polistes*) *chinensis* (Fabricius, 1793)**

Vespa chinensis Fabricius 1793: 261, ♀; "Japon ..., Konéy Tchéou".

Catalogues with Iranian records: Rahmani *et al.* (2020).

Distribution in Iran: Iran (no specific locality cited) (Antropov, 2017).

Distribution in the Middle East: Iran, Turkey (Antropov, 2017).

Extralimital distribution: China, Japan, Kazakhstan, Korea, Mongolia, New Zealand (introduced), Russia (Carpenter, 1996 as *P. chinensis chinensis*; Antropov, 2017), Hong Kong, Taiwan (Carpenter, 1996).

***Polistes* (*Polistes*) *dominula* (Christ, 1791)**

Vespa dominula Christ, 1791: 229, ♀, ♂; [Germany] "Kronenberg an der Höh" (type lost).

Catalogues with Iranian records: Ebrahimi and Carpenter (2008), Rahmani *et al.* (2020).

Distribution in Iran: East Azarbaijan, Kordestan, Markazi (Ebrahimi and Carpenter, 2008; Modarres Awal, 2012), Golestan (Abd-Rabou *et al.*, 2005; Ghahari *et al.*, 2007; Hayat *et al.*, 2008), Kerman (Pashaei Rad *et al.*, 2010), Razavi Khorasan (Sakenin *et al.*, 2010), Zanjan (Abbasi, 2009; Abbasi *et al.*, 2008a, 2009; Pashaei Rad *et al.*, 2010).

Distribution in the Middle East: Cyprus (Schmid-Egger *et al.*, 2017), Egypt (Guérin-Méneville, 1844 as *Polistes gallica* var. *lefebvrei*), Iran (see references above, and Carpenter, 1996; Gusenleitner, 2013; Ebrahimi, 2014; Kaplan and Efil, 2025), Israel-Palestine, Syria (Carpenter, 1996; Gusenleitner, 2013; Kaplan and Efil, 2025), Jordan (Haddad *et al.*, 2007), Turkey (Rudow, 1889 as *P. maculatus*; Gusenleitner, 1976 as *P. gallicus muchei* Gusenleitner, 1976, 2013; Özbay and Akbayin, 1993 as *P.* (*P.*) *dominulus bucharensis*; Yildirim and Özbek, 1993; Akbayin and Özbay, 1996 as *P.* (*P.*) *dominulus bucharensis*; Carpenter, 1996; Madl, 1997; Yildirim and Özbek, 1999 as *P.* (*P.*) *dominulus bucharensis*; Yildirim and Kojima, 1999; Yildirim and Gusenleitner, 2012; Schmid-Egger *et al.*, 2017; Kaplan and Efil, 2025).

Extralimital distribution: Afghanistan, Australia (introduced), Caucasus, Central Asia, China, Europe, North Africa, Kazakhstan, Mongolia, Pakistan, Russia, North America (introduced), India, South America (introduced), New Zealand (introduced) (Carpenter, 1996; Gusenleitner, 2013; Fateryga, 2018; Kaplan and Efil, 2025), Tunisia (Castro and Dvořák, 2010), Turkmenistan, Uzbekistan (Carpenter, 1996), Italy (Selis, 2023).

Comments: *Leptarthrus brevirostris* (Meigen) and *Promachus leoninus* Loew (Diptera: Asilidae) were recorded as the predators of *P. dominula* by Ghahari *et al.* (2007) and Hayat *et al.* (2008), respectively.

***Polistes* (*Polistella*) *fastidiosus* de Saussure, 1853**

Polistes fastidiosus de Saussure, 1853: 60, sex not stated.

Distribution in the Middle East: Saudi Arabia, Yemen (Richards, 1985b; Carpenter, 1996; Gusenleitner, 2013; Carpenter and Gadallah, 2020).

Extralimital distribution: Benin, Burkina Faso, Congo, Ivory Coast, Guinea-Bissau, Kenya, Mozambique, Nigeria, Senegal, Somalia, South Africa, Sudan, Tanzania, Zambia (Gusenleitner, 2013; Carpenter and Gadallah, 2020).

***Polistes* (*Polistes*) *foederatus* Kohl, 1898**

Polistes foederata (!) Kohl, 1898: 90, lectotype ♂; "Transkauk. Helenendorf".

Catalogues with Iranian records: Rahmani *et al.* (2020).

Distribution in Iran: Mazandaran (Guiglia, 1976a as *Polistes* (*Leptopolistes*) *foederatus* Kohl).

Distribution in the Middle East: Cyprus, Turkey (Schmid-Egger *et al.*, 2017; Selis *et al.*, 2025), Iran (Guiglia, 1976a), Iraq (Derwesh, 1965; Al-Ali, 1977; Selis *et al.*, 2025).

Extralimital distribution: Azerbaijan, Croatia, Greece, Italy (Schmid-Egger *et al.*, 2017; Selis *et al.*, 2025), Slovenia (Selis *et al.*, 2025).

***Polistes* (*Polistes*) *gallicus* (Linnaeus, 1767)**

Vespa gallica Linnaeus, 1767: 949, ♂; "Europe australi".

Distribution in the Middle East: Cyprus (Fateryga, 2018), Israel-Palestine (Carpenter, 1996; Fateryga, 2018), Jordan (Haddad *et al.*, 2007; Fateryga, 2018), Syria (Carpenter, 1996; Abo Kaf and Allewa, 2016), Turkey (Tanyolaç, 1964; Özeren, 1965; Gusenleitner, 1967 as both *P.* (*P.*) *gallicus* and *P.* (*Leptopolistes*) *foedratus*; Tüzün and Tanyolaç, 1987; Özbay and Akbayin, 1993; Yildirim and Özbek, 1993 as both *P. gallicus* and *P.* (*Leptopolistes*) *gallicus*; Akbayin and Özbay, 1996; Carpenter, 1996; Madl, 1997; Yildirim and Gusenleitner, 2012).

Extralimital distribution: Afghanistan, Algeria, Armenia, Azerbaijan, China, Croatia, Ethiopia, France, Greece, Italy, Kazakhstan, Kyrgyzstan, Malta, Mongolia, Morocco, Pakistan, Portugal, Russia, Spain, Switzerland, Tunisia, Turkmenistan, Ukraine (Carpenter, 1996; Schmid-Egger *et al.*, 2017; Fateryga, 2018).

***Polistes* (*Gyrostoma*) *indicus* Stolfa, 1934**

Polistes indicus Stolfa, 1934: 47, ♀, ♂; "India: Kewra, Salt Range, Punjab" [now Pakistan] (Kolkata).

Catalogues with Iranian records: Rahmani *et al.* (2020).

Distribution in Iran: Kerman (Pashaei Rad *et al.*, 2010; Schmid-Egger *et al.*, 2021), Semnan (Nosrati *et al.*, 2024), Sistan & Baluchestan (Blüthgen and Gusenleitner, 1970 as *Polistes* (*Leptopolistes*) *indicus* Stolfa, 1934).

Distribution in the Middle East: Iran (see references above, and Carpenter, 1996; Gusenleitner, 2013; Carpenter and Gadallah, 2020; Selis *et al.*, 2025), Iraq (Richards, 1985b; Carpenter, 1996; Gusenleitner, 2004, 2005; Kumar and Lambert, 2010; Carpenter and Gadallah, 2020; Selis *et al.*, 2025), Oman (Richards, 1985b; Carpenter, 1996; Gusenleitner, 2004, 2005, 2013; Carpenter and Gadallah, 2020; Selis *et al.*, 2025), United Arab Emirates (Richards, 1985b; Carpenter, 1996; Gillett and Howarth, 2004; Gillett and Gillett, 2005; Howarth and Gillett, 2008; Gusenleitner, 2013; Carpenter and Gadallah, 2020; Selis *et al.*, 2025).

Extralimital distribution: Afghanistan, Pakistan (Gusenleitner, 2013; Siddiqui *et al.*, 2015; Khan *et al.*, 2018; Selis *et al.*, 2025), India (Gusenleitner, 2013; Carpenter and Gadallah, 2020; Selis *et al.*, 2025).

***Polistes* (*Polistes*) *marginalis marginalis* (Fabricius, 1775)**

Vespa marginalis Fabricius, 1775: 367, sex not stated.

Distribution in the Middle East: Saudi Arabia, Yemen (Giordani Soika, 1957 as *P. marginalis arabicus*; Richards, 1985b; Carpenter, 1996; Wahis, 2000; Gusenleitner, 2013; Madl, 2018; Carpenter and Gadallah, 2020).

Extralimital distribution: Widely distributed in Africa from Cape of Good Hope, and Zanzibar to Ethiopia and Senegal (Gusenleitner, 2013; Carpenter and Gadallah, 2020).

***Polistes* (*Polistes*) *mongolicus* du Buysson, 1911**

Polistes gallicus var. *mongolicus* du Buysson, 1911: 218, ♂; [China] "Kachgarie, route de Koutchar à Karachar".

Catalogues with Iranian records: None.

Distribution in Iran: Ardabil, Chaharmahal & Bakhtiari, East Azarbaijan, Guilan, Hamadan, Hormozgan, Ilam, Isfahan, Kuhgiloyeh & Boyerahmad, Lorestan, Markazi, Northern Khorasan, Qom, (Ebrahimi and Carpenter, 2008 as *Polistes* (*Polistes*) *gallicus* (Linnaeus, 1767)), Fars, Golestan (Ebrahimi and Carpenter, 2008; Pashaei Rad *et al.*, 2010 both as *P.* (*P.*) *gallicus*; Schmid-Egger *et al.*, 2021), Kerman (Ebrahimi and Carpenter, 2008; Pashaei Rad *et al.*, 2010 both as *P.* (*P.*) *gallicus*; Schmid-Egger *et al.*, 2021), Kermanshah, Kordestan (du Buysson, 1912; Ebrahimi and Carpenter, 2008 both as *P.* (*P.*) *gallicus*), Khuzestan (Ebrahimi and Carpenter, 2008; Boyeri Shemi and Pashaei Rad, 2017 both as *P.* (*P.*) *gallicus*), Mazandaran (Guiglia, 1976a as both *P.* (*Leptopolistes*) *omissus*, and *P.* (*P.*) *gallicus*; Guiglia, 1976b; Ebrahimi and Carpenter, 2008 both as *P.* (*P.*) *gallicus*; Schmid-Egger *et al.*, 2021), Razavi Khorasan (Blüthgen and Gusenleitner, 1970 as both *Polistes* (*Leptopolistes*) *omissus* Weyrauch, 1939 and *P.* (*P.*) *gallicus*; Ebrahimi and Carpenter, 2008 as *P.* (*P.*) *gallicus*), Semnan (Schmid-Egger *et al.*, 2021), Sistan & Baluchestan (Blüthgen and Gusenleitner, 1970 as both *P.* (*Leptopolistes*) *omissus* and as *P.* (*P.*) *gallicus*; Ebrahimi and Carpenter, 2008 as *P.* (*P.*) *gallicus*), Southern Khorasan (Blüthgen and Gusenleitner, 1970 as both *P.* (*Leptopolistes*) *omissus*) and as *P.* (*P.*) *gallicus*, Tehran

(Ebrahimi and Carpenter, 2008 as *P.* (*P.*) *gallicus*; Schmid-Egger *et al.*, 2021), Zanjan (Abbasi, 2009; Abbasi *et al.*, 2008a, 2008b; Pashaei Rad *et al.*, 2010 all as *P.* (*P.*) *gallicus*); Generally distributed in the whole country (Modarres Awal, 1997, 2012 as *P.* (*P.*) *gallicus*; Schmid-Egger *et al.*, 2021).

Distribution in the Middle East: Cyprus (Schmid-Egger *et al.*, 2017; Selis *et al.*, 2025), Egypt, Israel-Palestine (Schmid-Egger *et al.*, 2017; Selis *et al.*, 2025), Turkey (Schmid-Egger *et al.*, 2017; Selis *et al.*, 2025), Iran (Schmid-Egger *et al.*, 2021; Selis *et al.*, 2025), Iraq (Selis *et al.*, 2025).

Extralimital distribution: Azerbaijan, China, Croatia, Greece, Mongolia, North Macedonia, Serbia (Schmid-Egger *et al.*, 2017; Selis *et al.*, 2025), Poland (Zyla *et al.*, 2023). Widespread from the Balkans to Central Asia, southwest Asia and northeast Africa (Schmid-Egger *et al.*, 2021).

Comments: The records of *Polistes gallicus* (Linnaeus, 1767) from Iran very likely refer to *Polistes mongolicus* du Buysson, 1911 or, very unlikely, it is *Polistes foederatus* Kohl, 1898 (see Schmid-Egger *et al.*, 2017, 2021; and Rainer Neumeyer personal communication). Re-examined specimens of some of the Iranian records prove such identification. *Polistes gallicus* was also excluded from the Iraqi fauna by Selis *et al.* (2025) for the following reasons: Augul (2017) cited Derwesh (1965) as a previous source for the presence of *P. gallicus* in Iraq. By examining photos of this species, Selis *et al.* (2025) attributed this to at least two different species apart from *P. gallicus*. Moreover, Derwesh (1965) has recorded *P. gallicus* var. *bucharensis*.

In Iran, *Polistes gallicus* has been reported as being a pest of date-palm and grapevine (Modarres Awal, 1997, 2012).

***Polistes* (*Polistes*) *nimpha* (Christ, 1791)**

Vespa nimpha Christ, 1791: 232, ♀; "Kronenberg an der Höh" (type lost).

Catalogues with Iranian records: Ebrahimi and Carpenter (2008), Rahmani *et al.* (2020).

Distribution in Iran: Guilan (Blüthgen and Gusenleitner, 1970), Mazandaran (Ebrahimi and Carpenter, 2008; Sakenin *et al.*, 2008), Razavi Khorasan (Abd-Rabou *et al.*, 2005; Sakenin *et al.*, 2010), Tehran (Blüthgen and Gusenleitner, 1970; Samin *et al.*, 2010), Zanjan (Abbasi *et al.*, 2008a, 2008b; Pashaei Rad *et al.*, 2010 all as *Polistes nimpha irakensis* Gusenleitner, 1976).

Distribution in the Middle East: Iran (see references above, and Carpenter, 1996; Gusenleitner, 2013; Schmid-Egger *et al.*, 2017; Selis *et al.*, 2025), Iraq (Derwesh, 1965; Gusenleitner, 1976; Al-Ali, 1977; Carpenter, 1996; Gusenleitner, 2013; Schmid-Egger *et al.*, 2017; Selis *et al.*, 2025), Israel-Palestine (Carpenter, 1996; Gusenleitner, 2013; Schmid-Egger *et al.*, 2017; Selis *et al.*, 2025), Jordan (Fateryga, 2018), Syria (Abo Kaf and Allewa, 2016), Turkey (Gusenleitner, 1967; Yildirim and Özbek, 1993, 1997, 1999 as *P.* (*P.*) *nimpha nimpha*; Carpenter, 1996; Yildirim and Gusenleitner, 2012; Gusenleitner, 2013; Schmid-Egger *et al.*, 2017; Fateryga, 2018; Kaplan and Efil, 2025; Selis *et al.*, 2025).

Extralimital distribution: Albania, Austria, Azerbaijan, Belgium, Czech Republic, Hungary, Latvia, Morocco, Poland, Serbia, Spain, Slovenia, Sweden (Schmid-Egger *et al.* 2017), Armenia (Castro and Dvořák, 2010), Bulgaria, China, Croatia, France, Germany, Greece, India, Italy, Kyrgyzstan, Kazakhstan, Mongolia, Pakistan, Portugal, Russia, Switzerland (Gusenleitner, 2013; Schmid-Egger *et al.*, 2017; Fateryga, 2018).

Comments: *Ancylorhynchus glaucus* (Rossi) (Diptera: Asilidae) has been recorded as being a predator of *P. nimpha* (Samin *et al.*, 2010).

***Polistes* (*Gyrostoma*) *olivaceus* (DeGeer, 1773)**

Vespa olivacea DeGeer, 1773: 582, sex not stated; "Amérique".

Catalogues with Iranian records: Ebrahimi and Carpenter (2008), Rahmani *et al.* (2020).

Distribution in Iran: Fars (Nemati and Pezhman, 2014 as *Polistes hebraeus* Fabricius), Hormozgan (Khoobdel *et al.*, 2014), Southern Iran (Behdad, 1984 as *P. hebraeus*).

Distribution in the Middle East: Egypt (Carpenter, 1997; Gusenleitner, 2013; Carpenter and Gadallah, 2020; Balachandar *et al.*, 2024), Iran (see references above, and Modarres Awal, 1997, 2012 both as *P. hebraeus*; Gusenleitner, 2013; Carpenter and Gadallah, 2020; Balachandar *et al.*, 2024), Oman (Richards, 1985b; Carpenter, 1996; Gusenleitner, 2013; Carpenter and Gadallah, 2020; Balachandar *et al.*, 2024).

Extralimital distribution: Afghanistan, Australia, Burma, China, India, Indonesia, Japan, Madagascar, Malaysia, Mauritius, New Zealand, Réunion, Seychelles, Singapore, Sri Lanka, Taiwan, Tanzania, Thailand (Carpenter, 1996; Gusenleitner, 2013; Carpenter and Gadallah, 2020), Bangladesh (Balachandar *et al.*, 2024), Cambodia, Chile, Fiji, French Polynesia, Hawaii, Laos, Marianas, Marquesas, Myanmar, Nepal, New Caledonia, Pakistan, Philippines, Samoa, Tonga, Tuamotu Archipelago, Vietnam (Carpenter, 1996).

Comments: Pashaei Rad *et al.* (2010) considered the record of *Polistes olivaceus* to actually belong to *P. wattii*, but we confirm and consider *P. olivaceus* as part of the fauna of Iran (southern regions) considering subsequent reports by Khoobdel *et al.* (2014) and Nemati and Pezhman (2014).

Given the confusion between *P. olivaceus* and *P. wattii*, the records of *P. olivaceus* by Morice (1921), Derwesh (1965), Khalaf and Al-Omar (1974), Abdul-Rassoul (1976), and Al-Ali (1977), are attributed to *P. wattii*, and so it was excluded from the Iraqi fauna (Selis *et al.*, 2025).

In Iran, it has been reported as being a pest of date palm by Behdad (1984) and Modarres Awal (1997, 2012) [both as *P. hebraeus* Fabricius].

***Polistes* (*Gyrostoma*) *rothneyi* Cameron, 1900**

Polistes rothneyi Cameron, 1900: 410, ♂; [India] "Barrackpore, Bengal".

Catalogues with Iranian records: Rahmani *et al.* (2020).

Distribution in Iran: Kerman (Nematy *et al.*, 2016).

Distribution in the Middle East: Iran (Nematy *et al.*, 2016).

Extralimital distribution: Pakistan (Khan *et al.*, 2018), India, Nepal (Carpenter, 1996 as *P.* (*Gyrostoma*) *rothneyi rothneyi*).

***Polistes* (*Polistes*) *semenowi* Morawitz, 1889**

Polistes semenowi Morawitz, 1889: 552, ♀, ♂; [Turkmenistan] "Copet-Dag im transcaspischen Gebiete".

Catalogues with Iranian records: Ebrahimi and Carpenter (2008 as both *Polistes semenowi* Morawitz, 1889 and *Polistes sulcifer* Zimmermann, 1931), Rahmani *et al.* (2020).

Distribution in Iran: Golestan (Schmid-Egger *et al.*, 2021), Mazandaran (Guiglia, 1976a as *Sulcopolistes semenowi* (Morawitz, 1889); Ebrahimi and Carpenter, 2008 as *P. sulcifer*), Tehran (Ebrahimi and Carpenter, 2008 as *Polistes sulcifer*).

Distribution in the Middle East: Egypt, Syria (Gusenleitner, 2013 as *P.* (*P.*) *sulcifer*; Schmid-Egger *et al.*, 2017), Iran (see references above, and Carpenter, 1996; Gusenleitner, 2013), Israel-Palestine (Carpenter, 1996 as *P.* (*Polistes*) *sulcifer*; Yildirim and Kojima, 1999 as *P. sulcifer*; Castro and Dvořák, 2009), Turkey (Blüthgen, 1961 as *Sulcopolistes semenowii*; Gusenleitner, 1967 as *Sulcopolistes sulcifer*; Guiglia, 1972 as *Sulcopolistes semenowii*; Yildirim and Özbek, 1993, 1999 both as *P.* (*Sulcopolistes*) *sulcifer*; Carpenter, 1996 as *P.* (*Polistes*) *sulcifer*; Madl, 1997 as *P.* (*Sulcopolistes*) *sulcifer*; Yildirim and Kojima, 1999 as both *P. sulcifer* and *P. semenowi*; Yildirim and Gusenleitner, 2012 as both *P. semenowi* and *P.* (*Sulcopolistes*) *sulcifer*; Gusenleitner, 2013; Schmid-Egger *et al.*, 2017).

Extralimital distribution: Azerbaijan, Caucasus, Croatia, France, Greece, Italy, Montenegro, North Macedonia, Serbia, Switzerland, Turkmenistan (Castro and Dvořák, 2009; Gusenleitner, 2013; Schmid-Egger *et al.*, 2017), North Africa (Carpenter, 1996; Gusenleitner, 2013), Russia (Carpenter, 1996).

***Polistes* (*Polistella*) *smithii neavi* von Schulthess, 1921**

Polistes smithii var. *neavi* von Schulthess, 1921: 126.

Distribution in the Middle East: Yemen (Dvořák and Carpenter, 2008; Gusenleitner, 2013).

Extralimital distribution: Zanzibar to Zambia, Sierra Leone (Gusenleitner, 2013).

***Polistes* (*Polistes*) *tenellus lahejensis* Giordani Soika, 1981**

Polistes tenellus lahejensis Giordani Soika, 1981: 175, ♀, ♂; "Arabia: Lahej".

Distribution in the Middle East: Known only from Yemen (Giordani Soika, 1981; Carpenter, 1996; Gusenleitner, 2013).

Extralimital distribution: None.

***Polistes* (*Gyrostoma*) *wattii* Cameron, 1900**

Polistes wattii Cameron, 1900: 416, ♀; [India] "Bengal".

Catalogues with Iranian records: Ebrahimi and Carpenter (2008), Rahmani *et al.* (2020).

Distribution in Iran: Bushehr, Fars, Ilam, Kermanshah (Ebrahimi and Carpenter, 2008; Modarres Awal, 2012), Hormozgan (Blüthgen and Gusenleitner, 1970 as *P.* (*Megapolistes*) *wattii* Cameron, 1900; Guiglia, 1976a; Ebrahimi and Carpenter, 2008; Modarres Awal, 2012), Kerman (Ebrahimi and Carpenter, 2008; Pashaei Rad *et al.*, 2010; Modarres Awal, 2012; Schmid-Egger *et al.*, 2021), Khuzestan (Blüthgen and Gusenleitner, 1970 as *P.* (*Megapolistes*) *wattii*; Ebrahimi and Carpenter, 2008; Modarres Awal, 2012; Boyeri Shemi and Pashaei Rad, 2017), Sistan & Baluchestan (Blüthgen and Gusenleitner, 1970 as *P.* (*Megapolistes*) *wattii*; Ebrahimi and Carpenter, 2008; Modarres Awal, 2012), Tehran (Ghahari *et al.*, 2007).

Distribution in the Middle East: Iran (see references above, and Castro and Dvořák, 2010; Gusenleitner, 2013; Ebrahimi, 2014; Selis *et al.*, 2025), Iraq (Morice, 1921 as *P. hebraeus*; Derwesh, 1965 as *P. hebraeus*; Khalaf and Al-Omar, 1974 as *P. olivaceus*; Swailem *et al.*, 1974; Abdul-Rassoul, 1976 as *P. macaensis*; Al-Ali, 1977 as *P. hebraeus*; Carpenter, 1996; Castro and Dvořák, 2010; Augul, 2017; Al-Saffar and Augul, 2021; Selis *et al.*, 2025), Kuwait (Richards, 1985b; Carpenter and Gadallah, 2020), Oman (Hamer, 1985 as *Polistes* Walt.; Richards, 1985b; Walker and Pittaway, 1987; Gusenleitner, 2013; Monks *et al.*, 2019; Carpenter and Gadallah, 2020; Selis *et al.*, 2025), Saudi Arabia (Richards, 1985b; Walker and Pittaway, 1987; Castro and Dvořák, 2010; Gusenleitner, 2013; Carpenter and Gadallah, 2020; Selis *et al.*, 2025), United Arab Emirates (Hamer, 1985 as *Polistes* Walt.; Richards, 1985b; Walker and Pittaway, 1987; Gillett and Howarth, 2004; Gillett and Gillett, 2005; Howarth and Gillett, 2008; Gusenleitner, 2010, 2013; Gess and Roosenschoon, 2016; Carpenter and Gadallah, 2020; Selis *et al.*, 2025).

Extralimital distribution: Afghanistan, China, India, Mauritius, Pakistan, Tajikistan, Turkmenistan (Castro and Dvořák, 2010; Gusenleitner, 2013; Khan *et al.*, 2018; Selis *et al.*, 2025).

Comments: *Saropogon pollinosus* Loew (Diptera: Asilidae) has been recorded by Ghahari *et al.* (2007) as a predator of *P. wattii.*

***Polistes* sp.**

Distribution in Iran: Isfahan (Ebadi, 1995), Iran (no specific locality cited) (Esmaili and Rastegar, 1974).

Comments: *Polistes* sp. was introduced as a pollinator insect by Ebadi (1995).

Tribe Ropaliini Guérin-Méneville, 1831

Genus *Belonogaster* de Saussure, 1854

Belonogaster de Saussure, 1854: 235. [Replacement name for *Raphigaster* de Saussure]

***Belonogaster adenensis adenensis* Giordani Soika, 1957**

Belonogaster abyssinicus adenensis Giordani Soika, 1957: 484, ♀; Southern Yemen: Jebel Jihaf, 7100ft [2165 m].

Distribution in the Middle East: Yemen (Giordani Soika, 1957; Richards, 1985b; Kojima and Carpenter, 1997; Wahis, 2000 as *B. adenensis somaliensis* Richards; Gusenleitner, 2007, 2013; Dvořák and Carpenter, 2010; Carpenter and Gadallah, 2020).

Extralimital distribution: Ethiopia, Somalia (Carpenter and Gadallah, 2020).

***Belonogaster arabica* Giordani Soika, 1957**

Belonogaster grisea arabicus Giordani Soika, 1957: 484, ♀; Southern Yemen: Dhala, 4800ft [1460 m].

Distribution in the Middle East: Saudi Arabia (Richards, 1982, 1985b; Gusenleitner, 2013; Carpenter and Gadallah, 2020), Yemen (Giordani Soika, 1957 as *B. grisea arabicus*; Richards, 1985a; Kojima and Carpenter, 1997; Wahis, 2000; Dvořák and Carpenter, 2010; Gusenleitner, 2013; Carpenter and Gadallah, 2020).

Extralimital distribution: None.

***Belonogaster filiformis* (de Saussure, 1853)**

Raphigaster filiformis de Saussure, 1853: 18, ♀; Saudi Arabia 'Arabie Djeddah'.

Distribution in the Middle East: Known only from Saudi Arabia (de Saussure, 1853 as *Raphigaster filiformis*; Richards, 1982, 1985b; Kojima and Carpenter, 1997; Gusenleitner, 2013; Carpenter and Gadallah, 2020).

Extralimital distribution: None.

***Belonogaster guichardi* Richards, 1982**

Belonogaster guichardi Richards, 1982: 94; ♂, ♀; Oman: Dhofar, Ayun Pools, 700 m.

Distribution in the Middle East: Oman (Richards, 1982, 1985b; Kojima and Carpenter, 1997; Gusenleitner, 2012, 2013; Carpenter and Gadallah, 2020), Saudi Arabia (Gusenleitner, 2013).

Extralimital distribution: None.

***Belonogaster juncea juncea* (Fabricius, 1781)**

Vespa juncea Fabricius, 1781: 468, ♀; Africa Aeqinoctiali.

Distribution in the Middle East: Saudi Arabia (Richards, 1985b; Gusenleitner, 2013; Carpenter and Gadallah, 2020), Yemen (Wahis, 2000; Carpenter and Gadallah, 2020).

Extralimital distribution: Entire Africa, India, Libya (Gusenleitner, 2013; Carpenter and Gadallah, 2020).

***Belonogaster saussurei* Kirby, 1881**

Belonogaster saussurei Kirby, 1881: 649, ♂; Southern Yemen: Socotra (Balfour).

Distribution in the Middle East: Known only from Yemen (including Socotra) (Kirby, 1881; Kohl, 1906; Richards, 1982, 1985b; Kojima and Carpenter, 1997; Carpenter and Gadallah, 2020).

Extralimital distribution: None.

Genus *Parapolybia* de Saussure, 1854

Parapolybia de Saussure, 1854: 207. Type species: *Polybia* (*Parapolybia*) *indica* de Saussure, 1854, by subsequent designation of Bingham, 1897.

***Parapolybia escalerae* (Meade-Waldo, 1911)**

Polybia escalerae Meade-Waldo, 1911: 109, ♀; "Dupuln Baktiari, S.W. Persia".

Catalogues with Iranian records: Ebrahimi and Carpenter (2008), Rahmani *et al.* (2020).

Distribution in Iran: Bushehr (Ebrahimi and Carpenter, 2008; Modarres Awal, 2012), Guilan (Ghahari *et al.*, 2007), Kerman (Ebrahimi and Carpenter, 2008; Shahreyari Nejad, 2022), Khuzestan (Meade-Waldo, 1911), Lorestan (Ebrahimi and Carpenter, 2008).

Distribution in the Middle East: Iran (see references above, and Kojima and Carpenter, 1997; Gusenleitner, 2013; Selis *et al.*, 2025), Iraq (Carpenter, 1999; Ali *et al.*, 2023; Selis *et al.*, 2025), Turkey (Gusenleitner, 1988, 2013; Kojima and Carpenter, 1997; Yildirim and Kojima, 1999; Selis *et al.*, 2025).

Extralimital distribution: Pakistan (Kojima and Carpenter, 1997; Gusenleitner, 2013; Maryam *et al.*, 2024; Selis *et al.*, 2025).

Comments: *Parapolybia escalerae* produces wild honey, and its honey is known as 'dry honey' or 'chocolate honey' and has medicinal uses too. The activity of *P. escalerae* is different from *Apis florea* Fabricius, 1781 (Hymenoptera: Apidae). The latter species works in open space and single comb but *P. escalerae* has activity in the dark space between mountains and rocks and produces several combs (Shahreyari Nejad, 2022).

Andrenosoma serratum Hermann, 1906 (Diptera: Asilidae) has been recorded by Ghahari *et al.* (2007) as being a predator of *Parapolybia escalerae*.

***Parapolybia persica* (Meade-Waldo, 1911)**

Polybia persica Meade-Waldo, 1911: 108, ♀; "Kuh Sefid, S.W. Persia".

Catalogues with Iranian records: Ebrahimi and Carpenter (2008), Rahmani *et al.* (2020).

Distribution in Iran: Khuzestan (Meade-Waldo, 1911 as *Polybia persica*).

Distribution in the Middle East: Known only from Iran (Meade-Waldo, 1911; von Schulthess, 1913; Kojima and Carpenter, 1997; Gusenleitner, 2013).

Extralimital distribution: None.

Genus *Ropalidia* Guérin-Méneville, 1831

Ropalidia Guérin-Méneville, 1831: pl. 9, fig. 8. Type species: *Ropalidia maculiventris* Guérin-Méneville, 1831, by monotypy.

***Ropalidia aethiopica aethiopica* (du Buysson, 1907)**

Icaria aethiopica du Buysson, 1907: 110, ♀, ♂; "Ethiopie meridionale: Dirouhoria".

Distribution in the Middle East: Saudi Arabia (Richards, 1985b; Kojima and Carpenter, 1997; Gusenleitner, 2013; Carpenter and Gadallah, 2020), Yemen (Giordani Soika, 1952; Kojima and Carpenter, 1997; Gusenleitner, 2007, 2013; Carpenter and Gadallah, 2020; Polašek *et al.*, 2025).

Extralimital distribution: Ethiopia, Somalia, South Africa, Sudan, Tanzania (Kojima and Carpenter, 1997; Gusenleitner, 2013; Carpenter and Gadallah, 2020; Polašek *et al.*, 2025).

***Ropalidia cincta* (Lepeletier de Saint Fargeau, 1836)**

Epipona cincta Lepeletier de Saint Fargeau, 1836: 541, ♀; "Sénégal, Afrique equinoxial".

Distribution in the Middle East: Yemen (Madl, 2018; Carpenter and Gadallah, 2020).

Extralimital distribution: Democratic Republic of Congo, Djibouti, Eritrea, Ethiopia, Gambia, Kenya, Mozambique, Namibia, Republic of Congo, Senegal, Somalia, South Africa, Sudan, Tanzania, Uganda, Zimbabwe (Kojima and Carpenter, 1997; Carpenter and Gadallah, 2020).

Comments: Madl (2018) recorded this species from Yemen without citing the examined specimens or a supporting reference. Therefore, presence of *R. cincta* in Yemen requires confirmation.

***Ropalidia irrequieta* (Kohl, 1906)**

Icaria irrequieta Kohl, 1906: 222, ♀; "Südarabien (Ras Farták)".

Distribution in the Middle East: Oman (Polašek *et al.*, 2025), Yemen (Kohl, 1906 as *Icaria irrequieta*; Kojima and Carpenter, 1997; Gusenleitner, 2013; Carpenter and Gadallah, 2020; Polašek *et al.*, 2025).

Extralimital distribution: Djibouti, Eritria, Somalia, Sudan (Polašek *et al.*, 2025).

Species Excluded from the Fauna of Iran and the Middle East

In total, two species are excluded from the fauna of the Middle East, both of which were erroneously recorded from Iran. Additionally, *Polistes* (*Polistes*) *gallicus* has been excluded from the fauna of Iraq by Selis *et al.* (2025), and also Iran (see comment under *Polistes mongolicus*). However, since it is distributed in other Middle Eastern countries, we retain it in the regional species list.

***Polistes caspica* Fabricius [nomen nudum]**

Records from Iran: Guilan, Mazandaran (Mossadegh, 1995), Khorasan, Caspian Sea area (Farahbakhsh, 1961; Modarres Awal, 1997).

Comments: This taxon has been recorded as a pest of grapevine, apricot and some other fruits (Farahbakhsh, 1961; Modarres Awal, 1997), as well as associated with hives of *Apis mellifera* Linnaeus (Hymenoptera: Apidae) (Mossadegh, 1995). *Polistes caspica* is probably a mistyping of *Vespa crabro caspica* Perez, 1910.

***Polistes* (*Fuscopolistes*) *carolina* (Linnaeus, 1767)**

Vespa carolina Linnaeus, 1767: 948.

Records from Iran: Hormozgan (Jahani *et al.*, 2020).

Comments: Distribution of *Polistes carolina* is restricted to USA, and adventive in Bermuda (Richards, 1978; Krombein *et al.*, 1979).

CONCLUSION

The Polistinae fauna of the Middle East comprises 32 species, approximately 3% of the world's known species, across four genera: *Belonogaster*, *Parapolybia*, *Polistes*, and *Ropalidia*. Among these, *Polistes* is the most diverse, with 21 recorded species, followed by *Belonogaster* (six species), *Ropalidia* (three species), and *Parapolybia* (two species). Two species, *Polistes caspica* and *Polistes carolina*, are excluded from the Middle Eastern fauna, as both were erroneously recorded from Iran.

A total of 17 species belonging to two genera, *Polistes* (15 species) and *Parapolybia* (two species), have been recorded from Iran to date. One species, *Parapolybia persica*, is known only from Iran (endemic). Iranian Polistinae have been recorded from 28 provinces, with the exceptions of Qazvin, West Azarbaijan, and Yazd. Among the provinces, Kerman has the highest species richness, with eight recorded species, followed by Tehran (six species), Golestan, Guilan, Khuzestan, Mazandaran (each with five species), Fars, Razavi Khorasan, Zanjan (each with four species), East Azarbaijan, Hormozgan, Kordestan, Lorestan, Markazi, Semnan, Sistan & Baluchestan (each with three species), Bushehr, Ilam, Isfahan, Kermanshah, Southern Khorasan (each with two species), Alborz, Ardabil, Chaharmahal & Bakhtiari, Hamadan, Kuhgiloyeh & Boyerahmad, Northern Khorasan, and Qom (each with one species).

The exact locality of one species, *Polistes chinensis*, is unknown, and its presence in Iran requires verification.

A comparison of the polistine wasp fauna across 17 countries in the Middle East suggests that Iran, with 17 recorded species, is exceptionally diverse. It is followed by Turkey (13 species), Yemen (11 species), Israel-Palestine (eight species), Saudi Arabia (seven species), Egypt and Iraq (each with six species), Cyprus and Syria (each with five species), Oman (four species), Jordan (three species), the United Arab Emirates (two species), and Kuwait (one species). No species have been recorded from Bahrain, Lebanon, and Qatar; this is very likely due to a lack of collection effort rather than true absence.

Among the Middle Eastern and adjacent countries, Turkey shares the highest number of species with Iran (12 species), followed by Pakistan (eight species), Israel-Palestine, Russia (each with seven species), Egypt, Iraq (each with six species), Azerbaijan, Kazakhstan, Syria (each with five species), Afghanistan, Cyprus (each with four species), Armenia, Oman, Turkmenistan (each with three species), Jordan, United Arab Emirates (each with two species), and Kuwait and Saudi Arabia (each with one species).

BIBLIOGRAPHY

Abbasi, R. 2009. Sexual dimorphism in the wing morphology of social vespid wasps – a case study on the genus *Polistes* Latreille using geometric morphometrics (Hymenoptera: Vespidae). *Zoology in the Middle East* 47: 73–84.

Abbasi, R., Mashhadikhan, M., Abbasi, M. and Kiabi, B. 2008a. Biodiversity of vespid wasps in spatial and temporal dimension in northern Zanjan province of Iran. *Chinese Journal of Ecology* 27(5): 797–802.

Abbasi, R., Pashaei Rad, S., Ebrahimi, E. and Sheidaei, M. 2008b. Faunistic study of vespid wasps in Zanjan province (Northwest of Iran) with some ecological measures. *Environmental Sciences* 6(1): 65–74.

Abbasi, R., Mashhadikhan, M., Abbasi, M. and Kiabi, B. 2009. Geometric morphometric study of populations of the social wasp, *Polistes dominulus* (Christ, 1791) from Zanjan province, north-west Iran. *New Zealand Journal of Zoology* 36: 41–46.

Abd-Rabou, S., Ghahari, H., Mauss, V. and Plant, J. 2005. New records of Apidae, Andrenidae, Sphecidae and Vespidae (Hymenoptera) from Iran. *Egyptian Journal of Agricultural Research* 83(4): 1613–1618.

Abdul-Rassoul, M.S. 1976. Checklist of Iraq Natural History Museum insects collections. *Iraq Natural History Research Center* 30: 1–41.

Abo Kaf, N.H. and Allewa, A.N. 2016. Identification of some species of wasps family (Vespidae; Hymneoptera) in Lattakia, Syria. *Tishreen University Journal for Research and Scientific Studies* 38 (4): 224–235.

Akbayin, H. and Özbay, C. 1996. Variability analysis on some Vespidae species from Diyarbakrr, Mardin and $anhurfa in South-east Anatolia. Doga. *Turkish Journal of Biology* 20(4): 351–358.

Al-Ali, A.S. 1977. Phytophagous and entomophagous insects and mites of Iraq. *Iraq Natural History Research Center* 33: 1–142.

Ali, W.K., Sirwan, M. and Ahmed, D.H. 2023. New record and observations of *Parapolybia escalerae* (Meade-Waldo, 1911) (Hymenoptera: Vespidae: Polistinae) in Kurdistan, Iraq. *Zootaxa* 5230(1): 97–100.

Al-Saffar, H.H. and Augul, R.S. 2021. Survey of insects in some Southern Iraqi marshes. *Bulletin of Iraq Natural History Museum* 16(4): 571–621.

Antropov, A.V. 2017. Subfamily Polistinae – Paper wasps, pp. 192–193. In: Belokobylskij, S.A. and Lelej, A.S. (eds), Annotated catalogue of the Hymenoptera of Russia. Volume I. Symphyta and Apocrita: Aculeata. *Proceedings of the Zoological Institute of the Russian Academy of Sciences*, Supplement No. 6, 475 pp.

Antropov A.V. and Proshchalykin M.Y. 2014. To the fauna of social wasps (Hymenoptera: Vespidae: Vespinae, Polistinae) of southern Siberia. *A.I. Kurentsov's Annual Memorial Meetings* 25: 127–133.

Arévalo, E., Zhu, Y., Carpenter, J.M. and Strassmann, J.E. 2004. The phylogeny of the social wasp subfamily Polistinae: Evidence from microsatellite flanking sequences, mitochondrial COI sequence, and morphological characters. *BMC Evolutionary Biology* 4(1): 8. DOI: 10.1186/1471-2148-4-8

Augul, R.S. 2017. Revision of vespid wasps (Hymenoptera; Vespidae) in Iraq. *Journal of Biodiversity and Environmental Sciences* 10(3): 99–111.

Balachandar, M., Malaiyapan, R. and Azhagu Raj, R. 2024. New record: Study on morphometric and molecular taxonomy of Vespidae species (Hymenoptera: Vespidae: Polistinae) in different districts of Tamil Nadu, India. https://doi.org/10.21203/rs.3.rs-4507993/v1

Behdad, A. 1984. *Pests of fruit crops*. Neshat Publication of Isfahan, Iran, 822 pp. [in Persian]

Bingham, C.T. 1897. The Fauna of British India, including Ceylon and Burma. In W.T. Blanford (ed.) *Hymenoptera*, Vol. I. Wasps and bees. Taylor and Francis, London, xxix + 579.

Blüthgen, P. 1956. Über einige *Polistes*-Arten der Zoolog. Staatssammlung in München. *Nachrichtenblatt der Bayerischen Entomologen* 5(9): 81–86.

Blüthgen, P. 1961. Die Faltenwespen Mitteleuropas. *Abhandlungen der Deutschen Akademie der Wissenschaften zu Berlin, Klasse für Chemie, Geologie und Biologie* 1961 (2): 251 pp.

Blüthgen, P. and Gusenleitner, J. 1970. Über Faltenwespen aus dem Iran (Hym., Diploptera), *Stuttgarter Beiträge zur Naturkunde* 223: 1–13.

Blüthgen, P. and Königsmann, E. 1969. Ergebnisse der Albanien-Expedition 1961 des Deutschen Entomologischen Institutes, 179: Hymenoptera: Vespidae, Eumenidae, Masaridae. *Beiträge zur Entomologie* 19 (7–8): 917–935.

Boyeri Shemi, F. and Pashaei Rad, S. 2017. Faunistic study of Vespidae (Hymenoptera) in Izeh, Khuzestan province. *Proceedings of the 3rd National Congress on Development and Promotion of Agricultural Engineering and Soil Sciences of Iran, 3 September 2017*, Tehran, 11 pp. [in Persian, English summary]

Brito, S., Oliveira, M.L., Carpenter, J.M. and Somavilla, A. 2024. The genus *Polistes* Latreille, 1802 (Hymenoptera: Vespidae) of Brazil: morphological diagnoses, identification key, and geographical distribution. *Zootaxa* 5496(4): 509–545.

Cameron, P. 1900. Descriptions of new genera and species of Hymenoptera. *Annals and Magazine of Natural History* (7) 6: 410–419, 495–506, 530–539.

Carpenter, J.M. 1991. Phylogenetic relationships and the origin of social behavior in the Vespidae, pp. 7–32. In: Ross, K.G. and Matthews, R.W. (eds), *The social biology of wasps*. Cornell University Press, Ithaca New York, xvii + 678 pp.

Carpenter, J.M. 1993. Biogeographic patterns in the Vespidae (Hymenoptera), two views of Africa and South America, pp. 139–155. In: Goldblatt, P. (ed.), Biological relationships between Africa and South America. Yale University Press, New Haven, CT, x + 630 pp.

Carpenter, J.M. 1996. Distributional checklist of species of the genus *Polistes* (Hymenoptera, Vespidae, Polistinae, Polistini). *American Museum Novitates* 3188: 1–39.

Carpenter, J.M. 1997. Phylogenetic relationships among European *Polistes* and the evolution of social parasitism. In: Grandcolas, P. (ed.), *The origin of biodiversity in insects; phylogenetic tests of evolutionary scenarios. Memoires du Museum national d'Histoire naturelle* 173: 135–161.

Carpenter, J.M. 1999. Taxonomic notes on paper wasps (Hymenoptera: Vespidae; Polistinae). *American Museum Novitates* 3259: 1–44.

Carpenter, J.M. and Gadallah, N.S. 2020. Biodiversity of aculeate wasps (Hymenoptera: Aculeata) of the Arabian Peninsula: Vespoidea, Vespidae. *Zootaxa* 4754(1): 191–216.

Carpenter, J.M. and Nguyen, L.P.T. 2003. Keys to the genera of social wasps of South-east Asia (Hymenoptera: Vespidae). *Entomological Science* 6: 183–192.

Carvalho, A.F., Menezes, R.S.T., Somavilla, A., Costa, M.A. and Del Lama, M.A. 2015. Polistinae biogeography in the Neotropics: history and prospects. *Journal of Hymenoptera Research* 42: 93–105.

Castro, L. and Dvořák, L. 2009. New and noteworthy records of vespid wasps (Hymenoptera: Vespidae) from the Palaearctic region (II). *Boletín de la Sociedad Entomológica Aragonesa* 44: 295–304.

Castro, L. and Dvořák, L. 2010. New and noteworthy records of vespid wasps (Hymenoptera: Vespidae) from the Palaearctic region (III). *Acta Musei Moraviae, Scientiae Biologicae* (Brno) 95(2): 37–53.

Christ, J.L. 1791. Naturgeschichte, Classification und Nomenclatur der Insekten vom Bienen-, Wespen, und Ameisengeschlecht, etc. 535 pp., 60 pls. Mainz.

de Saussure, H. 1852–1858. *Études sur la Famille des Vespides 1–3*. V. Masson, Paris and J. Cherbuliez, Genève, 350 + 470 + 364 pp., 22 + 39 + 16 pls.

de Saussure, H. 1853–1858. Monographie Des Guêpes Sociales Ou de la Tribu Des Vespiens. Masson, Paris and J. Cherbuliez, Genève.

de Saussure, H. 1854–1856. *Études sur la famille des Vespides III: La monographie des Masariens et un supplément a la monographie des Euméniens*. Kessmann & Cherbuliez Genève, 352 pp.

DeGeer, C. 1773. Memoires pour servir a l'histoire des insectes. Vol. 3, Stockholm: Pierre Hesselberg, 696 pp. + 44 pls.

Derwesh, A.I. 1965. A preliminary list of identified insects and arachnids of Iraq. *Director General Agriculture Research Projects Baghdad, Bulletin* 1965: p. 121, 123.

du Buysson, R. 1907. Nouvelles espèces d'Ischnogaster (Hyménoptères) appartenant au Musée de Leide. *Notes from the Leyden Musum* 29: 79–80.

du Buysson, M.R. 1911. Mission Pelliot-Vaillant dans l'Asie centrale; collections recueillis par M. le Dr. Vaillant, Hymenopteres. *Bulletin du Museum National d'Histoire Naturelle* 17(4): 217–219.

du Buysson, R. 1912. Les Vespides, Euménides et Scoliides. *Annales d'Histoire Naturelle Délégation en Perse II. Entomologie* 1: 77–83.

Dvořák, L. and Carpenter, J. 2008. The first record of the paper wasp *Polistes smithii* neavei from the Middle East (Hymenoptera: Vespidae), with notes on the social wasp fauna of Yemen. *Zoology in the Middle East*, 44(1): 119–120.

Dvořák, L. and Carpenter, J. M. 2010. New records of vespid wasps from Yemen with synonymy in *Belonogaster* (Hymenoptera: Vespidae: Polistinae and Eumeninae). *Linzer Biologische* Beiträge, 42 (1): 561–563.

Ebadi, R. 1995. Collection, identification and preliminary study of pollinator insects fauna in Isfahan province. *Proceedings of the 12th Iranian Plant Protection Congress, 2–7 September 1995, Karaj Junior College of Agriculture, Karaj*, p. 319.

Ebrahimi, E. 2014. Aculeate wasps and their role in natural equilibrium and pest control (with a review of the Iranian species). *Biocontrol in Plant Protection* 2(2): 59–68.

Ebrahimi, E. and Carpenter, J.M. 2008. Catalog of the vespid wasps of Iran (Hymenoptera, Vespidae). *Zootaxa* 1785: 1–42.

Erichson, W.F. 1849. Ménétriés E. Catalogue des insects recueillis par feu M. Lehman. *Mémoires de l'Académie Impériale des Sciences de Saint-Petersbourg* 8(6): 307.

Esmaili, M. and Rastegar, R. 1974. Identified species of aculeate Hymenoptera of Iran. *Journal of the Entomological Society of Iran* 2(2): 41–52.

Ezenwa, V.O., Peters, J.M., Hastings, M., Zhu, Y., Arévalo, E., Seppä, P., Pedersen, J.S., Zacchi, F., Queller, D.C. and Strassmann, J.E. 1998. Ancient conservation of microsatellite loci in polistine wasps. *Molecular Phylogenetics and Evolution* 9: 168–177.

Fabricius, J.C. 1775. *Systema entomologiae*. xxxii, Flensburgi & Lipsiae.
Fabricius, J.C. 1781. *Species insectorum exhibentes eorum differentias specificas*. 1. vii, 552 pp., Bohnii, Hamburgi & Kilonii [= Hamburg and Kiel]. I-VIII, 1–517 pp.
Fabricius, J.C. 1793. Entomologia systematica. Hafniae. 2: viii, 519 pp.
Farahbakhsh, G. 1961. Family Vespidae (Hymenoptera), p. 113. In: Farahbakhsh, G. (ed.), *A checklist of economically important insects and other enemies of plants and agricultural products in Iran*. The Ministry of Agriculture, Department Plant Protection, 153 pp.
Fateryga, A.V. 2018. Wasps of the family Vespidae (Hymenoptera) of the Crimean Peninsula. *Entomofauna* 39(1): 193–233.
Gadagkar, R. 1991. *Belonogaster*, *Mischocyttarus*, *Parapolybia*, and independent founding *Ropalidia*, pp. 149–190. In: Ross, K.G. and Matthews, R.W. (eds), *The social biology of wasps*. Cornell University Press, Ithaca New York, xvii + 678 pp.
Genise, J.F. 2000. The ichnofamily Celliformidae for *Celliforma* and allied ichnogenera. *Ichnos 7*: 267–282.
Gerstaecker, A. 1871. Beitrag zur Insektenfauna von Zanzibar. *Archiv für Naturgeschichte* 37: 345–363.
Gess, S.K. and Roosenschoon, P.A. 2016. A preliminary survey of flower visiting by aculeate wasps and bees in the Dubai Desert Conservation Reserve, United Arab Emirates. *Journal of Hymenoptera Research* 52: 81–141.
Ghahari, H., Lehr, P.A., Lavigne, R.J., Hayat, R. and Ostovan, H. 2007. New records of robber flies (Diptera, Asilidae) for the Iranian fauna with their prey records. *Far Eastern Entomologist* 179: 1–9.
Gillett, M. and Gillett, C. 2005. Insects and other arthropods. In: Hellyer, P. and Aspinall, S. (eds.), *The Emirates, a Natural History*. Trident Press Ltd., London, pp. 168–195, 352–356, 377–379.
Gillett, M. and Howarth, B. 2004. The insects of Jebel Hafit (including a short note on representatives of other arthropod groups), pp. 94–143. In: Aspinall, S. and Hellyer, P. (eds), *Jebel Hafit, a Natural History*. Emirates Natural History Group (ENHG), Abu Dhabi, 221 pp.
Giordani Soika, A. 1952. Hyménoptères récoltés par une mission Suisse au Maroc (1947). Eumeninae. *Bulletin de la Société Scientifique Naturelle de Maroc* 32: 235–267.
Giordani Soika, A. 1957. South-West Arabia Expedition 1937–38. 31 Hymenoptera: Vespidae. London, 1: 471–484.
Giordani Soika, A. 1981 [1980]. Eumenidi raccolti nell'Arabia meridionale da K. Guichard (Hym.). Nota II. *Bollettino del Museo Civico di Storia Naturale di Venezia* 31: 111–116.
Goulet, H. and Huber, J.T. 1993. Hymenoptera of the world: An identification guide to families. Agriculture Canada Publication 1894/E, Ottawa, 668 pp.
Guérin-Méneville, F.E. 1831. Insectes. In: Duperrey, L. (ed.), *Voyage autour du monde execute sur la corvette la Coquille, 1822–1825, par L. I. Duperrey, Zoologie, II (pt. 2, Div. 1), Atlas, pls*. Nabu Press, Paris, XX–XXI.
Guérin-Méneville, F.E. 1844. *Iconographie de règne animal nature de G. Cuvier, ou représentation, d'après nature et souvent non encore figurée, de chaqun des genres d'animaux Insectes*. J.B. Baillière, Paris, 576 pp.
Guiglia, D. 1972. *Les guepes sociales (Hymenoptera Vespidae) d'Europe occidentale et septentrionale. Faune de l'Europe et du Bassin Mediterraneen 6*. Masson et Cie, Paris.
Guiglia, D. 1976a. Contributo alla conoscenza dei Polistini dell'Iran (Hymenoptera, Vespidae). *Bollettino del Museo Civico di Storia Naturale di Venezia* 81: 195–201.
Guiglia, D. 1976b[1977]. Le vespe sociali (Missione A. Giordani Soika in Iran, 7). *Bollettino del Museo Civico di Storia Naturale di Venezia* 82: 99–104.
Gusenleitner, L. 1967. Vespidae, Eumenidae und Masaridae aus der Türkei, Teil II. *Polskie Pismo Entomologiczne* 37: 663–676.
Gusenleitner, J. 1976. Bemerkenswertes über Faltenwespen VI (Diploptera, Hymenoptera). *Nachrichtenblatt der Bayerischen Entomologen* 25(6): 112–119.
Gusenleitner, J. 1988. Neue und Bemerkenswerte Faltenwespen aus der Türkei (Hymenoptera, Vespoidea). *Linzer biologische Beiträge* 20(2): 713–737.
Gusenleitner, J. 2004. Über Faltenwespen aus dem Oman (Hymenoptera: Vespidae, Eumenidae, Masaridae). *Linzer biologische Beiträge* 36(1): 151–158.
Gusenleitner, J. 2005. Über Faltenwespen aus dem Oman Teil 2 (Hymenoptera: Vespidae, Eumenidae, Masaridae). *Linzer biologische Beiträge* 37(2): 1199–1201.
Gusenleitner, J. 2007. Vespidae aus Jemen in der Sammlung des Oberösterreichischen Landesmuseums (Hymenoptera, Vespidae). *Linzer biologische Beiträge* 39(1): 119–127.

Gusenleitner, J. 2010. Order Hymenoptera, family Vespidae. In: van Harten, A. (ed.), *Arthropod Fauna of the UAE* 3: 422–467.

Gusenleitner, J. 2012. Vespidae aus dem Oman und der Union der Arabischen Emirate (Hymenoptera: Vespidae: Eumeninae, Masarinae). Nachtrag. *Linzer biologische Beiträge*, 44(1): 307–312.

Gusenleitner, J. 2013. Die Gattungen der Vespidae im Nahen Osten, in Nordafrika und in Arabien (Hymenoptera: Vespidae). *Linzer biologische Beiträge* 45(2): 1025–1046.

Haddad, N.J., Dvořák, L., Adwan, O., Mdanat, H. and Batayneh, A. 2007. New data on vespid wasp fauna of Jordan (Hymenoptera, Vespidae). *Linzer biologische Beiträge* 39(1): 137–142.

Hamer, I. 1985. Bees and wasps of the Ruus al Jibal. *Bulletin of the Abu Dhabi Natural History Group* 25: 24.

Hayat, R., Ghahari, H., Lavigne, R.J. and Ostovan, H. 2008. Iranian Asilidae (Insecta: Diptera). *Turkish Journal of Zoology* 32: 175–195.

Howarth, B. and Gillett, M.P.T. 2008. The terrestrial and freshwater arthropods of Abu Dhabi Emirate, pp. 380–463. In: Perry, R.J. (ed.), *Terrestrial Environment of Abu Dhabi Emirate*. Environment Agency, Abu Dhabi, 463 pp.

Jahani, A., Hosseinzadeh Monfared, S., Ghasemi, M. and Shirvany, A. 2020. Mass trapping and identification of insect of mangrove forests (*Avicennia marina*) using cylindrical trap, yellow sticky trap and bucket-color trap (Case study: Minab city). *Journal of Animal Environment* 11(4): 271–280. [in Persian, English summary]

Jeanne, R.L. 1980. Evolution of social behavior in the Vespidae. *Annual Review of Entomology* 25: 371–396.

Jeanne, R.L. 1991. The swarm-founding Polistinae, pp. 191–231. In: Ross, K.G. and Matthews, R.W. (eds), *The Social Biology of Wasps*. Cornell University Press, Ithaca, NY, xvii + 678 pp.

Jeanne, R.L. 2003. Social complexity in the Hymenoptera, with special reference to the wasps, pp. 81–130. In: Kikuchi, T., Azuma, N. and Higashi, S. (eds.), *Genes, Behaviors and Evolution of Social Insects*. Hokkaido University Press, Sapporo, 293 pp.

Jeanne, R.L. 2021. Swarm-founding wasps, pp. 918–934. In: Starr, C.K. (ed.), *Encyclopedia of Social Insects*. Cham, Switzerland, Springer, 1075 pp.

Kaplan, E. and Efil, L. 2025. New contributions to families of Bembicidae, Megachilidae, Scoliidae, Tiphiidae, Mutillidae, and Vespidae (Hymenoptera: Aculeata) from Western Türkiye. *Munis Entomology and Zoology* 20(2): 2610–2618.

Keeping, M.G. 2002. Reproductive and worker castes in the primitively eusocial wasp *Belonogaster petiolata* (De Geer) (Hymenoptera: Vespidae): evidence for pre-imaginal differentiation. *Journal of Insect Physiology* 48: 867–879.

Khalaf, A.N. and Al-Omar, M.A. 1974. *A second list of insects from Iraq. Biological Research Centre Publication 2*. Biological Research Centre, Baghdad, 41 pp.

Khan, K., Rasool, M. and Zahid, M. 2018. Taxonomic study of Polistinae species (Hymenoptera: Vespidae) of Dir, Pakistan. *Journal of Entomology and Zoology Studies* 6(2): 791–794.

Khoobdel, M., Tavassoli, M., Salari, M. and Firozi, F. 2014. The stinging Apidae and Vespidae (Hymenoptera: Apocrita) in Iranian islands, Qeshm, Abu–Musa, Great Tunb and Lesser Tunb on the Persian Gulf. *Asian Pacific Journal of Tropical Biomedicine* 4: 258–262.

Kirby, W.F. 1881. On the Hymenoptera collected by Prof. I. Bayley Balfour in Socotra. *Proceedings of the Zoological Society of London* 1881: 649–650.

Kohl, F.F. 1898. Zur Kenntniss der europaischen *Polistes*-Arten. *Annalen der K.K. Naturhistorischen Hofmuseums in Wien* 13: 87–90.

Kohl, F.F. 1906. Zoologische Ergebnisse der Expedition der Kaiserlichen Akademie der Wissenschaften nach Südarabien und Sokotra in den Jahren, 1898-1899. Wien. Reprinted from *Denkschriften der Keiserlichen Akademie der Wissenschaften, Wien* 71: 169–301.

Kojima, J. and Carpenter, J.M. 1997. Catalog of species in the polistine tribe Ropalidiini (Hymenoptera: Vespidae). *American Museum Novitates* 3199: 1–96.

Krombein, K.V., Hurd, P.D., Smith, D.R. and Burks, B.D. 1979. *Catalog of Hymenoptera in America north of Mexico*. Vol. 1. Smithsonian Institution Press, Washington, pp. 1199–2209.

Kumar, P.G. and Lambert, K. 2010. New record of the *Polistes* (*Polistes*) *indicus* Stolfa from India (Hymenoptera: Vespidae: Polistinae). *Uttar Pradesh Journal of Zoology* 30(2): 193–196.

Latreille, P.A. 1802. *Histoire naturelle, générale et particulière, des Crustacés et des Insectes*. Tome troisième. Paris, 468 pp.

Latreille, P.A. 1810. *Considérations générales sur l'ordre naturel des animaux composant les classes des Crustacés, des Arachnides et des Insectes*. F. Schoell, Paris, 444 pp.

Lehr, P.A., Ghahari, H. and Ostovan, H. 2007. A contribution to the robber flies of subfamilies Stenopogoninae and Asilinae (Diptera: Asilidae) from Iran. *Far Eastern Entomologist* 173: 1–14.

Lepeletier, A. 1836. *Histoire Naturelle des Insectes. Hyménoptères 1*. Librairie Encyclopedique de Roret, Paris, 547 pp.

Linnaeus, C. 1758. *Systema naturae per regna tria naturae, secundum classes ordines, genera, species cum characteribus, differentiis, synonymis locis. Tomus I.* Editio decima, reformata. Laurnetii Salvii, Holmiae, 824 pp.

Linnaeus, C. 1767. Systema naturae per regna tria naturae secundum classes, ordines, genera species, cum characteribus, differentiis, synonymis, locis. Editio duodecima, reformata. Cum privilegio S:ae R:ae M: tis Sveciae & Electoris Saxon. Tomus I. Pars II. (Editio duodecima), Laurentii Salvii, Holmiae [= Stockholm]. pp. 533–1327 + 36 unnumbered.

Madl, M. 1997. Über Vespidae, Pompilidae, Scoliidae und Tiphiidae der Turkei. *Linzer biologische Beiträge* 29(2): 823–827.

Madl, M. 2018. A preliminary catalogue of the Hymenoptera (Insecta) of the Republic of Djibouti. *Linzer biologische Beiträge* 50(2): 907–967.

Maryam, Z., Ali, M., Shehbaz, M.H., Hussain, R., Qasim, M. and Hassan, M.A. 2024. An updated checklist of family Vespidae (Insecta: Hymenoptera) from Azad Kashmir and Gilgit-Baltistan, Pakistan. *Journal of the Entomological Research Society* 26(2): 321–336.

Meade-Waldo, G. 1911. X. New species of Diploptera in the collection of the British Museum. *Journal of Natural History* 7(37): 98–113.

Menezes, R.S.T., Lloyd, M.W. and Brady, S.G. 2020. Phylogenomics indicates Amazonia as the major source of Neotropical swarm-founding social wasp diversity. *Proceedings of the Royal Society B* 287: 20200480.

Modarres Awal, M. 1997. Family Vespidae (Hymenoptera), p. 284. In: Modarres Awal, M. (ed.), *List of agricultural pests and their natural enemies in Iran.* Second edition. Ferdowsi University of Mashhad Press, Iran, 429 pp.

Modarres Awal, M. 2012. Family Vespidae (Hymenoptera), pp. 525–527. In: Modarres Awal, M. (ed.), *List of agricultural pests and their natural enemies in Iran.* Third edition. Ferdowsi University of Mashhad Press, 759 pp.

Monks, J., Ross, S., Geiser, M., De Prins, J., Sharaf, M., Wyatt, N., Al Rijeibi, S. and Polaszek, A. 2019. A preliminary survey of the insect fauna of the Hajar Mountain Range, Oman. *Journal of Natural History* 53(15–16): 939–963.

Morawitz, F. 1889. Hymenopterologische Mittheilungen. *Horae Societatis Entomologicae Rossicae* 23: 540–554.

Morice, F. 1921. Annotated lists of aculeate Hymenoptera (except Heterogyna) and chrysids recently collected in Mesopotamia and north-west Persia. *The Journal of the Bombay Natural History Society* 28: 192–203.

Mossadegh, M. 1995. Insects associated with honey bee *Apis mellifera* L. hives in Iran. *Journal of Agricultural Sciences* 1(3–4): 22–27. [in Persian, English summary]

Nemati, S. and Pezhman, H. 2014. Comparing the pests and natural enemies fauna and determining the prevailing species in wheat fields under no-tillage and conventional tillage systems in Marvdasht region (Fars province, Iran). *Journal of Applied Researches in Plant Protection* 3(1): 117–127. [in Persian, English summary]

Nematy, A., Khayrandish, M., Ebrahimi, E. and Carpenter, J.M. 2016. First report of the paper wasp *Polistes rothneyi* Cameron (Hym.: Vespidae: Polistinae) from Iran. *Journal of the Entomological Society of Iran* 36(3): 237–239. [in Persian, English summary]

Neumeyer, R., Gereys, B., Castro, L. (2015) New data on the distribution of *Polistes bischoffi* Weyrauch, 1937 and *Polistes helveticus* Neumeyer, 2014, a synonym of *Polistes albellus* Giordani Soika, 1976 n. stat. *Boletín de la Sociedad Entomológica Aragonesa (S.E.A.)*, 57: 205–216.

Noll, F.B. 2013. Marimbondos: A review on the Neotropical swarm-founding Polistines. *Sociobiology* 60(4): 347–354.

Noll, F.B. and Wenzel, J.W. 2008. Caste in the swarming wasps: 'Queenless' Societies in Highly Social Insects. *Biological Journal of the Linnean Society* 93(3): 509–522.

Noll, F.B., da Silva, M., Soleman, R.A., Lopes, R.B., Grandinete, Y.C., Almeida, E.A.B., Wenzel, J.W. and Carpenter, J.M. 2021. Marimbondos: systematics, biogeography, and evolution of social behaviour of neotropical swarm-founding wasps (Hymenoptera: Vespidae: Epiponini). *Cladistics* 37(4): 423–441.

Nosrati, Z., Hakimitabar, M. and Rakhshani, E. 2024. First record of *Polistes indicus* Stolfa, 1934 (Hymenoptera: Vespidae) in the southern slopes of the Alborz mountains. *Proceedings of 25th Iranian Plant Protection Congress, 7-10 September 2024, IROST, Tehran*, p. 96.

Özbay, C. and Akbayin, H. 1993. Biometrical studies on four Vespidae (Hymenoptera) species from Diyarbaktr, Mardin and Sanhurfa. *Doga: Turkish Journal of Biology* 17: 75–92.

Özeren, M.H. 1965. Ankara Bölgesinde bazt meyve ve orrnan ağaçlari ile bitkilerine zarar veren Hymenoptera türleri, yayrhşr ve taksonomileri üzerinde ön araştrrmalar. *Ankara Üniversitesi Ziraat Fakultesi Yay*, No. 396, 73 pp.

Pashaei Rad, S., Abbasi, R., Soleimani, G. and Dvořák, L. 2010. New and supplementary information on the vespid fauna of Iran (Hymenoptera: Vespidae). *Zoology in the Middle East* 50(1): 95–100.

Perrard, A., Grimaldi, D. and Carpenter, J.M. 2017. Early lineages of Vespidae (Hymenoptera) in Cretaceous amber. *Systematic Entomology* 42(2): 379–386.

Pickett, K.M. and Carpenter, J.M. 2010. Simultaneous analysis and the origin of eusociality in the Vespidae (Insecta: Hymenoptera). *Arthropod Systematics and Phylogeny* 68(1): 3–33.

Piekarski, P.K., Carpenter, J.M., Lemmon, A.R., Lemmon, E.M. and Sharanowski, B.J. 2018. Phylogenomic evidence overturns current conceptions of social evolution in wasps (Vespidae). *Molecular Biology and Evolution* 35(9): 2097–2109.

Polašek, O., Onah, I., Kehinde, T., Rojo, V., van Noort, S. and Carpenter, J.M. 2025. Revision of the mainland African species of the Old World social wasp genus *Ropalidia* Guérin-Méneville 1831 (Hymenoptera; Vespidae). *Zootaxa* 5626(1): 1–142.

Qing-Bai, H., Fan-Ming, M., Sheng-Jie, M.A. and Xing-Cai, L. 2012. Nesting habits of *Polistes mandarinus* (Hymenoptera: Vespidae) in Bomi, Tibet. *Acta Entomologica Sinica* 55(2): 255–258.

Rahmani, Z., Rakhshani, E. and Carpenter, J.M. 2020. Updated checklist of Vespidae (Hymenoptera: Vespoidea) in Iran. *Journal of Insect Biodiversity and Systematics* 6(1): 27–86.

Reeve, H.K. 1991. Polistes, pp. 99–148. In: Ross, K.G. and Matthews, R.W. (eds), *The Social Biology of Wasps*. Cornell University Press, Ithaca, NY; xvii + 678 pp.

Richards, O.W. 1978. *The social wasps of the Americas, excluding the Vespinae*. British Museum (Natural History), London, 580 pp.

Richards, O.W. 1982. A revision of the genus *Belonogaster* de Saussure (Hymenoptera: Vespidae). *Bulletin of the British Museum (Natural History), Entomology* 44(2): 31–114.

Richards, O.W. 1985a [1984]. Insects of Saudi Arabia. Hymenoptera: Fam. Masaridae (The Arabian species). *Fauna of Saudi Arabia* 6: 413–422.

Richards, O.W. 1985b [1984]. Insects of Saudi Arabia. Hymenoptera: Fam. Vespidae (Social wasps of the Arabian Peninsula). *Fauna of Saudi Arabia* 6: 423–440.

Rudow, F. 1889. Einige entomologische Beobachtungen. *Societas Entomologica, Stuttgart* 4(11): 87–89.

Rusina, L.Y. 2015. Principles of organization of Polistinae (Hymenoptera, Vespidae) populations. *Entomological Review* 95(8): 1036–1050.

Sakenin, H., Raheb, J., Imani, S., Havaskary, M., Shirdel, F. and Mohseni, H. 2008. A preliminary survey on dipteran predators and parasitoids and Odonata in Iranian rice fields. *Proceedings of National Conference of Agronomical Rice Breeding, Young Researchers Club, Islamic Azad University of Ghaemshahr, 26-27 November, 2008*, 14 pp. [in Persian, English summary]

Sakenin, H., Samin, N. and Bagriacik, N. 2010. A contribution to the Aculeate Hymenoptera (Insecta) from Iran. *Efflatounia* 10: 15–20.

Samin, N., Sakenin, H., Imani, S. and Shojai, M. 2010. A contribution to the knowledge of robber flies (Diptera: Asilidae) from Tehran province and vicinity, Iran. *Journal of Biological Control* 24(1): 42–46.

Schmid-Egger, C., Achterberg, K.V., Neumeyer, R., Morinière, J. and Schmidt, S. 2017. Revision of the West Palaearctic *Polistes* Latreille, with the descriptions of two species–an integrative approach using morphology and DNA barcodes (Hymenoptera, Vespidae). *ZooKeys* 713: 53–112.

Schmid-Egger, C., Gusenleitner, J., Liebig, W.-H. and Jacobs, H.-J. 2021. New records of Vespidae (Hymenoptera) from Iran with descriptions of three species. *Linzer biologische Beiträge* 52(2): 1105–1121.

Selis, M. 2023. Illustrated key to the genera and a checklist of Italian Vespidae (Hymenoptera). *Fragmenta Entomologica* 55(1): 63–88.

Selis, M., Ismael, H.R., Hamad, M.I., Baiocchi, D. and Cilia, G. 2025. A first update to the vespid fauna of Iraq (Hymenoptera: Vespidae), with description of a new species. *Journal of Natural History* 59(13–16): 941–981.

Shahreyari Nejad, S. 2022. Introducing *Parapolybia escalerae* (Meade-Waldo, 1911) (Vespidae: Polistinae) as a paper wasp of the honey producer from Iran. *Plant Protection (Scientific Journal of Agriculture)* 45(3): 113–115.

Siddiqui, J.A., Bodlah, I., Carpenter, J.M., Naeem, M., Ahmad, M. and Bodlah, M.A. 2015. Vespidae (Hymenoptera) of the Pothwar region of Punjab, Pakistan. *Zootaxa* 3914(5): 501–524.

Silveira, O.T. 2008. Phylogeny of wasps of the genus *Mischocyttarus* de Saussure (Hymenoptera, Vespidae, Polistinae). *Revista Brasileira de Entomologia* 52: 510–549.

Silveira, O.T., Andena, S.R., Somavilla, A. and Carpenter, J.M. 2021. Phylogeny and classification of the Neotropical social wasps, pp. 267–291. In: Prezoto, F., Nascimento, F.S., Barbosa, B.C. and Somavilla, A. (eds.), *Neotropical social wasps: Basic and applied aspects*. Springer Nature, Switzerland, 472 pp.

Somavilla, A. and Carpenter, J.M. 2021. Key to the genera of social wasps (Polistinae) occurring in Neotropics, pp. 327–336. In: Prezoto, F., Nascimento, F.S., Barbosa, B.C. and Somavilla, A. (eds.), *Neotropical Social Wasps*. Springer Nature Switzerland, xiii + 472 pp.

Somavilla, A., Santos, B.F., Carpenter, J.M., Andena, S.R. and Oliveira, M.L. 2021. Total-evidence phylogeny of the New World *Polistes* Lepeletier, 1836, paper wasps (Vespidae, Polistinae, Polistini). *American Museum Novitates* 3973: 1–42.

Stolfa, E. 1934. Due nuovi Vespidi indiani. *Bollettino della Società veneziana di storia naturale* 1: 47–49, 4 figs.

Strassmann, J.E., Queller, D.C. and Hughes, C.R. 1987. Constraints on independent nesting by *Polistes foundresses* in Texas, pp. 379–380. In: Eder, J. and Rembold, H. (eds), *Chemistry and biology of social insects*. Verlag J Peperny, Munich, xxxv + 757 pp.

Strassmann, J.E., Nguyen, J.S., Arévalo, E., Cervo, R., Zacchi, F., Turillazzi, S. and Queller, D.C. 2003. Workers produce males only after queen death in *Polistes gallicus*, a Mediterranean social wasp. *Journal of Evolutionary Biology* 16: 254–259.

Swailem, S.M., Selim, A.A. and Amin, A.H. 1974. A contribution to the study of the insect fauna of Hammam Al-Alil, Part I. *Mesopotamia Journal of Agriculture* 9(1–2): 119–141.

Tanyolaç, T. 1964. Ankara gevresinde sosyal halde yaşayan Vespidae (Hymenoptera) türleri ve bunlardan *Polistes gallicus* L.'un biyolojisi tizerinde aragtrrmalar. Doktora tezi, 46 pp.

Tüzün, A. and Tanyolaç, T. 1987. Ege Btilgesinde Vespidae (Insecta: Hymenoptera) türlerinin saptanmasi. C. U. Fen Edebiyat Fakiiltesi, *Fen Bilimleri Dergisi* 5: 147–171.

von Schulthess, A. 1913. *Parapolybia* Saussure (Vespidae sociales). *Mitteilungen der Schweizerischen Entomologischen Gesellschaft* 12: 152–164.

von Schulthess, A. 1921. Zur Kenntnis athiopischer Vespiden (Hym). *Entomologische Mitteilungen* 10(4–5): 121–126, 143–149.

Wahis, R. 2000. Hyménoptères aculéates du Yemen (Pompilidae, Vespidae). *Notes Fauniques de Gembloux* 41: 73–100.

Walker, D.H. and Pittaway, A.R. 1987. *Insects of Eastern Arabia*. Macmillan, London, 191 pp.

Wenzel, J.W. 1990. A Social Wasp's Nest from the Cretaceous Period, Utah, USA, and its Biogeographical Significance. *Psyche* 97: 21–29.

Wenzel, J.W. 1998. A generic key to the nests of hornets, yellowjackets, and paper wasps worldwide (Vespidae: Vespinae, Polistinae). *American Museum Novitates* 3244: 1–39.

Wenzel, J.W. and Carpenter, J.M. 1994. Comparing Methods, pp. 97–101. In: Eggleton, P. and Vane-Wright, R. (eds.), *Phylogenetics and Ecology*. Academic Press, London, 376 pp.

Weyrauch, W. 1937. Zur Systematik und Biologie der Kuckuckswespen *Pseudovespa* und *Pseudopolistes*. *Zoologische Jahrbücher (Jena), Abteilung für Systematik, Ökologie und Geographie der Tiere* 70: 243–290.

Weyrauch, W. 1939. Zur Systematik der paläarktischen Polistinen auf biologischer Grundlage. *Archiv für Naturgeschichte* 8(2): 145–197.

Yildirim, E. and Gusenleitner, J. 2012. Contribution to the knowledge of the Vespidae (Hymenoptera, Aculeata) of Turkey, with a checklist to the Turkish species. *Turkish Journal of Zoology* 36(3): 361–374.

Yildirim E. and Kojima, J. 1999. Distributional checklist of the species of the family Vespidae (Insecta: Hymenoptera; Aculeata) of Turkey. *Natural History Bulletin of Ibaraki University* 3: 19–50.

Yildirim, E. and Özbek, H. 1993. Polistinae (Hymenoptera: Vespidae) of Turkey. *Turkish Entomology Dergisi* 17(3): 141–156.

Yildirim, E. and Özbek, H. 1997. Masaridae (Hymenoptera, Vespoidea) of Turkey. *Atatürk Üniversitesi Ziraat Fakültesi Dergisi* 28(E): 222–226.

Yildirim, E. and Özbek, H. 1999. s. *Turkish Journal of Zoology* 23(6): 591–604.

Zakharov, A.A. 1991. Organization of ant communities. Nauka, Moscow, 278 pp. [in Russian]

Zimmermann, K. 1931. Studien über individuelle und geographische Variabilität paläarktischer *Polistes* und verwandter Vespiden. *Zeitschrift für Morphologie und Ökologie der Tiere* 22(1): 173–230.

Zyła, W., Hilszczański, J. and Bystrowski, C. 2023. *Polistes mongolicus* du Buysson, 1911, a new species of paper wasp in Poland (Hymenoptera: Vespidae: Polistinae). *Annals of the Upper Silesian Museum in Bytom Entomology* 32: 1–6.

Vespula squamosa (Drury, 1773) (Vespinae) on white aster (Asteraceae). (Photo courtesy of Angella Moorehouse)

Vespula consobrina (de Saussure, 1854) (Vespinae) on early goldenrod (Asteraceae). (Photo courtesy of Angella Moorehouse)

Vespa orientalis (Linnaeus, 1771) (Vespinae). (Photo courtesy of Simon van Noort, Iziko Museums of South Africa; www.waspweb.org)

Vespa orientalis (Linnaeus, 1771) (Vespinae). (Photo courtesy of Josef Němec)

Dolichovespula sylvestris (Scopoli, 1763) (Vespinae). (Photo courtesy of Hana Křiženecká)

Vespula germanica (Fabricius, 1793) (Vespinae). (Photo courtesy of Alexander Fateryga)

5 Subfamily Vespinae Latreille, 1802

Libor Dvořák, Hassan Ghahari, Neveen S. Gadallah, and Patrick K. Piekarski

INTRODUCTION

The subfamily Vespinae (hornets and yellowjackets) includes the largest and best-known wasps, which are either social or social parasites of other congeners (Kimsey and Carpenter, 2012). It comprises the hornets (*Vespa* Linnaeus, 1758) and yellowjackets (*Dolichovespula* Rohwer, 1916 and *Vespula* Thomson, 1869). The remaining genus, *Provespa* Ashmead, 1902, is a small and poorly known group of nocturnal wasps restricted to Southeast Asia (Madl, 2012). The subfamily comprises more than 70 species (Kimsey and Carpenter 2012; Abels and Wedmann, 2022), predominantly distributed throughout the Holarctic and Oriental regions (Brothers and Finnamore, 1993).

The phylogenetic relationships among these four genera have been historically unstable and remain elusive (Persson, 2015; Lopez-Osorio *et al.*, 2017). Some studies have recovered *Vespa* as the sister-group to the remaining three genera (Carpenter, 1987; Saito and Kojima, 2011), and many have supported the monophyly of the yellowjackets (e.g., Carpenter, 1987; Pickett and Carpenter, 2010; Saito and Kojima, 2011; Perrard *et al.*, 2016). However, a phylogenetic study using data from nine genetic loci suggested that yellowjackets may not be monophyletic, finding some support that *Vespula* and *Dolichovespula* are not sister taxa (Lopez-Osorio *et al.*, 2014); this was further supported in a subsequent study (Lopez-Osorio *et al.*, 2015). A more recent phylogenomic analysis by Lopez-Osorio *et al.* (2017) recovered *Vespa* as sister to the *Dolichovespula*, further suggesting that yellowjackets are not monophyletic (also see text of Abels and Wedmann, 2022).

Phylogenies based on mitochondrial DNA, as presented by Fan *et al.* (2017) and Huang *et al.* (2019), supported *Dolichovespula* as sister to a clade formed by *Vespa* and *Vespula*. In contrast, phylogenomic data based on hundreds of nuclear DNA loci recovered relationships consistent with those of Lopez-Osorio *et al.* (2017), with strong support for *Vespula* as the basal vespine, and *Dolichovespula* forming a sister relationship with a clade comprised of *Provespa* + *Vespa* (Piekarski *et al.*, 2018).

Adults range in size from medium to very large, measuring 10–40 mm in total length, and have a compact body with a sessile gaster. In *Provespa* species, the ground color is brown, while in most other species, it is black marked with white, yellow, or orange markings (Yamane and Yamane, 2021). They are characterized by kidney-shaped eyes and strong mandibles. Unlike other social wasps, their first gastral segment is basally truncated, forming a vertical anterior face. Both queens and workers possess a well-developed venom apparatus adapted for stinging, which can pose medical and veterinary concerns (Edwards, 1980; Matsuura and Yamane, 1990; Martin, 2017).

Members of Vespinae can be distinguished from other Vespidae by the following combination of characteristics: the hind wing lacks an anal lobe; the marginal cell is pointed apically and terminates along the costal vein; the recurrent veins of the forewing are received by the same submarginal cell; the pronotal lobe is separated from the tegula by a distance equal to or shorter than its length; the tarsal claws are simple; the gaster is sessile, with the first gastral tergum truncate (Smith-Pardo *et al.*, 2020).

DOI: 10.1201/9781003740438-5

Among vespine wasps, there are both aerial and cavity/ground-nesting species. Some species of the genera *Vespa* and *Vespula* initiate their nests in early spring using abandoned rodent burrows or cavities in man-made structures or trees (Edwards, 1980; Matsuura and Yamane, 1990; Detoni *et al.*, 2021). Typically, they construct their nests using a mixture of plant fibers and salivary secretions. Nest sizes vary widely, ranging from small, baseball-sized structures with a few cells to massive nests containing hundreds of thousands of cells (Kimsey and Carpenter, 2012).

As the colony grows, workers expand the nesting cavity by excavating the surrounding soil, insulation, or decayed wood. The only external sign of such nests is often the steady traffic of foragers entering and exiting the entrance, offering camouflage from visually oriented predators, including humans. Other vespine species, particularly those of the genus *Dolichovespula*, construct aerial nests, which can be located as high as 40 meters above ground, making then inaccessible to most non-flying predators (Archer and Penney, 2012; Feás and Charles, 2019; Detoni *et al.*, 2021).

Some species, such as *Vespula germanica*, readily nest in human structures and scavenge a wide variety of food sources beyond live prey, leading to their classification as pests (Kimsey and Carpenter, 2012).

Due to the wide distribution of most Vespinae species and the considerable color variation exhibited by many of them, numerous subspecies have historically been recognized (see Archer, 1989; Carpenter and Kojima, 1997). However, these subspecies are now generally treated as synonyms of the nominate forms, or at best as informal local variants.

Notable examples include *Vespa analis* Fabricius, which ranges from India to Bali and extends through China to Siberia, Korea, and Japan; *Vespula orbata* (du Buysson), found in India and Southeast Asia; and *Vespula flaviceps* (Smith), which occurs from India and Southeast Asia through China to Siberia, Korea, and Japan (Carpenter *et al.*, 2011).

Vespines are highly eusocial, with a morphologically distinct worker and queen caste. They construct compact paper nests that are either annual or perennial, typically multi-combed and enclosed within a paper envelope. These nests may be suspended in open but sheltered locations or built underground. Larvae are fed masticated insects or, in some cases, flesh from dead vertebrates. A few are cleptoparasitic, invading and reproducing within the nests of other Vespinae species.

Several vespine wasps are invasive, having been introduced beyond their native ranges, and are considered major pests (Matsuura and Yamane, 1990; Ross and Matthews, 1991; Brothers and Finnamore, 1993). Most species are diurnal, beginning their activity one or two hours before sunrise and ceasing shortly after sunset. An exception is *Vespa crabro*, which continues foraging during the night (Yamane and Yamane, 2021).

A global checklist of the subfamily Vespinae was first compiled by Carpenter and Kojima (1997). Since then, numerous changes have been made to the taxonomy and nomenclature of the genera *Dolichovespula, Vespa*, and *Vespula*.

Checklists of the Iranian Vespinae: Farahbakhsh (1961) documented two species: *Vespa orientalis* Linnaeus, 1771 and *Vespula germanica* (Fabricius, 1793). Modarres Awal (1997 and 2012) listed four and five species and subspecies, respectively, in two and three genera. Ebrahimi and Carpenter (2008) reported six species in three genera (*Vespa* Linnaeus, *Vespula* Thomson, and *Dolichovespula* Rohwer). Bagriacik and Samin (2011) and Dvořák *et al.* (2012) catalogued nine and six species, respectively, across the same three genera. More recently, Rahmani *et al.* (2020) listed six species across three genera.

The present catalogue includes 13 species across three genera (*Dolichovespula* Rohwer, 1916, *Vespa* Linnaeus, 1758 and *Vespula* Thomson, 1869) for the Middle Eastern countries and eight species across three genera for Iran. The general distribution of species follows Carpenter and Kojima (1997).

Key to the species of the Middle Eastern Vespinae (modified from Carpenter and Nguyen, 2003, and Dvořák and Roberts, 2006)

1. Vertex long, distance between posterior ocelli and occiput greater than distance between posterior ocelli and compound eye (genus *Vespa* Linnaeus)....2

— Vertex short, distance between posterior ocelli and occiput shorter than or equal to distance between posterior ocelli and compound eye 4

2. Body brown and yellow; fifth gastral tergum in female, and sixth in male entirely brown .... *Vespa orientalis* (Linnaeus, 1771)

— Without this color combination....3

3. Body black, yellow and reddish; third and subsequent gastral terga yellow, often with two black spots; face yellow....*Vespa crabro* (Linnaeus, 1771)

— Body black, brown and orange; third and subsequent gastral terga orange to orange brown; face orange....*Vespa velutina* Lepeletier de Saint Fargeau, 1836

4. Pronotal carina absent; malar space shorter than last antennal segment (genus *Vespula* Thomson) 5

— Pronotal carina present laterally; malar space longer than last antennal segment (genus *Dolichovesoula* Rohwer)....8

5. Eye emargination entirely yellow; first gastral tergum with pale setae....6

— Eye emargination yellow beneath; first gastral tergum with black setae....7

6. Distance separating third and fourth mandibular teeth distinctly concave; male aedeagus emarginate at apex....*Vespula germanica* (Fabricius, 1793)

— Distance separating third and fourth mandibular teeth straight; male aedeagus rounded at apex .... *Vespula vulgaris* (Linnaeus, 1758)

7. Hind tibia with short setae along outer margin; clypeal free margin bluntly protruded in queens and workers; male aedeagus slightly emarginate at apex as well as preapically on both sides........ *Vespula rufa* (Linnaeus, 1758)

— Hind tibia with short yellow, and long black setae along outer margin; clypeal free margin sharply protruded in queens (no workers); male aedeagus straight at apex, but distinctly emarginate preapically on both sides ....*Vespula austriaca* (Panzer, 1799)

8. Lateral panel of pronotum ridged; eye emargination entirely yellow........ *Dolichovespula media* (Retzius, 1783)

— Lateral panel of pronotum nearly smooth (not ridged); eye emargination yellow at lower side only 9

9. Apical third of clypeus with dense, coarse punctures; clypeus yellow, rarely with 1–3 black spots in the middle....10

— Apical third of clypeus with sparse, fine punctures; clypeus usually with black longitudinal stripe, sometimes reduced into 1–3 black spots medially, exceptionally completely yellow....11

10. Clypeus sharply angled apico-laterally, pointed apico-medially in queens, rectangular in male; sting distinctly curved; no workers....*Dolichovespula omissa* (Bischoff, 1931)

— Clypeus bluntly angled apico-laterally in all casts; sting straight...*Dolichovespula sylvestris* (Scopoli, 1763)

11. Apico-lateral sides of clypeus sharply produced, pointed apico-medially in queens, rectangular in male; sting distinctly curved; no workers..........................*Dolichovespula adulterina* (du Buysson, 1904 [1905])

— Apico-lateral sides of clypeus produced in all casts; sting straight...12

12. Ocelli placed in equilateral triangle, distance from hind ocellus to vertex as long as or longer than distance between posterior ocelli; second gastral tergum without red; mesopleuron brightly setose ....*Dolichovespula saxonica* (Fabricius, 1793)

– Ocelli placed in distinctly obtuse ocellar triangle, distance between hind ocellus to vertex shorter than distance between posterior ocelli; second gastral tergum usually with red; mesopleuron nearly always with black setae...*Dolichovespula norwegica* (Fabricius, 1781)

List of Species of the Subfamily Vespinae in the Middle East

Subfamily Vespinae Latreille, 1802

Genus *Dolichovespula* Rohwer, 1916

Dolichovespula Rohwer, 1916: 642. Type species: *Vespa maculata* Linnaeus, 1763, by original designation.

Comments: The biology of the genus *Dolichovespula* is relatively poorly known compared to *Vespa* and *Vespula*, partly due to the global invasions and higher research attention focused on the latter two genera (Beggs *et al.*, 2011; Komonen, 2023). *Dolichovespula* and *Vespula* differ not only in their social organization but also in their colony and nest characteristics. However, variation also exists among species within *Dolichovespula* (Greene, 1991; Foster *et al.*, 2001; Archer, 2006, 2012).

Colony and nest characteristics are biologically significant, as they help interpret intra- and interspecific differences in the identity and prevalence of nest associates (parasites, parasitoids, predators, and scavengers), which can influence colony size, activity, and success (Greene *et al.*, 1976; Akre *et al.*, 1980; Archer, 2012; Komonen, 2023). A distinguishing trait of *Dolichovespula*, in contrast to its sister group *Vespula*, is that some workers are capable of producing haploid offspring that develop into males (Foster *et al.*, 2001).

Most *Dolichovespula* species typically construct aerial nests. To date, a total of 19 valid species of *Dolichovespula* have been described from the Palaearctic, Nearctic, and Oriental regions (Tan *et al.*, 2014; Daglio, 2020; Wang *et al.*, 2022).

Dolichovespula adulterina (du Buysson, 1904)

Dolichovespula norwegica var. *adulterina* du Buysson, 1905 (1904): 600, 628, ♀; «Europe».

Distribution in the Middle East: Turkey (Yildirim and Özbek, 1992 as *D.* (*Pseudovespula*) *adulterina*; Carpenter and Kojima, 1997; Yildirim and Kojima, 1999; Antropov, 2017).

Extralimital distribution: Canada, northern and western United States of America, Palaearctic (and adjacent parts of Oriental) from northern and western Europe to Sakhaline, Japan, northern China and Taiwan, Georgia, and Mongolia (Carpenter and Kojima, 1997; Archer, 1989; Gogala, 2022). Also, all records of this species from North America should be referred to as *D. arctica* (Rohwer, 1916) (= *Vespa borealis* Lewis, 1897) (Kimsey and Carpenter, 2012; Antropov, 2017).

Dolichovespula media (Retzius, 1783)

Vespa crabro var. *medius* Retzius, 1783: 63, sex not stated.

Distribution in the Middle East: Syria (Carpenter and Kojima, 1997; Castro and Dvořák, 2009; Antropov, 2017; Fateryga, 2018), Turkey (Carpenter and Kojima, 1997; Madl, 1997 as *D. media media*; Yildirim and Kojima, 1999; Yildirim and Gusenleitner, 2001; Castro and Dvořák, 2009; Antropov, 2017; Fateryga, 2018).

Extralimital distribution: Caucasus, China, Europe, Japan, Kazakhstan, Korea, Mongolia, North Africa, New Zealand (introduced), Russia (Carpenter and Kojima, 1997; Antropov, 2017; Fateryga, 2018; Gogala, 2022; Selis, 2023), Georgia, Morocco (Antropov, 2017).

***Dolichovespula norwegica* (Fabricius, 1781)**

Vespa norwegica Fabricius, 1781: 460, sex not stated; "Norwegia".

Distribution in the Middle East: Turkey (Carpenter and Kojima, 1997; Yildirim and Kojima, 1999; Yildirim and Özbek, 1992, 1999 as *D.* (*Boreovespula*) *norwegica*; Castro and Dvořák, 2010; Antropov, 2017).

Extralimital distribution: Canada, China, Europe, Georgia, Kazakhstan, Kyrgyzstan, Mongolia, Russia (Carpenter and Kojima, 1997; Gogala, 2022). Also, all records of this species from North America should be referred to as *D. albida* (Sladen, 1918) (Kimsey and Carpenter, 2012; Antropov, 2017).

***Dolichovespula omissa* (Bischoff, 1931)**

Vespa omissa Bischoff, 1931: 6, ♀; [Austria] "Thürengen".

Catalogues with Iranian records: Ebrahimi and Carpenter (2008), Bagriacik and Samin (2011), Dvořák *et al.* (2012), Rahmani *et al.* (2020).

Distribution in Iran: Tehran (Dvořák *et al.*, 2012).

Distribution in the Middle East: Iran (Dvořák *et al.*, 2012; Pekkarinen and Huldén, 1995; Carpenter and Kojima, 1997; Antropov, 2017), Turkey (Gusenleitner, 1966, 1988 both as *Pseudovespula omissa*; Yildirim and Özbek, 1992, 1999 both as *D.* (*Pseudovespula*) *omissa*; Carpenter and Kojima, 1997; Yildirim and Kojima, 1999; Antropov, 2017; Rahmani *et al.*, 2020).

Extralimital distribution: Europe, Georgia, Kazakhstan, Russia (Carpenter and Kojima, 1997; Antropov, 2017; Gogala, 2022).

***Dolichovespula saxonica* (Fabricius, 1793)**

Vespa saxonica Fabricius, 1793: 256, sex not stated; "in Germania".

Catalogues with Iranian records: Bagriacik and Samin (2011).

Distribution in Iran: East Azarbaijan (Abd-Rabou *et al.*, 2005; Ghahari *et al.*, 2007; Sakenin *et al.*, 2010a), Tehran (Samin *et al.*, 2010).

Distribution in the Middle East: Iran (see references above, and Antropov, 2017), Turkey (Carpenter and Kojima, 1997; Yildirim and Özbek, 1992 as *D.* (*Boreovespula*) *saxonica*; Yildirim and Kojima, 1999; Antropov, 2017).

Extralimital distribution: China, Europe, Japan, Kazakhstan, Korea, Mongolia, Russia, (Carpenter and Kojima, 1997; Gogala, 2022).

Comments: Re-examination of specimens reported by Abd-Rabou *et al.* (2005) and Sakenin *et al.* (2010a) confirms the presence of *D. saxonica* in Iran (Figure 5.1). Additionally, *Stichopogon elegantulus* (Wiedemann) and *Molobratia teutonus* (Linnaeus) (Diptera: Asilidae) were recorded as predators of *D. saxonica* by Ghahari *et al.* (2007) and Samin *et al.* (2010), respectively.

Nests are commonly found in elevated or aerial locations. They are often suspended from bush branches or beneath house eaves—even in densely built-up areas—as well as under rafters and roofs in barns and sheds, beneath park benches, on building exteriors, and within cavity walls (Falk, 1991; Colvin, 1992). The nest material is typically pale grey paper, constructed from fibers likely sourced from weathered, intact wood. The entrance to the nest typically features a prominent, lip-like rim (Allen and Archer, 1989; Else, 1991; Edwards, 1997).

***Dolichovespula sylvestris* (Scopoli, 1763)**

Vespa sylvestris Scopoli, 1763: 309, sex not stated; "Slovenia".

Catalogues with Iranian records: Ebrahimi and Carpenter (2008), Bagriacik and Samin (2011), Dvořák *et al.* (2012), Rahmani *et al.* (2020).

FIGURE 5.1 *Dolichovespula saxonica* (Fabricius, 1793) (Vespinae).

(Photo courtesy of Alexander Fateryga)

Distribution in Iran: Alborz (Ebrahimi and Carpenter, 2008), East Azarbaijan (Ebrahimi and Carpenter, 2008; Sakenin *et al.*, 2010a as *Dolichovespula media* (Retzius, 1783) – misidentification; Modarres Awal, 2012; Ghahari *et al.*, 2016), Golestan (Dvořák *et al.*, 2012; Schmid-Egger *et al.*, 2021), Guilan, Khuzestan (Archer, 1981 as *Dolichovespula sylvestris sumptuosa* (du Buysson, 1904)), Ilam (Abd-Rabou *et al.*, 2005 as *D. media* – misidentification), Kuhgiloyeh & Boyerahmad (Castro and Dvořák, 2009), Mazandaran (Abd-Rabou *et al.*, 2005; Ebrahimi and Carpenter, 2008; Dvořák *et al.*, 2012; Modarres Awal, 2012), Tehran (Archer, 1981 as *D. sylvestris sumptuosa*; Ebrahimi and Carpenter, 2008; Samin *et al.*, 2010 as *D. media* – misidentification; Dvořák *et al.*, 2012; Modarres Awal, 2012), West Azarbaijan (Blüthgen and Gusenleitner, 1970; Archer, 1981 both as *D. sylvestris sumptuosa*; Ghahari *et al.*, 2016).

Distribution in the Middle East: Iran (see references above, and Birula, 1930 as *D. sylvestris sumptuosa*; Carpenter and Kojima, 1997; Castro and Dvořák, 2009; Antropov, 2017; Fateryga, 2018), Syria (Carpenter and Kojima, 1997; Castro and Dvořák, 2009; Antropov, 2017; Fateryga, 2018), Turkey (Tanyolaç, 1964 as *D. sylvestris sylvestris*; Gusenleitner, 1966 as *D. sylvestris sumptuosa*; Eck, 1992 as *D. sylvestris sylvestris*; Yildirim and Özbek, 1992, 1998, 1999 all as *D.* (*Metavespula*) *sylvestris sylvestris*; Carpenter and Kojima, 1997; Madl, 1997 as *D. sylvestri sylvestri*; Yildirim and Kojima, 1999; Castro and Dvořák, 2009; Yildirim and Gusenleitner, 2012; Antropov, 2017; Fateryga, 2018).

Extralimital distribution: Afghanistan, Azerbaijan, China, Croatia, Europe, Georgia, India, Ireland, Kyrgyzstan, Mongolia, Morocco, Pakistan, Russia, Serbia, Sweden, Switzerland, Tajikistan,

Turkmenistan, Ukraine, Uzbekistan (Carpenter and Kojima, 1997; Castro and Dvořák, 2009; Antropov, 2017; Fateryga, 2018), Kazakhstan (Antropov, 2017).

Comments: *Jothopogon leucomallus* (Loew) (Diptera: Asilidae) has been recorded as a predator of *D. sylvestris* (Samin *et al.*, 2010 – misidentification as *Dolichovespula media*).

Genus *Vespa* Linnaeus, 1758

Vespa Linnaeus, 1758: 343, 572. Type species: "*Vespa crabro*, Fab." [= *Vespa crabro* Linnaeus, 1758], by subsequent designation of Latreille, 1810.

Comments: The hornet wasps of the genus *Vespa* are large, predatory, and eusocial wasps primarily distributed across the Oriental and Palaearctic Regions, though they are native to both Europe and Asia (Carpenter and Kojima, 1997; Smith-Pardo *et al.*, 2020). They prey on a wide variety of insects, including several agricultural, forest and sanitary pests, as well as honeybees (Smith-Pardo *et al.*, 2020); additionally, they may cause damage to fruits by feeding on them just before harvest.

In some parts of the world, their larvae and pupae are consumed as food by humans (Matsuura and Yamane, 1990; Girish Kumar and Srinivasan, 2010; Archer, 2012; Sorvari, 2019). These wasps exhibit a range of geographically variable color patterns (Smith-Pardo *et al.*, 2020). Although some authors have described several subspecies based solely on color variation, *Vespa* coloration differ significantly across regions (Perrard *et al.*, 2014; Smith-Pardo *et al.*, 2020). These subspecies were later synonymized by Carpenter and Kojima (1997).

A total of 22 valid species within this genus are currently known worldwide (Smith-Pardo *et al.*, 2020). Depending on the species, nests may be aerial, attached to tree branches or shrubs, or located in crevices, or underground (Smith-Pardo *et al.*, 2020). These nests are constructed using wood pulp and typically house large colonies consisting of a single female queen, numerous sterile workers, and males. Colonies often contain hundreds of workers, and in some cases, over 1,000 (Carpenter and Kojima, 1997; Archer, 2008).

Nest cycles vary with latitude: they can be annual, founded each spring by a newly emerged queen, or perennial, where younger queens replace older ones to maintain the colony (Matsuura and Yamane 1990; Archer, 2012). Colonies in warm tropical climates are more likely to be perennial (Smith-Pardo *et al.*, 2020).

Hornets prey on a wide range of organisms, including other social wasps, honeybees, flies, butterflies, moths (as they are capable of foraging by moonlight), and spiders (Pawlyszyn, 1994). Numerous insect species are commonly found within hornet nests, occupying various ecological roles: 1. **Scavengers**, such as *Fannia canicularis* (Diptera: Fanniidae) and *Ptinus fur* (Coleoptera: Ptinidae), which feed on decomposing organic material; 2. **Predators**, including *Velleius dilatatus* (Coleoptera: Staphylinidae) and *Korynetes caeruleus* (Coleoptera: Cleridae), which likely prey on hornet larvae or pupae; and 3. **Parasitoids**, such as *Sphecophaga vesparum* (Hymenoptera: Ichneumonidae), which parasitize the sealed brood stages.

In addition, fungus-feeding insects consume molds and other fungi that thrive on the moist materials within the nest (Edwards, 1980, 1997).

Vespa crabro Linnaeus, 1758

Vespa crabro Linnaeus, 1758: 572, ♀; "in Europae".

Catalogues with Iranian records: Ebrahimi and Carpenter (2008), Bagriacik and Samin (2011), Dvořák *et al.* (2012), Rahmani *et al.* (2020).

Distribution in Iran: Ardabil (Ghahari *et al.*, 2016), East Azarbaijan (Hayat *et al.*, 2008; Sakenin *et al.*, 2010a; Ghahari *et al.*, 2016), Fars (Mossadegh, 1995), Golestan (Ebrahimi and Carpenter, 2008, 2012; Dvořák *et al.*, 2012; Modarres Awal, 2012), Guilan (Perez, 1910 as *Vespa crabro caspica* Perez, 1910; Morice, 1921 as *Vespa crabro crabroniformis* Smith, 1852; Ebrahimi and Carpenter, 2008; Modarres Awal, 2012), Ilam (Hayat *et al.*, 2008; Dvořák *et al.*, 2012),

Khuzestan (Mossadegh, 1995; Abd-Rabou *et al*., 2005), Lorestan (Mossadegh, 1995), Mazandaran (Ebrahimi and Carpenter, 2008, 2012; Dvořák *et al*., 2012; Modarres Awal, 2012; Schmid-Egger *et al*., 2021), Northern Khorasan (Sakenin *et al*., 2010b), West Azarbaijan (Sakenin *et al*., 2010a; Ghahari *et al*., 2016), North of Iran (Esmaili and Rastegar, 1974; Modarres Awal, 1997, 2012 all as *V. crabro caspica*; Ebrahimi, 2014).

Distribution in the Middle East: Iran (see references above, and Ahmadi and Ebadi, 1986; Modarres Awal, 1997; Carpenter and Kojima, 1997; Antropov, 2017; Fateryga, 2018), Turkey (Gusenleitner, 1967; Tüzün and Tanyolaç, 1987; Yildirim and Özbek, 1992, 1999 both as *V. crabro germana*; Carpenter and Kojima, 1997; Yildirim and Kojima, 1999; Antropov, 2017; Fateryga, 2018; Kaplan and Efil, 2025).

Extralimital distribution: Algeria, Azerbaijan, Canada (introduced), China, Croatia, Czech Republic, Europe, Georgia, Guatemala, Japan, Kazakhstan, Korea, Mongolia, Russia, Sakhalin Islands, Taiwan, Turkmenistan, Ukraine (Carpenter and Kojima, 1997; Antropov, 2017; Fateryga, 2018).

Comments: *Vespa crabro* has been recorded in association with *Apis mellifera* (Hymenoptera: Apidae) hives (Mossadegh, 1995) and is also known to be a predator of honey bee (Ahmadi and Ebadi, 1986; Modarres Awal, 1997, 2012). *Choerades gilva* (Linnaeus) and *Dasypogon octonotatus* Loew (Diptera: Asilidae) have been identified as predators of *Vespa crabro* (Hayat *et al*., 2008).

Hornet nests are typically found in elevated sites, most commonly within hollow trees, but they may also be established in attics or outbuildings. In some instances, nests are constructed at the base of grass clumps or within cavities in the nests of *Lasius flavus* (Fabricius) (Hymenoptera: Formicidae). Occasionally, nests may be built entirely underground (Archer, 1993; Edwards, 1997). If the original nesting site becomes too confined, a colony may relocate to a more suitable cavity (Pawlyszyn, 1994). The nest structure is made from a pulp produced by collecting and processing decayed wood (Bunn, 1988).

Vespa orientalis Linnaeus, 1771

Vespa orientalis Linnaeus, 1771: 540, ♀; "Oriente".

Catalogues with Iranian records: Ebrahimi and Carpenter (2008), Bagriacik and Samin (2011), Dvořák *et al*. (2012), Rahmani *et al*. (2020).

Distribution in Iran: Alborz (Guiglia, 1976b; Ebrahimi and Carpenter, 2008), Ardabil (Ebrahimi and Carpenter, 2008, 2012; Dvořák *et al*., 2012), Bushehr (Ebrahimi and Carpenter, 2008, 2012; Ebrahimi and Carpenter, 2012; Dvořák *et al*., 2012), East Azarbaijan (Ebrahimi and Carpenter, 2008, 2012; Sakenin *et al*., 2010a; Dvořák *et al*., 2012; Ghahari *et al*., 2016), Fars (Guiglia, 1976b, 1977; Ostovan and Kamali, 1995; Ebrahimi and Carpenter, 2008, 2012; Dvořák *et al*., 2012; Nemati and Pezhman, 2014, in wheat field; Keshavarzi *et al*., 2019; Schmid-Egger *et al*., 2021), Golestan (Ebrahimi and Carpenter, 2008, 2012; Dvořák *et al*., 2012), Guilan (Hayat *et al*., 2008; Dvořák *et al*., 2012), Hamadan (Ebrahimi and Carpenter, 2008, 2012; Dvořák *et al*., 2012), Hormozgan (Ebrahimi and Carpenter, 2008, 2012; Dvořák *et al*., 2012; Khoobdel *et al*., 2014; Schmid-Egger *et al*., 2021), Ilam (Dvořák *et al*., 2012), Isfahan (Blüthgen and Gusenleitner, 1970; Dvořák *et al*., 2012; Dehghani *et al*., 2019), Kerman (Blüthgen and Gusenleitner, 1970; Ebrahimi and Carpenter, 2008, 2012; Dvořák *et al*., 2012 Schmid-Egger *et al*., 2021), Kermanshah (Ebrahimi and Carpenter, 2008, 2012; Dvořák *et al*., 2012; Azizkhani *et al*., 2016), Khuzestan (du Buysson, 1912; Ebrahimi and Carpenter, 2008, 2012, Dvořák *et al*., 2012; Boyeri Shemi and Pashaei Rad, 2017; Karam Kiani *et al*., 2025), Kordestan (Ebrahimi and Carpenter, 2008; Dvořák *et al*., 2012), Kuhgiloyeh & Boyerahmad (Ebrahimi and Carpenter, 2008; Dvořák *et al*., 2012), Lorestan (du Buysson, 1912; Ebrahimi and Carpenter, 2008, 2012; Dvořák *et al*., 2012), Markazi (Ebrahimi and Carpenter, 2008, 2012; Dvořák *et al*., 2012), Mazandaran (Dvořák *et al*., 2012), Northern Khorasan (Blüthgen and Gusenleitner, 1970; Ebrahimi and Carpenter, 2008), Qazvin (Morice, 1921 as *Vespa orientalis aegyptiaca* André, 1884; Ebrahimi and Carpenter, 2008, 2012), Qom (Ebrahimi and

Carpenter, 2008), Razavi Khorasan (Guiglia, 1976b, 1977; Ebrahimi and Carpenter, 2008, 2012; Dvořák *et al.*, 2012), Semnan (Dvořák *et al.*, 2012), Sistan & Baluchestan (Blüthgen and Gusenleitner, 1970; Ebrahimi and Carpenter, 2008, 2012; Dvořák *et al.*, 2012), Southern Khorasan (Dvořák *et al.*, 2012), Tehran (du Buysson, 1912; Blüthgen and Gusenleitner, 1970; Ebrahimi and Carpenter, 2008, 2012; Ebrahimi and Carpenter, 2012; Dvořák *et al.*, 2012), West Azarbaijan (Ebrahimi and Carpenter, 2012; Dvořák *et al.*, 2012), Yazd (Dvořák *et al.*, 2012), Zanjan (Abbasi *et al.*, 2008a, 2008b; Ebrahimi and Carpenter, 2012; Dvořák *et al.*, 2012).

Distribution in the Middle East: Bahrain (Giordani Soika as *Vespa orientalis arabica* Giordani Soika, 1957 and *V. orientalis somalica* Giordani Soika, 1934; Richards, 1985; Walker and Pittaway, 1987; Carpenter and Kojima, 1997; Antropov, 2017; Carpenter and Gadallah, 2020; Selis *et al.*, 2025), Cyprus (Carpenter and Kojima, 1997; Fateryga, 2018; Selis *et al.*, 2025), Egypt (Bytinski-Salz and Gusenleitner, 1971; Zalat, 1992; Carpenter and Kojima, 1997; Selis *et al.*, 2025), Iran (see references above, and Farahbakhsh, 1961; Esmaili and Rastegar, 1974; Mossadegh, 1995; Carpenter and Kojima, 1997; Modarres Awal, 1997, 2012; Ebrahimi, 2014; Antropov, 2017; Fateryga, 2018; Selis *et al.*, 2025), Iraq (Morice, 1921; Derwesh, 1965; Swailem *et al.*, 1974; Abdul-Rassoul, 1976; Al-Ali, 1977; Carpenter and Kojima, 1997; Augul, 2016, 2017; Antropov, 2017; Fateryga, 2018; Al-Saffar and Augul, 2021; Selis *et al.*, 2025), Israel-Palestine (Bytinski-Salz and Gusenleitner, 1971; Carpenter and Kojima, 1997; Antropov, 2017; Fateryga, 2018; Selis *et al.*, 2025), Jordan (Carpenter and Kojima, 1997; Antropov, 2017; Fateryga, 2018; Selis *et al.*, 2025), Kuwait (Walker and Pittaway, 1987; Al-Houty, 1989, 1997; Carpenter and Gadallah, 2020), Lebanon (Carpenter and Kojima, 1997; Antropov, 2017; Fateryga, 2018; Selis *et al.*, 2025), Oman (Giordani Soika, 1957 as *V. o. arabica* and *V. o. somalica*; Richards, 1985; Walker and Pittaway, 1987; Carpenter and Kojima, 1997; Antropov, 2017; Fateryga, 2018; Monks *et al.*, 2019; Carpenter and Gadallah, 2020; Selis *et al.*, 2025), Saudi Arabia, Yemen (Giordani Soika as *Vespa orientalis arabica* Giordani Soika, 1957 and *V. orientalis somalica* Giordani Soika, 1934; Richards, 1985; Walker and Pittaway, 1987; Carpenter and Kojima, 1997; Antropov, 2017; Fateryga, 2018; Carpenter and Gadallah, 2020; Selis *et al.*, 2025), Syria (Carpenter and Kojima, 1997; Abo Kaf and Allewa, 2016; Antropov, 2017; Fateryga, 2018; Selis *et al.*, 2025), Turkey (Bodenheimer, 1958; Tanyolaç, 1964; Özeren, 1965; Gusenleitner, 1966; Tüzün and Tanyolaç, 1987; Yildirim and Özbek, 1992; Özbay and Akbayrn, 1993; Akbayrn and Özbay, 1996; Carpenter and Kojima, 1997; Madl, 1997; Yildirim and Kojima, 1999; Yildirim and Gusenleitner, 2012; Antropov, 2017; Fateryga, 2018; Selis *et al.*, 2025), United Arab Emirates (Walker and Pittaway, 1987; Carpenter and Kojima, 1997; Gillett and Gillett, 2005 as *V. o. orientalis*; Gillett and Howarth, 2004as *V. o. orientalis*; Gusenleitner, 2010; Gess and Roosenschoon, 2016; Antropov, 2017; Fateryga, 2018; Carpenter and Gadallah, 2020; Selis *et al.*, 2025).

Extralimital distribution: Afghanistan, Algeria, Azerbaijan, Chad, China (introduced), Croatia, Ethiopia, Georgia, India, Kazakhstan, Kyrgyzstan, Libya, Madagascar (introduced), Nepal (introduced), Pakistan, Russia, Somalia, Sudan, Tajikistan, Turkmenistan, Uzbekistan, United States of America (not established) (Carpenter and Kojima, 1997; Fateryga, 2018; Carpenter and Gadallah, 2020; Gogala, 2022; Selis *et al.*, 2025), Albania, Belgium, Bosnia and Herzegovina, Bulgaria, Chile, Czech Republic, Greece, Italy, Malta, North Macedonia, Montenegro, Romania, Slovenia, Spain, United Kingdom (introduced) (Carpenter and Gadallah, 2020; Selis *et al.*, 2025), Mexico (Dvořák, 2006; Selis *et al.*, 2025).

Comments: *Vespa orientalis* is considered a pest of grapevine and apricots (Farahbakhsh, 1961; Modarres Awal, 1997, 2012), as well as date fruits in southern Iran (Ostovan and Kamali, 1995; Modarres Awal, 1997, 2012; Ebrahimi, 2014), and sweet prickly pear, *Opuntia ficus-indica* (L.) Mill. (Cactaceae) (Azizkhani *et al.*, 2016).

It has also been recorded in association with *Apis mellifera* (Hymenoptera: Apidae) hives (Mossadegh, 1995; Modarres Awal, 1997, 2012), and is a known predator of honey bee (Modarres Awal, 1997, 2012; Karam Kiani *et al.*, 2025).

The habitats and behavioral characteristics of *V. orientalis* have been studied by Ostovan and Kamali (1995) and Dehghani *et al.* (2019). *Philodicus ponticus* (Bigot) (Diptera: Asilidae) has been recorded as a predator of *Vespa orientalis* (Hayat *et al.*, 2008).

In Jordan, Haddad *et al.* (2005), and in Iran, Ghahari and Jussila (2015) recorded *Sphecophaga vesparum* Curtis (Ichneumonidae) as a parasitoid of *Vespa orientalis*.

***Vespa velutina* Lepeletier de Saint Fargeau, 1836**

Vespa velutina "De Hann" Lepeletier de Saint Fargeau, 1836: 507, ♀; "Indonesia".

Distribution in the Middle East: Yemen (introduced) (Richards, 1985 as *V. velutina auraria* Smith, 1852; Carpenter and Kojima, 1997 as *V. auraria*; Lien and Carpenter, 2002 as *V. auraria*; Carpenter and Gadallah, 2020).

Extralimital distribution: Afghanistan, Belgium (introduced), Bhutan, China, France, Germany, India, Indonesia, Italy, Japan, Kenya, Laos, Malaysia, Myanmar, Nepal, Pakistan, Portugal, Singapore, Spain, Taiwan, Thailand, United Kingdom, Vietnam (Carpenter and Kojima, 1997; Lien and Carpenter, 2002; Carpenter and Gadallah, 2020), Korea (Kim *et al.*, 2006).

***Vespa* sp.**

Distribution in Iran: Isfahan (Ebadi, 1995).

Comments: This unknown species, which probably is *Vespa crabro* or *V. orientalis*, was introduced as a pollinator insect by Ebadi (1995).

Genus *Vespula* Thomson, 1869

Vespula Thomson, 1869: 79. Type species: *Vespa austriaca* Panzer, 1799, by subsequent designation of Ashmead, 1902.

Comments: The yellowjacket genus *Vespula* can be distinguished from the other genera of the subfamily Vespinae by the absence of a pronotal carina and by having a malar space that is shorter than the last antennal segment. These wasps are eusocial in nature and typically build underground nests (Kumar and Carpenter, 2018; Sora *et al.*, 2019). In addition, the characteristics of their three castes, as well as nest architecture and larval morphology, are important for the phylogenetics and taxonomy of the genus (Greene, 1991; Kojima, 1998; Wenzel, 1998; Tan *et al.*, 2018).

A total of 25 valid *Vespula* species are currently recognized worldwide (Kumar and Carpenter, 2018; Tan *et al.*, 2018; Nguyen *et al.*, 2019). *Vespula* species are primarily found in temperate regions, and those occurring in subtropical areas are typically restricted to mountainous habitats (Akre *et al.*, 1980; Yamane *et al.*, 1980; Matsuura and Yamane, 1990). Most species have wide geographical distribution, and, as a result, exhibit considerable variation in coloration (Carpenter and Kojima, 1997; Carpenter *et al.*, 2011).

***Vespula austriaca* (Panzer, 1799)**

Vespa austriaca Panzer, 1799: 2, ♂; "Viennae Austr." (destroyed).

Distribution in the Middle East: Turkey (Yildirim and Özbek, 1992 as *V.* (*Allovespula*) *austriaca*; Carpenter and Kojima, 1997; Yildirim and Kojima, 1999; Antropov, 2017).

Extralimital distribution: Canada, China, Croatia, Europe, Georgia, India, Japan, Kazakhstan, Korea, Kyrgyzstan, Mongolia, Pakistan, Russia, Ukraine (Carpenter and Kojima, 1997), Azerbaijan (Antropov, 2017). Also, all records of this species from North America should refered to *Vespula infernalis* (de Saussure, 1854) (Antropov, 2017).

***Vespula germanica* (Fabricius, 1793)**

Vespa germanica Fabricius 1793: 256, sex not stated; [Germany] "Kiliae".

Catalogues with Iranian records: Ebrahimi and Carpenter (2008), Bagriacik and Samin (2011), Dvořák *et al.* (2012), Rahmani *et al.* (2020).

Distribution in Iran: Alborz (Ebrahimi and Carpenter, 2008), East Azarbaijan (Ebrahimi and Carpenter, 2008; Dvořák *et al.*, 2012), Fars (Ebrahimi and Carpenter, 2008; Dvořák *et al.*, 2012;

Keshavarzi *et al.*, 2019), Golestan (Hayat *et al.*, 2008; Dvořák *et al.*, 2012), Guilan (Morice, 1921; Blüthgen and Gusenleitner, 1970 as *Paravespula germanica* (Fabricius, 1793); Dvořák *et al.*, 2012), Hamadan (Dvořák *et al.*, 2012), Hormozgan (Khoobdel *et al.*, 2014), Isfahan (Lehr *et al.*, 2007; Ebrahimi and Carpenter, 2008; Dvořák *et al.*, 2012), Kerman (Ebrahimi and Carpenter, 2008; Dvořák *et al.*, 2012; Schmid-Egger *et al.*, 2021), Kermanshah (du Buysson, 1912; Morice, 1921; Dvořák *et al.*, 2012), Khuzestan (Dvořák *et al.*, 2012; Boyeri Shemi and Pashaei Rad, 2017), Kordestan (du Buysson, 1912; Ebrahimi and Carpenter, 2008; Dvořák *et al.*, 2012), Kuhgiloyeh & Boyerahmad (Ebrahimi and Carpenter, 2008; Dvořák *et al.*, 2012), Lorestan (Ebrahimi and Carpenter, 2008; Dvořák *et al.*, 2012), Markazi (Ebrahimi and Carpenter, 2008; Dvořák *et al.*, 2012), Mazandaran (Ebrahimi and Carpenter, 2008; Dvořák *et al.*, 2012), Northern Khorasan (Blüthgen and Gusenleitner, 1970 as *P. germanica*; Ebrahimi and Carpenter, 2008), Qazvin (Morice, 1921; Dvořák *et al.*, 2012), Qom (Ebrahimi and Carpenter, 2008), Razavi Khorasan (Guiglia, 1976b; Ebrahimi and Carpenter, 2008; Dvořák *et al.*, 2012), Semnan (Dvořák *et al.*, 2012), Sistan & Baluchestan (Ebrahimi and Carpenter, 2008), Southern Khorasan (Morice, 1921; Blüthgen and Gusenleitner, 1970 as *P. germanica*; Dvořák *et al.*, 2012), Tehran (Ebrahimi and Carpenter, 2008; Dvořák *et al.*, 2012), West Azarbaijan (Ebrahimi and Carpenter, 2008; Dvořák *et al.*, 2012; Ghahari *et al.*, 2016), Yazd (Dvořák *et al.*, 2012), Zanjan (Abbasi *et al.*, 2008a, 2008b; Dvořák *et al.*, 2012).

Distribution in the Middle East: Iran (see references above, and Farahbakhsh, 1961; Esmaili and Rastegar, 1974; Mossadegh, 1995; Modarres Awal, 1997, 2012; Ebrahimi, 2014; Carpenter and Kojima, 1997; Castro and Dvořák, 2010; Antropov, 2017; Fateryga, 2018; Selis *et al.*, 2025), Iraq (Khalaf, 1958; Derwesh, 1965; Al-Ali, 1977; Carpenter and Kojima, 1997; Castro and Dvořák, 2010; Augul, 2017; Antropov, 2017; Fateryga, 2018; Selis *et al.*, 2025), Israel-Palestine (Carpenter and Kojima, 1997; Braverman *et al.*, 1998; Antropov, 2017; Fateryga, 2018; Selis *et al.*, 2025), Jordan (Carpenter and Kojima, 1997; Haddad *et al.*, 2005, 2007; Antropov, 2017; Fateryga, 2018; Selis *et al.*, 2025), Lebanon (Carpenter and Kojima, 1997; Antropov, 2017; Fateryga, 2018; Selis *et al.*, 2025), Syria (Carpenter and Kojima, 1997; Abo Kaf and Allewa, 2016; Antropov, 2017; Fateryga, 2018; Selis *et al.*, 2025), Turkey (Bodenheimer, 1958; Tanyolaç, 1964; Özeren, 1965; Tüzün and Tanyolaç, 1987; Gusenleitner, 1967; Yildirim and Özbek, 1992 as *Vespula (Paravespula) germanica*; Özbay and Akbayrn, 1993 as *V. (Paravespula) germanica*; Akbayrn and Özbay, 1996 as *V. (Paravespula) germanica*; Carpenter and Kojima, 1997; Madl, 1997 as *V. (Paravespula) germanica*; Yildirim and Özbek, 1999 as *Vespula (Paravespula) germanica*; Castro and Dvořák, 2010; Antropov, 2017; Fateryga, 2018; Kaplan and Efil, 2025; Selis *et al.*, 2025).

Extralimital distribution: Afghanistan, Algeria, Argentina (introduced), Australia (introduced), Azerbaijan, Canada (introduced), Chile (introduced), Croatia, Mongolia, Morocco, Nepal, Pakistan, Russia, South Africa (introduced), Taiwan, Tajikistan, Tunisia, Turkmenistan, Ukraine, United States of America, Uzbekistan (Carpenter and Kojima, 1997; Castro and Dvořák, 2010; Antropov, 2017; Fateryga, 2018; Selis *et al.*, 2025); Armenia, Austria, Albania, Belarus, Belgium, Bosnia and Herzegovina, Bulgaria, Czech Republic, Denmark, Estonia, Finland, France, Georgia, Germany, Greece, Hungary, Iceland (introduced), India, Ireland, Italy, Kazakhstan, Korea, Kyrgyzstan, Lichtenstein, Lithuania, Luxembourg, Malta, Moldova, Netherlands, New Zealand, North Macedonia, Norway, Poland, Portugal, Romania, Serbia, Slovakia, Slovenia, Spain, Sweden, Switzerland, United Kingdom (Selis *et al.*, 2025).

Comments: *Vespula germanica* is considered a pest of grapevine, apricots, and other fruits (Farahbakhsh, 1961; Modarres Awal, 1997, 2012). It has also been recorded in association with hives of *Apis mellifera* (Hymenoptera: Apidae) (Mossadegh, 1995; Modarres Awal, 1997, 2012). *Stenopogon xanthotrichus* (Brullé) and *Philodicus ponticus* (Bigot) (Diptera: Asilidae) have been identified as predators of *V. germanica* (Lehr *et al.*, 2007; Hayat *et al.*, 2008).

Sphecophaga vesparum (Curtis) (Hymenoptera: Ichneumonidae) has also been collected from colonies of *V. germanica* in Mazandaran province, northern of Iran [6♀♀ (both brachypterous and winged females), Qaemshahr, km 2 Ahangarkola – Golafshan; 15 August 2020].

***Vespula rufa* (Linnaeus, 1758)**

Vespa rufa Linnaeus, 1758: 572, ♀; "in Europa".

Catalogues with Iranian records: Bagriacik and Samin (2011).

Distribution in Iran: Ardabil (Sakenin *et al.*, 2010a), Kordestan (Abd-Rabou *et al.*, 2005), Mazandaran (Sakenin *et al.*, 2010b), Semnan (Sakenin *et al.*, 2010c).

Distribution in the Middle East: Iran (see references above, and Antropov, 2017; Fateryga, 2018), Turkey (Yildirim and Özbek, 1992, 1999 both as *V.* (*Allovespula*) *rufa*; Carpenter and Kojima, 1997; Yildirim and Gusenleitner, 2012; Antropov, 2017; Fateryga, 2018).

Extralimital distribution: Afghanistan, Azerbaijan, Canada, China, Croatia, Europe, Georgia, Japan, Kazakhstan, Korea, Kyrgyzstan, Mongolia, Nepal, Russia, Serbia, Taiwan, Tajikistan, Turkmenistan, Ukraine, Uzbekistan (Carpenter and Kojima, 1997; Fateryga, 2018). Also, all records of this species from North America should referred to *V. intermedia* (du Buysson, 1904) (Antropov, 2017).

Comments: Re-examination of specimens reported of Abd-Rabou *et al.* (2005) and Sakenin *et al.* (2010b) confirms the presence of *Vespula rufa* in Iran (Figure 5.2). The specimens from other records were destroyed and are no longer available for re-examining. Additionally, *Hoplotriclis pallasii* (Wiedemann) (Diptera: Asilidae) has been recorded as a predator of *V. rufa* (Sakenin *et al.*, 2010c).

***Vespula vulgaris* (Linnaeus, 1758)**

Vespa vulgaris Linnaeus, 1758: 572, ♀; "in Europa".

Catalogues with Iranian records: Ebrahimi and Carpenter (2008), Bagriacik and Samin (2011), Dvořák *et al.* (2012), Rahmani *et al.* (2020).

FIGURE 5.2 *Vespula rufa* (Linnaeus, 1758) (Vespinae).

(Photo courtesy of Michael E. Archer)

Distribution in Iran: East Azarbaijan (Sakenin *et al.*, 2008, 2010b), Kermanshah (Abd-Rabou *et al.*, 2005), Northern Khorasan (Archer, 1989), Razavi Khorasan (Dvořák *et al.*, 2012), Semnan (Sakenin *et al.*, 2010c), Tehran (Samin *et al.*, 2010).

Distribution in the Middle East: Iran (see references above, and Antropov, 2017; Fateryga, 2018; Selis *et al.*, 2025), Iraq (Khalaf, 1961; Derwesh, 1965; Selis *et al.*, 2025), Israel-Palestine, Syria (Carpenter and Kojima, 1997; Selis *et al.*, 2025), Turkey (Tüzün and Tanyolaç, 1987; Yildirim and Özbek, 1992, 1999 both as *V.* (*Paravespula*) *vulgaris*; Carpenter and Kojima, 1997; Yildirim and Gusenleitner, 2012; Antropov, 2017; Fateryga, 2018; Kaplan and Efil, 2025; Selis *et al.*, 2025).

Extralimital distribution: Albania, Austria, Belarus, Belgium, Buthan, Bulgaria, Czech Republic, Denmark, Estonia, Finland, France, Germany, Greece, Iceland (introduced), Italy, Latvia, Lichtenstein, Lithuania, Luxembourg, Moldova, Netherlands, New Zealand, North Macedonia, Norway, Poland, Portugal, Romania, Serbia, Slovakia, Slovenia, Spain, Sweden, Switzerland, United Kingdom (Selis *et al.*, 2025), Argentina, Australia (introduced), Azerbaijan, Chile, China, Croatia, Georgia, Hawaii (introduced), Hungary, India, Japan, Kazakhstan, Korea, Kyrgyzstan, Mongolia, Pakistan, Russia, South Africa, Ukraine (Carpenter and Kojima, 1997; Antropov, 2017; Fateryga, 2018; Carpenter and Brown, 2022; Selis *et al.*, 2025), Tahiti (Carpenter and Brown, 2022).

Comments: All records of this species from North America should be referred to *V. alascensis* (Packard, 1870) (Antropov, 2017).

Hoplotriclis pallasii (Wiedemann) and *Leptogaster pubicornis* Loew (Diptera: Asilidae) were recorded as predators of *Vespula vulgaris* by Sakenin *et al.* (2010c) and Samin *et al.* (2010).

Species Excluded from the Fauna of Iran

Dolichovespula media (Retzius, 1783)

Vespa crabro medius Retzius, 1783: 63, sex not stated.

Records from Iran: East Azarbaijan (Sakenin *et al.*, 2010a), Ilam (Abd-Rabou *et al.*, 2005), Tehran (Samin *et al.*, 2010).

Comments: Re-examination of the specimen confirmed that the identification of *Dolichovespula media* by Sakenin *et al.* (2010a), Abd-Rabou *et al.* (2005), and Samin *et al.* (2010) was a misidentification of *Dolichovespula sylvestris*. Therefore, we exclude *D. media* from the fauna of Iran and reassign the records to *D. sylvestris*. Among the Middle Eastern countries, *D. sylvestris* has been recorded from Syria and Turkey.

CONCLUSION

The Vespinae fauna of the Middle East comprises 13 species (18.5% of the world species) within three genera, including *Dolichovespula* (with six species), *Vespa* (three species), and *Vespula* (four species). In Iran, a total of eight species across these three genera have been recorded to date.

Iranian Vespinae have been documented in 30 provinces, with the exception of Chaharmahal & Bakhtiari. Among the provinces, East Azarbaijan and Tehran have the highest species richness, each with seven recorded species, followed by Mazandaran (five species), Golestan, Guilan, Khuzestan, Northern Khorasan, Semnan, West Azarbaijan (each with four species), Alborz, Ardabil, Fars, Ilam, Isfahan, Kermanshah, Kordestan, Kuhgiloyeh & Boyerahmad, Lorestan, Razavi Khorasan (each with three species), Hamadan, Hormozgan, Kerman, Markazi, Qazvin, Qom, Southern Khorasan, Sistan & Baluchestan, Yazd, Zanjan (each with two species), and Bushehr (with one recorded species).

A comparison of the vespine fauna across 16 Middle Eastern countries suggests that Turkey, with 12 recorded species, has the highest diversity. This is followed by Iran (eight species), Syria (five species), Iraq and Israel-Palestine (each with three species), Jordan, Lebanon, and Yemen (each with two species), and Bahrain, Cyprus, Egypt, Kuwait, Oman, Saudi Arabia, and United Arab Emirates (each with one species). No species have been recorded from Qatar to date.

Additionally, among the Middle Eastern and neighboring countries adjacent to Iran, three countries, Kazakhstan, Russia, and Turkey, share the greatest number of vespine species with Iran (eight species each). These are followed by Azerbaijan (six species), Turkmenistan (five species), and Afghanistan, Pakistan, and Syria (each with four species), Iraq and Israel-Palestine (each with three species), Jordan and Lebanon (each with two species), Armenia, Bahrain, Cyprus, Egypt, Kuwait, Oman, Saudi Arabia, United Arab Emirates, and Yemen (each with one species).

REFERENCES

Abbasi, R., Mashhadikhan, M., Abbasi, M. and Kiabi, B. 2008a. Biodiversity of vespid wasps in spatial and temporal dimension in northern Zanjan province of Iran. *Chinese Journal of Ecology* 27(5): 797–802.

Abbasi, R., Pashaei Rad, S., Ebrahimi, E. and Sheidaei, M. 2008b. Faunistic study of vespid wasps in Zanjan province (Northwest of Iran) with some ecological measures. *Environmental Sciences* 6(1): 65–74.

Abd-Rabou, S., Ghahari, H., Mauss, V. and Plant, J. 2005. New records of Apidae, Andrenidae, Sphecidae and Vespidae (Hymenoptera) from Iran. *Egyptian Journal of Agricultural Research* 83(4): 1613–1618.

Abdul-Rassoul, M.S. 1976. Checklist of Iraq Natural History Museum insect collections. *Iraq Natural History Research Center* 30: 1–41.

Abels, J. and Wedmann, S. 2022. A fossil wasp (Hymenoptera: Vespidae: *Vespula? hassiaca*) from the Eocene of Messel. *Palaeontographica, Abteilung A: Palaeozoology – Stratigraphy* 323(1–3): 105–117.

Abo Kaf, N.H. and Allewa, A.N. 2016. Identification of some species from wasps family (Vespidae; Hymenoptera) in Lattakia, Syria. *Tishreen University Journal for Research and Scientific Studies* 38(4): 224–235.

Ahmadi, A.A. and Ebadi, R. 1986. *Diseases, pests and predators of honeybees*. Rahe Nadjat Publication, Esfahan, 219 pp. [in Persian]

Akbayrn, H. and Özbay, C. 1996. Variability analysis on some Vespidae species from Diyarbakrr, Mardin and Şanhurfa in South-east Anatolia. *Doga. Turkish Journal of Biology* 20: 35 1–358.

Akre, R.D., Greene, A., MacDonald, J.F., Landolt, P.J. and Davis, H.G. 1980. Yellowjackets of America north of Mexico. *U.S. Department of Agriculture, Agriculture Handbook* 552: 1–102.

Al-Ali, A.S. 1977. Phytophagous and entomophagous insects and mites of Iraq. *Iraq Natural History Research Center* 33: 1–142.

Al-Houty, W. 1989. *Insect fauna of Kuwait*. Fahad Al-Marzouk Printing and Publishing Establishment, University of Kuwait, Kuwait City, Kuwait, 186 pp.

Al-Houty, W. 1997. Checklist of the insect fauna of Kuwait. *Kuwait Journal of Science and Engineering* 24(1): 145–162.

Al-Saffar, H.H. and Augul, R.S. 2021. Survey of insects in some Southern Iraqi marshes. *Bulletin of Iraq Natural History Museum* 16(4): 571–621.

Antropov, A.V. 2017. Subfamily Vespinae – Hornets and Yellow Jackets, pp. 193–196. In: Belokobylskij, S.A. and Lelej, A.S. (eds), Annotated catalogue of the Hymenoptera of Russia. Volume I. Symphyta and Apocrita: Aculeata. *Proceedings of the Zoological Institute of the Russian Academy of Sciences*, Supplement No. 6, 475 pp.

Archer, M.E. 1981. Taxonomy of the *sylvestris* group (Hymenoptera: Vespidae, *Dolichovespula*) with the introduction of a new name and notes on distribution. *Insect Systematics and Evolution* 12(2): 187–193.

Archer, M.E. 1989. *A key to the world species of the Vespinae (Hymenoptera)*. Academic Board Research Committee, Research Monograph 2. University College of Ripon & York St. John, York, U.K., 94 pp.

Archer, M.E. 1993. The life history and colonial characteristics of the hornet, *Vespa crabro* L. (Hym., Vespinae). *Entomologist's Monthly Magazine* 129: 151–163.

Archer, M.E. 2006. Taxonomy, distribution and nesting biology of species of the genus *Dolichovespula* (Hymenoptera, Vespidae). *Entomological Science* 9: 281–293.

Archer, M.E. 2008. Taxonomy, distribution and nesting biology of species of the genera Provespa Ashmead and Vespa Linnaeus (Hymenoptera, Vespidae). *Entomologist's Monthly Magazine* 144: 69–101.

Archer, M.E. 2012. *Vespine wasps of the world: behaviour, ecology and taxonomy of the Vespinae*. Monograph Series. Vol. 4. Siri Scientific Press, Manchester, UK, 352 pp.

Archer, M.E. and Penney, D. 2012. *Vespine wasps of the world: behaviour, ecology and taxonomy of the Vespinae*. Siri Scientific Press, Monograph Series, Manchester, UK, 352 pp.

Ashmead, W.H. 1902. Papers from the Harriman Alaska expedition. XXVIII. Hymenoptera. *Proceedings of the Washington Academy of Sciences*, 4:117–274.

Augul, R.S. 2016. Insect pollinators in different regions of Iraq. *Journal of Entomology and Zoology Studies* 4 (2): 391–402.

Augul, R.S. 2017. Revision of the vespid wasps (Hymenoptera; Vespidae) in Iraq. *Journal of Biodiversity and Environmental Sciences* 10(3): 99–111.

Azizkhani, E., Omid, R., Moniri, V.R., Yarmand, H. and Zeynali, S. 2016. Investigation of pests of *Opuntia ficus-indica* in Bushehr, Ilam and Kermanshah provinces. *Iranian Journal of Forest and Range Protection Research* 13(2): 171–175. [in Persian, English summary]

Bagriacik, N. and Samin, N. 2011. A checklist of Iranian Vespinae (Hymenoptera: Vespoidea: Vespidae). *Archives of Biological Sciences, Belgrade* 63(2): 487–492.

Beggs, J.R., Brockerhoff, E.G., Corley, J.C., Kenis, M., Masciocchi, M., Müller, F., Rome, Q. and Villemant, C. 2011. Ecological effects and management of invasive alien Vespidae. *BioControl* 56: 505–526.

Birula, A. 1930. Über die russischen Wespen und ihre geographische Verbreitung (Dritter Beitrag). *Annuaire du Musée Zoologique de l'Académie des Sciences de l'URSS* 31: 291–339.

Bischoff, F. 1931. Hymenopteren aus dem nördlichen und östlichen Spanien (Evaniiden, Braconiden, Chalcididen, Proctotrupiden, Mutilliden, Scoliiden, Crabroniden und Vespiden), gesammelt von Dr. F. Haas 1914–1918 und von Prof. Dr. A. Seitz 1923. *Senckenbergiana Biologica* 13: 7–10.

Blüthgen, P. and Gusenleitner, J. 1970. Über Faltenwespen aus dem Iran (Hym., Diploptera), *Stuttgarter Beiträge zur Naturkunde* 223: 1–13.

Bodenheimer, P. 1958. *Türkiye'de ziraate ve ağaçlara zararlı olan böcekler ve bunlarla savaş hakkında bir etüt.* Bayur Matbaasr, Ankara, 319 pp.

Boyeri Shemi, F. and Pashaei Rad, S. 2017. Faunistic study of Vespidae (Hymenoptera) in Izeh, Khuzestan province. *Proceedings of the 3rd National Congress on Development and Promotion of Agricultural Engineering and Soil Sciences of Iran*, 3 September 2017, Tehran, 11 pp. [in Persian, English summary]

Braverman, Y., Ghizov-Ginzburg, A., Yeruham, I., Kolsky, O. and Saran, A. 1998. Control experiments with yellow jacket wasps (Hymenoptera: Vespidae) injuring cattle in Israel. *Journal of Economic Entomology* 91(2): 486–491.

Brothers, D.J. and Finnamore, A.T. 1993. Chapter 8. Superfamily Vespoidea, pp. 161–278. In: Goulet, H. and Huber, J.T. (eds.), *Hymenoptera of the world: An identification guide to families*. Research Branch, Agriculture Canada, Ottawa, 668 pp.

Bunn, D.S. 1988. Observations on the foraging habits of the hornet, *Vespa crabro* L. (Hym., Vespidae). *Entomologist's Monthly Magazine* 124: 187–193.

Bytinski-Salz, H. and Gusenleitner, J. 1971. The Vespoidea of Israel. *Israel Journal of Entomology* 6: 239–250.

Carpenter, J.M. 1987. Phylogenetic relationships and classification of the Vespinae (Hymenoptera: Vespidae). *Systematic Entomology* 12(4): 413–431.

Carpenter, J.M. and Brown, G.R. 2022. Catalogue of the Australian social wasps (Hymenoptera: Vespidae). *Zootaxa* 5214(4): 451–495.

Carpenter, J.M., Dvorak, L., Kojima, J.-I., Nguyen, L.T.P., Perrard, A. and Pickett, K.M. 2011. Taxonomic notes on the Vespinae of Yunnan (Hymenoptera: Vespidae). *American Museum Novitates* 3709: 1–10.

Carpenter, J.M. and Gadallah, N.S. 2020. Biodiversity of aculeate wasps (Hymenoptera: Aculeata) of the Arabian Peninsula: Vespoidea, *Vespidae. Zootaxa* 4754(1): 191–216.

Carpenter, J.M. and Kojima, J.-I. 1997. Checklist of the species in the subfamily Vespinae (Insecta: Hymenoptera: Vespidae). *Natural History Bulletin of Ibaraki University* 1: 51–92.

Carpenter, J.M. & Nguyen, L.P.T. 2003. Keys to the genera of social wasps of South-East Asia (Hymenoptera: Vespidae). *Entomological Science* 6: 183–192.

Castro, L. and Dvořák, L. 2009. New and noteworthy records of vespid wasps (Hymenoptera: Vespidae) from the Palaearctic region (II). *Boletín de la Sociedad Entomológica Aragonesa* 44: 295–304.

Castro, L. and Dvořák, L. 2010. New and noteworthy records of vespid wasps (Hymenoptera: Vespidae) from the Palaearctic region (III). *Acta Musei Moraviae, Scientiae Biologicae (Brno)* 95(2): 37–53.

Colvin, M. 1992. *Dolichovespula saxonica* (F.) found breeding in Britain, with nesting records of *D. media* (Retzius) (Hym., Vespidae). *Entomologist's Monthly Magazine* 128: 137–138.

Daglio, A. 2020. *Yellow jackets of the world*: *Taxonomy, distribution, and nesting biology*. LAP Lambert Academic Publishing, Mauritius, 376 pp.

de Saussure, H. 1854. Mélanges hyménoptérologiques. *Mémoires de la Société Physique et d'Histoire Naturelle de Genève* 14: 1–67.

Dehghani, R., Kassiri, H., Mazaheri-Tehrani, A., Hesam, M., Valazadi, N., and Mohammadzadeh, M. 2019. A study on habitats and behavioral characteristics of the hornet wasp (Hymenoptera: Vespidae: *Vespa orientalis*), an important medical-health pest. *Biomedical Research* 30(1): 61–66.

Derwesh, A.I. 1965. *A preliminary list of identified insects and some arachnids of Iraq*. The Government Press, Baghdad, 123 pp.

Detoni, D., Féas, X., Jeanne, R.L., Loope, K.J., O'Donnell, S., Santoro, D., Sumner, S. and Jandt, J.M. 2021. Evolutionary and ecological pressures shaping social wasps' collective defenses. *Annals of the Entomological Society of America* 114(5): 581–595.

du Buysson, R. 1904 [1905]. Monographie des Guêpes ou *Vespa*. *Annales de la Société Entomologique de France* 73(3): 485–634.

du Buysson, R. 1912. Les Vespides, Euménides et Scoliides. *Annales d'Histoire Naturelle Délégation en Perse II. Entomologie* 1: 77–83.

Dvořák, L. 2006. Oriental hornet *Vespa orientalis* Linnaeus, 1771 found in Mexico (Hymenoptera, Vespidae, Vespinae). *Entomological Problems* 36(1): 80.

Dvořák, L., Ghahari, H., Carpenter, J. and Abbasi, R. 2012. On the distribution and taxonomy of vespine wasps of Iran (Hymenoptera: Vespidae: Vespinae). *Acta Musei Moraviae, Scientiae Biologicae (Brno)* 97(2): 69–86.

Dvořák, L. & Roberts, S.P.M. 2006. Key to the paper and social wasps of Central Europe (Hymenoptera: Vespidae) *Acta Entomologica Musei Nationalis Pragae*, 46: 221–244.

Ebadi, R. 1995. Collection, identification and preliminary study of pollinator insect fauna in Isfahan province. *Proceedings of the 12th Iranian Plant Protection Congress*, 2–7 September 1995, Karaj Junior College of Agriculture, Karaj, p. 319.

Ebrahimi, E. 2014. Aculeate wasps and their role in natural equilibrium and pest control (with a review of the Iranian species). *Biocontrol in Plant Protection* 2(2): 59–68.

Ebrahimi, E. and Carpenter, J.M. 2008. Catalog of the vespid wasps of Iran (Hymenoptera, Vespidae). *Zootaxa* 1785: 1–42.

Ebrahimi, E. and Carpenter, J.M. 2012. Distribution pattern of the hornets *Vespa orientalis* and *V. crabro* in Iran: (Hymenoptera: Vespidae). *Zoology in the Middle East* 56(1): 63–66.

Eck, R. 1992. Morphologisch-zoogeographische Untersuchungen zur Populationsgliederung und Ausbreitungsgeschichte von *Dolichovespula sylvestris* und *Dolichovespula asiatica* (Insecta, Hymenoptera: Vespidae). *Entomologische Abhandlungen. Staatliches Museum für Tierkunde in Dresden* 54(8): 141–172.

Edwards, R. 1980. *Social wasps: Their biology and control*. Rentokil Unlimited, East Grinstead, UK, 398 pp.

Edwards, R. 1997. *Provisional atlas of the aculeate Hymenoptera of Britain and Ireland. Part 1. Bees, Wasps and Ants Recording Society*. Biological Records Centre; Dorset Press, Huntingdon; Dorchester, 139 pp.

Else, G.R. 1991. Wildlife reports: bees and wasps. *British Wildlife* 3: 51–53.

Esmaili, M. and Rastegar, R. 1974. Identified species of aculeate Hymenoptera of Iran. *Journal of the Entomological Society of Iran* 2(2): 41–52.

Fabricius, J.C. 1781. Species Insectorum exhibentes eorum differentias specificas, synonyma auctorum, loca natalia, metamorphosin adiectis observationibus, descriptionibus, 1. *Carol. Ernest. Bohnii, Hamburgi et Kilonii*. I-VIII: 1–517 pp.

Fabricius, J.C. 1793. *Entomologia systematica emendata et aucta. Secundum classes, ordines, genera, species adjectis synonymis, locis, observationibus, descriptionibus*. Vol. 2. Christ. Gottl. Proft, Hafniae [= Copenhagen]. VIII + 519 pp.

Fan, X.L., Gong, Y.J., Chen, P.Y., Tan, Q.Q., Tan, J.L. and Wen, S.J. 2017. Next-generation sequencing of the mitochondrial genome of *Dolichovespula panda* (Hymenoptera: Vespidae) with a phylogenetic analysis of Vespidae. *Journal of Asia-Pacific Entomology* 20: 971–976.

Farahbakhsh, G. 1961. Family Vespidae (Hymenoptera), p. 113. In: Farahbakhsh, G. (ed.), *A checklist of economically important insects and other enemies of plants and agricultural products in Iran*. The Ministry of Agriculture, Department of Plant Protection, Iran, 153 pp.

Fateryga, A.V. 2018. Wasps of the family Vespidae (Hymenoptera) of the Crimean Peninsula. *Entomofauna* 39(1): 193–233.

Feás, X. and Charles, R. 2019. Notes on the nest architecture and colony composition in winter of the yellow-legged Asian hornet, *Vespa velutina* Lepeletier 1836 (Hym.: Vespidae), in its introduced habitat in Galicia (NW Spain). *Insects* 10(8): 237. doi: 10.3390/insects10080237

Foster, K.R., Ratnieks, F.L.W., Gyllenstrand, N. and Thoren, P.A. 2001. Colony kin structure and male production in *Dolichovespula* wasps. *Molecular Ecology* 10: 1003–1010.

Gess, S.K. and Roosenschoon, P.A. 2016. A preliminary survey of flower visiting by aculeate wasps and bees in the Dubai Desert Conservation Reserve, United Arab Emirates. *Journal of Hymenoptera Research* 52: 81–141.

Ghahari, H., Gadallah, N.S. and Wahis, R. 2016. A faunistic study on Pompilidae and Vespidae (Hymenoptera: Vespoidea) from the west and northwest of Iran, with addition of new records. *Entomologie Faunistique – Faunistic Entomology* 69: 21–27.

Ghahari, H. and Jussila, R. 2015. A faunistic study on ichneumonid wasps (Hymenoptera: Ichneumonidae) from Ilam and Kermanshah provinces, western Iran. *Entomologia Croatica* 19(1–2): 55–66.

Ghahari, H., Lehr, P.A., Lavigne, R.J., Hayat, R. and Ostovan, H. 2007. New records of robber flies (Diptera, Asilidae) for the Iranian fauna with their prey records. *Far Eastern Entomologist* 179: 1–9.

Gillett, M. and Gillett, C. 2005. Insects and other arthropods. In: Hellyer, P. and Aspinall, S. (eds.), *The Emirates, a natural history*. Trident Press Ltd., London, pp. 168–195, 352–356, 377–379.

Gillett, M.P.T. and Howarth, B. 2004. The Insects of Jebel Hafit (including a short note on representatives of other arthropod groups), pp. 94–143. In: Aspinall, S. and Hellyer, P. (eds.), *Jebel Hafit, a natural history*, Emirates Natural History Group, Abu Dhabi, 220 pp.

Giordani Soika, A. 1934. *Labus* ed *Eumenes* nuovi o poco noti. *Memorie della Società Entomologica Italiana* 12: 215–228.

Giordani Soika, A. 1957. Hymenoptera: Vespidae. *Expedition to South-West Arabia 1937–8*. 1(31): 471–484.

Girish Kumar, P. and Srinivasan, G. 2010. Taxonomic studies of hornet wasps (Hymenoptera: Vespidae) *Vespa* Linnaeus of India. *Records of the Zoological Survey of India* 110(2): 57–80.

Gogala, A. 2022. Vespid wasps of Slovenia (Hymenoptera: Vespoidea: Vespidae). *Scopolia* 102: 1–79.

Greene, A. 1991. *Dolichovespula* and *Vespula*, pp. 263–305. In: Ross, K.G. and Matthews, R.W. (eds.), *The social biology of wasps*. Cornell University Press, Ithaca, NY, xviii + 678 pp.

Greene, A., Akre, R.D. and Landolt, P.J. 1976. The aerial yellowjacket, *Dolichovespula arenaria* (Fab.): nesting biology, reproductive production, and behavior (Hymenoptera: Vespidae). *Melanderia* 26: 1–34.

Guiglia, D. 1976a. Contributo alla conoscenza dei Polistini dell'Iran (Hymenoptera, Vespidae). *Bollettino del Museo Civico di Storia Naturale di Venezia* 81: 195–201.

Guiglia, D. 1976b [1977]. Le vespe sociali (Missione A. Giordani Soika in Iran, 7). *Bollettino del Museo Civico di Storia Naturale di Venezia* 82: 99–104.

Gusenleitner, J. 1966. Vespidae, Eumenidae und Masaridae aus der Türkei, Teil I. *Polskie Pismo Entomologiczne* 36: 343–363.

Gusenleitner, J. 1967. Vespidae, Eumenidae und Masaridae aus der Türkei, Teil II. *Polskie Pismo Entomologiczne* 37: 663–676.

Gusenleitner, J. 1988. Neue und bemerkenswerte Faltenwespen aus der Türkei (Hymenoptera, Vespoidea). *Linzer biologische Beiträge* 20(2): 713–737.

Gusenleitner, J. 2010. Order Hymenoptera, family Vespidae. *Arthropod Fauna of the UAE* 3: 422–467.

Haddad, N.J., Dvořák, L., Adwan, O., Mdanat, H. and Bataynah, A. 2007. New data on vespid wasp fauna of Jordan (Hymenoptera: Vespidae). *Linzer biologische Beiträge* 39(1): 137–142.

Haddad, N.J., Fuchs, S., Haddadin, J. and Kopelke, J.P. 2005. Record of *Sphecophaga vesparum* Curtis, a natural enemy of *Vespa orientalis* in northern Jordan. *Zoology in the Middle East* 35: 114–116.

Hayat, R., Ghahari, H., Lavigne, R.J. and Ostovan, H. 2008. Iranian Asilidae (Insecta: Diptera). *Turkish Journal of Zoology* 32: 175–195.

Huang, P., Carpenter, J.M., Chen, B. and Li, T.J. 2019. The first divergence time estimation of the subfamily Stenogastrinae (Hymenoptera: Vespidae) based on mitochondrial phylogenomics. *International Journal of Biological Macromolecules* 137: 767–773.

Kaplan, E. and Efil, L. 2025. New contributions to families of Bembecidae, Megachilidae, Scoliidae, Tiphiidae, Mutillidae, Vespidae (Hymenoptera: Aculeata) from Western Türkiye. *Munis Entomology and Zoology* 20(2): 2610–2618.

Karam Kiani, N., Rasekh, A., Shishehbor, P. and Tahmasbi, G. 2025. Annual population dynamics and daily activity rhythm of adult oriental hornets (*Vespa orientalis* L.), an important pest of Iranian honeybee in Ahvaz, Iran. *Journal of Agricultural Science and Technology* 27(4): 751–762.

Keshavarzi, S., Zaimy, M.A., Yusuf, M.A., Shahriarinamadi, M. and Parkhideh, S. 2019. Insect succession on carrion in Fars province, southwestern Iran. *Egyptian Journal of Forensic Sciences* 9: 18.

Khalaf, K.T. 1958. Some Hymenoptera and Coleoptera from Iraq. *Iraq Natural History Museum Publication* 14: 1–3.

Khalaf, K.T. 1961. Miscellaneous insects from Iraq. *Bulletin of the Iraq Natural History Museum* 1(7): 1–7.

Khoobdel, M., Tavassoli, M., Salari, M. and Firozi, F. 2014. The stinging Apidae and Vespidae (Hymenoptera: Apocrita) in Iranian islands, Qeshm, Abu–Musa, Great Tunb and Lesser Tunb on the Persian Gulf. *Asian Pacific Journal of Tropical Biomedicine* 4: 258–262.

Kim, J., Choi, M. and Moon, T. 2006. Occurrence of *Vespa velutina* Lepeletier from Korea, and a revised key for Korean *Vespa* species (Hymenoptera: Vespidae). *Entomological Research* 36: 112–115.

Kimsey, L.S. and Carpenter, J.M. 2012. The Vespinae of North America (Vespidae, Hymenoptera). *Journal of Hymenoptera Research* 28: 37–65.

Kojima, J. 1998. Larvae of social wasps (Hymenoptera: Vespidae). *Natural History Bulletin of Ibaraki University* 2: 7–227.

Komonen, A. 2023. Nest characteristics and associates of *Dolichovespula* (Hymenoptera, Vespidae) in Central Finland. *Journal of Hymenoptera Research* 96: 45–55.

Kumar, P.G. and Carpenter, J.M. 2018. A taxonomic review of the genus *Vespula* Thomson (Hymenoptera: Vespidae: Vespinae) from the Indian subcontinent. *Halteres* 9: 86–110.

Latreille, P.A. 1802. *Histoire naturelle*, Générale et particulière des Crustacés et des Insectes. Tome troisième. F. Schoell, Paris, 468 pp.

Latreille, P.A. 1810. *Considérations générales sur l'ordre naturel des animaux composant les classes des Crustacés, des Arachnides et des Insectes*. F. Schoell, Paris, 444 pp.

Lehr, P.A., Ghahari, H. and Ostovan, H. 2007. A contribution to the robber flies of subfamilies Stenopogoninae and Asilinae (Diptera: Asilidae) from Iran. *Far Eastern Entomologist* 173: 1–14.

Lepeletier de Saint Fargeau, A.L.M. 1836. *Histoire Naturelle des insectes. Hyménoptères*. Tome premier, Roret, Paris, pp. 1–547.

Lewis, H.W. 1897. Vespinae of the United States and Canada. *Transactions of the American Entomological Society* 24: 169–191.

Lien, N.T.P. and Carpenter, J. 2002. Vespidae of Vietnam (Insecta: Hymenoptera) 1. *Vespinae. Journal of New York Entomological Society* 110(2): 199–211.

Linnaeus, C. 1758. *Systema naturae per regna tria naturae, secundum classes ordines, genera, species cum characteribus, differentis, synonymis locis. Tomus I.* Editio decima, reformata. Laurentii Salvii, Holmiae, 824 pp.

Linnaeus, C. 1763. Amoenitates academicae; seu dissertationes variae physicae, medicae, botanicae, antehac seorsim editae. Nunc collecta et auctae cum tabulis aeneis. *Tomus* 6, 486 pp.

Linnaeus, C. 1771. Mantissa Plantarum altera. *Holmiae*, 143–588 pp.

Lopez-Osorio, F., Perrard, A., Pickett, K.M., Carpenter, J.M. and Agnarsson, I. 2015. Phylogenetic tests reject Emery's rule in the evolution of social parasitism in yellowjackets and hornets (Hymenoptera: Vespidae, Vespinae). *Royal Society Open Science* 2: 150159.

Lopez-Osorio, F., Pickett, K.M., Carpenter, J.M., Ballif, B.A. and Agnarsson, I. 2014. Phylogenetic relationships of yellowjackets inferred from nine loci (Hymenoptera: Vespidae, Vespinae, Vespula and Dolichovespula). *Molecular Phylogenetics and Evolution* 73: 190–201.

Lopez-Osorio, F., Pickett, K.M., Carpenter, J.M., Ballif, B.A. and Agnarsson, I. 2017. Phylogenomic analysis of yellowjackets and hornets (Hymenoptera: Vespidae, Vespinae). *Molecular Phylogenetics and Evolution* 107: 10–15.

Madl, M. 1997. Über Vespidae, Pompilidae und Tiphiidae der Türkei (Hymenoptera). *Linzer biologische Beiträge* 29(2): 823–827.

Madl, M. 2012. Notes on the genus *Provespa* Ashmead, 1903 (Insecta: Hymenoptera: Vespidae: Vespinae) based on the material of the Naturhistorisches Museum Wien (Austria). *Annalen des Naturhistorischen Museums in Wien, Serie B* 114: 27–35.

Martin, S.J. 2017. *The Asian hornet: Threats, biology and expansion*. University of Salford, International Bee Research Association, Manchester, 108pp.

Matsuura, M. and Yamane, S. 1990. *Biology of the vespine wasps*. Springer-Verlag, New York, 323 pp.

Modarres Awal, M. 1997. Family Vespidae (Hymenoptera), p. 284. In: Modarres Awal, M. (ed.), *List of agricultural pests and their natural enemies in Iran*. 2nd edition. Ferdowsi University of Mashhad Press, Iran, 429 pp.

Modarres Awal, M. 2012. Family Vespidae (Hymenoptera), pp. 525–527. In: Modarres Awal, M. (ed.), *List of agricultural pests and their natural enemies in Iran*. 3rd edition. Ferdowsi University of Mashhad Press, Iran, 759 pp.

Monks, J., Ross, S., Geiser, M., De Prins, J., Sharaf, M., Wyatt, N., Al Rijeibi, S. and Polaszek, A. 2019. A preliminary survey of the insect fauna of the Hajar Mountain Range, *Oman. Journal of Natural History* 53 (15–16): 939–963.

Morice, F. 1921. Annotated lists of aculeate Hymenoptera (except Heterogyna) and chrysids recently collected in Mesopotamia and north-west Persia. *The Journal of the Bombay Natural History Society* 28: 192–203.

Mossadegh, M. 1995. Insects associated with honey bee *Apis mellifera* L., hives in Iran. *Journal of agricultural Sciences* 1(3 and 4): 22–27. [in Persian, English summary]

Nemati, S. and Pezhman, H. 2014. Comparing the pests and natural enemies fauna and determining the prevailing species in wheat fields under no-tillage and conventional tillage systems in Marvdasht region (Fars province, Iran). *Journal of Applied Researches in Plant Protection* 3(1): 1–17. [in Persian, English summary]

Nguyen, L.T.P., Vu, T.T. and Wiśniowski, B. 2019. Taxonomic review and distribution of the genus *Vespula* Thomson, 1869 (Hymenoptera: Vespidae: Vespinae) from Vietnam. *Zootaxa* 4691(3): 279–285.

Ostovan, H. and Kamali, K. 1995. Some aspects of behavioural biology of hornet wasp, *Vespa orientalis* L. (Hym., Vespidae) a pest of date fruits in Kazerun, Iran. *Journal of Agricultural Sciences* 1(1): 33–43. [in Persian, English summary]

Özbay, C. and Akbayrn, H. 1993. Biometrical studies on four Vespidae (Hymenoptera) species from Diyarbaktr, Mardin and Şanhurfa. Doga. *Turkish Journal of Biology* 17: 75–92.

Özeren, M.H. 1965. Ankara Bölgesinde bazt meyve ve orrnan ağaçlari ile bitkilerine zarar veren Hymenoptera türleri, yayrhşr ve taksonomileri üzerinde ön araştrrmalar. *Ankara Üniversitesi Ziraat Fakultesi Yay*, No. 396, 73 pp.

Packard, A.S. 1870. List of hymenopterous and lepidopterous insects collected by the Smithsonian expedition to South America, under Prof. James Orton. *Annual Report of the Trustees of the Peabody Academy of Science* 1: 56–69.

Panzer, G.W.F. 1799. *Fauna Insectorum Germanicae initia oder Deutschlands Insecten. Felseckersche Buchhandlung, Nürnberg* 61–70: 1–2.

Pawlyszyn, B. 1994. Observations on the life cycle and behaviour of the hornet *Vespa crabro* L. (Hym., Vespidae) in Gloucestershire. *Entomologist's Monthly Magazine* 130: 159–163.

Pekkarinen, A. and Huldén, L. 1995. Distribution and phenology of the Vespinae and Polistinae species in eastern Fennoscandia (Hymenoptera: Vespidae). *Sahlbergia* 2: 99–111.

Perez, J. 1910. Notes sur les Vespides. *Actes de la Société Linnéenne de Bordeaux* 64: 1–20.

Perrard, A., Arca, M., Rome, Q., Müller, F., Tan, J., Bista, S., Nugroho, H., Baudoin, R., Baylac, M., Silvain, J.-F., Carpenter, J.M. and Villemant, C. 2014. Geographic variation of melanisation patterns in a hornet species: genetic differences, climatic pressures or aposematic constraints? *PLoS One* 9: e94162.

Perrard, A., Lopez-Osorio, F. and Carpenter, J.M. 2016. Phylogeny, landmark analysis and the use of wing venation to study the evolution of social wasps (Hymenoptera: Vespidae: Vespinae). *Cladistics* 32: 406–425.

Persson, S. 2015. Phylogeny and taxonomy of the subfamily Vespinae (Hymenoptera: Vespidae), based on five molecular markers. M. Sc. thesis, University of Gothenburg, Sweden, 44 pp.

Pickett, K.M. and Carpenter, J.M. 2010. Simultaneous analysis and the origin of eusociality in the Vespidae (Insecta: Hymenoptera). *Arthropod Systematics and Phylogeny* 68: 3–33.

Pickarski, P.K., Carpenter, J.M., Lemmon, A.R., Lemmon, E.M. and Sharanowski, B.J. 2018. Phylogenomic evidence overturns current conceptions of social evolution in wasps (Vespidae). *Molecular Biology and Evolution* 35(9): 2097–2109.

Rahmani, Z., Rakhshani, E. and Carpenter, J.M. 2020. Updated checklist of Vespidae (Hymenoptera: Vespoidea) in Iran. *Journal of Insect Biodiversity and Systematics* 6(1): 27–86.

Retzius, A.J. 1783. *Caroli de Geer. Genera et species insectorum et generosissimi auctoris extraxit, digessit. Latin quoad partem reddidit, et terminologiam insectorum Linneanam addidit*. Lipsiae, Cruse, 220 pp.

Richards, O.W. 1985. Insects of Saudi Arabia. Hymenoptera: Fam. Vespidae (Social wasps of the Arabian Peninsula). *Fauna of Saudi Arabia* 6: 423–440.

Rohwer, S.A. 1916. Vespidae. In: Guide to the Insects of Connecticut, Part III. The Hymenoptera, or wasp-like Insects of Connecticut. *State of Connecticut Geological and Natural History Survey Bulletin* 22: 640–644.

Ross, K.G. and Matthews, R.W. 1991. *The social biology of wasps*. Cornell University Press, Ithaca, New York, 696 pp.

Saito, F. and Kojima, J. 2011. Phylogenetic analysis and biogeography of the nocturnal hornets, *Provespa* (Insecta: Hymenoptera: Vespidae: Vespinae). *Species Diversity* 16: 65–74.

Sakenin, H., Havaskary, M., Samin, N., Shirdel, F. and Imani, S. 2010a. Odonata and Vespidae (Hymenoptera) from Arasbaran and Vicinity, Northwestern Iran. ISSAAS 2009 *International congress, Agriculture for Better Living and Global Economy*. Nong Nooch Tropical Botanical Garden and Resort, Pattaya, Thailand, 11–15 January, p. 184.

Sakenin, H., Raheb, J., Imani, S., Havaskary, M., Shirdel, F. and Mohseni, H. 2008. A preliminary survey on dipteran predators and parasitoids and Odonata in Iranian rice fields. *Proceedings of the National Conference of Agronomical Rice Breeding, Young Researchers Club of Islamic Azad University of Ghaemshahr*, 26–27 November, 2008, 14 pp. [in Persian, English Summary]

Sakenin, H., Samin, N. and Bagriacik, N. 2010b. A contribution to the Aculeate Hymenoptera (Insecta) from Iran. *Efflatounia* 10: 15–20.

Sakenin, H., Samin, N., Ghahari, H., Imani, S., Rastegari, J. and Jabbari, A. 2010c. A contribution to the knowledge of robber flies (Diptera: Asilidae) from Semnan province, Iran. *Linzer biologische Beiträge* 42(1): 833–841.

Samin, N., Sakenin, H., Imani, S. and Shojai, M. 2010. A contribution to the knowledge of robber flies (Diptera: Asilidae) from Tehran province and vicinity, *Iran. Journal of Biological Control* 24(1): 42–46.

Schmid-Egger, C., Gusenleitner, J., Liebig, W.-H. and Jacobs, H.-J. 2021. New records of Vespidae (Hymenoptera) from Iran with descriptions of three species. *Linzer biologische Beiträge* 52(2): 1105–1121.

Scopoli, J.A. 1763. *Eomologia Carniolica: exhibens insecta Carnioliae indigena et distributa in ordines, genera, species, varietates, methodo Linnaeana*. Typis Ioannis Thomae Trattner.

Selis, M. 2023. Illustrated key to the genera and a checklist of Italian Vespidae (Hymenoptera). *Fragmenta Entomologica* 55(1): 63–88.

Selis, M., Ismael, H.R., Hamad, M.I., Baiocchi, D. and Cilia, G. 2025. A first update to the vespid fauna of Iraq (Hymenoptera: Vespidae), with description of a new species. *Journal of Natural History* 59 (13–16): 941–981.

Sladen, F.W.N. 1918. The genus *Vespa* in Canada. *The Ottawa Naturalist* 32: 71–72.

Smith-Pardo, A.H., Carpenter, J.M. and Kimsey, L. 2020. The diversity of hornets in the genus *Vespa* (Hymenoptera: Vespidae; Vespinae), their importance and interceptions in the United States. *Insect Systematics and Diversity* 4(3): 1–27.

Sora, K., Gawas, S.M., Gupta, A. and Shaw, S.S. 2019. Checklist of vespid wasps (Hymenoptera: Vespidae) of Arunachal Pradesh, India, with a new distribution record of *Vespula flaviceps* (Smith). *Journal of Biological Control* 33(4): 336–342.

Sorvari, J. 2019. Yellow does not improve the efficiency of traps for capturing wasps of the genera *Vespula* and *Dolichovespula* (Hymenoptera: Vespidae). *European Journal of Entomology* 116(1): 240–243.

Swailem, S.M., Selim, A.A. and Amin, A.H. 1974. A contribution to the study of the insect fauna of Hammam Al-Alil, *Part I. Mesopotamia Journal of Agriculture* 9(1–2): 119–141.

Tan, J.L., Chen, X. and van Achterberg, C. 2014. Description of the male *Dolichovespula flora* Archer (Hymenoptera: Vespidae). *Entomotaxonomia* 36(1): 75–80.

Tan, Q.-Q., Tan, J.-L., Zhang, R.-N., Tian, X.-X. and Jian, J.-K. 2018. Discovery of the nest of the yellow jacket *Vespula structor* (Smith) (Hymenoptera: Vespidae) from China with description of its immature stages. *Journal of Hymenoptera Research* 67: 103–120.

Tanyolaç, T. 1964. Ankara gevresinde sosyal halde yaşayan Vespidae (Hymenoptera) türleri ve bunlardan Polistes gallicus L.'un biyolojisi tizerinde araştirmalar. Doktora tezt.

Tanyolaç, T. 1964. Ankara çevresinde sosyal halde yaşayan Vespidae (Hymenoptera) türleri ve bunlardan Polistes gallicus L.B'un biyolojisi üzerinde araştırmalar. Doktora tezi, Fen Fakültesi, Biyoloji Bölümü, Ankara Üniversitesi, 46 pp.

Thomson, C.G. 1869. Öfversigt af Sveriges Vesparnae. *Opuscula Entomologica* 1: 78–82.

Tüzün, A. and Tanyolaç, T. 1987. Ege Bölgesinde Vespidae (Insecta: Hymenoptera) türlerinin saptanmasi. *C. U. Fen Edebiyat Fakiiltesi, Fen Bilimleri Derg* 5: 147–171.

Walker, D.H. and Pittaway, A.R. 1987. *Insects of Eastern Arabia*. Macmillan, London, 191 pp.

Wang, H., Wen, Q., Wang, T., Ran, F., Wang, M., Fan, X., Wei, S., Li, Z. and Tan, J. 2022. Next-generation sequencing of four mitochondrial genomes of *Dolichovespula* (Hymenoptera: Vespidae) with a phylogenetic analysis and divergence time estimation of Vespidae. *Animals (Basel)* 12(21): 3004.

Wenzel, J.W. 1998. Generic key to the nests of hornets, yellow jackets, and paper wasps worldwide (Vespidae: Vespinae, Polistinae). *American Museum Novitates* 3224: 1–39.

Yamane, S., Wagner, R.E. and Yamane, S. 1980. A tentative revision of the subgenus *Paravespula* of eastern Asia (Hymenoptera: Vespidae). *Insecta Matsumurana, New Series* 19: 1–46.

Yamane, S. and Yamane, S. 2021. Vespinae, pp. 1000–1008. In: Starr, C.K. (ed.), *Encyclopedia of social insects*. Cham, Switzerland, Springer, 1075 pp.

Yildirim, E. and Gusenleitner, J. 2001. The present situation of the Vespidae-fauna (Hymenoptera, Aculeata) of Turkey. *Linzer biologische Beitraege*, 33 (1): 217–221.

Yildirim, E. and Gusenleitner, J. 2012. Contribution to the knowledge of the Vespidae (Hymenoptera, Aculeata) of Turkey, with a checklist of the Turkish species. *Turkish Journal of Zoology*, 36(3): 361–374.

Yildirim, E. and Kojima, J.-I. 1999. Distributional checklist of the species of the family Vespidae (Insecta: Hymenoptera; Aculeata) of Turkey. *Natural History Bulletin of Ibaraki University*, 3: 19–50.

Yildirim, E. and Özbek, H. 1992. Türkiye Vespinae (Hymenoptera: Vespoidea: Vespidae) türleri üzerinde sistematik ve faunistik gahqmalar. *Türkiye Entomoloji Dergisi* 16(4): 227–242.

Yildirim, E. and Özbek, H. 1998. Türkiye de *Dolichovespula sylvestris* (Scopoli, 1763) (Hymenoptera Vespoidea: Vespidae: Vespinae) in populasyonlanndaki varyasyonlar. *Atatürk Üniversitesi Ziraat Fakültesi Dergisi* 29: 772–776.

Yildirim, E. and Özbek, H. 1999. An evaluation on the fauna of Vespoidea (Hymenoptera, Aculeata) of Turkey, along with new records and new localities for some species. *Turkish Journal of Zoology* 23(6): 591–604.

Zalat, S.M. 1992. The Vespidae (social wasps) of Egypt. *Journal of the Egyptian-German Society of Zoology* 9(D): 247–258.

Katamenes sichelii (de Saussure, 1852) (Eumeninae). (Photo courtesy of Alexander Fateryga)

Katamenes sichelii (de Saussure, 1852) (Eumeninae). (Photo courtesy of Alexander Fateryga)

Polistes mongolicus (du Buysson, 1911) (Polistinae). (Photo courtesy of Alexander Fateryga)

6 Species Diversity of Vespidae in the Middle East with Special Emphasis on the Iranian Fauna

Hassan Ghahari

"Social wasps are among the least loved insects··· yet, where statistics will not alter a general impression, another approach might. Every schoolchild ··· ought to sit watching a Polistes wasp nest for just one hour ··· I think that few will be unaffected by what they see".

(William Donald Hamilton; 1 August 1936 – 7 March 2000; British evolutionary biologist)

"If we were to wipe out insects alone on this planet, the rest of life and humanity with it would mostly disappear from the land, within a few months."

(Edward Osborne Wilson; 10 June 1929 – 26 December 2021; American biologist, naturalist, ecologist, and entomologist at Harvard University)

INTRODUCTION

Vespidae is a large and cosmopolitan family of aculeate wasps, traditionally comprising six subfamilies: Eumeninae *s.l.*, Euparagiinae, Masarinae, Polistinae, Stenogastrinae, and Vespinae (Goulet and Huber, 1993; Pickett and Carpenter, 2010; Gawas *et al.*, 2020). The family exhibits a wide distribution, ranging from tropical Africa and Australia, through Asia and temperate Eurasia, and extending throughout South, Central, and North America (van der Vecht, 1966; Maryam *et al.*, 2024). Social wasps are primarily tropical in distribution and exhibit notable species richness (Dvořák and Straka, 2007). Fossil vespids are known from early Cretaceous sediments dating back more than 130 million years (Grimaldi and Engel, 2005). The family includes over 5,000 recognized species worldwide (Piekarski *et al.*, 2018; Silveira *et al.*, 2021), encompassing all eusocial and many solitary wasps (Goulet and Huber, 1993; Pham, 2014). Polistinae and Vespinae are strictly eusocial, while Eumeninae, Euparagiinae and Masarinae are primarily solitary wasps. The stenogastrine wasps are considered facultatively eusocial (Turillazzi, 2012). All eusocial wasps live in colonies typically composed of a single fertile queen, a few males, and many worker females. Generally, a single female is responsible for reproduction, while helper workers forgo reproduction and carry out tasks for the colony. However, some helpers may leave the parental nest to establish independent colonies or to join other foundresses (Varghese and Girish Kumar, 2023). Despite their often-fearsome reputation due to their ability to sting, wasps play vital roles in ecosystems as predators, pollinators, and agents of biological control. Their complex behaviors, morphological diversity, and ecological significance make them a fascinating subject for scientific studies (Sumner *et al.*, 2018; Maryam *et al.*, 2024).

DOI: 10.1201/9781003740438-6

STATUS OF THE FAMILY VESPIDAE IN THE MIDDLE EAST

The Vespidae of the Middle East have been relatively well studied; some subfamilies, and in some cases the entire family, have been revised by various authors (see the relevant chapters for each subfamily). In total, 716 valid species and subspecies in 69 genera across four subfamilies have been reported: 1) Eumeninae *s.l.* (57 genera, 605 species) is the most diverse subfamily, comprising 84% of the total listed species, 2) Masarinae (five genera, 66 species), 3) Polistinae (four genera, 32 species), and 4) Vespinae (three genera, 13 species). These represent 13.2% of all extant vespid species worldwide [with 5,407 extant species (Perrard *et al.*, 2017)], and 60.7% of the recorded species in the Palaearctic region [with 1,181 species in 90 genera (Antropov and Fateryga, 2017)].

SPECIES DIVERSITY OF VESPIDAE IN THE MIDDLE EAST COUNTRIES

Among the 16 Middle Eastern countries, Turkey has the most diverse vespid fauna, with 323 species and subspecies, representing 45% of the total number of species recorded in the Middle East (Eumeninae: 278 species; Masarinae: 20 species; Polistinae: 13 species; Vespinae: 12 species). This is followed by Iran, with 261 species (36.4%) in 55 genera (Eumeninae: 211 species in 45 genera; Masarinae: 25 species in 5 genera; Polistinae: 17 species in 2 genera; Vespinae: 8 species in 3 genera). Other countries include Israel-Palestine with 192 species (Eumeninae: 168 species; Masarinae: 13 species; Polistinae: 8 species; Vespinae: 3 species), and Egypt with 129 species (Eumeninae: 109 species; Masarinae: 13 species; Polistinae: 6 species; Vespinae: 1 species) (Table 6.1).

TABLE 6.1
The Fauna of Vespidae in the Countries of the Middle East

Country	Eumeninae	Masarinae	Polistinae	Vespinae	Total
Turkey	278	20	13	12	323
Iran	211	25	17	8	261
Israel-Palestine	168	13	8	3	192
Egypt	109	13	6	1	129
Saudi Arabia	84	7	8	1	100
Syria	80	9	5	5	99
Jordan	77	6	3	2	88
Yemen	61	4	12	2	79
Oman	61	6	5	1	73
Iraq	53	2	9	3	67
United Arab Emirates	47	11	2	1	61
Cyprus	31	7	5	1	44
Lebanon	33	4	0	2	39
Kuwait	3	0	1	1	5
Qatar	4	1	0	0	5
Bahrain	2	0	0	1	3

NEW RECORDS OF VESPIDAE FOR THE MIDDLE EAST

In this book, two species of Eumeninae collected from Iran are reported as new records for the fauna of the Middle East: *Alastor mocsaryi* (André, 1884), and *Eumenes pomiformis* (Fabricius, 1781). Given that the Middle East is a vast region with diverse geographical features and climates, it is expected that future regular and in-depth faunistic surveys will lead to additional discoveries, including new distributional data, new country records, and possibly new species.

SPECIES EXCLUDED FROM THE FAUNA OF THE MIDDLE EAST

A biogeographical analysis of vespid wasps is proposed in this book to provide a reliable and trustworthy species list, correcting several earlier misidentifications and erroneous records from the Middle Eastern countries. In total, 14 species of Vespidae in three subfamilies Eumeninae *s.s.* (three species), Masarinae (eight species), Polistinae (two species), and Vespinae (one species) are excluded from the fauna of the Middle East due to misidentifications or other reasons, as outlined below:

Subfamily Eumeninae *s.s.*: *Euodynerus aurantiacus* (Cam.); *Pseudepipona tricolor* Gusenleitner, 1976; and *Stenancistrocerus transcaspicus* (Kostylev, 1935).

Subfamily Masarinae: *Celonites* (*Eucelonites*) *crenulatus* Morawitz, 1888; *Celonites rothschildi* du Buysson, 1906; *Ceramius bicolor* (Thunberg, 1815); *Ceramius capicola* Brauns, 1902; *Ceramius hispanicus* Dusmet, 1909; *Jugurtia braunsi* (von Schulthess, 1922); *Masarina familiaris* Richards, 1962; and *Masaris carli* von Schulthess, 1922.

Subfamily Polistinae: *Polistes caspica* Fabricius [nomen nudum]; *Polistes* (*Fuscopolistes*) *carolina* (Linnaeus, 1767).

Subfamily Vespinae: *Dolichovespula media* (Retzius, 1783) (Vespinae) is excluded from the fauna of Iran because of misidentification.

DISTRIBUTION OF ENDEMIC SPECIES IN THE MIDDLE EASTERN COUNTRIES

Among the 16 countries of the Middle East, 12 countries (excluding Bahrain, Kuwait, Lebanon and Qatar) have species that are currently considered endemic. In total, 198 species and subspecies of Vespidae are confined to the Middle Eastern countries, with the greatest number of species being in Iran with 49 species and subspecies, followed by Turkey (43 species), Egypt (29 species), Israel-Palestine (21 species), Yemen (12 species), Oman (10), Saudi Arabia (eight species), Syria (seven species), Jordan, United Arab Emirates (each with six species), Cyprus (four species), and Iraq (three species). Details of endemic species in each country are provided below.

Cyprus. Four species of the subfamily Eumeninae are endemic to this country (9.3% of total species of Cyprus): *Eumenes mediterraneus cypricus* Blüthgen, 1938; *Euodynerus disconotatus laniensis* Giordani Soika, 1979; *Stenodynerus aequisculptus inaequisculptus* Giordani Soika, 1979; *Tachyancistrocerus cyprius* (Pittioni, 1950).

Egypt. In total, 29 species in two subfamilies Eumeninae (23 species) and Masarinae (six species) are endemic to this country (22.4% of total species of Egypt) (Table 6.2).

Iran. In total, 49 species of Vespidae in three subfamilies Eumeninae (38 species), Masarinae (10 species) and Polistinae (one species) are endemic to this country (19.2% of total species of Iran) (Table 6.3).

Iraq. Three species of the subfamily Eumeninae are endemic to this country (4.4% of total species of Iraq): *Antepipona irakensis* Giordani Soika, 1970; *Euodynerus garaensis* Selis, 2025; *Katamenes sichelii timanensis* Giordani Soika, 1986.

TABLE 6.2
List of Vespidae Species Only Known from Egypt (Endemic Species)

Subfamily	Species	Subfamily	Species
Eumeninae	*Allodynerus vinciguerrae pallidulus* Giordani Soika, 1970	Eumeninae	*Leptochilus flexilis* (Gioradni Soika, 1938)
Eumeninae	*Ancistrocerus impunctatus impunctatus* (Spinola, 1838)	Eumeninae	*Onychopterocheilus coctus* (Giordani Soika, 1970)
Eumeninae	*Chlorodynerus diglaensis* (Blüthgen, 1954)	Eumeninae	*Onychopterocheilus mochii* (Giordani Soika, 1942)
Eumeninae	*Chlorodynerus intricatus* Giordani Soika, 1958	Eumeninae	*Pterocheilus dives* Radoszkowski, 1876
Eumeninae	*Chlorodynerus mochii* Giordani Soika, 1958	Eumeninae	*Pterocheilus merpeba* Giordani Soika, 1943
Eumeninae	*Cyrtolabulus elbanus* Giordani Soika, 1977	Eumeninae	*Syneuodynerus fouadi fouadi* (Giordani Soika, 1939)
Eumeninae	*Cyrtolabulus sexspinosus* Giordani Soika, 1977	Eumeninae	*Xanthodynerus castrorum* (Blüthgen, 1954)
Eumeninae	*Euodynerus dantici tinctus* (Walker, 1871)	Eumeninae	*Xanthodynerus trichopygus* (Blüthgen, 1954)
Eumeninae	*Eustenancistrocerus nilensis* (Giordani Soika, 1935)	Masarinae	*Quartinia amiae* Carpenter, 2003
Eumeninae	*Katamenes dimidiativentris francisi* Zalat, 1992	Masarinae	*Quartinia halicticeps* Giordani Soika, 1939
Eumeninae	*Katamenes priesneri* (Giordani Soika, 1941)	Masarinae	*Quartinia mochii* Giordani Soika, 1939
Eumeninae	*Knemodynerus rhynchoides inclinans* (Giordani Soika, 1935)	Masarinae	*Quartinia nilotica* Fischer, 1964
Eumeninae	*Labochilus pulawskyi pulawskyi* Giordani Soika, 1970	Masarinae	*Quartinia tricolorata* Giordani Soika, 1954
Eumeninae	*Leptochilus camurus* (Giordani Soika, 1938)	Masarinae	*Quartinia tripolitana sinaitica* Riachards, 1964
Eumeninae	*Leptochilus errabundus* (Giordani Soika, 1938)	—	—

Israel-Palestine. In total, 21 species of two subfamilies Eumeninae (19 species) and Masarinae (two species) are endemic to this country (10.9% of total species of Israel-Palestine) (Table 6.4).

Jordan. Six species of the subfamily Eumeninae are endemic to this country (6.8% of total species of Jordan): *Allodynerus nutatus* Gusenleitner, 1991; *Antepipona jordanica* Gusenleitner, 2004; *Leptochilus bipunctatis* Gusenleitner, 2016; *Leptochilus longipilis* Gusenleitner, 1988; *Leptochilus villosus* Gusenleitner, 2001; *Xanthodynerus jordanicus* Gusenleitner, 1991.

Oman. Ten species in two subfamilies Eumeninae (nine species) and Masarinae (one species) are endemic to this country (13.7% of total species of Oman). Eumeninae: *Antepipona rotunda* Gusenleitner, 2014; *Antodynerus igneus* Gusenleitner, 2004; *Cyrtolabulus omanicus* Gusenleitner, 2014; *Knemodynerus familiaris muscatensis* (Giordani Soika, 1970); *Labochilus felix* Guichard, 1986; *Leptochilus arabicus* Giordani Soika, 1979; *Leptochilus masirahensis* Giordani Soika, 1981; *Leptochilus perterritus* Giordani Soika, 1970; *Pseudonortonia omanensis* Giordani Soika, 1979; Masarinae: *Quartinia monnerati* Gusenleitner, 2016.

Saudi Arabia. Eight species in three subfamilies Eumeninae (six species), Masarinae (one species) and Polistinae (one species) are endemic to this country (8% of total species of

TABLE 6.3
List of Vespidae Species Only Known from Iran (Endemic Species)

Subfamily	Species	Subfamily	Species
Eumeninae	*Alastor darius* Gusenleitner, 1986	Eumeninae	*Katamenes tauricus viratus* (Giordani Soika, 1949)
Eumeninae	*Alastor xerxes* Gusenleitner, 1986	Eumeninae	*Leptochilus iranus* Gusenleitner and Schmid-Egger, 2021
Eumeninae	*Alastor zoroaster* Gusenleitner, 1986	Eumeninae	*Leptochilus lorestanicus* Gusenleitner, 1995
Eumeninae	*Ancistrocerus tenebrosus* Gusenleitner, 2010	Eumeninae	*Onychopterocheilus spheciformis* (Gusenleitner, 1970)
Eumeninae	*Antepipona karadgensis* Giordani Soika, 1970	Eumeninae	*Paragymnomerus mammillatus* (Morawitz, 1885)
Eumeninae	*Antepipona levis* Gusenleitner, 2021	Eumeninae	*Parodontodynerus rufinus* (Kostylev, 1940)
Eumeninae	*Antepipona prompta* Giordani Soika, 1970	Eumeninae	*Pseudosymmorphus angulatus* Gusenleitner and Ockermüller, 2018
Eumeninae	*Brachypipona irana* Gusenleitner and Schmid-Egger, 2021	Eumeninae	*Pseudosymmorphus capillus* Gusenleitner and Ockermüller, 2018
Eumeninae	*Brachypipona montana* Gusenleitner and Schmid-Egger, 2021	Eumeninae	*Pterocheilus chobauti chlorodyneroides* Gusenleitner, 1970
Eumeninae	*Chlorodynerus araxanus* (Blüthgen, 1956)	Eumeninae	*Pterocheilus persicus* Gusenleitner and Fallahzadeh, 2013
Eumeninae	*Chlorodynerus flavoferrugineus* Giordani Soika, 1970	Eumeninae	*Pterocheilus tenebricosus* Gusenleitner, 1998
Eumeninae	*Chlorodynerus gratus* (Kostylev, 1940)	Eumeninae	*Stenodynerus cyrus* Gusenleitner, 2012
Eumeninae	*Chlorodynerus infrenis* Giordani Soika, 1970	Eumeninae	*Raphiglossa irana* Giordani Soika, 1970
Eumeninae	*Euodynerus amplus* Gusenleitner, 2011	Masarinae	*Celonites clarus* Gusenleitner, 1973
Eumeninae	*Euodynerus effrenatus* Giordani Soika, 1970	Masarinae	*Celonites discretus* Gusenleitner, 1973
Eumeninae	*Euodynerus khuzestanicus* Giordani Soika, 1970	Masarinae	*Celonites iranus* Gusenleitner, 2018
Eumeninae	*Euodynerus rufinus confictus* Giordani Soika, 1970	Masarinae	*Celonites longipilis* Gusenleitner, 1973
Eumeninae	*Euodynerus shirazensis* Giordani Soika, 1970	Masarinae	*Celonites persicus* Richards, 1962
Eumeninae	*Eustenancistrocerus malus* Giordani Soika, 1970	Masarinae	*Jugurtia zarudnyi* Kostylev, 1935
Eumeninae	*Eustenancistrocerus spinosissimus* Gusenleitner, 2000	Masarinae	*Masaris agnolii* Selis, 2019
Eumeninae	*Eustenancistrocerus sulcithorax* Giordani Soika, 1970	Masarinae	*Quartinia atrata* Gusenleitner, 1998
Eumeninae	*Extraepipona occulta* Gusenleitner, 2014	Masarinae	*Quartinia irana* Gusenleitner, 2012
Eumeninae	*Hemipterochilus punctiventris punctiventris* (Morawitz, 1885)	Masarinae	*Quartinia syriaca nitens* Gusenleitner, 1973
Eumeninae	*Hemipterochilus punctiventris subquadricolor* Gusenleitner, 1970	Polistinae	*Parapolybia persica* (Meade-Waldo, 1911)
Eumeninae	*Jucancistrocerus pareumeniformis* Giordani Soika, 1973	—	—

Saudi Arabia). Eumeninae: *Euodynerus soikai* Guichard, 1986; *Katamenes rauensis* (Giordani Soika, 1960); *Knemodynerus sauditus* (Giordani Soika, 1989); *Onychopterocheilus flaviventris* Gusenleitner, 1991; *Pseudonortonia bicarinatus* Guichard, 1986; *Pterocheilus nigrobilineolatus* Giordani Soika, 1942; Masarinae: *Masaris aegyptiacus arabicus* Giordani Soika, 1957; Polistinae: *Belonogaster filiformis* (de Saussure, 1853).

TABLE 6.4
List of Vespidae Species Only Known from Israel-Palestine (Endemic Species)

Subfamily	Species	Subfamily	Species
Eumeninae	*Knemodynerus familiaris judaicus* (Giordani Soika, 1952)	Eumeninae	*Pseudepipona herzi enslini* (Blüthgen, 1955)
Eumeninae	*Leptochilus amos amos* Gusenleitner, 1971	Eumeninae	*Pseudepipona rubidus* Gusenleitner, 1994
Eumeninae	*Leptochilus kurnubensis* Blüthgen, 1955	Eumeninae	*Pterocheilus varius* Gusenleitner, 2002
Eumeninae	*Leptochilus mauritanicus derufatus* Blüthgen, 1955	Eumeninae	*Stenodynerus fastidiosissimus judaicus* Gusenleitner, 1970
Eumeninae	*Leptochilus speciosus* Gusenleitner, 2002	Eumeninae	*Xanthodynerus pilosus* Gusenleitner, 1990
Eumeninae	*Leptochilus subtarsatellus* Gusenleitner, 2002	Eumeninae	*Rhaphiglossa bytinskii* Giordani Soika, 1974
Eumeninae	*Odynerus guichardi* Giordani Soika, 1977	Eumeninae	*Tachyancistrocerus komarowi derufata* Gusenleitner, 1971
Eumeninae	*Odynerus palaestinensis* Gusenleitner, 1971	Eumeninae	*Tropidodynerus flavus semiticus* (Giordani Soika, 1952)
Eumeninae	*Odynerus rotundigaster citripes* Gusenleitner, 1970	Masarinae	*Quartinia candida* Gusenleitner, 1997
Eumeninae	*Onychopterocheilus albidus aramaeus* (Blüthgen, 1955)	Masarinae	*Quartinia haemorrhoa* Gusenleitner, 1997
Eumeninae	*Onychopterocheilus fausti paravespoides* (Blüthgen, 1955)	—	—

Syria. Seven species in two subfamilies Eumeninae (six species) and Masarinae (one species) are endemic to this country (7% of total species of Syria). Eumeninae: *Alastor nitens* Gusenleitner, 2000; *Leptochilus mochianus mochianus* Giordani Soika, 1970; *Leptochilus nigrocitrinus* Giordani Soika, 1970; *Leptochilus tricolor* Gusenleitner, 2016; *Microdynerus syriacus* Gusenleitner, 2000; *Onychopterocheilus tertius* (Giordani Soika, 1970); Masarinae: *Jugurtia punctuosa* Gusenleitner, 2011.

Turkey. In total, 44 species in two subfamilies Eumeninae (41 species) and Masarinae (three species) are endemic to this country (13.6% of total species of Turkey) (Table 6.5).

United Arab Emirates. Six species in two subfamilies Eumeninae (three species) and Masarinae (three species) are endemic to this country (9.8% of total species of United Arab Emirates). Eumeninae: *Euodynerus setosus rubicundus* Gusenleitner, 1994; *Leptodynerus vanharteni* Gusenleitner, 2010; *Pseudonortonia emirata* Gusenleitner, 2012; Masarinae: *Quartinia arabica* Gusenleitner, 2010; *Quartinia longiceps* Gusenleitner, 2010; *Quartinia nitidiuscula* Gusenleiner, 2010.

Yemen. Twelve species in two subfamilies Eumeninae (nine species) and Polistinae (three species) are endemic to this country (15.2% of total species of Yemen). Eumeninae: *Afreumenes melanosoma yemenensis* Giordani Soika, 1996; *Ancistrocerus adenensis* Giordani Soika, 1952; *Antodynerus ignoratus* Gusenleitner, 2007; *Cyrtolabulus socotrae* (Kohl, 1906); *Cyrtolabulus yemenensis* Gusenleitner, 2007; *Euodynerus socotrae* (Kohl, 1906); *Knemodynerus lahijensis* Gusenleitner, 2002; *Leptochilus aterrimus* (Kirby, 1900); *Pseudonortonia scotti* Giordani Soika, 1957; Polistinae: *Belonogaster saussurei* Kirby, 1881; *Polistes tenellus lahejensis* Giordani Soika, 1981; *Ropalidia irrequieta* (Kohl, 1906).

TABLE 6.5
List of Vespidae Species Only Known from Turkey (Endemic Species)

Subfamily	Species	Subfamily	Species
Eumeninae	*Acanthodynerus giordanii* Gusenleitner, 1969	Eumeninae	*Leptochilus palandokenicus* Yildirim and Özbek, 1995
Eumeninae	*Alastor thymbrinus* Blüthgen, 1956	Eumeninae	*Leptochilus signatus gueruensis* Gusenleitner, 1995
Eumeninae	*Allodynerus floricola inaequalis* Giordani Soika, 1970	Eumeninae	*Microdynerus curdistanicus* Gusenleitner, 1988
Eumeninae	*Ancistrocerus claripennis ponticus* Giordani Soika, 1970	Eumeninae	*Microdynerus erzincanensis* Yildirim and Özbek, 1995
Eumeninae	*Antepipona albosignata* Gusenleitner, 1986	Eumeninae	*Onychopterocheilus anatolicus* (Blüthgen, 1955)
Eumeninae	*Antepipona tenuis* Gusenleitner, 1988	Eumeninae	*Onychpterocheilus fausti aurantiopictus* (Giordani Soika, 1970)
Eumeninae	*Brachyodynerus chloroticus* Gusenleitner, 2004	Eumeninae	*Onychopterocheilus pamirensis pentheri* (Blüthgen, 1956)
Eumeninae	*Cyrtolabulus anatolicus* Gusenleitner, 2006	Eumeninae	*Onychopterocheilus sarikamisensis* Yildirim, 2014
Eumeninae	*Cyrtolabulus mutinensis* Gusenleitner, 2006	Eumeninae	*Pseudepipona ankarensis* Giordani Soika, 1970
Eumeninae	*Cyrtolabulus yildirimi* Gusenleitner, 2006	Eumeninae	*Pseudosymmorphus adnexus* (Gusenleitner, 1969)
Eumeninae	*Euodynerus maximilianus* Gusenleitner, 1998	Eumeninae	*Pterocheilus kamanensis* Gusenleitner, 1967
Eumeninae	*Euodynerus quadrifasciatus rufipes* Gusenleitner, 1984	Eumeninae	*Pterocheilus multimaculatus* Gusenleitner, 2004
Eumeninae	*Eustenancistrocerus jerichoensis iconius* Blüthgen, 1957	Eumeninae	*Pterocheilus perpunctatus* Giordani Soika, 1970
Eumeninae	*Katamenes dimidiatus errectus* Giordani Soika, 1970	Eumeninae	*Pterocheilus schwarzi* Gusenleitner, 1994
Eumeninae	*Knemodynerus euodyneroides* Gusenleitner, 1997	Eumeninae	*Stenodynerus claviger* Gusenleitner, 1981
Eumeninae	*Leptochilus gusenleitneri* Yildirim and Özbek, 1995	Eumeninae	*Stenodynerus pannosus* Gusenleitner, 1985
Eumeninae	*Leptochilus hethiticus* Gusenleitner, 1985	Eumeninae	*Syneuodynerus erichi* (Gusenleitner, 1967)
Eumeninae	*Leptochilus kemali* Gusenleitner, 1977	Eumeninae	*Tachyancistrocerus aereus* Giordani Soika, 1952
Eumeninae	*Leptochilus montivagus* Gusenleitner, 2002	Eumeninae	*Tachyancistrocerus liliput* (Gusenleitner, 1993)
Eumeninae	*Leptochilus neutraliformis* Gusenleitner, 1977	Masarinae	*Celonites spinosus* Gusenleitner, 1966
Eumeninae	*Leptochilus nigroclypeus* Gusenleitner, 1985	Masarinae	*Celonites tenebricosus* Gusenleitner, 2012
Eumeninae	*Leptochilus osmanicus* Gusenleitner, 1988	—	—

FAUNAL SIMILARITY BETWEEN IRAN AND COUNTRIES IN THE MIDDLE EAST AND NEARBY TO IRAN

Among the countries of the Middle East as well as those adjacent to Iran, Turkey shares the highest number of species with Iran (141 species), followed by Russia (84 species), Kazakhstan (75 species), Azerbaijan (73 species), Israel-Palestine (72 species), Turkmenistan (63 species), Armenia

(57 species), Syria (51 species), Pakistan (45 species), Iraq (43 species), Afghanistan, Jordan (each with 34 species), Lebanon (28 species), United Arab Emirates (26 species), Egypt, Oman (each with 24 species), Cyprus, Saudi Arabia (each with 23 species), Yemen (10 species), Qatar (five species), Kuwait (four species), and Bahrain (three species).

A comparison of the vespid faunal similarity between Iran and the Middle East countries where the vespid fauna has been relatively well studied shows that 64% of Iraq's fauna is shared with Iran; followed by Syria (51.5%), Turkey (43.6%), United Arab Emirates (42.6%), Jordan (38.6%), Israel-Palestine (37.5%), Oman (32.8%), Saudi Arabia (23%), Egypt (18.6%), and Yemen (12.6%). These results are expected, as the countries with the highest percentage of faunal similarity (Iraq and Turkey) share significant land borders with Iran. The relatively high similarity with Syria, United Arab Emirates, Jordan, and Israel-Palestine likely reflects the more complete documentation of the vespid fauna in these countries. We anticipate that greater species overlap with neighboring countries will emerge as more faunistic studies are conducted. In contrast, Vespidae remain poorly studied in several other Middle Eastern countries, making reliable comparisons of faunal similarity with Iran currently difficult.

SPECIES DIVERSITY OF IRANIAN VESPIDAE BY PROVINCE

The distribution of Vespidae records across the 31 Iranian provinces (Figure 6.1) is shown in Figure 6.2. Among these provinces, Kerman has the highest reported species diversity, with 58

FIGURE 6.1 Map of Iran with provincial boundaries, and neighboring countries.

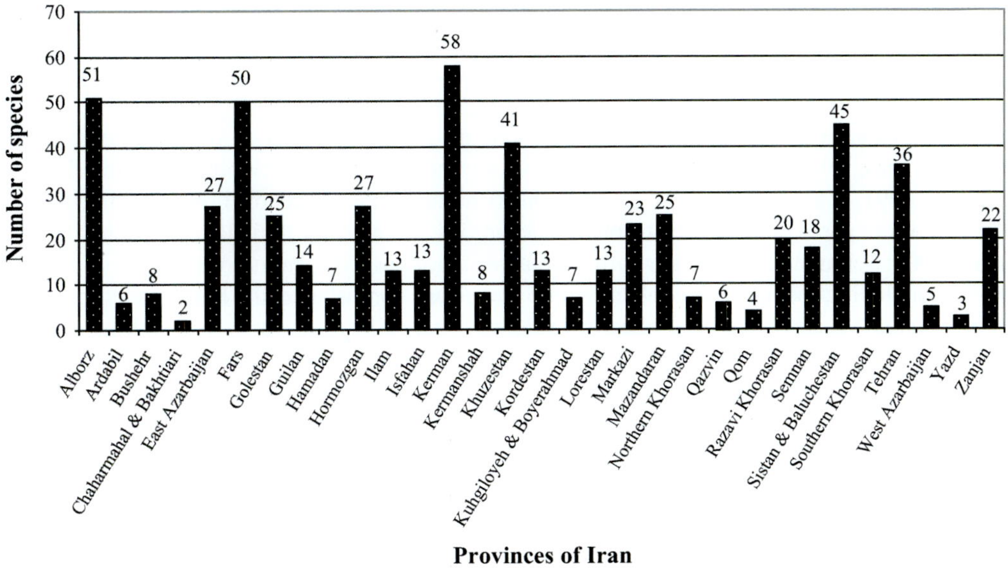

FIGURE 6.2 Number of reported species of Iranian Vespidae by province.

species, followed by Alborz (51 species), Fars (50 species), Sistan & Baluchestan (45 species), and Khuzestan (41 species). However, these results are likely biased toward provinces that have been more extensively sampled, as many other regions have not yet been sampled systematically.

Based on the known species diversity in countries adjacent to or near Iran, along with numerous specimens collected in Iran and preserved in various museums, insect collections of universities, and private collections, we estimate that the total number of Vespidae species in Iran may eventually reach 350 or more. More comprehensive surveys across all regions of the country, along with additional taxonomic research, are likely to reveal further species in the future.

PARASITOIDS AND PREDATORS OF VESPIDAE IN IRAN

Wasp colonies are preyed upon by a diverse array of natural enemies, including other social wasps (Jeanne and Hunt, 1992; Turillazzi, 1984; Gadagkar, 1991), ants (Jeanne, 1979; Tindo *et al.*, 2002), birds (McCann *et al.*, 2010; Huang *et al.*, 2004), mammals (Perry and Manson, 2008), and certain hymenopteran parasitoids (Johnson *et al.*, 2023). The prevalence of these various predator groups differs across geographic regions (Detoni *et al.*, 2021).

In total, five hymenopteran species are known parasitoids of Vespidae in Iran. Three of these were found within nests of *Delta dimidiatipenne*: 1) *Cotesia vanessae* (Reinhard) (Braconidae); 2) *Netelia fuscicornis* (Holmgren) (Ichneumonidae); and 3) *Ophion similis* (Szépligeti) (Ichneumonidae). Additionally, *Sphecophaga vesparum* (Curtis) (Ichneumonidae) was recorded as a parasitoid of *Vespa orientalis* by Ghahari and Jussila (2015) and was also collected from the nest of *Vespula germanica* in this study (see the related chapters).

Apart from these nest parasitoids, several studies have also reported the presence of parasitized prey and parasitoids of prey insects trapped within the nests of solitary wasps. This may occur when potter wasps collect and provision their nest with prey that has already been parasitized (Bohart *et al.*, 1982; Jennings and Houseweart, 1984; Auko *et al.*, 2014; Segoli *et al.*, 2020; Leduc *et al.*, 2022). In contrast to nest parasitoids, parasitoids of the prey species are unlikely to possess adaptations that allow them to emerge from the mud cells because they typically do not develop within a potter wasp nest (Yuan *et al.*, 2022; Johnson *et al.*, 2023).

Additionally, 28 species of robber flies (Diptera: Asilidae) have been reported by various authors as predators of Vespidae in Iran (Table 6.6), all of which were listed by Ghahari *et al.* (2014). In total, 69 species from 27 genera of Vespidae have been recorded as prey for Asilidae worldwide (Dennis and Lavigne, 2007).

Adult hornets, yellow jackets, and paper wasps prey on a wide range of arthropods, such as caterpillars, flies, spiders, grasshoppers, and cicadas, to nourish their developing offspring (Ross and Matthews, 1991; Yamazaki, 2025). In addition to hunting, they feed on flower nectar, tree sap, hemipteran honeydew, and sugary trophallactic liquids secreted by their larvae (Richards, 1971; Richter, 2000; Schmidt, 2009). Their carnivorous behavior has led to numerous reports of intraguild predation, including cases of cannibalism within Vespidae. For instance, the foundresses (queens) of certain polistine species have been observed raiding conspecific nests for larvae and pupae to feed their own brood during the early spring nesting period (Kasuya, 1980; Legan *et al.*, 2023). Hornet workers, such as *Vespa ducalis* Smith, 1852 and *V. mandarinia japonica* Radoszkowski, 1857, have been reported to steal larvae, and occasionally workers from paper wasp nests (Matsuura, 1984; Matsuura and Yamane, 1990; Yamazaki, 2025).

TABLE 6.6
Predator-Prey Relationships Between Asilidae and Vespidae in Iran

Predator Species (Asilidae)	Prey Species (Vespidae)
Ancylorhynchus glaucius (Rossi)	*Polistes nimpha* (Christ)
Andrenosoma serratum Hermann	*Allodynerus floricola* (de Saussure); *Parapolybia escalerae* (Meade-Waldo)
Antipalus truncatus (Loew)	*Delta unguiculatum* (Villers)
Antiphrisson trifarius (Loew)	*Delta unguiculatum* (Villers)
Apoclea femoralis (Wiedemann)	*Polistes* sp.
Apoclea micracantha Loew	*Ancistrocerus auctus* (Fabricius)
Choerades gilvus (Linnaeus)	*Delta dimidiatipenne* (de Saussure); *Vespa crabro* Linnaeus
Dasypogon octonotatus Loew	*Katamenes sichelii* (Giordani Soika); *Vespa crabro* Linnaeus
Dasypogon irinelae Weinberg	*Delta dimidiatipenne* (de Saussure)
Dioctria abdominalis Becker	*Euodynerus disconotatus* (Lichtenstein)
Dioctria linearis (Fabricius)	*Ancistrocerus auctus* (Fabricius); *Katamenes sichelii* (Giordani Soika)
Habropogon longiventris Loew	*Katamenes sichelii* (Giordani Soika)
Hoplotriclis pallasii (Wiedemann)	*Vespula rufa* (Linnaeus); *Vespula vulgaris* (Linnaeus)
Jothopogon leucomallus (Loew)	*Dolichovespula sylvestris* (Scopoli)
Leptarthrus brevirostris Meigen	*Polistes dominulus* (Christ)
Leptogaster pubicornis Loew	*Vespula vulgaris* (Linnaeus)
Machiremisca decipiens (Wiedemann)	*Ancistrocerus auctus* Fabricius
Machimus rusticus (Meigen)	*Delta unguiculatum* (Villers)
Molobratia teutonus (Linnaeus)	*Dolichovespula saxonica* (Fabricius)
Philodicus ponticus (Bigot)	*Vespula germanica* (Fabricius); *Vespa orientalis* Linnaeus
Pogonosoma maroccanum (Fabricius)	*Ischnogasteroides picteti picteti* (de Saussure)
Promachus leoninus Loew	*Polistes dominulus* (Christ)
Pycnopogon fasciculatus (Loew)	*Polistes atrimandibularis* Zimmermann
Pycnopogon leucostomus Engel	*Ischnogasteroides picteti tenuis* (de Saussure)
Saropogon pollinosus Loew	*Polistes wattii* Cameron
Stenopogon avus (Loew)	*Polistes bischoffi* Weyrauch
Stenopogon xanthotrichus Brullé	*Vespula germanica* (Fabricius)
Stichopogon elegantulus (Meigen & Wiedemann)	*Delta dimidiatipenne* (de Saussure); *Dolichovespula saxonica* (Fabricius)

Predation on wasp colonies has been recorded most frequently in Polistinae (52.7%), while predation on individual wasps is most often reported in Vespinae (42.0%). This is likely due to Polistinae nests often being exposed, whereas Vespinae typically build enveloped nests in cavities or underground (Spradbery, 1973; Detoni *et al.*, 2021). The main natural enemies of Eumeninae are birds, ants, bats, and parasitoids (West-Eberhard *et al.*, 1995). Some parasitoids of potter wasps are adapted to infest the nest while it is still under construction, while others do so after the nest has been completed and sealed, using their mandibles or ovipositors to puncture the nest cell walls (Krombein, 1979; West-Eberhard *et al.*, 1995). Regardless of when infestation occurs, parasitoids specialized to parasitize potter wasp nests typically possess morphological adaptations that allow them to break open the seal of a nests cell upon completing development (Martynova, 2020). Nevertheless, parasitoids occasionally fail to emerge from the nest cells (Buschini and Buss, 2010; Johnson *et al.*, 2023). Social wasps have evolved a variety of primary (e.g., nest construction) and secondary (e.g., coordinated behaviors and guarding) defenses to protect the colony and reduce the risk of predation and parasitism (Detoni *et al.*, 2021).

COMPARISON OF SPECIES DIVERSITY IN THE MIDDLE EASTERN COUNTRIES

Among the 16 countries of the Middle East, only the fauna of Turkey with 323 recorded species is more diverse than Iran. Although the number of species belonging to Eumeninae and Vespinae in Turkey is more than in Iran, there are more recorded species of Masarinae and Polistinae in Iran. It is likely that Turkey has more recorded diversity compared to Iran because the fauna of Turkey has been better studied.

Turkey, with a total surface area of 780,576 km^2, occupies Asia Minor between the Mediterranean Sea and the Black Sea and stretches into continental Europe. There are very different types of ecosystems, such as coastal and marine, agricultural, mountain, forests, steppes, and wetlands, but it is a mountainous country averaging about 1,000 meters in altitude. The ecosystems of Turkey are: 1) Agricultural Ecosystems; 2) Steppe Ecosystems; 3) Forest Ecosystems (There are two types according to the bio-geographical regions: Euro-Siberian bio-geographical region and Irano-Turanian bio-geographical region); 4) Mountain Ecosystems; 5) Inland Waters Ecosystems; and 6) Coastal and Marine Ecosystems. The topography and climatic diversity of the region are important preconditions for the development of a rich and diverse fauna. Turkey is generally divided into seven biogeographical regions. These are the Marmara Region, the Aegean Region, the Mediterranean Region, the Black Sea Region, and the Central, Eastern, and Southeastern Anatolian Regions (Yildirim, 2012b; Küçük and Ertürk, 2013; Muminjanov and Karagöz, 2018).

Iran is the 18th-largest country in the world (Hakimzadeh Khoei *et al.*, 2011), located in the Palaearctic region and naturally divided into six biomes and 14 ecoregions (Olson *et al.*, 2001). Climatologically, Iran is a predominantly arid and semi-arid country, but the northern slopes of the Alburz ranges and the Caspian lowland receive 800–2000 mm annual rainfall, making them the most humid parts of the country. The complex and varied climates, topography, and geological formations have led to a varied and unique biological diversity in the Iran Plateau (Hassan Zadeh *et al.*, 1993; Sabeti, 1994; Ghahreman and Attar, 1999; Zehzad *et al.*, 2002; Firouz, 2005; Mohammadi Fazel, 2010). Also, Iran is near the crossroads of three distinct biogeographical regions (the Palaearctic, Afrotropical, and Oriental regions), which undoubtedly positively impacts its diversity. Most Iranian species likely are Palaearctic in origin, but almost certainly there are faunal elements from all three regions, with Oriental species likely in the southeast. There are some Afrotropical species, although probably not too many because Iran is north of the Arabian Peninsula. Such a geographical position as well as its high environmental diversity promotes the diversity of the Iranian fauna not only with respect to Vespidae, but also other groups of Insecta, especially Hymenoptera (e.g., see Modarres Awal, 2012; Ghahari *et al.*, 2021, 2024; Gadallah *et al.*, 2022).

Comparison of land areas of the Middle Eastern countries (Table 6.7) indicates that Saudi Arabia has the largest area, but the number of recorded species (100 species) reflects how poorly the fauna of this family has been studied. On the other hand, Israel-Palestine is ranked 11th based on land

TABLE 6.7
Countries in the Middle East by Land Area, and Total Number of Vespidae Species

Country	Land Area (km^2)	Number of Recorded Species	Country	Land Area (km^2)	Number of Recorded Species
Saudi Arabia	2 149 690	100	Jordan	89 342	88
Iran	1 648 195	261	United Arab Emirates	83 600	61
Egypt	1 010 408	129	Israel-Palestine	28 092	192
Turkey	783 562	323	Kuwait	17 818	5
Yemen	555 000	79	Qatar	11 586	5
Iraq	438 317	67	Lebanon	10 452	39
Oman	309 500	72	Cyprus	9 251	44
Syria	185 180	99	Bahrain	760	3

area, while in terms of species diversity, it is only behind Turkey and Iran, meaning that its fauna is relatively well studied. The fauna of each country will be increased after systematic samplings are completed. However, regarding Saudi Arabia, Iran, and Egypt, we expect more species discoveries for these three countries than other countries of the Middle East.

CONCLUSIONS

Future studies of vespid wasps are essential to fully understand their life history, behavior, and habitat preference, as well as their role in the ecosystem. The high species diversity observed throughout the Middle East, many of which are endemic to the region, highlights the potential for significant findings for researchers working in this area. These discoveries may have important implications for agriculture, ecology, conservation, medicine, and animal behavior. Vespid wasps are beneficial insects, as they hunt caterpillars and other insects, including many agricultural pests, to feed their young (Figure 6.3) (Das and Gupta, 1989; Goulet and Huber, 1993). Despite their aggressive behavior and deadly stings, their foraging habits make them effective predators of insect pests in biological control programs (Iwata, 1967; Gould and Jeanne, 1984). Additionally, because they occupy a high trophic level in terrestrial arthropod food webs (or even animal) and frequently visit flowers for nectar as an energy source, they serve as both control agents of pests and pollinators for many plant species (Figure 6.4) (Kojima, 1993; Carpenter and Wenzel, 1999; Khuat *et al.*, 2004). These factors suggest that vespid wasps play critical roles in ecosystems and may serve as valuable bioindicators of environmental conditions and habitat perturbation (Itô, 1984; Carpenter, 1991). Although Aristole (384–322 BC), the ancient Greek philosopher and polymath, stated that "*Hornets and wasps are devoid of the extraordinary features which characterize bees; this we should expect, for they have nothing divine about them as the bees have*", their eusocial or social behaviors, such as highly coordinated and complex predatory strategies, demonstrate a remarkable level of evolutionary advancement that reflects a form of natural sophistication.

Regarding the importance and conservation value of Vespidae, there is ample evidence that their ecological role cannot be overlooked. Some key examples illustrating the significance of these insects are presented below.

1) The pollination of flowers should also be a direct concern to those managing areas of natural and semi-natural vegetation. Pollinator diversity has a significant impact on the composition of vegetation within a given area (Nicholls and Altieri, 2013; Ghahari, 2017). Vespid wasps can represent an important component of the ecological framework within ecosystems,

FIGURE 6.3 *Paravespa rex* (von Schulthess 1924) (Eumeninae) – carries the captured lepidopteran larva into the built nest.

(Photo courtesy of Alexander Fateryga)

contributing to inter-relationships among community units, that is by acting as pollinators and occupying various micro-habitats (Akre, 1982; Brock *et al.*, 2021; Prezoto *et al.*, 2021).

2) Vespid wasps play a natural and important role as predators in the biological control of pest insect populations in forest and agricultural crops (Picanço *et al.*, 2012; Southon *et al.*, 2019). Additionally, invasive vespids have the potential to disrupt native species assemblages, alter ecological dynamics, and degrade ecosystem services (Rojas-Nossa *et al.*, 2023). Social wasps are often recognized by their beneficial role in controlling various pests, although they can become aggressive when their nests are disturbed (Edwards, 1996). However, quantifying these effects in invaded areas is often challenging due to the complexity of ecological interactions and the lack of baseline from before the invasions occurred (Rojas-Nossa *et al.*, 2023). Solitary vespids perform similar ecological functions in a more inconspicuous manner, and most are incapable of stinging humans even when provoked. They are often selective in their prey choices, which may give them significant localized ecological importance (Matsuura and Yamane, 1990; Ross and Matthews, 1991; Southon *et al.*, 2019).
3) Both social and solitary vespid species offer valuable opportunities to study a wide range of highly advanced behavioral patterns. Consequently, wasps are significantly more important in studies of social behavior than most other insects. They are of considerable interest in various fields of study, including biogeography, ecology, evolution, behavioral genetics, and the neuroscience of learning and memory, especially in the context of facial recognition (Richter, 2000; O'Neill, 2001; Bourke, 2005; Legan *et al.*, 2021; Prezoto *et al.*, 2021; Jernigan *et al.*, 2023).

FIGURE 6.4 *Pterocheilus* sp. (Eumeninae) – pollinating while feeding on pollen and nectar.

(Photo courtesy of Alexander Fateryga)

4) Vespids are of interest in population genetics and zoogeography, as there are notable examples of local distinct populations and Middle Eastern forms that differ from those in other parts of the world (Ross and Matthews, 1991; Edwards, 1996; Schmitz and Moritz, 1998).
5) Members of Aculeata (which include the Vespidae) are particularly sensitive insects that respond to environmental changes and therefore serve as a valuable group for ecological monitoring (O'Neill, 2001; Brock *et al.*, 2021).

Several faunistic studies in Iran have shown that a diverse assemblage of Vespidae is present in both agricultural and natural ecosystems. Taxonomic research on vespid wasps has accelerated in Iran over the past 17 years, particularly following the work of Ebrahimi and Carpenter (2008). In this book, two species of Eumeninae from Iran are reported as new records for the fauna of the Middle East, representing 0.27% of the total number of species reported. Among Middle Eastern countries and those adjacent to Iran, the faunas of the Arabian Peninsula (Gadallah *et al.*, 2020), Iran (Ebrahimi and Carpenter, 2008; Rahmani *et al.*, 2020), Iraq (Selis *et al.*, 2025), Pakistan (Rafi *et al.*, 2017), Russia (Antropov and Fateryga, 2017), and Turkey (Yildirim, 2012a) have been catalogued. However, similar comprehensive cataloguing is still lacking for several Middle Eastern and neighbouring countries, and many additional species records are likely to be discovered across the region in the future.

The lack of comprehensive species cataloguing poses a significant taxonomic impediment to identifying and describing vespid biodiversity. The biology of relatively few species (e.g., *Vespa orientalis*) is adequately understood in the Middle East (see Ostovan and Kamali, 1995). Due to this gap in knowledge, it is difficult to assess their importance to human welfare beyond general terms.

We hope that this work will serve as a solid foundation for future research on the vespids throughout the entire region, not just in Iran. Many of the recorded species are likely to be beneficial in various ways. For example, some act as natural control agents against pests. Although an accurate assessment of the importance of Vespidae in Middle Eastern countries is still lacking, our general understanding of their life history suggests that conserving these important natural enemies of pests should be a fundamental goal for farmers and gardeners. This can be achieved, in part, by reducing the use of chemical and non-selective pesticides.

Everyone can dislike vespid wasps because of their extremely painful stings, but their nest-building behavior and the architectural beauty of nest-structure under the highest degree of cooperation and coordination inspire the utmost amazement and love in humans.

(Hassan Ghahari)

REFERENCES

Akre, R.D. 1982. Social wasps, pp. 1–105. In: Hermann, H.R. (ed.), *Social Insects*. Vol. 4. Academic Press, New York, 385 pp.

Antropov, A.V. and Fateryga, A.V. 2017. Family Vespidae, pp. 175–196. In: Belokobylskij, S.A. and Lelej, A.S. (eds), Annotated catalogue of the Hymenoptera of Russia. Volume I. Symphyta and Apocrita: Aculeata. *Proceedings of the Zoological Institute of the Russian Academy of Sciences*, Supplement No. 6, 475 pp.

Auko, T.H., Trad, B.M. and Silvestre, R. 2014. Five new associations of parasitoids in potter wasps (Vespidae, Eumeninae). *Revista Brasileira de Entomologia* 58: 376–378.

Bohart, G.E., Parker, F.D. and Tepedino, V.J. 1982. Notes on the biology of *Odynerus dilectus* [Hym.: Eumenidae], a predator of the alfalfa weevil, *Hypera postica* [Col.: Curculionidae]. *Entomophaga* 27: 23–31.

Bourke, A.F.G. 2005. Genetics, relatedness and social behaviour in insect societies. *Symposium - Royal Entomological Society of London* 22: 1–30.

Brock, R.E., Cini, A. and Sumner, S. 2021. Ecosystem services provided by aculeate wasps. *Biological Reviews of the Cambridge Philosophical Society* 96(4): 1645–1675.

Buschini, M.L.T. and Buss, C.E. 2010. Biologic aspects of different species of *Pachodynerus* (Hymenoptera; Vespidae; Eumeninae). *Brazilian Journal of Biology*, 70(3): 623–629.

Carpenter, J.M. 1991. Phylogenetic relationships and origin of social behavior in the Vespidae, pp. 7–32. In: Ross, K.G. and Matthews, R.W. (eds.), *The social biology of wasps*. Cornell University Press, Ithaca, New York, xviii + 678 pp.

Carpenter, J.M. and Wenzel, J.W. 1999. The relative abundance of swarm-founding social wasps in the Congo Basin (Insecta: Hymenoptera: Vespidae: Polistinae). *Natural History Bulletin of Ibaraki University* 3: 9–14.

Das, B.P. and Gupta, V.K. 1989. The social wasps of India and the Adjacent Countries (Hymenoptera: Vespidae). *Oriental Insects Monograph* 11: 176–189.

Dennis, D.S. and Lavigne, R.J. 2007. Hymenoptera as prey of robber flies (Diptera: Asilidae) with new prey records. *Journal of the Entomological Research Society* 9: 23–42.

Detoni, M., Feás, X., Jeanne, R.L., Loope, K.J., O'Donnell, S., Santoro, D., Sumner, S. and Jandt, J.M. 2021. Evolutionary and ecological pressures shaping social wasps collective defenses. *Annals of the Entomological Society of America* 114(5): 581–595.

Dvořák, L. and Straka, J. 2007. Vespoidea: Vespidae (vosovití). *Acta Entomologica Musei Nationalis Pragae*, Supplementum 11: 171–189.

Ebrahimi, E. and Carpenter, J.M. 2008. Catalog of the vespid wasps of Iran (Hymenoptera, Vespidae). *Zootaxa* 1785: 1–42.

Edwards, R. 1996. *BWAR. Bees, Wasps and Ants Recording Society*. 2nd edition. Biological Records Centre for Ecology and Hydrology, Natural Environment Research Council, Institute of Terrestrial Ecology, Monks Wood Huntingdon Cambs, 66 pp.

Firouz, E. 2005. *The Complete Fauna of Iran*. Tauris, I.B., New York, XIV + 322 pp.

Gadagkar, R. 1991. *Belonogaster*, *Mischocyttarus*, *Parapolybia*, and independent-founding *Ropalidia*, pp. 149–190. In: Ross, K.G. and Matthews, R.W. (eds.), *The social biology of wasps*. Cornell University Press, Ithaca, xvii, 678 pp.

Gadallah, N.S., Brothers, D.J. and Williams, K.A. 2020. Biodiversity of the aculeate wasps (Hymenoptera: Aculeata) of the Arabian Peninsula. *Zootaxa* 4754(1): 1–216.

Gadallah, N.S., Ghahari, H. and Shaw, S.R. 2022. *Braconidae of the Middle East*. Elsevier, London, 420 pp.

Gawas, S.M., Kumar, P.G., Pannure, A., Gupta, A. and Carpenter, J.M. 2020. An annotated distributional checklist of Vespidae (Hymenoptera: Vespoidea) of India. *Zootaxa* 4784(1): 1–87.

Ghahari, H. 2017. *Pollinator insects and honey bee*. Publication of Jame Elmi Karbordi University, Karaj, 312 pp. [in Persian]

Ghahari, H., Gibson, G.A.P. and Viggiani, G. 2021. *Chalcidoidea of Iran*. CABI, Wallingford, Oxfordshire, 432 pp.

Ghahari, H., Hayat, R., Lavigne, R.J. and Ostovan, H. 2014. An annotated checklist of Iranian Asilidae (Insecta: Diptera: Brachycera: Asiloidea). *Linzer biologische Beiträge* 46(2): 1379–1446.

Ghahari, H. and Jussila, R. 2015. A faunistic study on ichneumonid wasps (Hymenoptera: Ichneumonidae) from Ilam and Kermanshah provinces, western Iran. *Entomologia Croatica* 19(1–2): 55–66.

Ghahari, H., Moulet, P. and McPherson, J.E. 2024. *True bugs (Heteroptera) of the Middle-East*. Volumes 1 and 2. Springer, Singapore, 1471 pp.

Ghahreman, A. and Attar, F. 1999. *Biodiversity of plant species in Iran*. Vol. 1. Tehran University Press, Tehran, 1218 pp.

Gould, W.P. and Jeanne, R.L. 1984. Polistes wasps (Hymenoptera: Vespidae) as control agents for lepidopterous cabbage pest. *Environmental Entomology* 13(1): 150–156.

Goulet, H. and Huber, J.T. 1993. *Hymenoptera of the World: An Identification Guide to Families*. Agriculture Canada Publication, Ottawa, 668 pp.

Grimaldi, D. and Engel, M. 2005. Evolution of the insects. Cambridge University Press, New York, xv + 755 pp.

Hakimzadeh Khoei, M., Kaya, M. and Altindag, A. 2011. New records of Rotifers from Iran with biogeographic considerations. *Turkish Journal of Zoology* 35: 395–402.

Hassan Zadeh, B., Zehzad, B., Farhang, B., Majnounian, H. and Goshtasb, H. 1993. *Golestan National Park*. Department of Environment, Fardin Publication, Iran, 203 pp. [in Persian]

Huang, K.Y., Lin, Y.S. and Severinghaus, L.L. 2004. Nest provisioning of the Oriental Honey-buzzard (*Pernis ptilorhyncus*) in northern Taiwan. *Journal of Raptor Research* 38: 367–371.

Itô, Y. 1984. Social biology of tropical and subtropical paper wasps with special reference to theories of the evolution of eusociality in the Hymenoptera. *News Bulletin of the Entomological Society of Queensland* 12(6): 91–97.

Iwata, K. 1967. Report of the fundamental research on the biological control of insect pests in Thailand II. The report on the bionomics of Aculeate wasps – Bionomics of subsocial wasps of Stenogastrinae (Hymenoptera: Vespidae). *Nature and Life in Southeast Asia*, Tokyo 5: 259–293.

Jeanne, R.L. 1979. A latitudinal gradient in rates of ant predation. *Ecology* 60: 1211–1224.

Jeanne, R.L. and Hunt, J.H. 1992. Observations on the social wasp *Ropalidia montana* from peninsular India. *Journal of Biosciences* 17: 1–14.

Jennings, D.T. and Houseweart, M.W. 1984. Predation by eumenid wasps (Hymenoptera: Eumenidae) on spruce budworm (Lepidoptera: Tortricidae) and other lepidopterous larvae in spruce-fir forests of Maine. *Annals of the Entomological Society of America* 77(1): 39–45.

Jernigan, C.M., Stafstrom, J.A., Zaba, N.C., Vogt, C.C. and Sheehan, M.J. 2023. Color is necessary for face discrimination in the Northern paper wasp, *Polistes fuscatus*. *Animal Cognition* 26: 589–598.

Johnson, A.D., Rozenberg, T. and Segoli, M. 2023. Notes on the parasitoids found within the nests of *Delta dimidiatipenne* (Hymenoptera, Vespidae). *Journal of Hymenoptera Research* 96: 925–936.

Kasuya, E. 1980. Behavioral ecology of Japanese paper wasps, *Polistes* spp. (Hymenoptera: Vespidae) I. Extranidal activities of Polistes chinensis antennalis. *Researches on Population Ecology* 22: 242–254.

Khuat, D.L., Nguyen T.P.L. and Pham T.N. 2004. A survey of beneficial hymenopteran bees and wasps and their use of value in the buffer zone of Cuc Phuong National Park, North Vietnam. *Sylvatrop* 14 (1–2): 67–94.

Kojima, J. 1993. Relative abundance of swarm-founding *Ropalidia* in the eusocial wasps (Hymenoptera: Vespidae) in tropical Australia. *New Entomologist* 4: 4–7.

Küçük, M. and Ertürk, E. 2013. Biodiversity and protected areas in Turkey. *Sains Malaysiana* 42(10): 1455–1460.

Leduc, S., Rosenberg, T., Johnson, A.D. and Segoli, M. 2022. Nest provisioning with parasitized caterpillars by female potter wasps: costs and potential mechanisms. *Animal Behaviour* 188: 99–109.

Legan, A.W., Jernigan, C.M., Miller, S.E., Fuchs, M.F. and Sheehan, M.J. 2021. Expansion and accelerated evolution of 9-Exon odorant receptors in *Polistes* paper wasps. *Molecular Biology and Evolution* 38(9): 3832–3846.

Legan, A.W., Vogt, C.C. and Sheehan, M.J. 2023. Postural analysis reveals persistent changes in paper wasp foundress behavioral state after conspecific challenge. *Ecology and Evolution* 13: e10436.

Martynova, K.V. 2020. Cocoons of cuckoo wasps (Hymenoptera: Chrysididae): First overview of morphology reveals unexpected traces of taxonomy. *Zoologischer Anzeiger* 285: 122–138.

Maryam, Z., Ali, M., Shehbaz, M.H., Hussain, R., Qasim, M. and Hassan, M.A. 2024. An Updated Checklist of Family Vespidae (Insecta: Hymenoptera) from Azad Kashmir and Gilgit-Baltistan, Pakistan. *Journal of the Entomological Research Society* 26(2): 321–336.

Matsuura, M. 1984. Comparative biology of the five Japanese species of the genus *Vespa* (Hymenoptera, Vespidae). *Bulletin of the Faculty of Agriculture, Mie University* 69: 1–131.

Matsuura, M. and Yamane, S. 1990. *Biology of the vespine wasps*. Springer Verlag, Berlin, 323 pp.

McCann, S., Moeri, O., Jones, T., O'Donnell, S. and Gries, G. 2010. Nesting and nest-provisioning of the Red-throated Caracara (*Ibycter americanus*) in central French Guiana. *Journal of Raptor Research* 44: 236–240.

Modarres Awal, M. 2012. *List of agricultural pests and their natural enemies in Iran*. Third edition. Ferdowsi University of Mashhad Press, Iran, 759 pp.

Mohammadi Fazel, A. 2010. Iran's Fourth National Report to the Convention on Biological Diversity. Technical Report, Department of Environment, October 2010, 145 pp.

Muminjanov, H. and Karagöz, A. 2018. Biodiversity of Turkey. Contribution of genetic resources to sustainable agriculture and food systems. Food and Agriculture Organization of the United Nations, Ankara, 202 pp.

Naumann, I.D. 1983. The biology of mud nesting Hymenoptera (and their associates) and Isoptera in rock shelters of the Kakadu Region, Northern Territory. Australian National Parks and Wildlife Service. Spec. Pub. 10: 127–189.

Nicholls, C.I. and Altieri, M.A. 2013. Plant biodiversity enhances bees and other insect pollinators in agroecosystems. A review. *Agronomy for Sustainable Development* 33(2): 257–274.

O'Neill, K.M. 2001. *Solitary wasps: Behavior and natural history*. Cornell University Press, 406 pp.

Olson, D.M., Dinerstein, E., Wikramanayake, E.D., Burgess, N.D., Powell, G.V.N., Underwood, E.C., D'Amico, J.A., Itano, I., Strand, H.E., Morrison, J.C., Loucks, C.J., Allmutt, T.F., Ricketts, T.H., Kura, Y., Lamsreux, J.F., Wettengel, W.W., Hedao, P. and Kassem, K.R. 2001. Terrestrial ecoregions of the world: A new map of life on Earth. *BioScience* 51: 933–938.

Ostovan, H. and Kamali, K. 1995. Some aspects of behavioral biology of hornet wasp, *Vespa orientalis* L. (Hym., Vespidae), a pest of date fruits in Khuzestan, Iran. *Journal of Agricultural Sciences* (Islamic Azad University) 1(1): 33–43.

Perrard, A., Grimaldi, D. and Carpenter, J.M. 2017. Early lineages of Vespidae (Hymenoptera) in Cretaceous amber. *Systematic Entomology* 42: 379–386.

Perry, S. and Manson, J.H. 2008. *Manipulative monkeys: The Capuchins of Lomas Barbudal*. Harvard University Press, Cambridge, MA, 368 pp.

Pham, P.H. 2014. A checklist of Ropalidiini wasps (Hymenoptera: Vespidae: Polistinae) in Indochina. *Archives of Biological Sciences, Belgrade*, 66(3): 1061–1074.

Picanço, M.C., de Oliveira, I.R., Fernandes, F.L., Martinez, H.E.P., Bacci, L. and da Silva, E.M. 2012. Ecology of Vespidae (Hymenoptera) predators in *Coffea arabica* plantations. *Sociobiology* 59(4): 1269–1280.

Pickett, K.M. and Carpenter, J.M. 2010. Simultaneous analysis and the origin of eusociality in the Vespidae (Insecta: Hymenoptera). *Arthropod Systematics & Phylogeny* 68(1): 3–33.

Piekarski, P.K., Carpenter, J.M., Lemmon, A.R., Lemmon, E.M. and Sharanowski, B.J. 2018. Phylogenomic evidence overturns current conceptions of social evolution in wasps (Vespidae). *Molecular Biology and Evolution* 35: 2097–2109.

Prezoto, F., Nascimento, F.S., Barbosa, B.C. and Somavilla, A. 2021. *Neotropical social wasps: Basic and applied aspects*. Springer Nature Switzerland, xiii + 472 pp.

Rafi, M.A., Carpenter, J.M., Qasim, M., Shehzad, A., Zia, A., Khan, M.R., Mastoi, M.I., Naz, F., Ilyas, M., Shah, M. and Bhatti, A.R. 2017. The vespid fauna of Pakistan. *Zootaxa* 4362(1): 1–28.

Rahmani, Z., Rakhshani, E. and Carpenter, J.M. 2020. Updated checklist of Vespidae (Hymenoptera: Vespoidea) in Iran. *Journal of Insect Biodiversity and Systematics* 6(1): 27–86.

Richards, O.W. 1971. The biology of the social wasps (Hymenoptera, Vespidae). *Biological Reviews* 46: 483–528.

Richter, M.R. 2000. Social wasp (Hymenoptera: Vespidae) foraging behavior. *Annual Review of Entomology* 45: 121–150.

Rojas-Nossa, S.V., O'Shea-Wheller, T.A., Poidatz, J., Mato, S., Osborne, J. and Garrido, J. 2023. Predator and pollinator? An invasive hornet alters the pollination dynamics of a native plant. *Basic and Applied Ecology* 71: 119–128.

Ross, K.G. and Matthews, R.W. 1991. *The social biology of wasps*. Comstock Publishing Associates, Cornell University Press, Ithaca, xviii + 678 pp.

Sabeti, H. 1994. *Forests, trees and shrubs of Iran*, University of Yazd, Iran, 810 pp. [in Persian]

Schmidt, J.O. 2009. Wasps, pp. 1049–1052. In: Resh, V.H. and Cardé, R.T. (eds), *Encyclopedia of insects*. Second Edition. Academic Press, 1168 pp.

Schmitz, J. and Moritz, R.F.A. 1998. Molecular phylogeny of Vespidae (Hymenoptera) and the evolution of sociality in wasps. *Molecular Phylogenetics and Evolution* 9: 183–191.

Segoli, M., Leduc, S., Meng, F., Hoffmann, I., Kishinevsky, M. and Rozenberg, T. 2020. Frequency and consequences of the collection of already parasitized caterpillars by a potter wasp. *Scientific Reports* 10(1): 1–9.

Selis, M., Ismael, H.R., Hamad, M.I., Baiocchi, D. and Cilia, G. 2025. A first update to the vespid fauna of Iraq (Hymenoptera: Vespidae), with description of a new species. *Journal of Natural History* 59 (13–16): 941–981.

Silveira, O.T., Andena, S.R., Somavilla, A. and Carpenter, J.M. 2021. Phylogeny and classification of the Neotropical social wasps, pp. 267–291. In: Prezoto, F., Nascimento, F.S., Barbosa, B.C. and Somavilla, A. (eds.), *Neotropical social wasps: Basic and applied aspects*. Springer Nature, Switzerland, 472 pp.

Southon, R.J., Fernandes, O.A., Nascimento, F.S. and Sumner, S. 2019. Social wasps are effective biocontrol agents of key lepidopteran crop pests. *Proceedings of the Royal Society B: Biological Sciences* 286:20191676.

Spradbery, J.P. 1973. *Wasps: An account of the biology and natural history of solitary and social wasps*. University of Washington Press, Seattle, 408 pp.

Sumner, S., Law, G. and Cini, A. 2018. Why we love bees and hate wasps. *Ecological Entomology* 43(6): 836–845.

Tindo, M., Mony, R. and Dejean, A. 2002. Colony development and serial polygyny in the primitively eusocial wasp *Belonogaster juncea juncea* (Vespidae: Polistinae). *Journal of Insect Behavior* 15: 243–252.

Turillazzi, S. 1984. Defensive mechanisms in *Polistes* wasps, pp. 33–58. In: Hermann, H.H. (ed.), *Defensive mechanisms in social insects*. Praeger, New York, xii + 259 pp.

Turillazzi, S. 2012. *The biology of hover wasps*. Heidelberg: Springer. XVI+ 272 pp.

van der Vecht, J. 1966. The East-Asiatic and Indo-Australian species of *Polybioides* du Buysson and *Parapolybia* de Saussure (Hym., Vespidae). *Zoologische Verhandelingen* 82: 1–42.

Varghese, T. and Girish Kumar, P. 2023. A review of the taxonomy, biology and distribution of the social wasps (Hymenoptera: Vespidae) of the Indian subcontinent. *Journal of the Indian Institute of Science* 103: 1019–1047.

West-Eberhard, M.J., Carpenter, J.M. and Hanson, P.E. 1995. The vespid wasps (Vespidae), pp. 561–587. In: Hanson, P.E. and Gauld, I.D. (eds), *The hymenoptera of costa rica Oxford*, Oxford Science Publications, The Natural History Museum, London, 914 pp.

Yamazaki, K. 2025. Dangerous prey: a case of cannibalism between workers of *Polistes jokahamae* (Hymenoptera: Vespidae, Polistinae). *Fragmenta Entomologica* 57(1): 133–135.

Yildirim, E. 2012a. Contribution to the knowledge of the Vespidae (Hymenoptera, Aculeata) of Turkey, with a checklist of the Turkish species. *Turkish Journal of Zoology* 36(3): 361–374.

Yildirim, E. 2012b. The distribution and biogeography of Vespidae (Hymenoptera: Aculeata) in Turkey. *Türkiye Entomoloji Dergisi* 36(1): 23–42.

Yuan, D., Beckman, J., Fernandez, J.F. and Rodriguez, J. 2022. Nest ecology and prey preference of the mud dauber wasp *Sceliphron formosum* (Hymenoptera: Sphecidae). *Insects* 13(12): 1136.

Zehzad, B., Kiabi, B.H. and Madjnoonian, H. 2002. The natural areas and landscape of Iran: An overview. *Zoology in the Middle East* 26: 7–10.

Eumenes sareptanus André, 1884 (Eumeninae). (Photo courtesy of Alexander Fateryga)

"So great is the economy of nature, that most flowers which are fertilised by crepuscular or nocturnal insects emit their odour chiefly or exclusively in the evening."

(Charles Darwin; 12 February 1809 – 19 April 1882; English naturalist, geologist and biologist)

Celonites tauricus (Kostylev, 1935) (Masarinae). (Photo courtesy of Alexander Fateryga)

> *"Nature will bear the closest inspection. She invites us to lay our eye level with her smallest leaf, and take an insect view of its plain."*
>
> (Henry David Thoreau; 12 July 1817 – 6 May 1862; American naturalist, essayist, poet, and philosopher)

Polistes fuscatus (Fabricius, 1793) (Polistinae) on prairie willow (Salicaceae). (Photo courtesy of Angella Moorehouse)

> *"What happens in a world without wasps? We don't know for sure because we lack the basic science to tell us precisely what they do. But we know that wasps must be important for functioning ecosystems and a healthy planet. We know that they prey on the insects that would otherwise be a nuisance to us. In a world without wasps we would almost certainly need to use a lot of chemicals to keep other pests at bay. This, at least, is a good argument to forgive wasps for their sting just as we forgive bees."*
>
> (Seirian Sumner, November 2022; Interalia magazine, 74)

Vespula rufa (Linnaeus, 1758) (Vespinae). (Photo courtesy of Michael E. Archer)

"The yellow jacket is 1st cousin to the wasp and a close relation to the black hornet··· They are as yellow as gold; and when they get good and mad they blaze like forked lightning in a thunderstorm. Don't go near a yellow jacket's nest, especially in August, unless your life is heavily insured."

(Josh Billings; 21 April 1818 – 14 October 1885; American humorist and lecturer)

Appendix

List of Vespidae species (Hymenoptera) of the Middle East (Asterisk means fauna of Iran)

Subfamily Eumeninae Leach, 1815

(1) ***Acanthodynerus*** *giordanii* Gusenleitner, 1969
(2) *Acanthodynerus multimaculatus* Gusenleitner, 2000*
(3) ***Afreumenes*** *melanosoma yemenensis* Giordani Soika, 1996
(4) *Afreumenes nigrorufus* Giordani Soika, 1968
(5) ***Alastor*** *aegyptiacus* Blüthgen, 1955
(6) *Alastor arabicus* Giordani Soika, 1979
(7) *Alastor ardens* Kostylev, 1935
(8) *Alastor asiaticus* Morawitz, 1895
(9) *Alastor atropos* Lepeletier, 1841
(10) *Alastor dalyi* Giordani Soika, 1979
(11) *Alastor darius* Gusenleitner, 1986*
(12) *Alastor elisaei* von Schulthess, 1925
(13) *Alastor esfandiarii* Giordani Soika, 1970
(14) *Alastor iranus* Blüthgen, 1956*
(15) *Alastor mediomaculatus mediomaculatus* Giordani Soika, 1952*
(16) *Alastor mocsaryi* (André, 1884)*
(17) *Alastor nitens* Gusenleitner, 2000
(18) *Alastor pentheri* Kohl, 1905
(19) *Alastor ruficornis* Gusenleitner, 2001
(20) *Alastor sanctus* Blüthgen, 1956
(21) *Alastor savignyi* de Saussure, 1853
(22) *Alastor schwarzi* Gusenleitner, 1967
(23) *Alastor seidenstueckeri* Blüthgen, 1956
(24) *Alastor thymbrinus* Blüthgen, 1956
(25) *Alastor xerxes* Gusenleitner, 1986*
(26) *Alastor zoroaster* Gusenleitner, 1986*
(27) ***Allodynerus*** *delphinalis delphinalis* (Giraud, 1866)*
(28) *Allodynerus dignotus* (Morawitz, 1895)*
(29) *Allodynerus floricola floricola* (de Saussure, 1853)*
(30) *Allodynerus floricola inaequalis* Giordani Soika, 1970
(31) *Allodynerus nutatus* Gusenleitner, 1991
(32) *Allodynerus rossii* (Lepeletier, 1841)*
(33) *Allodynerus vinciguerrae pallidulus* Giordani Soika, 1970
(34) *Allodynerus vinciguerrae vinciguerrae* (Guiglia, 1929)
(35) ***Ancistrocerus*** *adenensis* Giordani Soika, 1952

(36) *Ancistrocerus antilope* (Panzer, 1798)
(37) *Ancistrocerus auctus* (Fabricius, 1793)*
(38) *Ancistrocerus biphaleratus biphaleratus* (de Saussure, 1852)
(39) *Ancistrocerus biphaleratus palaestinicus* Giordani Soika, 1952
(40) *Ancistrocerus biphaleratus triphaleratus* (de Saussure, 1855)
(41) *Ancistrocerus claripennis claripennis* Thomson, 1874*
(42) *Ancistrocerus claripennis ponticus* Giordani Soika, 1970
(43) *Ancistrocerus dusmetiolus* (Strand, 1914)
(44) *Ancistrocerus gazella* (Panzer, 1798)*
(45) *Ancistrocerus ichneumonideus ichneumonideus* (Ratzeburg, 1844)
(46) *Ancistrocerus impunctatus impunctatus* (Spinola, 1838)
(47) *Ancistrocerus longispinosus longispinosus* (de Saussure, 1855)
(48) *Ancistrocerus nigricornis* (Curtis, 1826)*
(49) *Ancistrocerus oviventris caucasicus* Gusenleitner, 1994
(50) *Ancistrocerus oviventris oviventris* (Wesmael, 1836)
(51) *Ancistrocerus oviventris siculus* Blüthgen, 1955
(52) *Ancistrocerus parietinus* (Linnaeus, 1761)*
(53) *Ancistrocerus parietum* (Linnaeus, 1758)*
(54) *Ancistrocerus renimacula* (Lepeletier, 1841)
(55) *Ancistrocerus scoticus scoticus* (Curtis, 1826)*
(56) *Ancistrocerus tenebrosus* Gusenleitner, 2010*
(57) *Ancistrocerus trifasciatus trifasciatus* (Müller, 1776)
(58) ***Antepipona** albosignata* Gusenleitner, 1986
(59) *Antepipona arabica* Giordani Soika, 1979*
(60) *Antepipona barrei barrei* (Radoszkovski, 1893)*
(61) *Antepipona barrei vastifica* (Morawitz, 1895)
(62) *Antepipona cingulifera* (Walker, 1871)
(63) *Antepipona cribrata* (Morawitz, 1885)
(64) *Antepipona curialis* (Morawitz, 1895)
(65) *Antepipona deflenda* (Saunders, 1853)*
(66) *Antepipona dentella* Gusenleitner, 1991
(67) *Antepipona iconia* (Blüthgen, 1951)
(68) *Antepipona insana clava* Gusenleitner, 1988
(69) *Antepipona insana insana* (Giordani Soika, 1943)
(70) *Antepipona irakensis* Giordani Soika, 1970
(71) *Antepipona jordanica* Gusenleitner, 2004
(72) *Antepipona karadgensis* Giordani Soika, 1970*
(73) *Antepipona kassalensis* (Giordani Soika, 1939)
(74) *Antepipona laevigata* (Blüthgen, 1951)*
(75) *Antepipona levis* Gusenleitner, 2021*
(76) *Antepipona nigricornis nigricornis* (Morawitz, 1885)
(77) *Antepipona omanensis* Giordani Soika, 1979*
(78) *Antepipona orbitalis ballioni* (Morawitz, 1867)
(79) *Antepipona orbitalis orbitalis* (Herrich-Schäffer, 1839)
(80) *Antepipona osmania* Gusenleitner, 1986*
(81) *Antepipona praeclara* Giordani Soika, 1970*
(82) *Antepipona prompta* Giordani Soika, 1970*
(83) *Antepipona rotunda* Gusenleitner, 2014
(84) *Antepipona specifica* (Morawitz, 1895)*
(85) *Antepipona tenuis* Gusenleitner, 1988
(86) *Antepipona vagabunda* (Dalla Torre, 1889)*

(87) *Antepipona varentzowi* (Morawitz, 1895)*
(88) *Antepipona vescovilis* Giordani Soika, 1970*
(89) *Antepipona yemenensis* (Giordani Soika, 1957)
(90) ***Antodynerus*** *ignaruris* (Kohl, 1906)
(91) *Antodynerus igneus* Gusenleitner, 2004
(92) *Antodynerus ignoratus* Gusenleitner, 2007
(93) *Antodynerus incomparabilis* Giordani Soika, 1970*
(94) *Antodynerus multicolor dhufariensis* (Giordani Soika, 1957)
(95) ***Brachyodynerus*** *binominatus* Blüthgen, 1939
(96) *Brachyodynerus chloroticus* Gusenleitner, 2004
(97) *Brachyodynerus djebaili* Giordani Soika, 1983
(98) *Brachyodynerus kopetdagicus* (Kostylev, 1940)
(99) *Brachyodynerus kusdasi* Gusenleitner, 1967
(100) *Brachyodynerus magnificus magnificus* (Morawitz, 1867)
(101) *Brachyodynerus quadrimaculatus* (André, 1884)
(102) *Brachyodynerus zhelochovtzevi* (Kostylev, 1929)*
(103) ***Brachypipona*** *irana* Gusenleitner and Schmid-Egger, 2021*
(104) *Brachypipona montana* Gusenleitner and Schmid-Egger, 2021*
(105) *Brachypipona orientalis* Gusenleitner, 2004*
(106) *Brachypipona schlaeflei* Gusenleitner, 2001
(107) *Brachypipona schmidti* (Gusenleitner, 1967)
(108) ***Cephalochilus*** *draco* Giordani Soika, 1970*
(109) ***Chlorodynerus*** *araxanus* (Blüthgen, 1956)*
(110) *Chlorodynerus chloroticus* (Spinola, 1838)
(111) *Chlorodynerus diglaensis* (Blüthgen, 1954)
(112) *Chlorodynerus flavoferrugineus* Giordani Soika, 1970*
(113) *Chlorodynerus gratus* (Kostylev, 1940)*
(114) *Chlorodynerus incisipes* (Kostylev, 1935)*
(115) *Chlorodynerus infrenis* Giordani Soika, 1970*
(116) *Chlorodynerus intricatus* Giordani Soika, 1958
(117) *Chlorodynerus kelidopterus* (Kohl, 1907)
(118) *Chlorodynerus loeffleri* Giordani Soika, 1958*
(119) *Chlorodynerus mochii* Giordani Soika, 1958
(120) *Chlorodynerus sanctus* (Blüthgen, 1954)
(121) *Chlorodynerus schulthessianus* (Kostylev, 1937)*
(122) *Chlorodynerus ypsilon rhodius* (Blüthgen, 1954)
(123) *Chlorodynerus ypsilon ypsilon* (Kostylev, 1929)
(124) ***Cyphodynerus*** *canaliculatus* (de Saussure, 1855)
(125) *Cyphodynerus sculpturatus* (Dover, 1925)*
(126) ***Cyrtolabulus*** *anatolicus* Gusenleitner, 2006
(127) *Cyrtolabulus caputabnormis* Gusenleitner, 2000
(128) *Cyrtolabulus chudeaui* (du Buysson, 1908)
(129) *Cyrtolabulus elbanus* Giordani Soika, 1977
(130) *Cyrtolabulus eremicus* (Giordani Soika, 1952)
(131) *Cyrtolabulus exiguus* (de Saussure, 1853)
(132) *Cyrtolabulus gracilis* (Kohl, 1906)*
(133) *Cyrtolabulus grossepunctatus* (Kirby, 1900)
(134) *Cyrtolabulus iranus* (Giordani Soika, 1968)*
(135) *Cyrtolabulus karachiensis* Gusenleitner, 2006*
(136) *Cyrtolabulus mutinensis* Gusenleitner, 2006
(137) *Cyrtolabulus omanicus* Gusenleitner, 2014

(138) *Cyrtolabulus sexspinosus* Giordani Soika, 1977
(139) *Cyrtolabulus socotrae* (Kohl, 1906)
(140) *Cyrtolabulus spinithorax* (Giordani Soika, 1968)
(141) *Cyrtolabulus syriacus* (Giordani Soika, 1968)*
(142) *Cyrtolabulus yemenensis* Gusenleitner, 2007
(143) *Cyrtolabulus yildirimi* Gusenleitner, 2006
(144) *Cyrtolabulus zarudnyi* (Kostylev, 1939)*
(145) ***Delta*** *anomala* (Zavattari, 1909)*
(146) *Delta asina mixtum* (Giordani Soika, 1944)
(147) *Delta campaniforme chloroticum* Giordani Soika, 1979*
(148) *Delta dimidiatipenne* (de Saussure, 1852)*
(149) *Delta emarginatum* (Linnaeus, 1758)
(150) *Delta esuriens esuriens* (Fabricius, 1787)*
(151) *Delta esuriens gracile* (de Saussure, 1852)*
(152) *Delta fenestrale* (de Saussure, 1852)
(153) *Delta hottentotta elegans* (de Saussure, 1852)*
(154) *Delta lepeleterii lepeleterii* (de Saussure, 1852)
(155) *Delta lepeleterii pilosellum* Giordani Soika, 1982
(156) *Delta unguiculatum libanicum* (Giordani Soika, 1941)
(157) *Delta unguiculatum sporadense* (von Schulthess, 1929)
(158) *Delta unguiculatum unguiculatum* (Villers, 1789)*
(159) *Delta viatrix* (Nurse, 1903)
(160) ***Discoelius*** *dufourii dufourii* Lepeletier, 1841*
(161) *Discoelius pictus* Kostylev, 1940*
(162) *Discoelius zonalis* (Panzer, 1801)*
(163) ***Eumenes*** *coarctatus coarctatus* (Linnaeus, 1758)*
(164) *Eumenes coarctatus lunulatus* Fabricius, 1804
(165) *Eumenes comberi* Dover, 1925*
(166) *Eumenes coronatus* (Panzer, 1799)*
(167) *Eumenes crimensis* Blüthgen, 1938*
(168) *Eumenes cyrenaicus* Blüthgen, 1938
(169) *Eumenes dubius dubius* de Saussure, 1852*
(170) *Eumenes dubius palaestinensis* Blüthgen, 1938*
(171) *Eumenes interpositus* Gusenleitner, 1972*
(172) *Eumenes jarkandensis* Blüthgen, 1938*
(173) *Eumenes mediterraneus cypricus* Blüthgen, 1938
(174) *Eumenes mediterraneus mediterraneus* Kriechbaumer, 1879*
(175) *Eumenes mediterraneus quettaensis* Cameron, 1907
(176) *Eumenes modestus* Gusenleitner, 2006*
(177) *Eumenes papillarius papillarius* (Christ, 1791)*
(178) *Eumenes pedunculatus pedunculatus* (Panzer, 1799)*
(179) *Eumenes pomiformis* (Fabricius, 1781)*
(180) *Eumenes pseudubius* Gusenleitner, 1972
(181) *Eumenes punctaticlypeus* Giordani Soika, 1943*
(182) *Eumenes sareptanus* André, 1884
(183) *Eumenes subpomiformis* Blüthgen, 1938
(184) *Eumenes tripunctatus* (Christ, 1791)*
(185) ***Eumicrodynerus*** *europaeus* (Giordani Soika, 1942)
(186) ***Euodynerus*** *amplus* Gusenleitner, 2011*
(187) *Euodynerus annae* (Kostylev, 1937)*
(188) *Euodynerus breviventris* (Giordani Soika, 1952)*

(189) *Euodynerus caspicus caspicus* (Morawitz, 1873)*
(190) *Euodynerus cherkensis* (Giordani Soika, 1942)
(191) *Euodynerus clatratus* Blüthgen, 1951
(192) *Euodynerus curictensis curictensis* Blüthgen, 1940*
(193) *Euodynerus dantici dantici* (Rossi, 1790)*
(194) *Euodynerus dantici tinctus* (Walker, 1871)
(195) *Euodynerus disconotatus disconotatus* (Lichtenstein, 1884)*
(196) *Euodynerus disconotatus laniensis* Giordani Soika, 1979
(197) *Euodynerus diversus* (Walker, 1871)
(198) *Euodynerus effrenatus* Giordani Soika, 1970*
(199) *Euodynerus enodatus* (Kostylev, 1940)*
(200) *Euodynerus fastidiosus* (de Saussure, 1853)*
(201) *Euodynerus garaensis* Selis, 2025
(202) *Euodynerus hellenicus* Blüthgen, 1942
(203) *Euodynerus khuzestanicus* Giordani Soika, 1970*
(204) *Euodynerus maximilianus* Gusenleitner, 1998
(205) *Euodynerus niloticus ebneri* (von Schulthess, 1921)
(206) *Euodynerus niloticus niloticus* (de Saussure, 1855)
(207) *Euodynerus niloticus sauditus* (Giordani Soika, 1957)
(208) *Euodynerus notatus notatus* (Jurine, 1807)*
(209) *Euodynerus posticus* (Herrich-Schäffer, 1841)*
(210) *Euodynerus quadrifasciatus quadrifasciatus* (Fabricius, 1793)
(211) *Euodynerus quadrifasciatus rufipes* Gusenleitner, 1984
(212) *Euodynerus rufinus confictus* Giordani Soika, 1970*
(213) *Euodynerus rufinus rufinus* Blüthgen, 1942*
(214) *Euodynerus salzi* (Giordani Soika, 1952)
(215) *Euodynerus semidantici semidantici* (Giordani Soika, 1952)*
(216) *Euodynerus semisaecularis macedonicus* Blüthgen, 1951*
(217) *Euodynerus semisaecularis semisaecularis* (Dalla Torre, 1889)*
(218) *Euodynerus setosus rubicundus* Gusenleitner, 1994
(219) *Euodynerus setosus setosus* Gusenleitner, 1970*
(220) *Euodynerus shirazensis* Giordani Soika, 1970*
(221) *Euodynerus socotrae* (Kohl, 1906)
(222) *Euodynerus soikai* Guichard, 1986
(223) *Euodynerus variegatus kruegeri* (von Schulthess, 1928)
(224) *Euodynerus velutinus* Blüthgen, 1951
(225) ***Eustenancistrocerus*** *amadanensis amadanensis* (de Saussure, 1855)*
(226) *Eustenancistrocerus askhabadensis askhabadensis* (Radoszkowski, 1886)*
(227) *Eustenancistrocerus inconstans* (de Saussure, 1863)*
(228) *Eustenancistrocerus iranicus* Gusenleitner, 2013*
(229) *Eustenancistrocerus jerichoensis iconius* Blüthgen, 1957
(230) *Eustenancistrocerus jerichoensis jerichoensis* (von Schulthess, 1928)*
(231) *Eustenancistrocerus khuzestanicus* Giordani Soika, 1970*
(232) *Eustenancistrocerus malus* Giordani Soika, 1970*
(233) *Eustenancistrocerus nilensis* (Giordani Soika, 1935)
(234) *Eustenancistrocerus pharao* (de Saussure, 1863)
(235) *Eustenancistrocerus spinosissimus* Gusenleitner, 2000*
(236) *Eustenancistrocerus sulcithorax* Giordani Soika, 1970*
(237) *Eustenancistrocerus tegularis* (Morawitz, 1885)*
(238) ***Extraepipona*** *occulta* Gusenleitner, 2014*
(239) ***Gymnomerus*** *laevipes laevipes* (Shuckard, 1837)*

(240) ***Hemipterochilus*** *aberrans* (Morawitz, 1885)
(241) *Hemipterochilus bembeciformis bembeciformis* (Morawitz, 1867)
(242) *Hemipterochilus bembeciformis terricola* (Mocsáry, 1883)
(243) *Hemipterochilus bicoloricornis* (Giordani Soika, 1952)
(244) *Hemipterochilus hohlbecki* (Kostylev, 1935)*
(245) *Hemipterochilus punctiventris punctiventris* (Morawitz, 1885)*
(246) *Hemipterochilus punctiventris subquadricolor* Gusenleitner, 1970
(247) *Hemipterochilus rubrosignatus* (André, 1884)*
(248) *Hemipterochilus simplex* Gusenleitner, 2000*
(249) ***Ischnogasteroides*** *annae* (Kostylev, 1939)
(250) *Ischnogasteroides carinatus* (Kostylev, 1939)
(251) *Ischnogasteroides leptogaster* (Walker, 1871)
(252) *Ischnogasteroides picteti picteti* (de Saussure, 1852)
(253) *Ischnogasteroides picteti tenuis* (Morawitz, 1888)*
(254) *Ischnogasteroides tenuissimus* (Giordani Soika, 1941)
(255) *Ischnogasteroides zarudnyi* (Kostylev, 1939)*
(256) ***Jucancistrocerus*** *atrofasciatus* (Morawitz, 1885)*
(257) *Jucancistrocerus caspicus* Giordani Soika, 1970*
(258) *Jucancistrocerus citreodecoratus* Giordani Soika, 1970*
(259) *Jucancistrocerus consimilis* (Morawitz, 1895)*
(260) *Jucancistrocerus jucundus jucundus* (Mocsáry, 1883)*
(261) *Jucancistrocerus pareumeniformis* Giordani Soika, 1973*
(262) *Jucancistrocerus saharensis* (Giordani Soika, 1934)
(263) ***Katamenes*** *dimidiativentris dimidiativentris* (Giordani Soika, 1941)
(264) *Katamenes dimidiativentris francisi* Zalat, 1992
(265) *Katamenes dimidiatus dimidiatus* (Brullé, 1832)*
(266) *Katamenes dimidiatus arrectus* Giordani Soika, 1970
(267) *Katamenes dimidiatus montanus* (Nurse, 1904)
(268) *Katamenes flavigularis* (Blüthgen, 1951)*
(269) *Katamenes jenjouristei jenjouristei* (Kostylev, 1939)
(270) *Katamenes jenjouristie rubroniger* (Giordani Soika, 1941)
(271) *Katamenes lybicus lybicus* (Giordani Soika, 1941)
(272) *Katamenes niger* (Brullé, 1839)*
(273) *Katamenes priesneri* (Giordani Soika, 1941)
(274) *Katamenes rauensis* (Giordani Soika, 1960)
(275) *Katamenes sichelii baerii* (Radoszkowski, 1865)*
(276) *Katamenes sichelii biblicus* (Giordani Soika, 1935)
(277) *Katamenes sichelii coranicus* Giordani Soika, 1970
(278) *Katamenes sichelii fulvus* (Eversmann, 1854)*
(279) *Katamenes sichelii sichelii* (de Saussure, 1852)*
(280) *Katamenes sichelii tauriae* (Giordani Soika, 1960)*
(281) *Katamenes sichelii timanensis* Giordani Soika, 1986
(282) *Katamenes tauricus tauricus* (de Saussure, 1855)*
(283) *Katamenes tauricus viratus* (Giordani Soika, 1949)
(284) ***Knemodynerus*** *djarabubensis djarabubensis* (von Schulthess, 1928)
(285) *Knemodynerus euodyneroides* Gusenleitner, 1997
(286) *Knemodynerus excellens* (Pérez, 1907)*
(287) *Knemodynerus familiaris familiaris* (Giordani Soika, 1939)
(288) *Knemodynerus familiaris judaicus* (Giordani Soika, 1952)
(289) *Knemodynerus familiaris muscatensis* (Giordani Soika, 1970)
(290) *Knemodynerus lahijensis* Gusenleitner, 2002

(291) *Knemodynerus pseudolateralis* (Meade-Waldo, 1915)
(292) *Knemodynerus rhynchoides inclinans* (Giordani Soika, 1935)
(293) *Knemodynerus rhynchoides rhynchoides* (de Saussure, 1853)
(294) *Knemodynerus sauditus* (Giordani Soika, 1989)
(295) *Knemodynerus sinaiticus* (Giordani Soika, 1939)
(296) *Knemodynerus stigma* (de Saussure, 1863)
(297) ***Labochilus*** *felix* Guichard, 1986
(298) *Labochilus orientalis* Gusenleitner, 2001*
(299) *Labochilus pulawskyi pulawskyi* Giordani Soika, 1970
(300) *Labochilus pulawskyi soharensis* Giordani Soika, 1979
(301) ***Leptochilus*** *aegineticus* Gusenleitner, 1970
(302) *Leptochilus alborufulus* Gusenleitner, 1977
(303) *Leptochilus alpestris alpestris* (de Saussure, 1855)
(304) *Leptochilus alterego* Gusenleitner, 1970
(305) *Leptochilus ambitiosus* Giordani Soika, 1986
(306) *Leptochilus amos amos* Gusenleitner, 1971
(307) *Leptochilus arabicus* Giordani Soika, 1979
(308) *Leptochilus aterrimus* (Kirby, 1900)
(309) *Leptochilus ayunensis* Giordani Soika, 1981
(310) *Leptochilus bellus* Blüthgen, 1955
(311) *Leptochilus bipunctatus* Gusenleitner, 2016
(312) *Leptochilus bytinskii* Giordani Soika, 1952
(313) *Leptochilus callidus* (Kostylev, 1940)*
(314) *Leptochilus camurus* (Giordani Soika, 1938)
(315) *Leptochilus errabundus* (Giordani Soika, 1938)
(316) *Leptochilus flegias* (Giordani Soika, 1938)
(317) *Leptochilus flexilis* (Giordani Soika, 1938)
(318) *Leptochilus fuscipes* Gusenleitner, 1985*
(319) *Leptochilus gemma* Giordani Soika, 1970*
(320) *Leptochilus gusenleitneri* Yildirim and Özbek, 1995
(321) *Leptochilus hermon* Gusenleitner, 1971
(322) *Leptochilus hethiticus* Gusenleitner, 1985
(323) *Leptochilus iranus* Gusenleitner and Schmid-Egger, 2021*
(324) *Leptochilus jordaneus* Gusenleitner, 1994
(325) *Leptochilus josephi* Giordani Soika, 1947
(326) *Leptochilus kemali* Gusenleitner, 1977
(327) *Leptochilus kurnubensis* Blüthgen, 1955
(328) *Leptochilus limbiferoides* (Giordani Soika, 1938)
(329) *Leptochilus limbiferus anatolicus* Blüthgen, 1955
(330) *Leptochilus longipilis* Gusenleitner, 1988
(331) *Leptochilus lorestanicus* Gusenleitner, 1995*
(332) *Leptochilus masirahensis* Giordani Soika, 1981
(333) *Leptochilus medanae falkenhayni* (Dusmet, 1917)
(334) *Leptochilus medanae medanae* (Gribodo, 1886)
(335) *Leptochilus membranaceus luxuriosus* Blüthgen, 1955
(336) *Leptochilus membranaceus membranaceus* (Morawitz, 1867)*
(337) *Leptochilus mauritanicus derufatus* Blüthgen, 1955
(338) *Leptochilus mochianus mochianus* Giordani Soika, 1970
(339) *Leptochilus modestus* de Saussure, 1853
(340) *Leptochilus montivagus* Gusenleitner, 2002
(341) *Leptochilus muscatensis* Giordani Soika, 1979

(342) *Leptochilus nabataeus* Gusenleitner, 1990
(343) *Leptochilus neutraliformis* Gusenleitner, 1977
(344) *Leptochilus neutralis neutralis* (Giordani Soika, 1943)*
(345) *Leptochilus neutralis rufior* Giordani Soika, 1952
(346) *Leptochilus nigrocitrinus* Giordani Soika, 1970
(347) *Leptochilus nigroclypeus* Gusenleitner, 1985
(348) *Leptochilus ornatulus* Gusenleitner, 1977
(349) *Leptochilus osiris* (Schmiedeknecht, 1900)
(350) *Leptochilus osmanicus* Gusenleitner, 1988
(351) *Leptochilus palandokenicus* Yildirim and Özbek, 1995
(352) *Leptochilus perterritus* Giordani Soika, 1970
(353) *Leptochilus praestans* Giordani Soika, 1970
(354) *Leptochilus pseudojosephi* Giordani Soika, 1952
(355) *Leptochilus regulus* (de Saussure, 1855)*
(356) *Leptochilus ressli* Gusenleitner, 1985*
(357) *Leptochilus salomon* Gusenleitner, 1971
(358) *Leptochilus schatzmayri* (Giordani Soika, 1938)
(359) *Leptochilus schindleri* (Guiglia, 1933)
(360) *Leptochilus signatus gueruensis* Gusenleitner, 1995
(361) *Leptochilus signatus signatus* Gusenleitner, 1995
(362) *Leptochilus simurgh* Gusenleitner, 2016*
(363) *Leptochilus speciosus* Gusenleitner, 2002
(364) *Leptochilus subtarsatellus* Gusenleitner, 2002
(365) *Leptochilus tarsatellus* Giordani Soika, 1970
(366) *Leptochilus tarsatiformis* (Giordani Soika, 1943)
(367) *Leptochilus tarsatus* (de Saussure, 1855)
(368) *Leptochilus torretassoi* (Giordani Soika, 1938)
(369) *Leptochilus tricolor* Gusenleitner, 2016
(370) *Leptochilus turcicus* Gusenleitner, 1971
(371) *Leptochilus tussaci* Giordani Soika, 1986
(372) *Leptochilus villosus* Gusenleitner, 2001
(373) ***Leptodynerus*** *arabicus* Giordani Soika, 1970
(374) *Leptodynerus vanharteni* Gusenleitner, 2010
(375) ***Micreumenes*** *arabicus* Giordani Soika, 1979
(376) *Micreumenes glaber glaber* Giordani Soika, 1983
(377) *Microdynerus abdelkader abdelkader* (de Saussure, 1856)
(378) *Microdynerus abdelkader pharaonum* Giordani Soika, 1960
(379) *Microdynerus aegaeicus* Gusenleitner, 1998
(380) *Microdynerus anatolicus* (Blüthgen, 1938)*
(381) *Microdynerus appenninicus* Giordani Soika, 1960
(382) *Microdynerus atriceps* Morawitz, 1895*
(383) *Microdynerus confinis* Gusenleitner, 1979
(384) *Microdynerus curdistanicus* Gusenleitner, 1988
(385) *Microdynerus erzincanensis* Yildirim and Özbek, 1995
(386) *Microdynerus eurasius* (Blüthgen, 1938)
(387) *Microdynerus exilis* (Herrich-Schäffer, 1839)
(388) *Microdynerus hoetzendorfi* (Dusmet, 1917)
(389) *Microdynerus latro* Blüthgen, 1955
(390) *Microdynerus longicollis longicollis* Morawitz, 1895
(391) *Microdynerus microdynerus* (Dalla Torre, 1889)*
(392) *Microdynerus mirandus* (Giordani Soika, 1947)

(393) *Microdynerus nugdunensis nugdunensis* (de Saussure, 1855)
(394) *Microdynerus parvulus* (Herrich-Schäffer, 1838)*
(395) *Microdynerus rubronotatus* rubiculus Gusenleitner, 2000
(396) *Microdynerus saundersi saundersi* Blüthgen, 1955
(397) *Microdynerus syriacus* Gusenleitner, 2000
(398) *Microdynerus urdunus* Gusenleitner, 2004
(399) ***Odynerus*** *albopictus* de Saussure, 1856*
(400) *Odynerus alexandrinus* de Saussure, 1853
(401) *Odynerus alpinus* von Schulthess, 1897
(402) *Odynerus armeniacus* (Morawitz, 1885)*
(403) *Odynerus consobrinus* consobrinus Dufour, 1839
(404) *Odynerus ezechiae* von Schulthess, 1923*
(405) *Odynerus femoratus* de Saussure, 1856*
(406) *Odynerus fulvitarsis* (Morawitz, 1895)
(407) *Odynerus guichardi* Giordani Soika, 1977
(408) *Odynerus laticinctus laticinctus* (Bialynicki-Birula, 1926)*
(409) *Odynerus leucopus* (Blüthgen, 1941)*
(410) *Odynerus melanocephalus melanocephalus* (Gmelin, 1790)
(411) *Odynerus nigrospinosus* (Morawitz, 1895)*
(412) *Odynerus palaestinensis* Gusenleitner, 1971
(413) *Odynerus poecilus* de Saussure, 1856
(414) *Odynerus reniformis* (Gmelin, 1790)
(415) *Odynerus rotundigaster citripes* Gusenleitner, 1970
(416) *Odynerus rotundigaster rotundigaster* de Saussure, 1853
(417) *Odynerus serricrus* (Blüthgen, 1963)
(418) *Odynerus shestakovae* (Kostylev, 1935)*
(419) *Odynerus simillimus* Morawitz, 1867
(420) *Odynerus spinipes spinipes* (Linnaeus, 1758)
(421) ***Omicrabulus*** *arabicus* Gusenleitner, 2002
(422) ***Onychopterocheilus*** *albidus albidus* (Giordani Soika, 1941)
(423) *Onychopterocheilus albidus aramaeus* (Blüthgen, 1955)
(424) *Onychopterocheilus albopictus* (Kriechbaumer, 1869)
(425) *Onychopterocheilus anatolicus* (Blüthgen, 1955)
(426) *Onychopterocheilus atrohirtus* (Morawitz, 1885)
(427) *Onychopterocheilus bytinskii* (Gusenleitner, 1970)
(428) *Onychopterocheilus coctus* (Giordani Soika, 1970)
(429) *Onychopterocheilus cyanipennis* (André, 1895)
(430) *Onychopterocheilus ecarinatus* (Morawitz, 1895)
(431) *Onychopterocheilus fausti aurantiopictus* (Giordani Soika, 1970)
(432) *Onychopterocheilus fausti fausti* (Morawitz, 1873)
(433) *Onychopterocheilus fausti paravespoides* (Blüthgen, 1955)
(434) *Onychopterocheilus fereniger* (Giordani Soika, 1952)
(435) *Onychopterocheilus flaviventris* Gusenleitner, 1991
(436) *Onychopterocheilus grandiceps* (Blüthgen, 1955)
(437) *Onychopterocheilus hellenicus hellenicus* (Morawitz, 1885)
(438) *Onychopterocheilus hellenicus syriacus* (Blüthgen, 1952)*
(439) *Onychopterocheilus mavromoustakisi* (Giordani Soika, 1970)*
(440) *Onychopterocheilus mena* (Giordani Soika, 1943)
(441) *Onychopterocheilus mochii* (Giordani Soika, 1942)
(442) *Onychopterocheilus pallasii* (Klug, 1805)
(443) *Onychopterocheilus pamirensis pentheri* (Blüthgen, 1956)

(444) *Onychopterocheilus rectus teta* (Giordani Soika, 1943)
(445) *Onychopterocheilus rothi* (Dusmet, 1928)
(446) *Onychopterocheilus sarikamisensis* Yildirim, 2014
(447) *Onychopterocheilus spheciformis* (Gusenleitner, 1970)*
(448) *Onychopterocheilus stiziformis* (Gusenleitner, 1970)*
(449) *Onychopterocheilus tertius* (Giordani Soika, 1970)
(450) ***Paragymnomerus*** *amitinorum* Blüthgen, 1952*
(451) *Paragymnomerus discoidalis* (de Saussure, 1856)*
(452) *Paragymnomerus mammillatus* (Morawitz, 1885)*
(453) *Paragymnomerus signaticollis tauricus* (Kostylev, 1940)
(454) *Paragymnomerus spiricornis spiricornis* (Spinola, 1808)*
(455) ***Paraleptomenes*** *miniatus miniatus* (de Saussure, 1855)*
(456) ***Paramischocyttarus*** *subtilis* Magretti, 1884
(457) ***Paravespa*** *grandis caucasica* (Kokujev, 1912)*
(458) *Paravespa grandis grandis* (Morawitz, 1885)*
(459) *Paravespa mimetica* (von Schulthess, 1923)*
(460) *Paravespa quadricolor* (Morawitz, 1885)*
(461) *Paravespa rex* (von Schulthess, 1923)*
(462) ***Pareumenes*** *arabicus* Giordani Soika, 1981
(463) *Pareumenes enslini* (von Schulthess, 1931)
(464) *Pareumenes laminatus laminatus* (Kriechbaumer, 1879)
(465) *Pareumenes laminatus palaestinensis* Giordani Soika, 1952
(466) ***Parodontodynerus*** *aramaeus* Blüthgen, 1955*
(467) *Parodontodynerus ephippium ephippium* (Klug, 1817)*
(468) *Parodontodynerus rufinus* (Kostylev, 1940)*
(469) ***Pseudepipona*** *ankarensis* Giordani Soika, 1970
(470) *Pseudepipona augusta* (Morawitz, 1867)
(471) *Pseudepipona herrichii herrichii* (de Saussure, 1856)*
(472) *Pseudepipona herzi enslini* (Blüthgen, 1955)
(473) *Pseudepipona herzi herzi* (Morawitz, 1895)
(474) *Pseudepipona ionia clausa* (Giordani Soika, 1943)
(475) *Pseudepipona ionia ionia* (de Saussure, 1855)
(476) *Pseudepipona lativentris cypria* Blüthgen, 1942
(477) *Pseudepipona lativentris lativentris* (de Saussure, 1855)*
(478) *Pseudepipona niveopicta* Giordani Soika, 1970
(479) *Pseudepipona peculiaris* (Morawitz, 1895)*
(480) *Pseudepipona priesneri* Gusenleitner, 1970*
(481) *Pseudepipona przewalskyi* (Morawitz, 1885)
(482) *Pseudepipona pseudominuta* Gusenleitner, 1971
(483) *Pseudepipona rubidus* Gusenleitner, 1994
(484) *Pseudepipona sellata* (Morawitz, 1885)*
(485) *Pseudepipona tricarinata* (Kokujev, 1912)
(486) *Pseudepipona tripunctata* (Fabricius, 1787)
(487) ***Pseudodontodynerus*** *gambiensis gambiensis* (Meade-Waldo, 1915)
(488) *Pseudodontodynerus peculiariventris* (Giordani Soika, 1952)*
(489) *Pseudodontodynerus pretiosus* (Dusmet, 1928)
(490) ***Pseudonortonia*** *aberratica aberratica* (Morice, 1903)
(491) *Pseudonortonia aberratica arabica* Giordani Soika, 1957
(492) *Pseudonortonia aegyptiaca aegyptiaca* (de Saussure, 1863)
(493) *Pseudonortonia bicarinatus* Guichard, 1986
(494) *Pseudonortonia bisuturalis* (de Saussure, 1852)

(495) *Pseudonortonia difformis* (de Saussure, 1853)
(496) *Pseudonortonia emirata* Gusenleitner, 2012
(497) *Pseudonortonia omanensis* Giordani Soika, 1979
(498) *Pseudonortonia scotti* Giordani Soika, 1957
(499) *Pseudonortonia sudanensis* (von Schulthess, 1921)
(500) *Pseudonortonia tilkiani* Guichard, 1986
(501) ***Pseudosymmorphus*** *adnexus* (Gusenleitner, 1969)
(502) *Pseudosymmorphus angulatus* Gusenleitner and Ockermüller, 2018*
(503) *Pseudosymmorphus capillus* Gusenleitner and Ockermüller, 2018*
(504) *Pseudosymmorphus moricei* (von Schulthess, 1925)
(505) *Pseudosymmorphus prophetus* (Giordani Soika, 1952)
(506) ***Pterocheilus*** *arabicus* Giordani Soika, 1970
(507) *Pterocheilus chobauti calefactus* Giordani Soika, 1970
(508) *Pterocheilus chobauti chlorodyneroides* Gusenleitner, 1970*
(509) *Pterocheilus chobauti chobauti* Dusmet, 1928
(510) *Pterocheilus coccineus* André, 1884
(511) *Pterocheilus dives* Radoszkowski, 1876
(512) *Pterocheilus kamanensis* Gusenleitner, 1967
(513) *Pterocheilus merpeba* Giordani Soika, 1943
(514) *Pterocheilus multimaculatus* Gusenleitner, 2004
(515) *Pterocheilus nigrobilineolatus* Giordani Soika, 1942
(516) *Pterocheilus perpunctatus* Giordani Soika, 1970
(517) *Pterocheilus persicus* Gusenleitner and Fallahzadeh, 2013*
(518) *Pterocheilus phaleratus phaleratus* (Panzer, 1797)
(519) *Pterocheilus schwarzi* Gusenleitner, 1994
(520) *Pterocheilus tenebricosus* Gusenleitner, 1998*
(521) *Pterocheilus unipunctatus* (Lepeletier, 1841)
(522) *Pterocheilus varius* Gusenleitner, 2002
(523) ***Rhynchium*** *acromum* Giordani Soika, 1952*
(524) *Rhynchium cyanopterum cyanopterum* de Saussure, 1852*
(525) *Rhynchium oculatum adenense* Giordani Soika, 1952
(526) *Rhynchium oculatum distinguendum* du Buysson, 1913*
(527) *Rhynchium oculatum hebraeum* Giordani Soika, 1952
(528) *Rhynchium oculatum oculatum* (Fabricius, 1781)*
(529) ***Stenancistrocerus*** *biblicus* (Giordani Soika, 1952)*
(530) *Stenancistrocerus leonhardi* Gusenleitner, 2015*
(531) *Stenancistrocerus obstrictus* (Morawitz, 1895)*
(532) ***Stenodynerus*** *aequisculptus aequisculptus* (Kostylev, 1940)
(533) *Stenodynerus aequisculptus inaequisculptus* Giordani Soika, 1979
(534) *Stenodynerus aequisculptus taurus* (Gusenleitner, 1966)
(535) *Stenodynerus bluethgeni* van der Vecht, 1971*
(536) *Stenodynerus chevrieranus* (de Saussure, 1855)*
(537) *Stenodynerus chitgarensis* Giordani Soika, 1970*
(538) *Stenodynerus claviger* Gusenleitner, 1981
(539) *Stenodynerus clypeopictus* (Kostylev, 1940)
(540) *Stenodynerus cyrus* Gusenleitner, 2012*
(541) *Stenodynerus difficilis* (Morawitz, 1867)*
(542) *Stenodynerus fastidiosissimus fastidiosissimus* (de Saussure, 1855)
(543) *Stenodynerus fastidiosissimus judaicus* Gusenleitner, 1970
(544) *Stenodynerus haladaorum* Gusenleitner, 1999
(545) *Stenodynerus orenburgensis* (André, 1884)

(546) *Stenodynerus pannosus* Gusenleitner, 1985
(547) *Stenodynerus picticrus* (Thomson, 1874)
(548) *Stenodynerus pullus* Gusenleitner, 1981
(549) *Stenodynerus punctifrons* (Thomson, 1874)*
(550) *Stenodynerus sapidus* Giordani Soika, 1970*
(551) *Stenodynerus simulatus* Gusenleitner, 1981
(552) *Stenodynerus steckianus* (von Schulthess, 1897)
(553) *Stenodynerus trotzinai* (Morawitz, 1895)*
(554) *Stenodynerus xanthomelas* (Herrich-Schäffer, 1839)*
(555) ***Symmorphus*** *allobrogus* (de Saussure, 1855)
(556) *Symmorphus angustatus* (Zetterstedt, 1838)
(557) *Symmorphus bifasciatus* (Linnaeus, 1761)
(558) *Symmorphus connexus* (Curtis, 1826)
(559) *Symmorphus crassicornis* (Panzer, 1798)*
(560) *Symmorphus debilitatus* (de Saussure, 1855)
(561) *Symmorphus declivis* Harttig, 1932
(562) *Symmorphus gracilis* (Brullé, 1832)*
(563) *Symmorphus murarius* (Linnaeus, 1758)*
(564) ***Synagris*** *rubescens* (Giordani Soika, 1989)
(565) *Synagris spiniventris* (Illiger, 1802)
(566) *Synagris spinosuscula* de Saussure, 1852
(567) ***Syneuodynerus*** *egregius egregior* (Gusenleitner, 1970)*
(568) *Syneuodynerus egregius egregius* (Herrich-Schäffer, 1839)
(569) *Syneuodynerus erichi* (Gusenleitner, 1967)
(570) *Syneuodynerus fouadi dhofariensis* Giordani Soika, 1979
(571) *Syneuodynerus fouadi fouadi* (Giordani Soika, 1939)
(572) *Syneuodynerus fritzi* (Gusenleitner, 1997)
(573) *Syneuodynerus gebi* (Gusenleitner, 1970)
(574) *Syneuodynerus insolitus* Gusenleitner, 1994
(575) *Syneuodynerus siegberti* (Gusenleitner, 1967)
(576) ***Tachyancistrocerus*** *aereus* Giordani Soika, 1952
(577) *Tachyancistrocerus cyprius* (Pittioni, 1950)
(578) *Tachyancistrocerus komarowi derufata* Gusenleitner, 1971
(579) *Tachyancistrocerus komarowi komarowi* (Morawitz, 1885)*
(580) *Tachyancistrocerus liliput* (Gusenleitner, 1993)
(581) *Tachyancistrocerus pallidenotatus* Giordani Soika, 1970*
(582) *Tachyancistrocerus quabosi* Giordani Soika, 1979*
(583) *Tachyancistrocerus rhodensis* (de Saussure, 1855)*
(584) *Tachyancistrocerus schmidti* (Kokujev, 1913)*
(585) *Tachyancistrocerus serenus* (Giordani Soika, 1935)
(586) *Tachyancistrocerus syriacus* Giordani Soika, 1970
(587) ***Tricarinodynerus*** *arabicus* Guichard, 1986
(588) *Tricarinodynerus magrettii* (Gribodo, 1884)
(589) ***Tropidodynerus*** *flavus semiticus* (Giordani Soika, 1952)
(590) *Tropidodynerus interruptus interruptus* (Brullé, 1832)
(591) ***Xanthodynerus*** *castrorum* (Blüthgen, 1954)
(592) *Xanthodynerus jordanicus* Gusenleitner, 1990
(593) *Xanthodynerus octavus* (Giordani Soika, 1943)
(594) *Xanthodynerus pilosus* Gusenleitner, 1990
(595) *Xanthodynerus trichopygus* (Blüthgen, 1954)
(596) ***Psiliglossa*** *algeriensis* E. Saunders, 1905*

(597) *Psiliglossa odyneroides odyneroides* (S.S. Saunders, 1850)
(598) *Psiliglossa pulchra zhelochovtsevi* Panfilov, 1968*
(599) ***Raphiglossa*** *bytinskii* Giordani Soika, 1974
(600) *Raphiglossa eumenoides caucasica* Giordani Soika, 1970*
(601) *Raphiglossa eumenoides eumenoides* S. S. Saunders, 1850
(602) *Raphiglossa formosa* Kostylev, 1940*
(603) *Raphiglossa irana* Giordani Soika, 1970*
(604) *Raphiglossa spinosa rufescens* (E. Saunders, 1905)
(605) ***Zethus*** *favillaceus* Walker, 1871

Subfamily Masarinae Latreille, 1802

(606) ***Celonites*** *abbreviatus abbreviatus* (Villers, 1789)*
(607) *Celonites andreasmuelleri* Mauss, 2013
(608) *Celonites cagrii* Mauss and Yildirim, 2022
(609) *Celonites clarus* Gusenleitner, 1973*
(610) *Celonites cyprius* de Saussure, 1854*
(611) *Celonites discretus* Gusenleitner, 1973*
(612) *Celonites fischeri* Spinola, 1838
(613) *Celonites foveolatus nigrior* Richards, 1962
(614) *Celonites hamanni* Gusenleitner, 1973
(615) *Celonites hermon* Gusenleitner, 2002
(616) *Celonites iranus* Gusenleitner, 2018*
(617) *Celonites jousseaumei jousseaumei* du Buysson, 1906*
(618) *Celonites laetus* Panfilov, 1968
(619) *Celonites longipilis* Gusenleitner, 1973*
(620) *Celonites osseus* Morawitz, 1888*
(621) *Celonites persicus* Richards, 1962*
(622) *Celonites pictus* Richards, 1962*
(623) *Celonites rugipes* Bischoff, 1928
(624) *Celonites semenovi* Kostylev, 1935*
(625) *Celonites spinosus* Gusenleitner, 1966
(626) *Celonites syriacus* Gusenleitner, 2012
(627) *Celonites tauricus* Kostylev, 1935*
(628) *Celonites tenebricosus* Gusenleitner, 2012
(629) *Celonites yemenensis* yemenensis Giordani Soika, 1957
(630) ***Ceramius*** *bureschi* Atanassov, 1938
(631) *Ceramius caucasicus* André, 1884*
(632) *Ceramius fonscolombei fonscolombei* Latreille, 1810*
(633) *Ceramius palaestinensis* (Giordani Soika, 1957)
(634) *Ceramius tuberculifer* de Saussure, 1854*
(635) ***Jugurtia*** *escalerae* Meade-Waldo, 1910*
(636) *Jugurtia eurycara* Kostylev, 1935*
(637) *Jugurtia irana* Kostylev, 1935*
(638) *Jugurtia jemenensis* Kostylev, 1935
(639) *Jugurtia punctuosa* Gusenleitner, 2011
(640) *Jugurtia zarudnyi* Kostylev, 1935*
(641) ***Masaris*** *aegyptiacus aegyptiacus* Meade-Waldo, 1911
(642) *Masaris aegyptiacus arabicus* Giordani Soika, 1957
(643) *Masaris agnolii* Selis, 2019*
(644) *Masaris vespiformis* Fabricius, 1793

(645) ***Quartinia*** *amiae* Carpenter, 2003
(646) *Quartinia arabica* Gusenleitner, 2010
(647) *Quartinia araxana* Giordani Soika, 1960*
(648) *Quartinia atrata* Gusenleitner, 1998*
(649) *Quartinia candida* Gusenleitner, 1997
(650) *Quartinia gusenleitneri* Schmid-Egger, 2017
(651) *Quartinia haemorrhoa* Gusenleitner, 1997
(652) *Quartinia halicticeps* Giordani Soika, 1939
(653) *Quartinia irana* Gusenleitner, 2012*
(654) *Quartinia libanica* Richards, 1964 [1962]
(655) *Quartinia longiceps* Gusenleitner, 2010
(656) *Quartinia mochii* Giordani Soika, 1939
(657) *Quartinia monnerati* Gusenleitner, 2016
(658) *Quartinia nilotica* Fischer, 1964
(659) *Quartinia nitidiuscula* Gusenleitner, 2010
(660) *Quartinia nubiana* Richards, 1962
(661) *Quartinia orientalis* Gusenleitner, 1973
(662) *Quartinia popovi* Richards, 1972
(663) *Quartinia separata* Gusenleitner, 1997
(664) *Quartinia soikai* Richards, 1964 [1962]*
(665) *Quartinia syriaca nitens* Gusenleitner, 1973*
(666) *Quartinia syriaca syriaca* Richards, 1964 (1962)
(667) *Quartinia thebaica* du Buysson, 1902
(668) *Quartinia tricolorata* Giordani Soika, 1954
(669) *Quartinia tripolitana sinaitica* Richards, 1964 (1962)
(670) *Quartinia tripolitana tripolitana* Richards, 1962
(671) *Quartinia tuareg* Giordani Soika, 1954

Subfamily Polistinae Latreille, 1802

(672) ***Polistes*** *associus* Kohl, 1898*
(673) *Polistes atrimandibularis* Zimmermann, 1931*
(674) *Polistes badius* Gerstaecker, 1871
(675) *Polistes biglumis* (Linnaeus, 1758)*
(676) *Polistes bischoffi* Weyrauch, 1937*
(677) *Polistes bucharensis* Erichson, 1849*
(678) *Polistes chinensis* (Fabricius, 1793)*
(679) *Polistes dominula* (Christ, 1791)*
(680) *Polistes fastidiosus* de Saussure, 1853
(681) *Polistes foederatus* Kohl, 1898*
(682) *Polistes gallicus* (Linnaeus, 1767)
(683) *Polistes indicus* Stolfa, 1934*
(684) *Polistes marginalis marginalis* (Fabricius, 1775)
(685) *Polistes mongolicus* du Buysson, 1911*
(686) *Polistes nimpha* (Christ, 1791)*
(687) *Polistes olivaceus* (DeGeer, 1773)*
(688) *Polistes rothneyi* Cameron, 1900*
(689) *Polistes semenowi* Morawitz, 1889*
(690) *Polistes smithii neavi* von Schulthess, 1921
(691) *Polistes tenellus lahejensis* Giordani Soika, 1981
(692) *Polistes wattii* Cameron, 1900*

(693) ***Belonogaster*** *adenensis adenensis* Giordani Soika, 1957
(694) *Belonogaster arabica* Giordani Soika, 1957
(695) *Belonogaster filiformis* (de Saussure, 1853)
(696) *Belonogaster guichardi* Richards, 1982
(697) *Belonogaster juncea juncea* (Fabricius, 1781)
(698) *Belonogaster saussurei* Kirby, 1881
(699) ***Parapolybia*** *escalerae* (Meade-Waldo, 1911)*
(700) *Parapolybia persica* (Meade-Waldo, 1911)*
(701) ***Ropalidia*** *aethiopica aethiopica* (du Buysson, 1907)
(702) *Ropalidia cincta* (Lepeletier de Saint Fargeau, 1836)
(703) *Ropalidia irrequieta* (Kohl, 1906)

Subfamily Vespinae Latreille, 1802

(704) ***Dolichovespula*** *adulterina* (de Buysson, 1905)
(705) *Dolichovespula media* (Retzius, 1783)
(706) *Dolichovespula norwegica* (Fabricius, 1781)
(707) *Dolichovespula omissa* (Bischoff, 1931)*
(708) *Dolichovespula saxonica* (Fabricius, 1793)*
(709) *Dolichovespula sylvestris* (Scopoli, 1763)*
(710) ***Vespa*** *crabro* Linnaeus, 1758*
(711) *Vespa orientalis* Linnaeus, 1771*
(712) *Vespa velutina* Lepeletier de Saint Fargeau, 1836
(713) ***Vespula*** *austriaca* (Panzer, 1799)
(714) *Vespula germanica* (Fabricius, 1793)*
(715) *Vespula rufa* (Linnaeus, 1758)*
(716) *Vespula vulgaris* (Linnaeus, 1758)*

Index

Pages in *italics* refer to figures, pages in **bold** refer to tables, and pages followed by n refer to notes.

For Product Safety Concerns and Information please contact our EU representative GPSR@taylorandfrancis.com Taylor & Francis Verlag GmbH, Kaufingerstraße 24, 80331 München, Germany

Batch number: 10748241

Printed by Printforce, the Netherlands